CONVENTIONAL AND ORGANIC FARMING

Nitrogen

McCarrison
Organic matter   Teragram   Cisgenic
Agroecology

Ammonia   Preeclampsia

Genome   Nitrification   Greenhouse gases

Green Revolution

Bones

Pesticides

Methane

Hunzas   Cover crops
        Phenolics   MRL
Howard   Anthroposophy

Leaching

Haber-Bosch

NUE   Eutrophication

ADI   Denitrification

Balfour   Carbon

Guano   Fertile crescent

Manure

BNF

Transgenic

Mutagenesis

Steiner

IPM

Phosphorus

Living Soil

CRISPR-Cas9

Gene editing

Broadbalk

LD_{50}

$PM_{2.5}$ HERP

Compost   Alkaloids

Nitrates

Biodiversity

Nitrous oxide   Antioxidant

Mycorrhizae   Planetary Boundaries

Rhizosphere   Sequestration

GM crops

CONVENTIONAL AND ORGANIC FARMING

A Comprehensive Review through the
Lens of Agricultural Science

Victor M. Shorrocks

First published 2017

Published by
5M Publishing Ltd,
Benchmark House,
8 Smithy Wood Drive,
Sheffield, S35 1QN, UK
Tel: +44 (0) 1234 81 81 80
www.5mpublishing.com

A Catalogue record for this book is available from the British Library

ISBN 9781910455999

Book layout by Servis Filmsetting Ltd, Stockport, Cheshire
Printed by Replika Press Pvt. Ltd, India
Photos by Vic Shorrocks

Contents

Author's Foreword

The origin of this book can be traced back to the 1950s when I became aware of organic farming as a student studying plant nutrition and learning that plants absorb minerals by active processes as ions and not as organic compounds. The opposition to the use of fertilizers, which on dissolution release the same ions that are ultimately generated when organic matter is mineralized, was difficult to understand. It was not then apparent to me that the organic movements were fundamentally opposed to modern farming, to agricultural science and especially to field experiments with small plots. Their dislike of conventional farming would subsequently extend from fertilizers to pesticides and to genetically modified crops. I mistakenly thought, in the 1950s, that the organic movements were looking for solutions to the problems of crop nutrition when the raw materials for fertilizers would no longer be available; for example would efficient recycling of nutrients provide the answer?

The classic text on organic farming was *The Living Soil* by Lady Eve Balfour published in 1943, which we now know influenced farmers and thinking in many countries. The complete title of the book was *The Living Soil – Evidence of the Importance to Human Health of Soil Vitality, with Special Reference to National Planning*. It was thus important here to examine the origin of the idea that human health was directly associated with soil vitality and the use of manure/compost, and also to consider what medical research has revealed over the years about any connection.

As a boy I recall being excited on hearing about the long-living Hunzas, in what is now Pakistan, with their magnificent physiques. So I was fascinated later to learn about the work of McCarrison on the Hunzas and of Howard in India (in the 1930s) which was interpreted by some as showing that food produced using compost, later called organic farming, could have special health-promoting qualities. This aspect is still important to organic farmers but for reasons different to those suggested by Howard; now it is the virtues of the beneficial compounds found and the absence of pesticides in organic food that are extolled. Two basic questions about organic farming are — has it a monopoly on the production of inherently healthy food and are there any basic and meaningful differences in the composition of organic and conventionally produced food?

Manure played no part in my research in Malaysia on the mineral nutrition of the rubber tree (*Hevea brasiliensis*) but fertilizer use and the nutrition of associated legume cover crops grown between the tree rows did. I became familiar with aspects of farming across the world later when initiating research into the agricultural use of the micronutrient boron and the development of a cotton/soy bean herbicide in many countries.

A revealing connection with organic agriculture came in 1965 when I corrected *The Scotsman* newspaper after it wrote that work at Rothamsted (Broadbalk) had shown that soil fertility could only be maintained by the use of fertilizers; my published letter, which pointed out that manure could be as effective as fertilizers, prompted an organic supporter to write and suggest that I should visit the organic experiment at Haughley in Suffolk; I was told: "We know what we think should be, or should like to be, the truth about the questions being asked there, but we try to accept the answers and go on probing, whether we like them or no." This was when the Haughley experiment had been running for about twenty years.

In the 1990s, after retiring, I became intrigued by the opposition of the organic movements to genetically modified crops (GM) when I was bringing myself up to date with what was involved in genetic engineering. It seemed strange that the potential crop improvements that could be achieved by this new plant breeding tool were not appreciated by the organic movements as a way of benefiting from biology rather than chemistry. I wanted to find the evidence behind the anti-agricultural science thinking as well as that against genetic engineering as evident in the Reith Lecture given by the Prince of Wales in May 2000 and also the activities of the green and organic movements in trashing field trials of GM crops in 1999. It was at this time that I began a study of the history of the organic movements in the hope that I would find the evidence to support their opposition to GM as well as their objections to fertilizers and pesticides. The result was eventually this book.

A curious development has been the use of the word organic, which has been hallowed by repetition to suggest the superiority not only of a farming system and food but also of many items with no obvious connection with farming (e.g. cosmetics). The uncritical repetition of claims can easily give the impression of validity.

Initially I envisaged looking at the practices and claims about organic farming and food as seen through the lens of agricultural science but it quickly became apparent that it was necessary to establish how agricultural science had developed and is still developing and then assess where organic farming fitted. An example was the need to consider the existence and the toxicology of natural pesticides (phytoalexins) and how they compared with man-made pesticides. The distinction between hazard, risk and vulnerability is particularly important with regard to phytoalexins and man-made pesticides.

The information in this book is derived from the reported work of a large number of researchers in many countries, several of whom have been helpful in providing supplementary material. Much use has been made of the internet to find papers of interest. Such literature searches as well as reports in the media led to follow-up enquiries for evidence and confirmation of reported claims; unfortunately a significant minority, usually concerning excessive claims, did not reply.

Over the years the academic papers, especially those by Trewavas in Scotland, have been strong motivators. His papers and those by Kirchmann and Bergstrom in Sweden have variously discussed and commented critically about specific aspects of organic farming and food. Inevitably some adverse comments have been refuted by organic supporters, especially when reports reached a generally sympathetic media; organic proponents would often refer to work in progress to support their position. The hope here is that by presenting in one place the available evidence covering all the aspects of farming relevant to the comparisons of organic with conventional it is possible for the whole story to be appreciated and "cherry-picking" of topics avoided.

We are living in an age of increasing specialization so it is not surprising that this book is written by a generalist. My aim has been to be as comprehensive as possible but it is inevitable that I will have oversimplified some aspects and will have been out of date with others; corrections, suggestions and additions will be gratefully received.

Having been retired for more than twenty years I am hopeful that I will not be maliciously accused of being in the pocket of agribusiness.

When I was approached by 5m Publishing about writing this book I hesitated, saying that as long as I enjoyed the investigative process I would continue writing ...

Victor M. Shorrocks MA, DPhil, CBiol, MRSB.

Acknowledgements

A book such as this that covers many aspects of agriculture and nutrition is inevitably dependent on the research of very many across the world and the now vast literature. The contribution of these workers is gratefully acknowledged. Their work has been consulted and cited. Without their endeavours it would not have been possible to attempt this survey.

Special thanks are due to several who encouraged and advised me in the early days and supported me by ensuring that I was kept up to date; particular thanks go to Lord Dick Taverne; Professor Chis Leaver; and Professor Anthony J. Trewavas in the UK, with the latter's motivating papers on organic food and farming; to Professor Sir Colin Berry and Dr John V. Possingham in Australia; and Professor James S. Bus in the USA.

Professor Vaclav Smil, an expert in several fields relating to the environment, food and population, helped with information on provision of N and kindly gave permission to quote from his many publications. Professor Dr. Jan Willem Erisman is thanked for providing data on global N fixation.

Particular thanks go to Ken Gilbert for discussions on fertilizers and their availability and for commenting on the early chapters. The valuable discussions on the Classical experiments at Rothamsted with A.E. Johnston are much appreciated. Aaron Kinsman and Emmanuel Omondi of the Rodale Institute are thanked for help and comments on the Rodale trials. The help of Markus Arbenz at IFOAM for providing information on IFOAM over the years is gratefully acknowledged. Dr Pierre Lagoda of FAO/IAEA kindly supplied up-to-date information on mutagenesis. Dr Richard Cooper is thanked for information on elemental S behaving as a phytoalexin. Dr Lauren Ponisio very kindly provided explanatory information on her studies.

Special acknowledgement is required of the pioneering work of Professor Bruce N. Ames and his many co-workers, especially to the late Dr Lois Gold who, in the words of Ames, instilled sanity into the controversy about human exposure to trace amounts of chemicals. It was their work, starting in the 1980s, that opened the door to our understanding of the similar toxicology of man–made and natural plant pesticides, the phytoalexins. Having found many natural plant components were both mutagens and carcinogens, they questioned the interpretation of rodent toxicology tests regarding the risks of low doses; their work remains very relevant but is seldom considered. Unfortunately they were confronted by the anti-pesticide lobby and although the scientific battle was won the public relations battle was lost and remains so. Since the 1990s, Professor Ames has devoted his attention to studying the nutritional causes of cancer and aging, and is still working at 88 years young.

Permissions have been given for using specific quotations by Duchy Home Farm and by Pesticide Action Network, for data by European Food Safety Authority and by IFOAM, for a figure by Leslie Corrice and for using books as sources by Rudolf Steiner Press and Floris Books.

It is believed that all reasonable attempts have been made to contact the owners of copyrighted material used in this book. However, if you are the copyright owner of any source used that is not credited, please notify the publisher and this will be corrected in any new edition.

Biographical Notes

VICTOR M. SHORROCKS

Vic Shorrocks is a specialist in the mineral nutrition of crops, gaining his D.Phil. at Oxford with work on the uptake of phosphorus and potassium by barley and sunflower in 1957. Research on the nutrition of the rubber tree and associated cover crops in Malaysia was followed by a period with RTZ Borax that involved research on the micronutrient boron and the development of a herbicide around the world. In the latter part of his career he concentrated on the mineral micro-nutrition of plants, animals and man with his quarterly publication *Micronutrient News*, consulting worldwide. Since retiring he has campaigned for a better understanding of the importance of iodine and selenium in human health and also for an appreciation of the way that risks associated with pesticides are assessed in Europe (by MEPs). He has continued to keep up with developments in organic farming, a topic he has followed for more than sixty years and which led to him study plant breeding.

Part 1

From Hunter Gatherers to Agricultural Science

Origin of Farming and the Historical Appreciation of the Value of Manure and Legumes

"The fairest thing we have, a flower, still has its roots in earth and in manure"
D.H. Lawrence

In early times there was much specu-lation about how plants grow and the general conclusion was that plants feed on humus or soil particles in much the same way as animals feed on plants (Anon, 1960). The earliest appreciation of the value of manure for crops may have been in one place but more likely in the various places where man first prac-tised farming and observed that animal manure, dead animals and possibly blood increased plant growth. This probably followed hunter-gatherers abandoning their way of life and the domestication of animals and plants.

Domestication of animals and crops 10,000–8,000 BC

Fertile Crescent

Farming began with the domestication first of animals, initially dogs, about 15,000 years ago, then reindeer, sheep, goats and cattle, probably in that order (Branford Oltenacu, 2005). So-called domesticated wheat, barley and the pulses, lentils and peas appeared later, some 10,000 to 12,000 years ago, in the Fertile Crescent in the Middle East (Zohary *et al.*, 2012). There are some doubts about timing. Suggestions that cereals were being cultivated at a 23,000 year old site in Galilee in Israel indicates that man may have been experimenting with farming some 11,000 years earlier than is generally thought. The evidence for this is based on the presence of domestic-types of wheat and barley, of weeds found in wheat and of stone harvesting sickles (Nadel and Sternberg, 2015). It is curious that consumption of chickens in Israel has also been dated to 23,000 years ago at an archaeological site at Tel Maresha, south of Jerusalem (Perry-Gal, 2015).

The Fertile Crescent is the region in the Middle East where the civilizations of the Middle East and the Mediterranean basin originated. It includes a roughly crescent-shaped area across Mesopotamia from the Persian Gulf and southwards from Syria through Palestine into northern Egypt. Irrigation water provided by the Tigris, Euphrates and Nile was absolutely essential for agriculture. In the case of the Nile, the clay and silt provided by the annual flood did supply about 17,000 t N, 6,000 t K_2O and 7,500 t P_2O_5 to the soils in the valley and the delta (Abu Zeid, 1989). The deposit of silt behind the Aswan High Dam since the 1960s has reduced the deposit of sediment via the irrigation water but the detrimental effect is likely to have been minimal because the amounts formerly supplied only amounted to about 5 kg/ha N, 3 kg/ha P_2O_5 and 2 kg/ha K_2O based on current arable land in Egypt.

The selection from emmer and einkorn grasses of plants whose seed did not shatter and spill on to the ground was the key factor in the origin of farming. Natural selection would have favoured the grasses with a weak rachis that easily breaks and permits seed dispersal. A single mutation that resulted in a tough rachis caused the grain to remain on the plant. Artificial selection would have favoured large seeds, which would have been more likely to survive after being planted in the top soil (Allaby, 2008).

It is currently thought that a domesticated variant of an einkorn wheat originated around the Karacadag Hills 30 km from Gobekli Tepe (south-east Turkey) near an apparently religious centre built 11,500 years ago by hunter-gatherers (Harari, 2014).The unanswered question is: was farming promoted by the desire to build and remain close to Gobekli Tepe and by the manpower required to construct the temple or is it just a coincidence that the Karacadag Hills are so close to Gobekli Tepe? The sowing of the saved seed of what we now know as wheat necessitated the farmers staying close to the areas chosen for cultivation, and so farming began. Vermin attracted to the stored cereals were preyed on by wild cats and so began domestication of cats, possibly in Egypt some 7,000 years ago.

Several leguminous crops also originated in the Fertile Crescent, being variants that did not readily shed their seed. Peas and lentils have been found in archaeological sites in the Middle East together with barley and wheat (Kislev and Bar-Yosef (1988). Wild cattle, sheep and goats were also present in the Fertile

Crescent. A creationist could be forgiven for thinking that cereals, legumes and animals were placed together on purpose.

Another curiosity of the Fertile Crescent was the development there of lactose tolerance in adults; this followed mutations promoting lactase production. Most babies can digest milk thanks to the enzyme lactase but up until several thousand years ago that enzyme was turned off when a person grew into adulthood. Lactose tolerance was selected for primarily in north-west Europe when migrating farmers, with the low-yielding wheat and barley they took from the Fertile Crescent, consumed milk from their domestic cattle, especially in times of famine. Where people were less dependent on cattle, such as in China, Thailand and the American South-west, milk was not drunk and lactose tolerance did not evolve (National Geographic, 2014). Today about 35% of the global population – mostly with European ancestry – can digest lactose in adulthood without a problem (Peeples, 2009).

Other areas of plant domestication

Olives were domesticated about 10,000 years ago in the Middle East. Other crops originated at about the same period in different parts of the world, with rice and millet in China and beans and squash in Mexico and North America. Maize had to wait for mutants from its wild ancestor, teosinte, which ultimately gave rise to plants producing corn cobs more than 6,000 years ago. Sunflowers in North America date back 5,000 years, about the same time that potatoes were grown in the Andes. Sugar cane and bananas originated about 8,000 years ago in Papua New Guinea. Pumpkins and squashes were probably selected by native North Americans about 8,000 years ago. Some 5,000 years ago West African farmers were selecting African millet, rice and sorghum.

With the domestication of grapevines about 5,500 years ago the main wave of domestication was over in the Middle East. It is salutary to realize that our diet today is based very largely on that of the earliest farmers across the world. On the other hand, much of our physiology is still that of our hunter-gatherer forbears. Man made use of edible plants where he found them and domesticated animals that were not the fiercest.

And so ended plant and animal domestication or, was it man who was domesticated by his crops and animals and prevented from leading a nomadic way of life?

From a total of about 150 plant species used globally as food, today man is mostly reliant on about ten species of crops, namely maize, rice, wheat, potatoes, cassava, soybeans, sweet potatoes, sorghum, yams and plantains.

Hunter-gatherers and a natural way of life

Organic proponents often cite many practices of modern farming as being unacceptable because they are not natural. It is important therefore when considering the progression from hunter-gatherers to modern man to decide at what point in the history of mankind it is felt that living and farming became unnatural. We readily accept the transition from hunter-gatherer to resident farmer, but why? Regulated hunting and gathering could be sustainable but probably only for a small global population. It is thought that the global hunter-gatherer population rose to about 5 million, which suggests it was not an unsuccessful lifestyle (McKevitt and Ryan, 2013). It can be argued that the hunter-gatherer way of life was preferable to that of the farmers (Harari, 2014) because it was more satisfying, less time-demanding and healthier. The size of the hunter-gatherer population was limited by the availability of their food. In contrast, the population that can be supported by farming is largely limited by the ability of farmers. The result of the move to successful farming was a population explosion of non-farmers in developed but not in developing countries, the effects of which are still being felt today. So did the transition from hunter-gatherer to farmer mark an improvement in living for mankind?

Should we also question whether it is more natural to allow the wild predecessors of the domesticated animals to roam at will and be hunted or to confine them in fields and pens? Most domesticated animals will be slaughtered after an unnaturally very short life or, in the case of dairy cows, after they have produced far more milk than required by their calves. The ecosystem of grasslands would be more naturally maintained by grazing animals depositing their dung directly on the land rather than in cow sheds or feedlots.

Early Europeans lived as hunter-gatherers for more than 35,000 years. What has happened since farmers reached Europe some 8,500 years ago, coming first from the Near East and later, about 4,500 years ago, the nomadic sheep herders from Yamnaya in western Russia (Zimmer, 2015)? They lived alongside the hunter-gatherers for possibly 2,000 years (Marshall, 2013). The hunter-gatherers' diet would have consisted of wild animals, fish, nuts and fruit, whereas early farmers would have mostly eaten domestic animals, cereals and pulses. Modern dieticians would probably prefer the former diet to one dominated by mineral and vitamin-poor cereals.

Recent genetic studies have indicated that after agriculture arrived in Europe people's DNA underwent widespread changes, altering height, digestion, immune system and skin colour – very probably attributes brought with them but not developed by natural selection before their arrival in Europe (Zimmer, 2015). Early farmers were likely to be pastoralists tending their animals on grassland, with crops taking a minor position. When they migrated from the Fertile Crescent they would have taken their crops with them. Following the adoption of farming as a way of life the global population grew to 200–330 million by about 1000 BC. There is no general agreement as to why farming ousted the hunter-gatherer way of life but it was most likely due to it being preferred as a more secure and safe strategy. Higher birth and survival rates in resident farmers would probably have been contributory factors. Moreover, it is probable that farmers adopted some of the customs of their predecessors and so benefited from the best of both worlds. Man still fishes and collects fruit and nuts from the wild.

We now have a divided world. On one hand there are the developed countries where a very small number of farmers produce food for the rest and on the other hand, in the developing countries, there are many subsistence farmers, particularly in Africa and Asia with a lifestyle not too different from the earliest farmers.

There is possibly just one remaining truly hunter-gatherer group (of about 400) in the world, the Hadza living in the central Rift valley in Tanzania. Other groups that live very close to nature and are very reliant on local plants and animals include the Tsimane in Amazonian Bolivia, the Kyrgyz in the Pamir mountains of north Afghanistan, the Arctic Inuit and the Bajau of Malaysia (National Geographic, 2014).

Short of a nuclear holocaust or population-destroying plagues there is no likelihood of returning to a hunter-gatherer way of life, which effectively means man will continue to rely on farming. The part that farming according to organic principles could play in

feeding mankind is one of the basic questions addressed in this book.

Appreciation of the value of manure and legumes

When did man first realize that animal dung had the power to improve crop growth? Probably the benefits were so obvious and occurred in pre-historic times, so were not recorded. We should be grateful to the first farmers. But who were the first to write about the use of animal manure, often historically translated as dung? Answers are sought from Sumerian tablets, Egyptian caves, tombs and temples, the Bible and Greek and Roman literature. It is likely that animal and human excrement would have been disposed of for hygienic reasons so references to the handling of manure does not necessarily imply an appreciation of its value. Maybe the benefits of human faeces were noticed before those of dung?

Dried bison manure was used as fuel by native Indians on the tree-less plains of America. It is still used by many cultures, e.g. Tibetans with their yaks. Dried cowpats were used as fuel by peoples transported to Siberian and Kazakhstan work camps by the Russians in the Second World War (H. Shaw, Oxford, England, 2014, personal communication). Burning results in the loss of nitrogen and also the potential for the organic matter in manure to improve soil conditions.

It is likely that the value of animal dung

would only have been appreciated when it was necessary to move it from where it was deposited to where crops were grown, when improved plant growth may have been observed. Such disposal of dung would have followed the corralling of cattle. Grazing sheep and foraging goats played a small part and droppings from dovecotes were valued.

Mesopotamia

There is a noticeable absence of any reference to dung in the cuneiform records of the Sumerians in Mesopotamia (the Tigris–Euphrates river system) – although it is clear archaeologically that dried manure was used extensively as a fuel. If manure had been used on the frequently grown wheat and barley it would have been expected that scribes would have mentioned it (P. Collins, Oxford, England, 2014, personal communication). As many as 100,000 cuneiform tablets have so far been translated, dating back 5,000 years.

China, Japan and Korea

In China settled agriculture can be dated to around 5000 BC, when rice and millet were being grown. According to Davy (c.1810), the Chinese mixed night soil with one-third of its weight of marl (calcareous clay), made it into cakes and dried them in the sun. However, the cakes were most likely used for fuel, not as a soil amendment.

In the early years of the 20th century an eminent American soil scientist, F.H. King, visited China, Japan and Korea and wrote about what he called permanent agriculture in these countries (King, 1911). His book *Farmers of Forty Centuries*, was later to influence the pioneers of mainstream organic farmers in England. King maintained that cereal yields in China exceeded those in the USA and were achieved by the use of animal manure, night soil, multiple cropping and by growing legumes.

King wrote that for thousands of years Chinese farmers used human manure – night soil – as a fertilizer for most crops, including rice and wheat. By the time of King's visit the use of night soil had become restricted to vegetables grown close to home because it was handled as a liquid, making transport difficult. King commented on the ability of night soil to improve both the yield and flavour of vegetables, especially pak choi. Use of night soil in Japan only started in the 17th century (Tanaka, 1998) after the authorities had banned the discharging of human excreta into rivers and moats.

In China and Japan night soil was applied annually at about 4,000 kg/ha. King also reported that during the fifth and sixth centuries AD, with the expansion and intensification of agriculture, cattle began to be used for ploughing and to supply manure to fertilize crops. King's book was perceived by Paull (2009, 2011) as the precursor of organic farming. According to Pfeiffer (1947), the Chinese had for hundreds of years made compost heaps by mixing exhausted topsoil with organic refuse and farmyard manure, so extending the manure. Calcined bones were also used in China as a soil amendment 2,000 years ago. Smil (2001) commented that it would only have been in densely populated areas that night soil would have contributed more than 20 kg N/ha; total annual N output in human faeces and urine would typically be about 3.3 kg N/capita.

In China, Japan and Korea it had been the practice for centuries to grow leguminous crops in rotation with cereals, notably rice, in order to maintain soil fertility. Written records of the use of leguminous cover crops in China date from the fifth century BC (Smil, 2001). While leguminous plants were known by many farmers over the centuries to be beneficial, it was not until the end of the 19th century that it became known that legumes have the capacity to capture nitrogen from the air.

Egypt

Cattle probably reached Egypt more than 14,000 years ago, grazing peripheral land well away from the annually flooded land close to the Nile. Cattle provided meat, milk and hides. Oxen were beasts of burden and bull fighting was practised as long ago as 2000 BC.

At Nabta Playa, in the eastern desert, herds of cattle are depicted on rock paintings dating back 7,000 years, It is thought that most Egyptian cattle were herded on the open grasslands, so there may have been little opportunity for disposing of manure on crops such as barley and emmer wheat (used to make bread and beer). The main vegetables grown in ancient Egypt were onions, leeks, beans, lentils, garlic, radish, cabbage, cucumbers, and lettuce. The fruit grown consisted of dates, figs, grapes, pomegranates, and melon (Harris, 2001). The inclusion of the nitrogen-fixing beans and lentils was valuable. The absence of trees in the Nile valley explains why dried dung was used as a fuel for cooking.

Murray (2000) considers that it was unlikely that animal dung was used extensively to improve soil fertility but that it was sometimes used and therefore appreciated. On the other hand, there is evidence that bird droppings collected from dovecotes were used on crops.

South America

Archaeological evidence suggests that Andean people collected guano from small islands located off the desert coast of Peru for use as a soil amendment for well over 1,500 years (Rau, 1878). The Incas apparently assigned great value to guano, restricted access to it, and punished any disturbance of the birds by execution (Cushman, 2013).

Bible

Oxen and cattle are mentioned in the Torah (Genesis 30.32), Exodus (9.4, 20.10, 21.28, 22.1, 23.4 in 5th and 6th century BC) and Deuteronomy (5.14, 22.1, 25.4). There are a few mostly derogatory references to dung and dunghills in

1 Samuel 2.8 (630–640 BC) and Psalm 113, Philippians 3.8 and Luke 14.35). Leaving fields fallow every seventh year was appreciated by the children of Israel (Exodus 23.11), which would suggest that the benefits of dung might also have been appreciated.

There is one appreciative mention (Luke 13.8, written 80–100 AD) about giving dung to a poorly grown fig tree in the hope of reviving it.

Carthage

It was reported that the Carthaginians used bird droppings as a fertilizer in North Africa in 200 BC.

Greece

Cattle and dung feature in Greek mythology, notably the sixth labour of Hercules and the story of his cleaning the Augean stables. It is believed that the myth dates back to Mycenean times (1600–1100 BC) and may be based on someone living in Argos. However, documents on Greek agriculture are few and far between. Hesiod in *Works and Days*, a farmer's almanac (8th century BC), gave details about living and farming, including the use of oxen for ploughing, but there is no mention of manure. In the Odyssey Homer tells that Odysseus, when returning to Ithaca from Troy, found his dog Argus lying on dung outside the stable door and that the manure was to be used on his estate (Book 17 c.800 BC).

Xenophon (430–354 BC) in his *Economy* gives information about living off the land in the form of a Socratic dialogue. In a debate about how many are not aware that it is good to mix manure with the soil it is said: "No wonder a man does not get wheat from his farm when he takes no pains to have it properly manured." Manure is said to be "the best thing in the world for agriculture, and every one can see how naturally it is produced."

The Greeks also appreciated the value of leaving land fallow every other year and of crop rotations, indicating the breadth of their agricultural knowledge. At the time of the Trojan wars (c.1200 BC) Greek agriculture and diet was based on barley and to a lesser extent on the more demanding wheat. Arable land was limited. The rocky soils were better suited for olive trees, fruit orchards and vines. Lentils, chick pea and beans were also grown and there was an appreciation of their value to improve soils for the next crop. Theophrastus (370–285 BC) wrote that beans best invigorate the ground and that they rot easily (Smil, 2001). Grassland was limited which meant that only the wealthy reared cattle. Oxen were used as work animals. Sheep and goats were better adapted to the hilly terrain.

Rome

The Romans appear to be the first to write at length about the use of manure and of legumes. Early accounts of Roman agriculture come from the writings of Cato the Elder (234–149 BC),

Virgil (70–19 BC) and Columella (4–70 AD).

It is not possible to say whether the Romans were self-taught about manure or whether they learnt from the Greeks, the Egyptians or the Phoenicians. With their extensive trading activities around the Mediterranean the Phoenicians could have been likely transmitters of agricultural knowledge from the Fertile Crescent during the period 1500 to 300 BC.

Cato's manual on running a farm, De Agri Cultura or *On Agriculture*, is the oldest surviving work of Latin prose. It covers farm management as well as crop and animal husbandry. It became a textbook at a time when Romans were expanding their agricultural activities and when the cultivation of grain was less important than that of the vine, the olive and domestic vegetables, or the rearing of cattle.

Cato paid particular attention to varieties of olives and vines as well as the cultivation of a wide range of crops including wheat, barley, lentils, rape, turnips, radishes, millet, panic-grass, lupines, asparagus, figs, fruit and nut trees. Grafting, preferably around a new moon, of apples, figs, olives, pears, quinces and vines was described in detail. It is curious that Cato talked of hauling the manure and of applying it to meadows "in the dark of the Moon". Did Steiner get some of his ideas about the supposed influence of the Moon from Cato?

The application of urine from pigs was recommended for pomegranates "to serve as food for the fruit". Clover, vetch, fenugreek, beans and bitter-vetch were grown as forage for cattle.

Cato advocated having a large dunghill and that all manure, including that from pigeons, goats, and sheep, should be carefully saved and any foreign matter removed. Half the manure should be applied to forage crops and grassland, preferably in autumn. The remaining manure was applied in trenches dug around the olives, vines and fruit trees; vegetables were then grown on top of the trenches.

Manure hampers and baskets were included in the detailed lists of required equipment for olive groves and vineyards. The beneficial effects of cattle dung on crops were thus clearly appreciated by the Romans more than 2,100 years ago. Cato also understood that lupines, beans, and vetch "were crops which fertilize land". The preparation of kilns and the burning of limestone with wood were described by Cato. The lime was used to make mortar but it is not clear whether any was also applied to soils.

Virgil wrote four instructive poems on farming activities, *The Georgics* (Agricultural Things, 29 BC). He wrote about wheat, fruit, olives and vines and about grafting, ploughing, seeding, about bees and about animal breeding, but there was no mention of manure.

In contrast, Columella (3–70 AD) in his twelve volumes on all aspects of Roman agriculture, wrote at length in Book II of *De re Rustica* about dung and manuring of a wide range of crops. On the one hand he recommended that organic wastes had to be matured before being used as soil amendments and that weed infestations could be minimized by using of old manure. On the other hand he also pointed out that incorporating manure on the day of its production maximized its effect; he also compared the fertilizer value of different legume species and advocated the growing of lucerne to improve soil fertility. He valued its longevity as well as its productivity.

Varro, a prolific Roman scholar (116–27 BC), gave many details about the use not only of manure but also the use of leguminous crops and of clover to improve subsequent crops in his three books, *De re Rustica*. The Romans used human excrement and considered it superior to the dung of sheep and goats but inferior to bird droppings. Little was mentioned about the quantities of manure applied.

Europe

Chorley (1981) reported on the marked increase in the cultivation of leguminous crops between 1750 and 1880 in North Europe. He pointed out that the legumes, with their ability to fix atmospheric nitrogen, were very valuable and compared favourably with manure, which was thought at the time to be the main input that improved yields. He concluded that the widespread growing of leguminous crops and the consequent increase in the nitrogen supply was a major driver of agricultural development during this period.

Poore, G.V. (1843–1904)

Poore deserves a mention because it is thought that his books (e.g. Poore, 1902) and articles on the use of organic matter, vegetable and human waste influenced Howard and others (Conford, 2001). Poore was a doctor, an authority on sanitation and a keen gardener. He composted all organic wastes, to which he added human excreta from an outdoor earth closet. He believed that soil could handle human waste and that humus had powerful disinfecting properties. He postulated that man-made fertilizers could be harmful but the only one available when he wrote was superphosphate.

Summary

1. Domestication of some animals may have preceded or was more or less contemporaneous with that of plants.
2. The realization that the application of animal manure can improve crops inevitably followed the domestication and corralling of cattle. The benefits of human excrement on crops may have been appreciated in China and Japan in pre-historic times well before the benefits of animal dung.
3. The Sumerians in Mesopotamia did not mention dung on any of their thousands of tablets.
4. The ancient Egyptians would appear to have valued bird droppings more than animal manure.
5. Cattle and dung are mentioned in the Torah but without any indication of

its value. The Carthaginians used bird droppings as fertilizer.

6. The Greeks and particularly the Romans were the earliest to write about the value of manure. They were clearly aware that several kinds of dung were beneficial and should be carefully used. The Greeks also knew that it was beneficial to let fields lie fallow.

7. The Chinese, Japanese and Koreans were probably the first to appreciate the value of legumes, followed by the Romans who understood that leguminous crops helped to maintain soil fertility.

8. The Chinese, Japanese, Koreans and Romans employed the two key elements of organic farming – the recycling of animal manure and the growing of legumes. As they were not tempted by man-made fertilizers or pesticides they cannot be thought of as the first organic farmers.

9. Farmers in the 18th and early 19th centuries began to grow more legumes but they had no reason to know that legumes were able to take nitrogen from the air and make it available for crops or that the manure they appreciated was not generating nutrients only recycling them.

References

Abu Zeid, M.A. (1989) Environmental impacts of the High Dam. *Water Resources Development* 5, 3, 156.

Allaby, R.G. (2008) The rise of plant domestication: life in the slow lane. *Biologist*, 55, 94–99.

Anon (1960) Crop Nutrition and the World food supply. *Journal of the Royal Institute of Chemistry* 84, 47–88.

Branford Oltenacu, E.A. (2005) Domestication of animals. In: Pond, W.G. and Bell, A.W. (eds.) *Encyclopaedia of Animal Science*. Marcel Dekker, New York.

Cato (234–149 BC), De Agri Cultura (on Agriculture). Available at http://penelope.uchicago.edu/Thayer/E/Roman/Texts/Cato/De_Agricultura/A*.html

Chorley, G.P.H. (1981) The Agricultural Revolution in Northern Europe, 1750-1880: Nitrogen, Legumes, and Crop Productivity. *The Economic History Review New Series*, 34, 1, 71–93.

Columella (3–70 AD) De re rustica. Available at http://penelope.uchicago.edu/Thayer/E/Roman/Texts/Columella/de_Re_Rustica/2*.html

Conford, P. (2001) *The Origins of the Organic Movement*. Floris Books, Great Britain.

Cushman, G.T. (2013) *Guano and the Opening of the Pacific World: A Global Ecological History*, Review 1589. Cambridge and New York: Cambridge Univ. Press. doi: 10.14296/RiH/2014/1589

Davy, H. (1813, 1844) *Elements of Agricultural Chemistry*. Richard Griffin and Company, Glasgow.

Harari, Y.N. (2014) *Sapiens – A Brief History of Humankind*. Harvil Secker, London.

Harris, C.C. (2001) Ancient Egyptian Agriculture. Available at www.touregypt.net/egypt-info/magazine-mag07012001-magf5.htm#ixzz38EDRPoYj

Hesiod (8th century BC), *Works and Days* Translated by Evelyn-White, H.G. Available at www.sacred-texts.com/cla/hesiod/works.htm

King, F.H. (1911) *Farmers of Forty Centuries, or Permanent Agriculture in China, Korea and Japan*. Mrs F.H. King, Madison, Wisconsin, USA. Available at www.gutenberg.org

Kislev, M.E. and Bar-Yosef, O. (1988) The Legumes: The Earliest Domesticated Plants in the Near East? *Current Anthropology* 29, 1, 175–178. doi: 10.1086/203623.

Marshall, M. (2013) *New Scientist* October. Hunter-gatherers got on fine with Europe's first farmers. Available at www.newscientist.com/article/dn24385-huntergatherers-go-on-line-with-europes-first-farmers.html

McKevitt, S. and Ryan, T. (2013) *Project Sunshine: How Science Can Use the Sun to Fuel and Feed the World*. Icon Books Ltd, London.

Murray, M. A. (2000) Cereal production and processing. In: Nicholson, T and Shaw, I. (Eds.) *Ancient Egyptian Materials and Technologies*. Cambridge University Press, Great Britain.

Nadel, D. and Sternberg, M. (2015) Discovery in Israel pushes back dawn of agriculture to 23,000 years ago. Available at www.haaretz.com/jewish/archaeology/1.667258

National Geographic (2014) Evolution of diet. September. Available at www.nationalgeographic.com/foodfeatures/evolution-of-diet

Paull, J. (2009) Permanent Agriculture: Precursor to Organic Farming. *Elementals, Journal of Bio-Dynamics Tasmania*, 83, 19–21.

Paull, J. (2011) The making of an agricultural classic: farmers of forty centuries or permanent agriculture in China, Korea and Japan, 1911–2011. *Agricultural Sciences* 2, 3, 175–180. doi: 10.4236/as.2011.23024.

Peeples, L. (2009) Did Lactose Tolerance First Evolve in Central, Rather Than Northern Europe? *Scientific American* August 28. Available at www.scientificamerican.com/article/lactose-toleraence/

Perry-Gal, L. (2015) Ancient Israelites first to eat chickens rather than make them fight. Available at www.haaretz.com/jewish/archaeology/.premium-1.667089

Pfeiffer, E. (1947) Soil Fertility, Renewal and Preservation. *Biodynamic Farming and Gardening* (original edition 1938). Faber and Faber. London. Available at www.soilandhealth.org

Poore, G.V. (1902) *The Earth in Relation to the Preservation and Destruction of Contagia* (1899 lecture). Longmans. Green and Co. London.

Rau, W.R. (1878) Genuine Peruvian Guano compared with other manures. East Anglian Handbook 1878, 73. Available at www.bernardoconnor.org.uk/Coprolites/Guano.htm

Smil, V. (2001) *Enriching the Earth, Fritz Haber, Carl Bosch and the Transformation of World Food Production*. MIT Press. Cambridge, Massachusetts.

Tanaka, Y. (1998) The cyclical sensibility of the Edo period Japan. *Japan Echo* 25, 2, 12–16.

Varro (116–27 BC) *De re Rustica*. Available at http://penelope.uchicago.edu/Thayer/E/Roman/Texts/Varro/de_Re_Rustica/home.html

Virgil (70–19 BC) *Georgics* (four poems on agricultural things).

Xenophon (430–354 BC) *Oeconomicus* (the Economy) Translated by Dakyns, H.G. The Project Gutenberg. Available at www.gutenberg.org/ebooks/search/?query=dakyns&go=Go

Zimmer, C. (2015) Agriculture Linked to DNA Changes in Ancient

Europe. Available at www.nytimes.
com/2015/11/24/science/agriculture-
linked-to-dna-changes-in-ancient-europe.
html?_r=0

Zohary, D., Hopf, M. and Weiss, E. (2012)
Domestication of Crops in the Old World.
4th edition. Oxford University Press,
Oxford.

CHAPTER 2

The Fertilizer and Pesticide Roads – from Manure to Man-made Fertilizers and from Sulphur to Synthetics

"All roads lead from and to Rome"

The road to fertilizers is paved with farmyard manure and droppings. It proceeds from the historical use of manure to the promotion of improved agricultural practices facilitated by the printing press and the writings of wealthy farmers. Here the story will be told, from a European perspective. With the advent of agricultural chemistry and increasing world trade, farmers were able to supplement animal manure with nutrients from new materials such as guano and Chilean nitrates. After the basic research of Haber, the door was opened to the fertilizer industry.

From Romans to the printing press (1476)

Much would have been learnt about agriculture from the several translations of the book *Ruralia Commoda* by Petrus de Crescentius (1230–1321) of the University in Bologna. It is likely that Crescentius passed on the Romans' appreciation of dung. His book was widely read, passing through many editions. Many books on agriculture appeared in Europe, particularly in Italy and France in the 15th century.

The Romans would have brought their knowledge of the value of manure with them to the UK but it is doubtful whether dung would have been applied to crops because most cattle would have been permanently on pastures; many cattle would have been slaughtered and eaten every winter because of the shortage of winter feed, leaving little manure to be moved. It is understood that the Romans brought turnips to Britain together with several vegetables, including onions, leeks, cabbage, peas, celery, radishes and asparagus (Cross, 2006).

According to Pliny, the Celts applied lime and marl to their fields. Saxon chiefs probably realized the value of animal droppings as their tenants were obliged to put their sheep and other stock on the chief's fields at night (Porteus, 1948). The tenants may have applied other organic materials such as the straw, ferns and bracken used as carpets when combined with night soil.

From Roman times to the 16th century farming expertize would have been passed on by word of mouth. Here the experience, largely in Great Britain, as reported by Porteus (1948), will be used as an example of the developments that probably took place in countries across Europe but with different actors.

Medieval farmers were well aware of the need to maintain soil fertility, both by fallowing and the application of manure applied directly on the fallows by sheep or after collection (Baugh, 1989). As previously, the lord of the manor would often compel his tenants to fold their sheep on his land.

The creation of fields from common land – the enclosures – was necessary before it was worthwhile devoting time and effort on cultivation and manuring. Before enclosure much farmland existed in the form of numerous, dispersed strips only under the control of individual cultivators during the growing season. After harvesting, the land reverted to communal use. Enclosing, which prevented the exercise of common grazing, was instigated by the landowners as a means of increasing their own grazing land for sheep. In England the movement for enclosure began in the 12th century and proceeded rapidly so that by 1500 45% of land was enclosed; this number rose to 71% by 1700, and to 95% by 1914 (Turner, 1986). In the rest of Europe enclosure made little progress until the 19th century, which might possibly have delayed the appreciation and use of manure.

Before agricultural chemistry (1477–1800)

Many people played a part in furthering the progress of agricultural science on the road to fertilizers; here we highlight a few authors on agriculture mostly from Great Britain. It is likely that similar progress was being made at the same time in other European countries. Widespread dissemination of agricultural knowledge was greatly facilitated by the introduction of printing and the writings of progressive landowners keen to pass on their experience.

Fitzherbert of Norbury (1470–1538)

The first book on agriculture in English was *The Boke of Husbandry* by Fitzherbert of Norbury in 1523 (Porteus, 1948). Fitzherbert recommended fallowing to keep up soil fertility and advised against applying manure before winter rains but that it should be spread evenly in the late spring. Fitzherbert, in common with the Romans, expressed a preference for dove and pigeon dung – this was at a time when eating pigeons was common.

Bernard Palissy (1510–1590)

Palissy was a Huguenot who achieved fame for discoveries regarding geology, hydrology and fossil formation. In agricultural circles Palissy is remembered for his prescient statement in 1563 (Russell, 1973): "When you bring dung into a field it is to return to the soil something that has been taken away". Likewise, he realized that the ash of burnt straw was beneficial as it returned salts to the soil that the crop had taken away.

Thomas Tusser (1524–1580)

It is curious that the second agricultural book in English was farming poetry by Tusser in 1557 entitled *Five Hundred Points of Good Husbandry* (Porteus, 1948). Did Tusser know that he was following in the footsteps of Virgil's *Georgics*? The poem is essentially an agricultural calendar. He was a great believer in the value of well-rotted farmyard manure, which he recommended should be quickly ploughed in "before the hot sun could dry, or rain wash away its goodness". He advised that land should be fallowed every third year. He also appreciated that the benefit of folding sheep on pastures was due to their droppings. Like Fitzherbert, he approved of pigeon manure and advised that night soil was buried in trenches at night. The only crops mentioned by Tusser in his comprehensive book are cereals, peas, hops and saffron. Crop rotations and turnips are not mentioned.

Walter Blith (1605–1654)

Blith's book *The English Improver Improved*, published in 1653, carried a very strong message on yield promoting properties of manure. Blith also had a great faith in the fertilizing power of water, known as floating, which involved a slow flooding of the land from nearby streams. The success of this method may have been partly due to the streams being polluted with sewage. A modern parallel is the application of the contents of drains in Cairo on centre pivot farms on desert soils. Blith was one of the first to champion crop rotations and temporary grass crops.

John Houghton (d. 1705)

Houghton wrote more than 500 weekly letters on husbandry known as *The Collection*. He wondered how manure benefited crops. His problem was that he believed crops got their nutriment from nitre or other salts in the air that was washed into the soil by rain (Porteus, 1948). His nitre was potassium nitrate, which was known from its use in gunpowder. Perhaps if he had applied potassium nitrate to crops he would, today, be credited with being the originator of agricultural chemistry. At the same time he was a firm supporter of manure and pigeon droppings.

J. Mortimer (1656–1707)

Mortimer, in his book, *The Whole Art of Husbandry* (1707), wrote of the need to have a balance on the farm between the areas of pasture and arable crops whereby there is sufficient pasture to support the cattle needed to provide manure for the arable crops. Today this is a key aspect of most biodynamic farming. He recommended how to prepare dung and also the use of various waste organic materials.

Rotations, turnips and Charles Townshend (1674–1738)

Early farming developed either by the practice of shifting cultivation, in which fields were abandoned after some years of cultivation and virgin land occupied instead, or by farming in the valleys where annual floods deposited nutrient-rich sediments to replace the nutrients removed in the harvests. The practice of growing a series of different types of crops in the same area in sequence over the seasons evolved over the years

alongside the realization of the benefits of fallow, of animal grazing and of legumes.

In the simplest two-course rotation of the Romans a cereal crop was followed by a fallow year. From the middle ages to the 18th century a three-year rotation was practised in much of Europe. A winter-sown wheat or rye crop would be followed by spring-sown oats or barley and then a fallow; alternatively a legume crop could have been grown in the three-course rotation with a cereal and a fallow. Livestock may have grazed the spring-sown crops as well as weeds in the fallow year.

The growing of turnips was pioneered by farmers, in Belgium (Waasland) in the early 16th century and popularized by the British agriculturist Townshend in the 18th century.

Viscount Townshend enjoyed a distinguished political life before resigning in 1730 and devoting himself to farming in the last few years of his life. He promoted growing turnips for stock and became so obsessed with turnips that he gained the title "Turnip Townshend". Turnips were valuable in two ways. First, they could be grown as a winter feed and so permit more animals to be kept over the winter. The manure produced in the winter was used on arable crops, possibly as much as 25 t/ha once every four years (Johnston and Poulton, 2005). Second, it led to subsequent crops benefiting from the manure that was deposited when animals were folded on the turnips in the field. Animal breeding was considerably facilitated by using turnips as a winter feed. Townshend's studies on crop rotations led to the Norfolk four-course rotation, which typically consisted of wheat, turnips, barley or oats and a clover/grass ley. The grazed turnip crop prepared the soil for the following season's cereal, after which a one year clover/grass ley restored soil nitrogen ready for the next crop of wheat. Such a system eliminated the need for fallowing and kept the soil in good condition. Townshend found that his four-course rotation gave higher yields than when the land was fallowed for one year (Porteus, 1948). The adoption of this rotation was a key development in farming.

Jethro Tull (1674–1741)

Tull is famous for inventing a turnip seed drill designed to replace the broadcasting of massive amounts of seed that was not only wasteful but also resulted in uneven, difficult to weed, crops. Experiences in Shropshire (Baugh, 1989) revealed that in the 1320s wheat yields exceeded the amount sown by a factor of up to 5:1, whereas by the 1850s the ratio could be as high as 16:1. Tull was sadly mistaken about how plants obtain nourishment from the soil. He thought that plants feed on small particles of soil, which led him into developing hoes and ploughs that would break up the soil into small edible particles.

Arthur Young (1741–1820)

Young was very influential through his fifteen books on agricultural matters, although it is acknowledged that he was not a successful farmer. He travelled widely and was a campaigner for the rights of agricultural workers. He was an enthusiastic advocate of the use of farmyard manure, of turnips and of crop rotations and he helped to introduce potatoes, carrots, cabbages and onions. In 1793 he became the first secretary of the Board of Agriculture.

Coke of Holkham, Norfolk (1752–1842)

Coke is famous for his many annual gatherings held at shearing time known as "Holkham Clippings". The gatherings can be thought of as a cross between a modern agricultural show and the Oxford farming conference. He valued manure, clover and turnips and promoted seed drilling.

Johann Rudolf Glauber (1604–1670)

Glauber was a Dutch alchemist and chemist who handled two chemicals that were to figure as fertilizers. They were potassium nitrate, which he found in the earth from cattle sheds, and potassium sulphate, which he made from reacting potassium nitrate with sulphuric acid. He also discovered sodium sulphate, which led to the laxative being named after him (Glauber's salt). Having found that potassium nitrate increased crop growth, he thought that the value of dung and animal material such as horns and bones was entirely due to potassium nitrate.

William Ellis (c.1700–1758)

By the early 18th century many materials were being used as soil amendments. For example, Ellis (1732), a Hertfordshire (Little Gaddesden) farmer, wrote (1732) that much of the success of local farming was due to manuring with soot, ashes, horn shavings, rabbit clippings and rags. He advised that wheat grew better after a clover crop, preferably of two years. His popular book, *The Practical Farmer*, is available on Amazon.

Agricultural chemistry from 1800

Despite the fact that many people around the world had appreciated the value of manure and night soil for many hundreds of years it was not until the 19th century that any progress was made in understanding the reasons for the benefits as well as how crops grew. Eminent natural philosophers (now known as scientists) such as Francis Bacon (1561–1626) and Robert Boyle (1627–1691) in England and van Helmont (1580–1644) in the Netherlands all believed that water was the principal nourishment of plants (Russell, 1973). The advent of what we would now call agricultural chemistry provided the breakthrough, starting in the 18th century. Bacon and Boyle would have not been called scientists – William Whewell, an English polymath, coined the term scientist in 1832.

Joseph Priestley (1733–1804)

The first evidence that gases participate in photosynthesis was provided by Joseph Priestley in 1772 when he showed if a plant was placed in a closed glass container lacking oxygen that oxygen was generated. He had earlier demonstrated that a burning candle in a sealed chamber soon goes out and that similarly a mouse in a sealed chamber soon suffocates; in both case all the oxygen in the air had been used up.

Sir Humphry Davy (1778–1829)

Davy, a renowned chemist, better known for his invention of a safety lamp for miners, can claim to be the first agricultural chemist. Analytical chemistry had taken off in Europe and chemists such as Davy were motivated to analyse all the substances they could. He used chemistry to further the understanding of farming practices. His analytical studies and eight lectures (1802 to 1812) were published as a book, *Elements of Agricultural Chemistry*, in 1812. He relied on the findings of previous workers but did not accept de Saussure's demonstration that plant carbon came from the air as he thought it came from the soil. The recommendations he made regarding the preparation and use of manure in order to minimize loss of nutrients and its weed seed content would be readily accepted today. Davy may also be thought of as the father of the man-made fertilizer industry with his assessments of the potential value of what he called manures of mineral origin, or fossil manures. He analysed many organic materials, including blubber, bones, farmyard manure, fish refuse, guano, hair, horns, peat, night soil, seaweed, soot, straw and urine, to assess their value as soil amendments. He also analysed ammonium sulphate, common salt, gypsum, lime, magnesium limestone, peat ashes, potassium sulphate and sodium sulphate to determine their value.

Davy pre-dated both Boussingault and Liebig; however, their work is discussed in the edited version of Davy's lectures published in 1844. His analysis of pigeon droppings led Davy to forecast that Peruvian guano would be of great benefit; guano was first imported in 1840 into Great Britain.

John Joseph Mechi (1802–1879)

Mechi was primarily a businessman who lived during the time when Liebig had published his book on soil chemistry, and when Lawes was making superphosphate of lime and sodium nitrate and Peruvian guano were on the market. In his book *How to Farm Profitably* he forecast that one day analytical chemists would solve problems of plant nutrition.

Théodore de Saussure (1767–1845)

De Saussure was a Swiss chemist who made basic advances in plant physiology in the early 19th century, principally by using quantitative methods and chemical analysis. By measuring the weight of plants grown in air or in known mixtures of air and carbon dioxide he showed that the increase in dry mass of a plant as it grows was due almost entirely to its use of CO_2 (in the process we now know as photosynthesis) and only marginally to the absorption of minerals from the soil (Russell, 1973). De Saussure had effectively disproved the humus theory that decaying organic matter in the soil was the direct source of carbon in plants. Nevertheless, the humus theory held sway for many years until the work of Sprengel and Liebig. He appreciated that nitrogen was very important and he showed that it came from the soil and not from the air, a finding that was not accepted by Liebig. He analysed plant ash, finding it contained alkalis and phosphates as found in soil organic matter. It was fortunate that his teachings were accepted by Boussingault.

Jean Baptiste Boussingault (1802–1887)

Boussingault is a largely forgotten pioneer who after a career in the mining industry and academia devoted himself, from 1836, to agricultural chemistry. He spent ten years in South America. Having witnessed the remarkable effects of guano on crops growing on sandy Peruvian soils he decided to study guano when he returned to France in the early 1830s (Gorman, 2013). Boussingault demonstrated that legumes were able to utilize nitrogen from the air, something that Liebig declared was impossible. He was the first researcher to show that clovers and peas grown in nitrogen-free sterilized sand were able to increase the nitrogen content of the sand by harvest time – something wheat and oats were unable to do (Smil, 2001).

Boussingault established the world's first agricultural experiment station on his wife's property in Alsace in 1836, some seven years ahead of Rothamsted and the German equivalent in Moeckern (Saxony) in 1852. Sadly his experimental station did not survive him due largely to the Franco–Prussian war (1870). However, his discoveries were built on by others, including his better-known contemporary, Liebig – who acknowledged Boussingault as a pioneer and great discoverer in the field of soil and plant chemistry. Boussingault reintroduced the quantitative methods employed by de Saussure and showed that the nutritional value of fertilizer materials was proportional to their nitrogen content. His demonstration that legume crops can supply nitrogen to the soil was an important advance.

The spotlight was now on nitrogen, where it was to remain with a minor interruption caused by Liebig not accepting that plant nitrogen did not come directly from the atmosphere.

Carl Sprengel (1787–1859)

Sprengel was an agronomist and chemist who conducted pioneering research in agricultural chemistry during the first half of the 19th century; he deserves more than a brief mention. He predated Liebig, who it would appear from the studies of van der Ploeg *et al.* (1999), adopted much of Sprengel's findings and presented them as his own. By his paper

Sprengel (1826) refuted the then popular humus theory, whereby plants derive all their mineral matter from humus-derived organic compounds, consisting of C, H, O and N, some of which were transmuted into the mineral constituents found by plant analysis. Soil minerals were not thought to play a part.

Sprengel (1828) formulated the concept of the Law of the Minimum, which explains how a deficiency of only one essential nutrient would limit crop growth even if all other nutrients were present in adequate amounts. This was at a time when it was speculated that all minerals found in plants were essential. It is now known that plants have a limited capability to be selective in the mineral nutrients they absorb. Elements that may be toxic are absorbed.

Sprengel played a major but largely forgotten, and therefore unappreciated, part in the early days of plant mineral nutrition and soil chemistry. It would appear that Liebig appropriated Sprengel's findings in his books published in 1840 and 1855.

The contributions made by Sprengel are now recognized in Germany, where the Association of German Agricultural Experimental and Research Stations have created the Sprengel–Liebig Medal to recognize and commemorate the achievements of both pioneering scientists and so avoid disputes. It is hoped (van der Ploeg *et al.*, 1999) that the international community of agronomists will act similarly by recognizing Sprengel as a co-founder of agricultural chemistry and that the Law of the Minimum be called the Sprengel–Liebig Law of the Minimum.

Justus von Liebig (1803–1873)

Justus von Liebig was a German chemist who made major contributions to agricultural, biological and organic chemistry particularly by his book and papers. He was a great chemistry teacher and is widely considered to be the father of the fertilizer industry for his work on plant nutrients. He is frequently, but incorrectly, thought to be the originator of the Law of the Minimum. He was responsible, with Sprengel, for debunking the commonly held view that plants derived their carbon entirely from the humus in the soil – the humus theory.

However his claim to fame needs to be tempered by the realization (van der Ploeg *et al.* (1999) that he got several of the concepts now attributed to him from Sprengel, to whom he does not appear to have given credit; Sprengel is not mentioned in Liebig (1840). It is noteworthy that Liebig did not write about agricultural chemistry until 1840.

In his wide-ranging book he wrote about the inorganic contents of plants and soils, the origin and action of humus and rotations, as well as on manure. He pioneered the use of plant analysis, using a dry ash procedure for determining the nutrients that were in short supply. On the basis of ash content he divided crops up into potash crops (beets, turnips, potatoes and maize), lime crops (legumes and tobacco) and silica crops (wheat, oats, rye and barley). A separate wet ashing method is required for analysis of nitrogen in plant tissues, a procedure that perhaps Liebig seldom used and may have resulted in him being less aware of the variation in crop nitrogen contents than he would otherwise have been.

Liebig believed that plants get their all-important nitrogen from ammonia brought down in rain and that there was no opportunity for man to intervene. So it came about that Liebig's patent manures comprised plant ashes, gypsum, calcined bones, potassium silicate and magnesium sulphate. Not surprisingly, they did not produce heavy crops. Others, notably Lawes and Gilbert (1895), challenged the efficacy of Liebig's manure after finding in 1846 how it was important to supply nitrogen to wheat on Broadbalk, Rothamsted (Table 2.1).

Despite Lawes gaining similar results year after year Liebig disputed them and the two men were never reconciled. It is strange that Liebig was happy to use the work of plant physiologists when formulating his theories but was not prepared to accept the experimental evidence of Lawes. Henry Gilbert and Augustus Voelcker both studied and attended Liebig's lectures at his laboratory in Giessen where, after the Second World War, the university was given his name. Liebig is credited with inventing beef extracts from carcasses (OXO) and also of Marmite.

Sir John Bennet Lawes (1814–1900)

In 1822 Lawes succeeded to the Rothamsted estate at around 1837 and

Table 2.1 Comparison of wheat yields (1846) on Broadbalk treated with Liebig's manure with and without nitrogen and with farmyard manure (Lawes and Gilbert, 1895)

	Wheat kg/ha
Liebig's mineral manure	1361
Liebig's manure + 135 kg/ha $(NH_4)_2SO_4$	1966
35 t/ha farmyard manure	1832

began to experiment on the effects of various materials, particularly of ground bones on plants growing in pots. Having found that crushed bones were only effective in acid soils he treated bones (and rock phosphate) with sulphuric acid to solubilize the phosphate content. The product later called single superphosphate gave good results, especially on turnips, and was patented in 1842. Imported rock phosphate and coprolites (fossilized facces, particularly dinosaur) replaced the more expensive bones. The experiments were quickly extended to crops in the field. Apart from testing Liebig's theory that plants obtained their nitrogen from the air – a problem that was to dominate the rest of his life – Lawes's aim was to find ways of making farmers less reliant on farmyard manure because he realized the supply was insufficient to meet requirements.

Wheat in the Broadbalk experiment at Rothamsted was found to require abundant nitrogen, initially applied as either ammonium sulphate or sodium nitrate, in order to produce good yields. Similar results were found in parallel studies on a sandy loam soil at Woburn, much to the chagrin of Leibig, who had claimed that

the clay loam soils on Broadbalk were responsible for Lawes's results.

Lawes enlisted the services of Joseph Henry Gilbert in 1843 and the two worked together for more than half a century. By the 1850s they had proved the value of chemical salts supplying N, P, K and Mg. Lawes and Gilbert were puzzled by the ability of clovers to grow well even on poor soils whether they were given nitrogen or not and at the same time were rich in nitrogen. They had no knowledge of the nitrogen–fixing capacity of the nodule bacteria on the roots of clover and so never solved the problem.

Lawes also challenged Liebig's theory about ash analysis being an infallible guide to the nutrient requirement of crops, having shown with turnips that responsiveness was not necessarily related to crop composition.

Sir Henry Gilbert (1817–1901)

Gilbert, a trained chemist, worked with Lawes at Rothamsted and is to be remembered for the contributions he made to all the experiments carried out there. It was Gilbert who persuaded Lawes to continue the long-term Broadbalk experiment (and others) when Lawes wanted to discontinue it after he had disproved Liebig's theory about the source of nitrogen.

It had become clear by the second half of the 19th century that nitrogen occupied a central position in agriculture from its presence in all plant and animal tissues, its requirement for vigorous growth and its yield promoting properties.

Dr Augustus Voelcker (1822–1884)

Voelcker was the consulting chemist to the Royal Agricultural Society of England. His main interest was how farmyard manure should be made and stored to minimize deterioration. He also concerned himself with the loss of nutrients in drainage water from the different plots in the Broadbalk experiment. He found that phosphates and potassium are largely retained by the soil whereas considerable amounts of nitrogen, calcium, magnesium and sulphates are leached through soil.

Théophile Schloesing (1824–1919)

Schloesing, working with Pasteur, proved the bacterial origins of nitrification, in 1885 by experiments involving soil sterilization. He also showed that plants can absorb NH_4 through their leaves.

Winogradsky (1856–1953)

Winogradsky was the first scientist to isolate the two nitrifying bacteria (*Nitrosomonas* and *Nitrobacter*) in 1889.

Hermann Hellriegel (1831–1895) and Hermann Wilfarth (1853–1904)

Although the value of growing legumes has been appreciated by farmers from ancient times and in several parts of the world including China, Japan, Korea Greece and Rome, it was not until the end of the 19th century that anything was known about the nature of the benefit of legumes. Even after Boussingault had shown that clover and peas grown in sterilized sand could increase the nitrogen content of the sand, how this came about remained a mystery.

The breakthrough came with the classic work of Hellriegel and Wilfarth (1888) in Germany who demonstrated that soil microbes when existing in association with peas, beans and clovers fixed nitrogen. They also showed that the nodules on the roots were not just storage organs of nitrogen but also the place where the nitrogen fixation occurred (Smil, 2001).

Martinus Beijerinck (1851–1931)

Beijerinck, a Dutch microbiologist, was the first to isolate the nitrogen-fixing Rhizobia bacteria in pea nodules.

Albert Frank (1839–1900)

Frank was a German botanist who discovered that blue-green algae (cyanobacteria) were able to fix nitrogen. The cyanobacterium *Anabaena azollae* forms a symbiotic relationship with *Azolla*, an aquatic fern commonly used in paddy rice cultivation to provide nitrogen. Frank is also credited with coining the term mycorrhiza.

Ulysse Gayon (1845–1929)

Gayon and his assistants demonstrated that denitrifying bacteria can reduce nitrates and, via NO and N_2O, return N_2 to the atmosphere.

Dr Robert Warrington (1837–1907)

Work at Rothamsted had shown that the ammonia applied in farmyard manure or as ammonium sulphate did not stay as ammonia in the soil for long but was converted to nitrate, in which form the nitrogen was absorbed by the plant. How this came about had to wait until Pasteur had discovered microorganisms and Schloesing had shown that they were able to generate nitrate during the purification of sewage (Porteus, 1948). It was Warrington at Rothamsted who proved that microorganisms in the soil had a similar capacity. Both organic and inorganic nitrogenous substances can act as raw materials for the soil microorganisms, indicating that the microorganisms were not adversely affected by the application of soluble inorganic nitrogen compounds.

Warrington also advocated that soil physical conditions and soil microorganisms should not be forgotten when considering how to improve crop growth; this was important at a time when inorganic fertilizers were taking centre stage and being seen as a universal panacea. He championed the value of organic materials not only to increase the humus content of soils but also to improve the physical condition of both clay and sandy soils. It is possible that his promotion of soil physics alienated Warrington from Gilbert the chemist and led to his departure from Rothamsted. The improvements of soil physical conditions by massive amounts of manure on the growth of spring-sown crops at Rothamsted have taken more than 100 years to become apparent (see Chapter 13).

Samuel William Johnson (1830–1909)

Johnson was an American agricultural chemist who by his book *How Crops Grow* provided some of the first insights into plant nutrition in USA in 1868. He had studied under Liebig and probably as a result of using dry-ashing techniques for multi-element plant analysis tended to ignore nitrogen requirements. The early fertilizer industry in the USA was largely based on superphosphate in the latter years of the 19th century.

Early soil amendments

Up to the beginning of the 19th century farmers had developed three ways to supply plant nutrients in order to maintain soil fertility. First, they recycled as much as possible of the nutrients originally removed by the crops as farm-yard manure and crushed animal bones. Second, they applied litter collected from meadows and forests. Third, they grew leguminous crops as part of the rotation. Over the centuries farmyard manure had played a dominant role in maintaining soil fertility but by the late 18th century there was a growing interest in the use of and experiments with other materials, starting with bones and later with an impure sodium nitrate and guano. By the latter half of the 19th century a considerable trade of these new materials had developed, mainly from South America to Europe and the USA (Table 2.2).

Bones

Some 2,000 years ago the Chinese applied calcined bones, probably a by-product in the production of bone China. In Europe the fertilizing property of bones was probably first realized in the early 18th century by farmers in the Sheffield area, who collected night soil containing entire bones and horn shavings (Perkins, 1994). Bones were then considered simply as another animal waste. The bones were repeatedly placed in the middle of a dung heap until they disintegrated. The phosphate content of bone was not appreciated until Liebig's work. Lawes solved the problem as to why bones were satisfactory on some soils and not on others. He recognized that differences in soil acidity determined whether or not the bones became soluble. The practice of applying crushed bone usually mixed in with farmyard manure became popular

and imports were required to augment domestic supply. In 1837 more than 230,000 tons of bones were imported from as far as Argentina and India into the UK. In the 1850s 300,000 cat mummies were imported from a Pharaonic cat graveyard in Beni Hassan (Perkins, 1994). Coprolites from Gloucestershire and Cambridgeshire were also used for their phosphate content. Crushed bone meal is still promoted for use by the home gardener.

In the USA, buffalo bones, possibly from more than 30 million carcasses that had been left to rot, were collected from the Great Plains and shipped east to be ground up for fertilizer. This was after the near elimination of buffalo hunted for their skins in the mid-19th century.

Guano

Guano was first imported into Europe in about 1840. Although some of the deposits were very deep (up to 60m) these were quickly exhausted and exports fell after 1870.

The guano of commerce is the excrement of seabirds that can only build up under arid conditions where very numerous birds have access to huge fish stocks, as in regions adjacent to the Humboldt Current in the Pacific adjacent to South America. Guano is a highly effective fertilizer due to its nitrogen, phosphate, and potassium content.

Guano is still bought by some organic farmers, for example seabird and bat guano is offered for sale online in the USA and the UK. The best Peruvian guano typically contained 8 to 16% N (mainly uric acid), 8 to 12% P_2O_5 and 2 to 3% K_2O. However, the nitrogen content can vary widely and could be as low as 1% N where it has been heavily leached. In Shropshire it was applied to wheat at 250–375 kg/ha (Baugh, 1989). Von Humboldt is credited with being the first European to encounter Peruvian guano in 1802 when he investigated its fertilizing properties. His writings later alerted Europeans, particularly Boussingault, to its potential. During the guano boom of the 19th century, the vast majority of seabird guano was harvested from deep deposits on Peruvian islands. Guano imports into the UK peaked around 1870 at about 300,000 tons per year and fell to 25,000 tons in 1888 when supplies were running out. By this time about 20 million tons had been exported, of which 12 million tons went to Europe. After 1870 Peruvian guano lost its position as a fertilizer to sodium nitrate mined in the Atacama Desert. During the war with Peru (1879–1883) Chile seized much of the guano as well as Peru's nitrate-producing area. Fraudulent marketing of adulterated guano led to the development of a test to determine its chemical composition.

Guano is approved for use by IFOAM but Codex Alimentarius advises that the permission of the certifying body is required.

Chilean sodium nitrate

Sodium nitrate had originally been used in gunpowder. The benefits of Chilean sodium nitrate ($NaNO_3$) as a fertilizer were appreciated in the early 19th century.

After the rapid depletion of the guano deposits, European agriculture turned increasingly to Chilean sodium nitrate as a nitrogen source and later to the by-product ammonium sulphate.

The mineral deposit caliche contains 6–10% $NaNO_3$, together with sulphates and chlorides of sodium, potassium, calcium, magnesium and small amounts of various micronutrients including borates and iodates. Davy (1813) found Chilean nitrates to be very variable in composition.

To manufacture Chilean sodium nitrate, caliche is crushed and then the soluble nitrates are extracted by leaching. The sodium and potassium nitrates are recovered separately by crystallization. Chilean sodium nitrate contains about 98% $NaNO_3$. Large scale production of $NaNO_3$ began around 1850. The first shipment of Chilean sodium nitrate to Europe arrived in England in 1820 and between 1830 and 1880 annual imports increased from 8,300 to 226,000 t.

Chile became embroiled in the First World War when Germany was the major importer of its nitrates for use in explosives. After an initial defeat, off the coast of Chile, in late 1914 the British navy regrouped and defeated the German fleet. Thereafter a massive blockade prevented German access to its nitrate holdings.

IFOAM, Codex Alimentarius (1999) and the European regulations do not permit organic farmers to use Chilean sodium nitrate. Although it is a natural product it is a mineral and as such is in the same category as man-made fertilizers. It is not permitted because it is in accord with the conventional concept of plant nutrition and contradicts the organic concepts (FiBL, 2005).

Birth of the modern fertilizer industry

The story of the birth of the fertilizer industry is essentially one about the manufacture of nitrogenous fertilizers based on the ground-breaking Haber–Bosch process. Developments with phosphorus and potassium fertilizers depended on more mundane activities – mostly of mining and refining of ores.

Sir William Crookes (1832–1919)

Crookes, a renowned chemist, warned in a British Association lecture in 1898 that conventional agriculture could not feed the growing population without fixing nitrogen from the air (Smil, 2001). This was at a time when the benefits of Chilean sodium nitrate and of guano were appreciated but supplies were diminishing. He called on all chemists to find a way to fix nitrogen chemically from the unlimited reserves in the air. He said that without a new source of nitrogen fertilizer famine would be inevitable. A few years before, in 1895, Frank and Caro had shown how calcium cyanamide ($CaCN_2$, 20% N) could be manufactured from calcium carbide and nitrogen. Apart from its use

as a fertilizer calcium cyanamide was used as a defoliant, herbicide and a fungicide before the development of specific pesticides. The economics of calcium cyanamide production depended on the availability of cheap energy.

Birkeland and Eyde (1902)

The natural phenomenon, in which the heat generated by lightning causes gaseous nitrogen in the air to be oxidized to NO and NO_2, which after oxidation in the atmosphere are deposited as NO_3, had been understood since the 18th century. In 1902 Birkeland and Eyde in Norway simulated nature by passing an electric arc through air at very high temperatures to make first nitric acid and then calcium nitrate. This was seven years before Haber's work in Germany but it is Haber who is usually remembered and the Norwegians forgotten. Eyde, the industrialist, had realized there was a growing need for nitrogenous fertilizers. Although large-scale manufacture was subsequently achieved by Norsk Hydro, the high costs of production meant that the process was largely restricted to Norway, where hydroelectric power provided a relatively cheap source of electricity. In the 1920s, it was replaced in Norway by a combination of the Haber process and the Ostwald process for making nitric acid from ammonia.

Fritz Haber (1868–1934)

There is no unequivocal evidence that Haber was motivated by developments in agricultural chemistry. It would appear that initially he studied the combustion of hydrocarbons and electrochemistry, a far remove from Liebig. His association with Bunsen possibly led to his work on the chemistry of gases and his finding, in 1909, when working at the University of Karlsruhe, that he had made ammonia by reacting nitrogen and hydrogen at 1,000° C with iron as a catalyst. Realizing the potential, he attempted to synthesize ammonia at lower temperatures using different catalysts and by circulating nitrogen and hydrogen over the catalysts at 150–200 atmospheres at 500° C. At this stage Bosch of BASF became involved and by 1913 an ammonia plant was operating at Oppau near Ludwigshafen – and the rest, as they say, is history. Ammonia is the basic raw material for the production of nitric acid, and thence of nitrates, that would be used not only as fertilizers but also as explosives. Haber's reputation is sullied by the fact that he was also closely involved in developing and using poison gases (chlorine and phosgene) that were used in the First World War and of Zyklon-B used in the extermination camps in the Second World War.

The importance of Haber's discovery on how to fix atmospheric nitrogen for mankind cannot be overstated. It is now known that about half of man's nitrogen needs comes from the man-made fixation of ammonia pioneered by Haber: the other half being derived from the biological fixation of nitrogen by microorganisms (Chapter 16). Haber received the Nobel Prize for Chemistry in 1918.

Carl Bosch (1874–1940)

The Haber process for fixing nitrogen was purchased by the German chemical company BASF, who assigned Carl Bosch to scale up Haber's laboratory procedure to an industrial level. He succeeded and was awarded a Nobel Prize for Chemistry, in 1931. BASF started industrial production in 1913 in time for the war-extending production of explosives. It was not until 1924 that the first Haber plants were built in France and Britain after Bosch had given the French government the required technical details at the treaty of Versailles after the First World War.

Global consumption of N fertilizers

In 1920 only 16% of consumed N fertilizers was derived by the Haber–Bosch process but by 1940 the contribution had risen to 69%, to 82% in 1955 and to 99% by 1990 (Smil, 2001). The contributions by Chilean nitrates and ammonium sulphate became minor (Table 2.2).

The global consumption of N fertilizers has grown rapidly since the 1950s and seems to have plateaued at around 110 million t N in 2014 (Table 2.3).

Fertilizers – the end of the road

A basic attribute of most modern fertilizers is that they should be readily available, which often equates with products being water soluble. This property does not have the approval of the organic movements who, with special

Table 2.2 Global nitrogen production (1,000 t N). Synthetic ammonia = Haber–Bosch process (Smil, 2001)

	Chilean nitrate	Guano	Ammonium sulphate	Calcium cyanamide	Synthetic ammonia
1850	5	–	0	0	0
1860	10	70	0	0	0
1870	30	70	0	0	0
1880	50	30	0	0	0
1890	130	20	–	0	0
1900	220	20	120	0	0
1905	260	10	130	0	0
1910	360	10	230	10	0
1920	410	10	290	70	150
1929	510	10	425	255	930
1940	200	10	450	290	2,150
1950	270	–	500	310	3,700
1960	200	–	950	300	9,540
1970	120	–	950	300	30,230
1980	90	–	970	250	59,290
1990	120	–	550	110	76,320
2000	120	–	370	80	85,130

Table 2.3 Global consumption of N fertilizers (1,000 t N). Data 1850–2000 (Smil, 2001), from 2002 (FAOSTAT, 2016)

1850	5	1960	11,290
1860	80	1970	31,600
1870	100	1980	60,600
1880	80	1990	77,100
1890	150	2000	85,700
1900	360	2002	82,588
1905	400	2004	89,023
1910	610	2006	91,963
1920	930	2008	95,567
1930	2,130	2010	100,791
1940	3,100	2012	106,282
1950	4,780	2014	108,937

exceptions, only accept a few inorganic materials if they are water insoluble. It is likely that the reasons for the objection to water-soluble fertilizers is based on them being counter to the organic concepts (e.g. disapproval of man-made materials) and concerns over possible detrimental effects of soluble salts on soil microorganisms.

A high active ingredient content is a valuable property of fertilizers in order to cut shipping and handling costs, e.g. di-ammonium phosphate with 18% N and 46% P_2O_5 and triple superphosphate with 44–48% P_2O_5 are preferred to single superphosphate with 16–22% P_2O_5; urea with 46% N is preferred to ammonium sulphate with 21% N and to sodium nitrate with 16% N.

The importance of N relative to P and K is reflected in the dominance of the fertilizer market by N, which accounts for 60% of the more than 170 million t nutrients compared with P (24%) and K (16%) in 2010/2011. Wheat, rice and maize consume about half of all fertilizer used globally in agriculture, followed by cash crops such as vegetables, fruit, flowers and vines (YARA, 2012).

Nitrogen fertilizers

In the 19th century farmers came to appreciate the value of nitrogenous fertilizers such as guano and Chilean nitrates, which required less handling than farmyard manures – often applied at rates of more than 25 t/ha.

With the exception of Chilean sodium and potassium nitrates (also available from the Dead Sea) and ammonium sulphate made by reacting the ammonia in coal gas with sulphuric acid, all nitrogenous fertilizers are ultimately dependent on fixation of ammonia in factories. Initially hydrogen derived from the electrolysis of water was combined with nitrogen from the air. Today hydrogen is generated from natural gas by steam reforming. Trinidad has benefited from its massive reserves of natural gas by becoming the largest exporter of ammonia, most of which goes to North America. The ammonia is used to produce nitric acid and thence for all nitrogenous fertilizers, such as ammonium nitrate (NH_4NO_3), urea ($CO(NH_2)_2$), di- and mono-ammonium phosphates and

calcium ammonium nitrate. When natural gas supplies run out it is to be hoped that new cheap forms of power generation (possibly fusion energy) will permit the use of the electrolytic production of hydrogen to be used instead of methane.

The global nitrogen fertilizer market is dominated by urea (Table 2.4) despite it not being as efficient as other N fertilizers and it being a polluter (via NH_3 and NO_2, Chapter 17). Up to 20% more N is needed to give the same yields as ammonium nitrate (YARA, 2011). Nevertheless, urea is likely to continue to be the main N fertilizer because of its use in China, India and Brazil. The preference for urea is fundamentally because of the economics of production and transport. Although ammonium nitrate (33% N) is a very efficient fertilizer there are also security considerations because of its use as an explosive.

Whilst urea has a lower carbon footprint at the production stage than ammonium nitrate, this benefit is offset by the loss of NH_3 and generation of more NO_2. Plants do not directly absorb urea

Table 2.4   Main nitrogenous fertilizers. Proportions used based on N (YARA, 2012)

	%
Urea	57
Ammonium/Calcium Nitrate	8
Urea/Ammonium Nitrate solutions	5
NPK	9
Ammonium phosphates	6
Ammonia	4
Other	11

in significant quantities. Urea needs to be first hydrolysed to ammonium by soil enzymes, which takes between a day and a week, depending on temperature and soil moisture content.

Although urea could be thought of as a natural substance, being found in urine, and capable of being synthesized in leaves, organic farmers are not permitted to use it, probably because it is man-made.

Not surprisingly, optimum rates of N application exceed national average rates (Table 2.5). For winter wheat in Europe about 200kg N/ha as ammonium nitrate is required for optimal yields (in France 183 and Germany 210 kg N/ha, YARA, 2011).

The lower average rates of N application in India and Brazil indicates that, subject to water availability, there is a considerable potential to increase productivity of the main crops.

The increasing global demand for meat is the major driver of the application of N to wheat and maize that are mostly used to feed livestock. Pork and poultry are gaining in popularity. To produce 1 t of poultry meat, feed corresponding to 2 t grain is needed, for pork about 4 t and for beef 7 t. About 20–30% of total N fertilizer

Table 2.5 Average N application rates kg N/ha/year to main crops (YARA, 2012)

	EU	USA	China	India	Brazil
Wheat	130	75	182	110	
Maize	166	155	173		83
Rice			191	100	

consumption is understood to be required for meat production (YARA, 2012).

The application of 192 kg N/ha to winter wheat in Europe, can result in yields of 9.3 t of grain/ha in soils where the yield without N application would be 2.07 t/ha (YARA, 2012). Massive amounts of manure would be required to supply such amounts of N.

Phosphorus and potassium fertilizers

Liebig is credited with being the first to show that bones treated with sulphuric acid produced a more effective fertilizer than crushed bone. The first man-made phosphatic fertilizer was superphosphate initially made from treating crushed bone with sulphuric acid and later by using imported rock phosphate. The reaction results in a fertilizer with 60% calcium sulphate (gypsum) and 25–30% water-soluble mono-calcium phosphate, most of which is converted to an insoluble form in the soil. Lawes set up a superphosphate factory in 1843 using coprolites as the main raw material. James Fison and associates founded companies in the 1850s to make superphosphate using bones and coprolites from East Anglia and by 1871 there were about eighty factories making superphosphate in Great Britain.

Phosphoric acid, produced by reacting rock phosphates with sulphuric acid, is the starting material for the manufacture of mono- and di-ammonium phosphate (46% P_2O_5 and 18% N). Triple superphosphate is made by reacting rock phosphate with phosphoric acid. Organic farmers are not allowed to use synthetic

P fertilizers but they can use basic slag (c. 2% P_2O_5) and rock phosphate containing less than 90 mg Cd/kg P_2O_5. In the UK Tunisian rock phosphate seems to be preferred, which according to Oosterhuis *et al.* (2000), is likely to contain 94–173 mg Cd/kg P_2O_5. Food grade phosphates with 13 mg Cd/kg P_2O_5 are also manufactured in Tunisia.

Consumption of potassium and phosphorus is much less than that of nitrogen (Table 2.6).

Potassium nitrate was the first potassium salt known to improve plant growth. In Elizabethan times the production of KNO_3 for use in gunpowder was small scale and would have precluded it being used as a soil amendment. It was perhaps not appreciated that the impure Chilean nitrates were partially beneficial because of their potassium content.

It was the opening of the German potash mines in 1860 that effectively

Table 2.6 Global consumption (million t) of potassium and phosphorus fertilizers (FAOSTAT, 2016)

	K_2O	P_2O_5
1961	8.7	10.9
1965	12.1	15.8
1970	16.4	21.1
1975	21.4	25.6
1980	24.2	31.7
1985	25.7	32.5
1990	24.7	36.0
1995	20.7	30.7
2000	21.7	32.4
2005	25.3	34.5
2010	33.2	42.9
2014	37.6	46.7

kick-started the potassium fertilizer industry, which today is dominated by potassium chloride, potassium sulphate and potassium nitrate. Before potash minerals such as carnallite, sylvinite and kainite had become available the nutrient potassium was obtained from wood ashes – hence the name potash. Organic farmers can use several potassium salts, such as potassium chloride and potassium sulphate, provided they are produced by physical processes and not by any chemical procedure. The ores kainite and sylvinite are also permitted. Some variation with regard to potassium is apparent. In the UK potassium sulphate can be used after approval by the Soil Association (2014); justification has to be provided for the use of natural rock potash and ores. The main potassium fertilizer is muriate of potash (KCl), which contains 50% K.

A wide range of compound fertilizers containing N, P, and K, can be made simply by mixing straight fertilizers. Other compound fertilizers involve chemical reactions e.g. the production of nitrophosphates by acidifying phosphate rock with nitric acid to produce phosphoric acid and calcium nitrate which is then ammoniated. Other nutrients can be added to form compound fertilizers containing several nutrients in a single pellet.

Secondary nutrients and micronutrients

The regulations governing the use of secondary nutrients such as Mg, Ca and S by organic farmers appear to be relaxed. Provided there is justification, dolomitic limestone, gypsum (calcium sulphate), ground chalk and limestone, Epsom salts and kieserite can all be used (IFOAM, 2014 and Soil Association, 2014).

Historically manure has been an important source of micronutrients and will have prevented deficiencies in low yielding crops. The discovery of the micronutrients that are essential for plants, namely Fe (1860), Mn (1922), B (1923), Zn (1926), Cu (1931), Mo (1938), Cl (1954) and Ni (1987), followed the growth in the use of N, P and K. Micronutrient deficiencies are largely restricted to a limited number of crops and soils are treated by conventional farmers by soil or foliar application of appropriate inorganic salts (often metal sulphates) or organic chelates.

Organic farmers are allowed to treat micronutrient and sulphur deficiencies after diagnosis and in some cases approval by the certifying body. In contrast to the organic objections to manmade N and P fertilizers and to the use of chemical processes to make K fertilizers, where it is the way the fertilizer is produced that contradicts the concepts of organic farming, no such prohibition seems to exist with regard to the micronutrients.

Use of fertilizers

In 2010–11 total world fertilizer consumption reached 171 million t (Mt) of nutrients, comprising 104 Mt N, 40 Mt P_2O_5 and 27 Mt K_2O. About 60% of this total is estimated to have been applied to cereals (Heffer, 2013). Consumption of $N+P_2O_5 + K_2O$ in China, USA and India together accounted for 30 Mt (per year) in the period 2006–2009 (FAOSTAT, 2016). The FAO forecasts that the demand for nitrogen, phosphate, and potash will grow annually by about 1.3–2.7% in the near future (FAO, 2012).

The influence of farming practices and of fertilizers on agricultural production can be seen from the progression of wheat yields over the last 800 years before the advent of plant breeding and better varieties (see Table 18.1).

The ability of American farmers to produce food for urban populations has increased markedly over the years (Table 2.7).

At the end of the road to fertilizers we find the traditional farmer on a mixed farm, rotating his crops and supplementing the nutrients in farmyard manure with inorganic fertilizers in order to improve yields. Such traditional farming existed in Europe between the wars. It was a transient phenomenon – the age of synthetic herbicides, insecticides and fungicides was yet to dawn and industrial agriculture was some way off.

Table 2.7 Persons fed by one American farmer (USDA, 2010)

	Number
1940	19
1950	27
1960	61
1970	72
1980	112
1990	129
2000	139
2010	155

The pesticide road from sulphur to synthetics

The finding of insect pests, of diseases and of suffocating weeds would have motivated agricultural chemists to search for solutions.

The road to pesticides, which here includes insecticides, fungicides and herbicides, followed a path similar to that of fertilizers. Farmers were faced with the problem of finding ways to prevent or limit the damage caused by pests and diseases. Inevitably they looked for natural, often noxious, materials for the solution. With fertilizers the breakthrough came with the factory fixation of nitrogen in the early 20th century. Advances in synthetic chemistry around the Second World War led to the discovery of the insecticidal properties of DDT and the selective weed-killing properties of the phenoxyacetic acid compounds 2, 4-D and 2, 4, 5-T.

Insecticides

The first use of insecticides was probably about 2400 BC by the Sumerians, who rubbed foul-smelling sulphur compounds on their bodies to deter insects and mites. In Greece, Homer, around 800 BC, described how Odysseus fumigated the hall, house and court with burning sulphur to control pests. At this time the Chinese were using mercury and arsenic compounds to control body lice. The Chinese were also practising perhaps the earliest form of integrated pest management when using predatory ants to protect citrus groves from caterpillars and wood-boring beetles; they even used ropes or bamboo sticks tied between adjacent trees to help the ants move around easily (Taylor, 2007). The Greeks and Romans used olive oil, ash and sulphur to deter insects.

The Persians used a powder derived from grinding the dried flowers of *Chrysanthemum cinerariaefolium* to protect stored grain. The Crusaders brought the knowledge back to Europe that dried pyrethrum controlled head lice. By the 19th century pyrethrum was sometimes used on crops but its instability in sunlight and high cost limited its use. The synthetic pyrethroids are stable. Nicotine sulphate, extracted from tobacco leaves, has been used as an insecticide since the 17th century. It is effective against most types of insect pests, but is used particularly against aphids and caterpillars – soft-bodied insects.

Rotenone can be obtained from tissues of several members of the *Fabaceae* family but particularly from the roots of the two members *Derris* and *Lonchocarpus* grown in South America. It was used there as early as 1649 to paralyse fish, causing them to surface. The insecticidal properties of rotenone were realized later in the middle of the 19th century. It acts both as a stomach and a contact insecticide and has been used for about 150 years to control leaf-eating caterpillars (Taylor, 2007).

Insects can be controlled by materials acting as stomach poisons and for many years arsenicals (copper arsenate and calcium arsenate) were so used; from the mid-19th century the less soluble and less harmful, but more expensive, lead arsenate was used.

Carbon disulphide was used, from 1858, to great effect in the fight against the American phylloxera (an aphid-like insect) that attacks the roots of grapes. The only successful means of controlling phylloxera had been the grafting of European grape varieties on phylloxera-resistant American rootstocks.

Hydrogen cyanide was used for treating citrus trees (under tents) against scale insects in California at the end of the 19th century (Sherwood Taylor, 1957). It was still used for the fumigation of the dormant nursery stock of deciduous trees up to the middle of the 20th century and for the quarantine treatment of bananas and pineapples to control aphids and mealy bugs FAO (1969).

Cryolite (Na_3AlF_6, sodium hexafluoroaluminate), studied because of its fluorine content, was found in the 1930s to be an effective insecticide by acting as a stomach poison. Until 1987 cryolite was sourced from the west coast of Greenland but has been replaced by synthetic cryolite, which is more efficaceous. Cryolite is used in the US, not only on grapes, citrus and potatoes but also on other crops including broccoli, Brussels sprouts, cauliflower, melons, peppers, and squash as a wettable powder sometimes in combination with other insecticides.

The insecticide armoury in 1940 was limited to several arsenicals, petroleum oils, turpentine, nicotine, pyrethrum, rotenone, sulphur and cryolite. But the situation was about to change.

Discovered in 1939, dichloro-diphenyl-trichloroethane (DDT) was, by 1945, being widely used in agriculture and also to combat insect-borne illness such as malaria and typhus. DDT was particularly effective in killing Colorado potato beetle, a very serious pest in North America and Europe. Its success was a big incentive to agricultural chemists.

The synthesis of the organophosphate insecticide parathion was the result of German chemists experimenting with nerve gases during the Second World War. Its extreme toxicity led it to be banned in 1972 in the USA. Other insecticides developed during the Second World War included BHC, Aldrin, dieldrin and endrin, all of which were banned following the Stockholm Convention in 2001 largely because of their persistency and their effects on birds. The bio-accumulation of these insecticides down the food chain resulted in the loss of birds of prey due to the thinning and breakage of their egg shells.

Bacillus thuringiensis (Bt) and the synthetic pyrethroid insecticides were released in the 1970s and became very popular in the 1980s. The spores and crystalline insecticidal proteins produced by *B. thuringiensis* had been studied since the 1920s. The insecticidal activity of *B. thuringiensis* has recently been bred, by genetic modification, into Bt maize and Bt cotton.

The wide spectrum activity of many early insecticides was a drawback because beneficial organisms and natural enemies of pests were also killed. The development of resistance has been another problem.

Fungicides

Diseases such as blights and mildews (and also of locusts and caterpillars) are mentioned in the Old Testament (First Kings 8:37, about 550 BC).

During Roman times smoke was used against mildew and blights. Material such as straw, dung, and animal horn was burnt windward of the orchard or vineyard. The Romans dealt with cereal rust diseases by invoking a special rust god (Taylor, 2007). In 1807 a copper sulphate solution was used to control bunt disease in wheat. Bordeaux mixture, based on copper sulphate and lime, has been used to control downy mildew on grapes from the late 19th century. It remains one of the most widely used fungicides worldwide. Sulphur was also used to treat fungal diseases on grapes. The successful development of Thiram, in 1934, was followed by other fungicides. Organic mercury compounds were used to protect seeds from rust and other diseases from 1913 to the 1960s, when they were banned due to their toxicity.

Herbicides

Olive oil may have been the first herbicide. It is said that Theophrastus (371–287 BC), who is considered the father of botany, killed young, undesirable trees by pouring olive oil on their roots.

Historically weeds were controlled mainly by hand weeding but various chemical methods have also been described, such as the use of salt and sea water. Varro (116–27 BC) described how amurca, an olive oil waste product, could be used either on its own or mixed with other materials both as a herbicide and as an insecticide; its activity may have been because it was prepared in copper vessels.

During the 19th and early 20th century, European and the American broadleaf weeds in cereal crops were controlled with dilute sulfuric acid, iron sulphate, copper sulphate, copper nitrate, sodium chlorate and sodium arsenate; all acted non-selectively.

The Second World War saw the introduction of the phenoxyacetate herbicides. 2, 4-D and 2, 4, 5-T. Agent Orange, which consisted of these two herbicides, was used to destroy vegetation and crops on a massive scale in the Vietnam War during 1961–1972. The presence of contaminant highly toxic dioxins in 2, 4, 5-T caused serious health issues, including tumours and birth defects in exposed citizens. Herbicides atrazine, paraquat, and picloram were developed in the late 1950s and the broad spectrum herbicide glyphosate in the 1970s.

The ability of some crops to withstand glyphosate and very recently 2, 4-D has been achieved by introducing the appropriate genes by genetic engineering. This has facilitated weed control, particularly in maize and soybeans.

The global pesticide market is dominated (in value) by herbicides, which in 2009 accounted for about 50% of the total with insecticides and fungicides both at about 25% (UNEP, 2013).

Integrated Pest Management

Integrated Pest Management (IPM) aims to keep pests at economically insignificant levels by using crop production methods that discourage pests and encourage beneficial predators that attack pests (e.g. ladybirds that eat aphids and the tiny wasps such as *Aphelinus abdominalis* that lay their eggs in caterpillars and aphids). Limited use of pesticides is allowed in IPM. Insecticides that are less harmful to beneficial insects may be required to assist IPM.

Summary

1. Before the advent of agricultural chemistry many wrote about improving crop production, which usually involved the application of manure and the cultivation of clover.

2. It was appreciated many years ago that manure should be carefully stored and quickly incorporated into the soil.

3. The introduction of turnips and the adoption of the Norfolk four-course rotation, which also included a nitrogen-fixing grass-clover ley, marked a major advance in farming. Soil fertility was improved by grazing sheep on turnips in the winter and by the presence of clover in the rotation.

4. The big breakthrough in the understanding of the nutrient requirements of plants came with the work of physiologists such as de Saussure and Boussingault and analytical chemists, particularly Liebig.

5. De Saussure was responsible for demonstrating that plants use atmospheric CO_2 in order to grow.

6. Boussingault, using quantitative methods, showed that the nutritional value of fertilizer materials was mainly due to their nitrogen content.

7. Sprengel, an agronomist and chemist, played a major but largely unappreciated part in the early days of plant mineral nutrition and soil chemistry. He formulated the concept of the Law of the Minimum, whereby plant growth is limited by the element or compound that is present in the soil in the least adequate amount. His work appears to have been plagiarized by Liebig.

8. Liebig mistakenly believed that the nitrogen he found in plants came from ammonia in the air, which was at odds with the work of Boussingault.

9. Lawes at Rothamsted proved that Liebig's fertilizer, which did not contain nitrogen, was ineffective.

10. Some of Liebig's beliefs were presented as facts, which was unfortunate for him as others were able to design experiments to test his beliefs.

11. Liebig's emphasis and studies on phosphates led to increased use of bones and eventually to the manufacture of water-soluble phosphatic fertilizers both from bones and rock phosphates.

12. The finding of guano in Peru and the nearby Chilean saltpetre triggered exports to Europe and North America in the 19th century and led to improved crop yields.

13. By the end of the 19th century agriculture in Europe and North America was benefiting from nitrogenous and phosphatic fertilizers that were supplementing the nutrients in farmyard manure.

14. Much had been learnt about the value of applying nitrogen to crops and by the end of the 19th century it was realized

that guano and Chilean nitrates would soon run out. In a prescient lecture, Crookes called on chemists to find a way to capture nitrogen chemically from the unlimited reserves in the air. He said that without a new source of nitrogen fertilizer famine would be inevitable.

15. Atmospheric nitrogen was fixed by man, firstly in Norway in 1902 and then in Germany by Haber in 1909. So began in 1923, after the scaling up the process by Bosch of BASF, the commercial production of ammonia, nitric acid and nitrates.

16. Developments in the production of fertilizers containing potassium and phosphorus and of the micronutrients were relatively simple, mainly involving mining and refining.

17. The development of effective pesticides over the years has been largely fortuitous, depending on the local availability of noxious substances.

18. The insecticidal activity of pyrethrum and of rotenone has been appreciated for a few hundred years. Carbon disulphide played a valuable part in controlling the aphid-causing phyloxera on European grapes in the 19th century.

19. The discovery, in 1939, of the ability of DDT to control mosquitoes led to its widespread use in agriculture and at the same time stimulated the synthesis of compounds with insecticidal properties.

20. The Romans employed smoke when attempting to control mildew and blights in orchard and vineyards but it was not until the early 19th century that copper sulphate was found to be effective. Bordeaux mixture, based on copper sulphate and lime, was used for controlling downy mildew on grapes and it remains one of the most popular fungicides.

21. Olive oil was sometimes used as a herbicide in ancient Greece and Rome but farmers would have relied on hand weeding until it was found that various inorganic salts acted as non-selective herbicides. The Second World War saw the introduction of the phenoxyacetate herbicides, e.g. 2, 4-D, which could be used selectively.

22. The latter half of the 20th century saw the development of a very large number of several types of pesticides by a burgeoning agrochemical industry.

23. Weed control and insect control has entered a new phase with the use of genetic engineering. Some crops, notably maize and soybean, have now been genetically modified to be resistant to the non-selective herbicide glyphosate in order to facilitate weed control. Other crops, notably cotton, have been engineered to contain the Bt crystal protein and so permit selective insect control.

References

Apostolides, A., Broadberry, S., Campbell, B., Overton, M. and van Leeuwen (2008) English agricultural output and labour productivity, 1250–1850: some preliminary estimates. From the Project: Reconstructing the National Income of Britain and Holland, Available at www.basvanleeuwen.net/bestanden/agriclongrun1250to1850.pdf

Baugh, B.C. (1989) (Ed.) *A History of Shropshire, Volume 4 Agriculture*. The Victoria history of the counties of England, Oxford University Press, Oxford.

Codex Alimentarius (1999) Codex Alimentarius, GL 32–1999, Guidelines for the Production, Processing, Labelling and Marketing of Organically Produced

Foods Available at www.fao.org/fao-who-codexalimentarius/standards/list-of-standards/en/

Crescentius, P. (1306) *Ruralia Commoda*. Augsburg 1471 (in latin).

Cross, N. (2006) *Food in Romano Britain*. Available at http://resourcesforhistory.com/Roman_Food_in_Britain.htm

Davy, H. (1813, 1844) *Elements of agriculture*. R. Griffin, Glasgow (1844 edition contains notes on findings of Boussingault and Liebig).

Ellis, W. (1732) *The Practical Farmer: or the Hertfordshire Husbandman Containing Many New Improvements in Husbandry*.

FiBL (2005) Research Institute of Organic Agriculture. *Chilean nitrate and organic farming*, Frick, Switzerland. Available at www.betriebsmittelliste.ch/fileadmin/documents/de/hifu/stellungnahmen/0503-chilean-nitrate.pdf

FAO (1969) Manual of fumigation for insect control. Available at http://www.fao.org/docrep/x5042e/x5042E0b.htm#Hydrogen cyanide (HCN)

FAO (2012) *Current world fertilizer trends and outlook to 2016*. Available at ftp://ftp.fao.org/ag/agp/docs/cwfto16.pdf

FAOSTAT (2016) FAO Statistics Division www.fao.org/faostat/en/#data/RF

Gorman, H.S. (2013) *The Story of N: a Social History of the Nitrogen Cycle and the Challenge of Sustainability (Studies in Modern Science, Technology, and the Environment)*. Rutgers University Press New Brunswick, New Jersey.

Heffer, P. (2013) Assessment of Fertilizer Use by Crop at the Global Level *2010–2011*. *International Fertilizer Industry Association*. Available at www.fertilizer.org

Hellriegel, H. and Wilfarth, H. (1888) Untersuchungen über die Stickstoff-nahrung der Gramineen und Leguminosen.

Beilageheft zu der Zeitschrift des Vereins für die Rübenzucker-Industrie des Deutschen Reiches, Buchdruckerei der Post. Kayssler & Co., Berlin, Germany

IFOAM (2014) *IFOAM norms for organic production and processing Version 2014)*. Available at http://infohub.ifoam.bio/sites/default/files/ifoam_norms_version_july_2014.pdf

Johnston, A.E. and Poulton, P.R. (2005) Soil Organic Matter: Its Importance in Sustainable Agricultural Systems. Proceedings 565 of the International Fertiliser Society, York 1–48.

Lawes, J.B. and Gilbert, J.H. (1895) *The Rothamsted Experiments Over 50 Years*. Edinburgh and London: Blackwood and Sons, Edinburgh and London, UK.

Liebig (1840) *Organic Chemistry in its Application to Agriculture and Physiology*. Taylor and Walton London. Available at https://archive.org/stream/organicchemistry00liebrich#page/n5/mode/2up

Oosterhuis, F.H. Brouwer, F.M. and Wijnants, H.J. (2000). *A possible EU wide charge on cadmium in phosphatic fertilizers: economic and environmental implications*. Report 000/02 EU Commission. Available at http://ec.europa.eu/environment/enveco/taxation/pdf/cadium.pdf

Perkins, J. (1994) On bones and English agriculture 1750 to 1850, *Journal of the Royal Agricultural Society of England* 155, 199.

Ploeg, R. R. van der, Bőhm, W. and Kirkham, M. B. (1999) History of soil science. On the Origin of the theory of mineral nutrition of plants and the law of the minimum. *Soil Science Society of America Journal* 63: 5, 1055–1062.

Porteus, C. (1948) *Pioneers of Fertility*. Clareville Press, London.

Russell, E.W. (1973) *Soil Conditions and Plant Growth*. 10th edition. Longman, London.

Sherwood Taylor, F. (1957) *A History of Industrial Chemistry*. Heinemann, London.

Smil, V. (2001) *Enriching the Earth: Fritz Haber, Carl Bosch and the Transformation of World Food Production*. MIT Press, Cambridge, Massachusetts.

Soil Association (2014) *Soil Association organic standards farming and growing Revision 17*. 3 November. Available at www.soilassociation.org/organicstandards

Sprengel, C. (1826) Ueber Pflanzenhumus, Humussaure und Salze (About plant humus, humic acids and salts of humic acids). *Archiv für die Gesammte Naturlehre* 8: 145–220.

Sprengel, C. (1828) Von den Substanzen der Ackerkrume und des Unter- grundes (About the substances in the plow layer and the subsoil). *Journal für Technische und Ökonomische Chemie* 2, 423–474, 3, 42–99, 313–352, and 397–421.

Taylor, E.L. (2007) Pesticide Development a Brief Look at the History. Southern Regional Extension Forestry SREF-FM-010

Turner, M. (1986) English Open Fields and Enclosures: *Journal of Economic History* 46, 669–692.

UNEP (2013) *Global chemical outlook*. Available at www.unep.org/chemicalsandwaste/Portals/9/Mainstreaming/GCO/Rapport_GCO_calibri_greendot_20131211_web.pdf

Varro (116–27 BC) *De Re Rustica*. Available at http://penelope.uchicago.edu/Thayer/E/Roman/Texts/Varro/de_Re_Rustica/home.html

USDA (2010) *Briefing on the status of rural America*. Available at www.usda.gov/documents/Briefing_on_the_Status_of_Rural_America_Low_Res_Cover_update_map.pdf

YARA (2011) *Nitrate fertilizer Optimizing yield, preserving the environment*. Available at. http://yara.com/doc/33521_Nitrate_-_Pure_Nutrient.pdf

YARA (2012) *Yara Fertilizer Industry Handbook February 2012*.Available at http://yara.com/doc/37694_2012%20fertilizer%20industry%20handbook%20wfp.pdf

CHAPTER 3

Social and Agricultural Scene in 1920s and 1930s

"Study the past if you would define the future" Confucius

The 1920s and 1930s saw the birth in Europe of both biodynamic and mainstream organic farming. What were the social, economic and agricultural circumstances that provided fertile ground for the development of these organic movements?

Whereas biodynamic farming originated with the eight lectures by just one man, Rudolf Steiner, several people and groups were involved either actively or as supporters with the development of mainstream organic farming, which was initially and principally in the UK. Steiner lectured to a small audience in Koberwitz in Lower Silesia in 1924. The way in which biodynamic and mainstream organic farming developed will be discussed in Chapters 5 and 6.

Social, political and economic background

Several of the founder members of the Soil Association were of an age to have either experienced first-hand, or at least been well aware of, the horrors of the First World War. They would have been aware of the use of chemical warfare in what was perhaps the first industrial war. There would have been few families and communities in Western Europe who were unaffected by the conflict.

1920s

The 1920s was mostly a period of economic prosperity that accompanied the rebuilding after the ravages of the war.

The young generation from the aristocracy and wealthier classes who had been too young to fight were led into a hedonistic lifestyle, sometimes referred to as the Roaring Twenties, at clubs in the cities.

Poverty among the unemployed contrasted strikingly with the affluence of the middle and upper classes (Pugh, 2009). By the mid-1920s the post-war period of prosperity was over and unemployment had risen to more than 2 million in the UK. In some parts of the country it reached 70%. This led in the UK to the General Strike of 1926. In the USA the Wall Street crash of 1929 was followed by the Great Depression of the 1930s. In Germany economic crises in the early 1920s led to hyperinflation of the currency and to political instability. Steiner was approached for his advice on farming in 1923.

Apart from the Turkish wars of independence, Europe was at peace in the decade. However, across Europe there was a rise in popularity of extreme political movements with fascists setting up to counteract communism.

The women's suffrage movement advanced in the UK, with all women aged over 21 gaining the right to vote in 1928. There is no record that Lady Balfour was a suffragette. In the USA women, having been given the vote in 1920, began to enter the workplace in large numbers. Women took part in the Olympics for the first time in 1928.

Consumerism began. Technological advances led to the manufacture of many new consumer goods in the 1920s, including hair dryers (1920), hearing aids (1923), Clarence Birdseye invented a frozen food process (1924), television (1925), Geiger counter (1925), silent films gave way to sound films (1927), sliced bread in the USA (1928), and home refrigerators and dishwashers appeared in the USA (*The Telegraph*, 2000). The BBC began radio broadcasting in the UK (1922), insulin was discovered (1922), Lindbergh made the first solo flight across the Atlantic (1927), penicillin was discovered by Fleming (1928) (History of learning). Were there too many changes in lifestyle for those who longed for a simpler more natural life?

Plastics began to be used in consumer products. Nylon was first produced in the 1930s. The demand for household goods continued to grow, now prompted by advertising, the power of which had been appreciated at least in the USA, where by 1930 more than half of families had a radio set. The term 'soap opera' came into being when soap manufacturers sponsored domestic radio dramas and in return got frequent plugs for their products. In the UK the BBC's charter stressed education and the provision of news but advertisements were not allowed. Advertising on the radio had to wait for the arrival of Radio Luxembourg in 1933.

Transport by motor car had been popularized first by Ford, especially with its Model T, just before the First World War and later by Morris, in Oxford, who by the mid-1920s dominated the

UK market. Howard Carter had discovered the tomb of Tutankhamun in 1922, stimulating interest in foreign lands and cultures.

1930s

The story of the 1930s is dominated by the Great Depression (USA) and Great Slump (UK) starting in 1929 and the effects it had on living standards in the USA and Europe. There was massive unemployment, particularly in industrial and mining areas.

In Europe reform groups were formed whose main concerns were electoral, educational and social. Only one, the New English Weekly in the UK, advocated changes to farming practices and it provided the supporters of the organic movement with a mouthpiece. Food reform movements in Germany and the USA disapproved of industrialization, urbanization and the growing dominance of technology in the modern world; was this one of the legacies of the First World War? They called for a natural way of living consisting of vegetarian diets, physical training, natural medicines and a return to living on the land (Vogt, 2007).

Scientific advances and inventions in the 1930s included the electric razor (1931), radio telescope (1932), electron microscope (1933), cats' eyes (1934), ballpoint pen (1938), the helicopter (1939) and cloud seeding to trigger rain (1939) (*The Telegraph*, 2000, and History of Learning).

Agricultural background

Cereal prices in the UK had been guaranteed by the government during the First World War in order to stimulate production but after 1921 this ended and farmers suffered from cheap cereal imports. The wheat and barley acreage declined and remained low until the Second World War. There were some moves to specialize, e.g. dairying, motivated by efficiency and profitability. The Milk Marketing Board set up in 1933 helped to stabilize prices and to see many mixed farmers through the depression.

The dependence in the UK on imported sugar during the First World War stimulated the growing of sugar beet, with the first processing factory being built by the Dutch in Norfolk in 1912. This was followed by a further seventeen factories during the 1920s. It would appear that the agriculturally depressed areas were those where mixed farming predominated (Orwin, 1930).

There is little evidence that the yearning for living in harmony with nature popularized by the writings of Wrench had been taken on board by the fledgling organic movement, which seems to have been primarily motivated by a dislike of the way farming was developing. The revulsion to the use of science and of industrialization in the First World War had spread to agriculture.

Man-made fertilizers were being used, notably ammonium sulphate, nitro chalk, ammonium phosphate and superphosphate. The very limited use of pesticides before the Second World War meant that they would have attracted little attention. Plants provided several important non-synthetic insecticides, notably nicotine extracts from tobacco leaves, pyrethrums from chrysanthemums and rotenone extracted from the root of *Derris*. Synthetic chemistry advanced rapidly in the 1930s and the insecticide DDT (1937) and the first organophosphate compounds were synthesized. Nothing was made public about these compounds at the time in view of their potential as chemical warfare agents.

The first herbicides were developed during the Second World War. In 1946, the first commercially available chlorine-based herbicides were marketed to kill broadleaf plants. The compounds included 2, 4-D (2, 4-dichlorophenoxyacetic acid) and 2, 4, 5-T (2, 4, 5-trichlorophenoxyacetic acid).

The North American scene in the 1930s was dominated by the devastating soil erosion on the prairies, mostly by wind but exacerbated by repeated droughts, monoculture and basic farming ignorance as to how to minimize wind erosion. This was despite Americans being earlier alerted to the dangers and ravages of soil erosion by Bennett and Chapline (1928).

The story of a poor Oklahoma farming family engulfed by the Great Depression was told by John Steinbeck in his novel *The Grapes of Wrath* (1939). The experiences were those of the many small tenant farmers who were driven from their homes having suffered frequent droughts and economic hardship. They were trapped in the Dust Bowl and only escaped by going to California. This was the reality that has since been portrayed as the classic example of the lunacy of modern farming methods and incorrectly said to be due to the misuse of fertilizers. Instead, deep ploughing and the loss of the protection previously provided by deep-rooted grasses was the principal farming practice that was responsible. By 1940 more than 2 million people had fled from the Great Plains of Kansas, the Oklahoma panhandle and Texas.

At the same time floods in the basin of the Mississippi River led to water erosion and by 1933 the Tennessee Valley Authority had taken control of the valley, building dams that combined flood control and electricity generation. Perhaps the TVA was seen as further evidence of the industrialization of agriculture?

Increased use of tractors in the UK is evident from the decline in the numbers of agricultural horses, which had varied between 800,000 and 980,000 (1871 to 1910) and which then fell to 680,000 in 1930 and 290,000 in 1950 (Anon).

Public health

Malnutrition was a major cause of death in the UK in the interwar years (Pugh, 2009). Despite advances in treating diseases, it was felt that insufficient attention was being paid to the prevention of disease and particularly to human nutrition. It was such concerns that led the doctors and dentists in Cheshire to hold a meeting and to publish their Medical

Testament in 1939. The apparent links between the health of isolated communities and their natural methods of food production resonated well with those seeking solutions to poor health.

In London two doctors, Williamson and his wife, Innes Pears, had established the Pioneer Health Centre in Peckham in 1926 with the objective of improving health by combining nutrition and physical well-being with better social conditions. They farmed organically, probably under the influence of Howard, on their own farm to produce food for their canteen. Almost 1,000 families benefited from what became known as the Peckham experiment. The centre was finally closed in 1950 after the formation of the National Health Service (1948). The Pioneer Health Centre was ahead of the game in the 1920s (Pioneer Health Foundation, 2015).

The discovery of thirteen vitamins (including the eight B complex vitamins) between 1913 and 1931 and the realization of their importance could well have led to people hypothesising that other health-related factors in food were waiting to be found. Was it against such a background that the potential of arbuscular mycorrhizae were seen after their discovery by Rayner (1927)?

Summary

1. The horrors of the First World War, which would have been experienced by many who were to become involved with organic farming, are seen as the backcloth to developments in society and in farming in the post-war years.
2. The rise in consumerism, especially by the people enjoying the Roaring Twenties lifestyle, contrasted with the poverty of the unemployed.
3. The Wall Street crash, the Great Depression, the strikes in the UK and the economic crises in Germany provided fertile ground for reformers of society, including those looking for a simple way of living close to nature.
4. It is likely that the industrialization of the First World War and the continuing scientific developments were also anathema to reform groups and those who would form the organic movement in the UK and in Europe.
5. Cereal prices, which had been high during the First World War, were depressed by cheap imports post-war and resulted in a major reduction in wheat and barley production the UK.
6. In the USA the wind erosion in the Great Plains and the water erosion in the Mississippi basin led to them being cited as examples of the folly of mechanized/industrial agriculture and, incorrectly, to the use of fertilizers.
7. The building of dams to generate electricity and provide irrigation water by the Tennessee Valley Authority saw the start of the regeneration of agriculture in the USA.
8. There was a growing awareness of the importance of diet for good health.

References

Anon. A vision of Britain through time (1801–2001) Available at www.

visionofbritain.org.uk/unit/10001043/
cube/AGCEN_HORSES_GEN

Bennett, H.H. and Chapline, W.R. (1928)
Soil erosion a national menace. USDA,
Circular 33.

History of Learning. Available at https://
naldc.nal.usda.gov/naldc/download.xhtml
?id=CAT87212611&content=PDF

Orwin, C.S. (1930) *The Future of Farming*.
Clarendon Press, Oxford.

Pioneer Health Foundation (2015) The
Peckham experiment in the 21st century.
Available at http://thephf.org

Pugh, M. (2009) *We Danced All Night.
A Social History of Britain Between the
Wars*. Bodley Head Vintage Books,
London.

Rayner, M.C. (1927) Mycorrhiza. *New
Phytologist* 26, 2, 85–114. doi: 10.1111/
j.1469-8137.1927.tb06710.x.

The Telegraph (2000) A timeline of
inventions. An overview of the inventions
which have changed the course of history
(1900–1944). Available at www.telegraph.
co.uk/news/science/science-news/
4751183/A-timeline-of-inventions.html

Vogt, G. (2007) The origins of organic
farming. 9–29. In: Lockeretz, W. (Ed.)
Organic Farming, An International History.
CAB International, Oxford.

Part 2

The Nature, Origin and Benefits of Organic Farming

CHAPTER 4

Organic Farming – Principles, Promotion and Perceptions

"When I use a word, Humpty Dumpty said, it means just what I choose it to mean – neither more nor less. The question is, said Alice, whether you can make words mean so many different things? The question is, said Humpty Dumpty, who is to be master – that's all?" Lewis Carroll

Organic farming is here understood to be farming according to the methods and procedures certified by an approved body and complying with national regulations. Worldwide there are now more than 570 certifying bodies. What constitutes organic farming and what are the basic principles are considered using the views of international, national and non-governmental agencies. Proponents of organic agriculture sometimes broaden the organic umbrella so it is also important to establish what does not constitute organic farming. The basic nature of the two types of organic farming, namely biodynamic and mainstream, are compared as well as the promotion of organic food and the public perceptions that lead to purchases.

There are signs of an increasing influence of politics and bureaucracy in the

European and US organic markets, which may lead to organic farming being defined by lawyers far removed from farming and experts in the organic field.

Apart from the many claims that are made about the superiority of organic food, organic farming is said to be the route to sustainable food production (e.g. Duchy Home Farm, 2014). Whilst the matter of sustainability is unlikely to attract much public attention and so affect consumption of organic food, it is nevertheless clear that organic systems by using legumes, crop rotations, recycling, biological pest control measures and by minimizing external inputs can indicate ways to improved land management. However, these practices are not unique to organic farming but are also followed by conventional farmers.

Lord Northbourne (Walter James, 1896–1982)

It was Lord Northbourne who gifted the world the term "organic farming" in his book *Look to the Land* (Northbourne, 1940), which is a manifesto of organic agriculture (Paull, 2014). In it he foresaw a contest of organic versus chemical farming and a clash of world views that may last for generations. Rather than dealing with the practicalities of organic farming, Lord Northbourne's book presents the philosophy, the rationale, and the imperative of adopting organic methods. His ideas on organic farming spread quickly and were taken up in the USA and Australia in particular.

Some authors, mainly in the USA, mistakenly attribute the first use of the term organic farming to the American publisher Jerome Rodale, citing his use of the term in his periodical, *Organic Farming and Gardening*, which dates from 1942, two years after *Look to the Land* was published.

Northbourne was influenced by Steiner's ideas, and he implemented them on his own estate in Kent. He can thus be thought of as a biodynamic farmer, which is confirmed by the fact that Pfeiffer advised him. He was not a follower of Howard (Conford, 2001). However, Northbourne did have many ideas that were very similar to Howard's, such as the dislike of the industrialization of agriculture and especially of man-made fertilizers. Northbourne did not approve of experimentation and was not prepared to wait for scientific proof of every idea or hypothesis before taking action; otherwise he said we would hardly ever decide to act at all (Paull, 2014). He was of the opinion that decisions about most things that really matter have to be taken on impressions, or on intuition – an opinion that would have been readily accepted by Howard.

Farmers and growers

Roman and Peasant Farmers

It could be argued that the Romans, with their appreciation of the use of manure and of leguminous crops, were organic farmers. Likewise, the millions of farmers in the underdeveloped world who have no access to fertilizers and man-made pesticides could be said to be organic farmers, but that is only when looking back with some of the many self-imposed rules of the mainstream organic movement. The low productivity of Roman and African peasant farmers suggests their methods left room for improvement. However, if we wish to learn what is the best way of returning to a subsistence level of farming then it might be instructive to learn from peasant farmers in underdeveloped countries.

McKevitt and Ryan (2013) were perceptive when they said that images of rural living in the 19th century were responsible for the enduring myth that prior to the Industrial Revolution the lives of ordinary people living in rural communities were simpler, better and happier. There is little evidence that the peasants of the time would have considered their lives as idyllic. The notion of a noble peasant community completely in tune with nature and eking out a subsistence existence based on organic farming was a fallacy according to Moore-Colyer (2001). Peasant, self-sufficiency farming is sometimes referred to as organic by default because it is practised by the millions of small-scale farmers who cannot afford to buy chemical fertilizers.

Cuba

Farming in Cuba was changed abruptly following the collapse of the Soviet Union in 1991. Before then fertilizers and pesticides were purchased using the support provided by the USSR, which had bought Cuban sugar at inflated prices. Imports of both fertilizers and pesticide were stopped and Cuban agriculture effectively became organic. There are said to be 10,000 urban farmers and gardeners, particularly in and around Havana, who make a significant contribution to vegetable production by employing a growing system known as Organopónicos. The crops, mainly vegetables, are grown in narrow beds consisting of concrete or wooden walls filled with organic matter and soil; but they were difficult to find on a visit in 2016.

Field-grown organic coffee, bananas and mangoes are exported to the USA. Conventional farmers growing such crops as sugar cane, cereals, potatoes and beans are now able to obtain fertilizers and pesticides but pressure to satisfy the American organic market may mean the survival of some organic farming in Cuba. GM crops are now accepted in Cuba.

African farmers and UNEP-UNCTD report (2008)

From time to time supporters of organic farming include farmers in developing countries as fellow travellers, especially when yield improvements have been achieved. A typical example is afforded by the report (UNEP-UNCTAD, 2008) on 286 projects during the 1990s–2000s in several African countries, mostly in East Africa, designed to improve agricultural productivity: this work was reported by Pretty *et al.* (2006, 2011).

The UNEP–UNCTAD report, which extols the virtues of organic and near-organic farming, attracted attention when it was interpreted as demonstrating that organic agriculture is superior to conventional/traditional agriculture in Africa. There were claims that yields had been more than doubled by introducing organic or near-organic practices. The story was taken up by the media (e.g. Lean, 2009), which reported that many studies had shown organic techniques can be more productive in developing countries than conventional chemical-based agriculture. Both the Soil Association (2009) and the UK Green party (Lucas, 2008) used the studies to support their understanding about the superiority of organic farming.

However, the conclusions were not based on yield comparisons of organic farming with low-input farming using fertilizers and pesticides in any of the projects, at least in East Africa.

On seventy-one projects in East Africa average yield increases of more than 128% were derived from comparing the yields at the end of the projects with those at the beginning. In no project was the yield differences measured in the same field or on the same farm or in the same season. On all the 286 projects reported by Pretty *et al.* (2006) the average yield difference was 79%.

The better yields were the result of farmers being trained and advised on land and water management, on crop varieties, on conservation tillage, agroforestry and on integrated pest management. It was disappointing that no data was given of actual yields of the various crops either at the beginning or end of the projects.

Pretty *et al.* (2006, 2011) attributed the yield improvements to several factors associated with the training and advice given. The introduction of legumes and of new varieties played important parts. They mentioned that reduced pesticide use allowed the farmers to buy fertilizers and better seed, which indicated that the farmers could not all be described as organic. In reality the improvements in productivity had nothing to do with the introduction of an organic system in place of a traditional system but with the provision of advice and materials.

Increasing attention is being paid to organic farming practices in Africa, with conferences being held every few years under the auspices of the FAO, IFOAM and UNCTAD with assistance from ISOFAR and the International Centre for Research in Organic Food Systems (ICROFS) in Denmark. Confidence was expressed in the report on the second conference in Zambia in 2012 (FAO, 2013) that organic management can benefit people, the economy and ecosystems in Africa. The potential benefits of the use of fertilizers and pesticides seem to be largely ignored so the assessments of successes in terms of food production and living standards are made by comparing yields before and after introduction of improved organic techniques. The very low initial maize yields were exemplified by results from KwaZulu–Natal, where yields in poorly farmed semi-arid areas were as low as 0.4 t/ha. Fertilizers brought the yield up to 3 t/ha. To raise yields above this level by using organic methods could only be achieved with massive inputs of manure (20 t/ha) and/or composting, and by mixed cropping systems.

At the third conference in 2015 in Nigeria there was a call to expand the scope of organic agriculture beyond the use of manure, compost or organic fertilizers into other aspects such as organic livestock and aquaculture (ISOFAR, 2015).

Gardeners

Growing vegetables on allotments and in gardens is often described as being organic, which is clearly not the case. Garden centres sell inorganic fertilizers, herbicides and pesticides to meet demand.

A recent study by Sheffield University (Edmondson *et al.*, 2014) compared soils from small-scale organic gardening on urban allotments with soils from the surrounding agricultural region and other urban green spaces. Gardeners used manure and most made composts. It is not surprising that the allotment soils had more organic carbon and total nitrogen than nearby pastures and arable land and had a better physical structure. The potential of allotments to contribute to food production without the penalty of soil degradation led to a call for the promotion of allotments rather than further intensification of conventional agriculture.

Such calls to consider reducing food production on farms and to emphasize environmental concerns can only be contemplated in developed countries that have the ability to import much of their food. It also ignores the potential for pollution of garden soils. A study several years ago by Purves (1985) on garden soils in the Edinburgh area found them to be heavily contaminated with toxic heavy metals when compared with nearby farms on the same soil type. The reason for the contamination was that the composting and burning of vegetable and especially wood wastes by the home gardener had unwittingly concentrated the heavy metals (cadmium, copper, lead, nickel and zinc). Studies in the US (Baltimore) also showed elevated levels of the same metals in inner-city garden soils (Mielke *et al.*, 1983).

Biodynamic and mainstream organic farming

Globally there are basically two organic farming systems, each with their own set of rules and regulations. They are distinguished here as firstly biodynamic farming and secondly mainstream organic farming. Whereas biodynamic farming was very largely dependent on the ideas of one man, Steiner, mainstream organic farming benefited from the activities of a wide range of people not just farmers (see Appendix for Chapter 6).

Biodynamic farmers, the first to form a group, produced their first regulations in 1928. They will be members of a national association that is registered to Demeter International, which is responsible for worldwide certification. This results in biodynamic certification being more uniform globally than mainstream organic certification.

The first rules and regulations governing mainstream organic farming were published by the Soil Association in the UK in 1967. Since then standards have been drawn up by organic farming groups in different countries, by national authorities as well as by international bodies such as IFOAM and FAO.

Although biodynamic farming had at least a twenty-year start it is now overshadowed by mainstream organic farming. The latter benefited from the activities of several people, notably in the UK of Sir Albert Howard, Sir Robert McCarrison, Lord Northbourne, Lord Lymington and Lady Eve Balfour. Balfour was the principal driving force for mainstream organic farming with her two books (Balfour, 1943, 1975) and her lectures. She garnered huge support in many countries. It is revealing that her first book ran to at least eight editions whereas her second book, which gave the first results of the unsuccessful Haughley experiment, was less well received and ran to only one edition. In 2006 the Soil Association reprinted the Balfour (1943) edition, altering the subtitle to refer to post–war planning instead of "national planning" and omitting chapter 9 of Balfour (1943, 1949) entitled "The Haughley Research Farm Project". In America it was mainly the activities of Rodale and Pfeiffer (1938, 1947) that spread the organic story. Rodale popularized the term "organic".

The two organic farming systems have much in common but there are significant differences (Table 4.1).

Common features of biodynamic and mainstream organic farming

Both groups highly value farmyard manure and composts, which are seen as life-giving food for the soil rather

Table 4.1   Comparison of practices of conventional, biodynamic and mainstream organic farmers

Practice	Conventional	Biodynamic	Mainstream organic
Use of crop rotations	Yes	Yes	Yes
Use of animal manure	Yes	Yes	Yes
Use of legumes	Yes	Yes	Yes
Use of man-made fertilizers	Yes	No	No
Use of man-made pesticides	Yes	No	No
Use of antibiotics	Yes	No	No
Choosing pest- and disease-resistant varieties	Yes	Yes	Yes
Mandatory free-range and open-air livestock systems	No	Yes	Yes
Use of GM crops	Yes	No	No
Use of homeopathic preparations			
On soil and crops	No	Yes	No
With manure	No	Yes	No

than for the crop. Both groups use legumes for their nitrogen-fixing ability and crop rotations as a means of controlling/minimizing weeds, insect pests and plant diseases. Mixed farming is generally considered to be essential in order that the farmyard manure is produced on the farm where it is used.

Both groups abhorred the way society was developing. It was probably because of their instinctive dislike of modern (early 20th century) agriculture, which they called industrial, that both groups prohibited the use of man-made fertilizers and more recently, man-made pesticides, antibiotics, genetically modified (GM) crops and food irradiation.

Distinguishing features of biodynamic agriculture and mainstream organic farming

According to the Demeter Biodynamic Trade Association (Demeter, 2014), the biodynamic method is fundamentally distinguished from mainstream organic and other farming methods by four key attributes:

i.   Biodynamic farmers think in terms of forces and processes, whereas organic and conventional farmers think in terms of substances.
ii. Biodynamic farmers use nine biodynamic preparations in homeopathic quantities to enhance manures and composts.
iii. By understanding and working in harmony with the earthly and cosmic rhythms/cycles the biodynamic farmer is attuned to and works with the daily, monthly and seasonal rhythms of nature.
iv. The biodynamic farmer strives to maintain the farm as a self-contained and self-sustaining unit importing the minimum of nutrients.

The origin of the key aspects of working with cosmic forces and using homeopathy are discussed in Chapter 5 (also see Appendix). Some biodynamic farmers use homeopathic remedies when treating livestock, for example for mastitis.

Mainstream organic farming is essentially a collection of positive and negative actions or features. Here the term "negative" refers to practices that are not carried out, not that the actions are detrimental.

The use of farmyard manure, of leguminous crops and crop rotations are clearly positive actions. Not only have these farming practices been valued for many centuries but they are also valued today by conventional modern farmers.

It is the negative aspects of mainstream organic farming that largely distinguish it from conventional farming (Table 4.1). The Soil Association in the UK has become aware of being typecast as being against things (Soil Association, 2013).

The prohibiting of the use of soluble inorganic fertilizers, man-made pesticides, antibiotics and GM crops are the key practices with which the organic movement identifies itself. It is the prohibition that permits these practices to be labelled negative here. The organic movement is convinced that crops grown without fertilizers and pesticides contain more beneficial nutrients and fewer harmful pesticides, additives and nitrates, and are accordingly better for both animal and human health.

The Soil Association (2012) in its document *The Road to 2020* refreshed its strategy with the objective of reaching out to a wider audience. It wants to discuss with conventional farmers how ways can be found to get the best out of organic and conventional farming practices.

There is little doubt that the success of the organic food and farming business has been dependent on the negative aspects. How mainstream organic farming came to champion the negative features is discussed in Chapter 6. The question that can be asked about the negatives is what would the future for organic farming be if it was shown that soluble fertilizers and man-made pesticides were not harmful and that GM crops posed no risk to man or to the environment –would there still be a valid case for promoting organic food?

The inevitable diminishing supply of oil and raw materials means that ideally we need to eliminate, or at least drastically reduce, the use of man-made fertilizers and pesticides and work towards more sustainable forms of agriculture. On the other hand, the increasing global demand for food and particularly of animal products will limit such moves. Mankind is unlikely to turn vegan voluntarily so it will be difficult to reduce the demand for livestock products, despite the fact that they are very inefficient producers of protein.

Conventional and organic farming share several features (Table 4.1). Organic farming is increasingly science based and innovative, especially with regard to animal welfare and ecology.

Organic farms utilize much of the same technology that conventional farmers employ. Precision cropping equipment and hi-tech field mapping in crops are used by conventional and organic farmers alike. The assertion of superior animal welfare practised by organic farmers is difficult to accept, especially when it is realized that conventional farmers do their best to ensure that their animals are well cared for. All farmers know that animals that are cared for well will be more productive.

International and national agencies

Organic agriculture is variously defined and described by different organizations.

IFOAM

The International Federation of Organic Agriculture Movements (IFOAM) consider that the term "Organic Agriculture" is all-embracing and means more than agriculture certified as organic. IFOAM's mission is to work for the day when there is worldwide adoption of organic agriculture, including various forms of non-certified organic agriculture. IFOAM regards any system that uses organic methods and is based on the principles of organic agriculture, as organic agriculture and any farmer practising such a system as an organic farmer. IFOAM considers the collection of berries, medicinal herbs and beekeeping as organic under the heading "wild collection areas". This approach can make it difficult to interpret statistics on organically farmed areas and crops. Willer and Lernoud (2016) report for IFOAM that globally there are about 31 million ha of wild collection areas and 37.5 million ha of agricultural organic land.

IFOAM considers that organic agriculture follows several principles, namely:

Health: Organic agriculture sustains and enhances the health of soil, plant, animal, human and the planet.
Ecology: Organic agriculture is based on ecological systems and cycles with which it works and helps sustain.
Fairness: Organic agriculture builds on relationships that ensure fairness with regard to the environment and life opportunities.
Care: Organic agriculture is managed in a precautionary and responsible manner to protect the health and well-being of current and future generations as well as the environment.

IFOAM is now closely linked to the Sustainable Organic Agriculture Action Network in developing what are called sustainable holistic systems, which cover issues relating to the environment, biodiversity, climate change, energy, gender equality, social justice, workers' and farmers' rights and responsibilities, land ownership, indigenous rights and animal welfare (IFOAM, 2013). In 2014 IFOAM joined the coalition opposed to GM organisms. IFOAM's belief is that as technology of GM crops is not fully understood, organic agriculture should take a precautionary approach. IFOAM takes a similar position regarding nano-technology (IFOAM, 2013).

Jahanban and Davari (2014) considered that little is known about the possible toxicological or ecological effects of nanoparticles in crop production and in food; they see the absence of national and international regulation and licensing requirements as impediments to the organic movement being able to accept nanotechnology.

FAO and WHO

The FAO took a positive position on organic agriculture at a meeting in Rome FAO (1999) when it acknowledged that organic agriculture should have a place in sustainable agriculture programmes in view of its potential contribution. FAO felt that organic farmers may discover new and innovative production technologies that may apply to other agricultural systems as well as their own and may provide market opportunities for farmers to meet consumer demands.

FAO (1999, 2013) defines organic agriculture as a system that relies on ecosystem management rather than external inputs. By eliminating the use of synthetic fertilizers and pesticides, veterinary drugs, GM crops, preservatives, additives and irradiation organic farming avoids potential damage to the environment.

The Codex Alimentarius Commission (1999/2001) established by the FAO and WHO describes organic agriculture as a holistic production system that promotes and enhances agro-ecosystem health, including biodiversity biological cycles and soil biological activity. Organic agriculture is identified as being based on minimizing the use of external inputs and avoiding the use of synthetic fertilizers and pesticides.

ISOFAR

The International Society of Organic Agriculture Research (ISOFAR) promotes and supports research in all areas of organic agriculture by facilitating global co-operation in research, education and knowledge exchange. It is based in Germany and appears to be particularly active in developing countries.

Europe, UK and USA Europe

The European Commission (Agriculture and Rural Development) takes a positive line with regard to organic farming in its document *What is Organic Farming?* (EU, 2007, 2014). Its somewhat vague definition of organic farming is that it is a system that seeks to provide the consumer with fresh, tasty and authentic food while respecting natural life-cycle systems. The essential features include:

i.   Rotating crops.
ii. Limiting the use of synthetic pesticides and fertilizers, antibiotics and food additives.

iii. Prohibiting the use of GM organisms.
iv. Choosing plant and animal varieties that are resistant to disease and are adapted to local conditions.
v.   Taking advantage of on-site production of animal feed and manure.
vi. Raising livestock in free-range, open-air systems and providing them with organically grown feed.
vii. Using appropriate animal husbandry practices.

UK

The Food Standards Agency (FSA, 2006), which is neither for nor against organic food, commented that the agency recognizes:

i.   The availability of organic food increases choice.
ii. Consumers may choose to buy organic food because of concern about the environment and animal welfare.
iii. Eating organic food can help to minimize people's intake of pesticide residues and additives.
iv. Consumers may also choose to buy organic food because they believe that it is the healthier option.

USA

The National Organic Standards Board (NOSB), an advisory body for the US government (USDA, 2009), defines organic agriculture as an ecological production management system that promotes and enhances biodiversity, biological cycles and soil biological activity. It is based on minimal use of off-farm inputs and on management practices that restore, maintain and enhance ecological harmony. Synthetic fertilizers, sewage sludge, irradiation and GM crops cannot be used. A product labelled organic in the USA must contain at least 95% organically produced material.

Non-governmental bodies

Greenpeace

Greenpeace (2010) considers that the definition of what constitutes organic agriculture should be under constant revision in order to reflect the latest science and changing market conditions. In view of its opposition to GM crops such revisions could allow for the championing of new causes against developments in agricultural science.

Greenpeace sees several benefits of organic agriculture, including the contribution to social, economic and ecologically sustainable development notably by using local resources (e.g. seed and dung) and by increasing farmers' income and improving their well-being. It understands that studies have shown the strong benefits of organic production on soil fertility, biodiversity, reduction of toxic inputs, water management, provision of healthier and more diverse food, improved livelihood opportunities,

access to markets for small farmers and social equality. Greenpeace feels that the future of organic agriculture is uncertain in the absence of sufficient research and investment, both public and private.

Friends of the earth

Friends of the Earth (FoE, 2004) champions organic farming in the UK and elsewhere. It feels that over the past fifty years farming has become over mechanized, too dependent on agro-chemicals and very wasteful of energy. Such intensive farming is thought to create widespread environmental and health problems. Pesticides, fertilizers and animal waste pollute the land and water supplies while destroying and degrading wildlife habitats. The FoE comprehensively admonishes traditional agriculture, citing food health scares and animal diseases, e.g. *E. coli, Salmonella,* bovine spongiform encephalopathy and foot-and-mouth disease, which it sees as symptoms of an ailing system that is not sustainable, economically or environmentally.

Pesticides action network (PAN)

PAN is an international network of like-minded groups that are focused primarily on addressing the harm caused by chemical pesticides, so it is not surprising that it supports organic farming. PAN considers pesticides to be highly hazardous to human health, causing severe and sometimes fatal illness among farm families and workers, as well as damaging wildlife, the environment and putting toxins into the food chain (PAN, 2015).

PAN North America takes a much broader view, condemning industrial agriculture because it exhausts the natural resources on which we all depend at the same time as making us sick and undermining the world's capacity to feed itself. Moreover, current agricultural practices are said to be unsustainable. Pesticide exposure undermines public health by increasing risks of cancer, auto-immune disease (e.g. diabetes, lupus, rheumatoid arthritis and asthma), non-Hodgkin's lymphoma and Parkinson's disease. It is confident that organic and other non-industrial farming systems are more than capable of feeding every person on earth.

Organic facts

Organic Facts, an information service based in Bangalore, India, makes all-embracing claims about the benefits of organic food on its website (www.organicfacts.net). Organic food is said to prevent premature aging, to reduce the risk of heart disease and to boost the immune system. It also promotes animal welfare, ensures a safe and healthy world for future generations, reduces the presence of pesticides, tastes better and prevents cancer. Organic Facts covers itself by saying that as the information it publishes comes from secondary sources it does not assume any liability relating to the accuracy, usefulness, completeness and adequacy of the information!

Promotion by mainstream organic groups

Health

It is natural that organic groups should do their best to promote and protect their niche markets and the way they have done this needs to be scrutinized.

Internet searches for organic farming and food readily reveal statements extolling the virtues of organic farming as being much healthier for the ecosystem, that it boosts biodiversity, protects nearby waterways from chemical pollution and that it helps soils retain water and nutrients. It also reduces human exposure to harmful chemicals and toxic residues, which have been linked to a variety of illnesses (e.g. Reynolds, 2013); the consumption of non-organic food is also linked to developmental and learning problems such as attention deficit hyperactivity disorder, high blood pressure, type 2 diabetes, depression and cancer. Promotion of the merits of organic farming usually concentrates on a few aspects relating to health, beneficial nutrients in food, avoidance of pesticides and GM crops, biodiversity, livestock management, the sequestration of carbon in the soil and the conservation of diminishing stocks of raw materials.

Balfour's belief that organic food was going to be very important for the health of the nation was central to her thinking and the reason for her setting up the Haughley farm experiment, which after twenty-four years failed to demonstrate any health benefits to crops or livestock.

In 2000 the Advertising Standards Authority (ASA) in the UK required the Soil Association in the UK to withdraw leaflets claiming that organic food tastes better, is healthier and better for the environment, because it found the claims could not be substantiated (ASA, 2000). Evidence was not available to show that organically produced food possessed noticeable health benefits over and above the same food when conventionally produced or that a diet of organic food could guarantee no harmful effects.

This has not stopped the practice altogether, as was evident by the reporting by organic supporters (Watson, 2014) and the media (BBC, 2014) following the publication of Baranski *et al.* (2014). Health benefits were ascribed to the composition of organic food, although causal relationships had not been demonstrated between any components of organic food and health.

Detailed advice has recently been given by the Soil Association (2014a), advising promoters of organic food how they should proceed by getting approval from "Copy Advice" (part of ASA) and what, from a long list they are permitted to say in support of their products, e.g. the best way to reduce your exposure to pesticides in all food is to buy organic.

Superior nutrient composition of organic food

There have been many studies comparing the composition of organically and conventionally grown food (Chapter 14). Differences in the composition of organic

and conventional food are found but currently there is no convincing evidence that organic food is inherently superior in its composition, e.g. Dangour *et al.* (2009). Nevertheless, it is expected that attempts will continue to be made by organic supporters to establish the superiority of organic food. The literature study by Baranski *et al.* (2014) is a recent example.

Avoidance of GM crops and products

Opposition to GM crops is universal among organic movements. In the UK the Soil Association was quick in the 1990s to mount opposition to GM crops, a position it maintains today. On its website the association (2014) presents its reasons, namely:

i.   Genetically modified crops will make it more difficult to feed the world because they drive out and destroy the systems that are needed worldwide, particularly by poor African and Asian farmers.

ii.   Farming that helps Bayer, Syngenta and Monsanto is not needed.

iii.   Higher yields have not been achieved despite the promises.

Genetically modified crops were said to be worse for wildlife in the UK on the basis of the results of the only large-scale, farm-based research project in the world to look at the impact of GM oilseed rape, maize and sugar beet on wildlife (1999–2003) – the Farm Scale Evaluation. The finding of fewer insects in and under the GM crops was misinterpreted as being due to the technology of genetic engineering. The fewer insects were actually due to the fact that the use of the more effective herbicide (glyphosate) had killed more weeds, thereby leaving less vegetation for insects to eat. The greater number of weeds in conventional sugar beet and spring rape crops than in their genetically modified counterparts meant that there were more weed seeds for animals, particularly farmland birds (FSE, 2003).

The objection that farmers will have to buy costly seed and herbicides from the major agrochemical companies does not seem to be valid. Why would farmers continue to buy such seed and herbicides if they were not beneficial and did not increase their own profits?

Public perceptions and the reasons for choosing organic food

It is important to consider why people buy organic food and what they think organic farming is. The success of the organic food market is dependent on what the public perceive and how they respond to promotion.

The promotion of organic farming has been so effective that even the word "organic" has been hijacked. The adjective is commonly used to give the impression of superiority, perfection and purity of a product, whether it is carrots or cotton lingerie, melons or mascara.

Organic compounds are chemicals essentially based on carbon, hydrogen and oxygen in which the bonding between the elements involves the sharing of electrons rather than the donation of electrons, as is the case in inorganic compounds. Organic compounds are generally produced by a biological organism.

It is likely that the early buyers of organic food appreciated and supported the organic production methods but today the term organic means many different things to people. As Humpty Dumpty said: "A word can mean just what I choose it to mean."

Bruce (2013) examined the way the word "organic" has been used, concentrating on a) *New Scientist* (8 million words), b) *The Times* (52 million words), which was seen as a target market for organic produce and where the 'ideology' of organic farming was most likely to be represented, and c) the BBC. This usage was then compared to the general use of the term in the 434 million word Bank of English corpus (a multi-source compendium of written and spoken words). The objective was to highlight differences in usage of the word organic and to reveal how and where the adjective is used as part of an ideological discourse.

The frequency of use of the word organic varied widely, with the *New Scientist* having the highest usage at ninety times per million words, the comparable number for *The Times* was thirty-six, for the BBC thirteen and overall twenty-one. It was found that the word natural, which is often used as a synonym for organic, occurred much more often, with the *New Scientist* at 324 times per million words, *The Times* with eighty-nine and the BBC with seventy-one. In the *New Scientist*, organic most commonly modified nouns relating to chemistry, ideology and agriculture (in that order), whereas in *The Times* organic modified nouns describing agriculture and gardening, food and drink. It was found that, apart from its use in a chemical context, the word organic was used adjectivally to give nouns a positive value.

It is unfortunate that too few people, including many in the media, have a basic knowledge or even understanding of chemistry. The result is that words such as chemicals and organic are misused. The public have been led to believe that chemicals are by definition nasty and harmful and that anything labelled organic is likely to be wholesome and good. It is necessary to go back to basics and to try to explain that all substances are made up of chemicals and that they all have the propensity to be toxic depending on the amounts involved. As Paracelcus, the Swiss alchemist and botanist, said in the early 16th century: "The dose makes the poison."

Quiet, repetitive promotion that all things labelled organic are to be preferred is now the norm and judging by the absence of opposition it would seem to be generally accepted. For example, a twelve-page insert in *The Times* (9 August 2016) published by Raconteur in association with the Soil Association, Natural Products News and the Organic Trade Board extolled the virtues of natural and organic living. Several suggestions and claims were made, including:

i.   There is a growing liking for things organic.

ii.   In order to have a healthy life we should be careful about the chemicals we eat and put on our skin.

iii. There is much concern about chemicals in items such as cosmetics, clothing and cleaning products.

iv.   Natural and organic beauty is obtainable at an affordable price.

v.   Organic farmers arc in the forefront of agricultural innovation and conventional farmers are looking to learn from them.

vi.   Growth in sales of organic food is just beginning.

vii. When marketing organic food it is most important to get across the message that the product tastes good.

Evidence of the research behind most of these suggestions and claims, which is seldom considered by the popular press, will be discussed in later chapters.

Surveys

Consumers of organic food generally believe that it is healthier than conventional food despite the lack of supporting evidence, as demonstrated by the studies of Dangour *et al.* (2009) and by

Smith-Spangler *et al.* (2012) whose work partly stimulated a survey of consumers of organic food about their perceptions regarding their health (Oates *et al.*, 2014). Not surprisingly, the Organic Health and Wellness Survey found that 76% of respondents perceived their overall health to be better after moving to an organic diet.

Top of the list of the reasons people give for eating organic food relate to health and the avoidance of man-made pesticides, food additives and hormones. The Organic Research Centre (formally the Progressive Farming Trust Ltd) whose business is to develop and support sustainable land use, presented survey data (Padel, 2011) on shoppers' attitudes to organic food using data collected by Beaufort Research in Wales and Scotland (number of respondents 1407) in 2011 (Table 4.2 and 4.3).

A survey in Belgium of consumer perceptions and the scientific evidence related to food quality (Hoefkens *et al.*, 2009) found that organic vegetables are

Table 4.2 The main motivating factors for the purchase of organic food (Padel, 2011)

Motivation	%
Healthier/better for you	21
Better taste/flavour	21
If on offer or price acceptable	16
No or fewer chemicals	11
Better for animals	6
Better quality	6
Fresher	6
Buy for children	5
All other reasons	8

Table 4.3 Responses of shoppers to specific question (Padel, 2011)

Agreed	%
Organic farming means better standards of animal welfare	74
Producing food organically allows wildlife to flourish	71
Organic produce is healthier for you	60
Producing food organically helps reduce the carbon footprint	57
Organic produce tends to be better quality	50
Organic food represents good value for money	28

perceived as containing fewer contaminants and more nutrients, and as such are healthier and safer compared to conventional vegetables. However, those behind the survey concluded that not enough evidence is available to support or refute such perceptions.

A revealing difference in attitudes across the North Sea to organic food was revealed on BBC Farming Today (BBC, 2009) when organic sales were declining in the UK but were booming in Germany. The suggested explanation was that in Germany people who believe in organic food would only eat organic whereas in the UK a lot of people eat organic only when they can afford it. When they can't afford it they do not.

Avoidance of pesticides

Pesticides and their presence in fruit and vegetables are sometimes vilified on websites, which bolster consumer perceptions. It is likely that health concerns are the main reasons for choosing organic food. Whether or not there is evidence to support this perception is considered in Chapter 10.

Smallwood (2014), a Rodale director in the US, could perhaps be accused of exaggerating about pesticide use when he preferred blemished apples that have been attacked by fungi over the flawless apples grown conventionally and treated with more than seventy-five different toxic pesticides and fungicides. He is happy to say that the black spots on his apples are simply cosmetic and to ignore any effects the infection might have had on the composition of the fruit. An alternative view regarding the presence, production and activity of potentially harmful natural pesticides has been presented by Ames and Gold (1997) and Gold *et al.* (2002), and is discussed in Chapter 15.

Moreover, it is not correct to say that organic food will be pesticide-free. Apart from the naturally occurring pesticides that all plants contain, organic farmers are allowed to use certain pesticides – mostly used before synthetic pesticides became available. These include copper sulphate on vines and potatoes, sodium hypochlorite, lime sulphur, *Bacillus thuringiensis*-based preparations, pyrethrums and nicotine sulphate.

Further all-embracing claims were made at a Rodale Institute meeting about "Why organic farming could save the world" (Rodale News, 2014). Organic farming was said to be the complete solution to health, unemployment, poverty, biodiversity and water problems.

Better for the environment and livestock

The Duchy Farm website proclaims: "Organic farming isn't only better for the environment – thanks to the restricted use of artificial chemical fertilizers and pesticides – but it also assures greater wildlife protection and animal welfare standards" (Permission to quote kindly given by Duchy Home Farm, 2014). However, garden birds do not seem to have heeded the story of organic superiority. McKenzie and Whittingham (2010) found in a three-year study that wild birds were not swayed by the organic label, but instead preferred the more protein-rich, conventional food that helped them to survive the winter. Claims of the merits of organic food for birds should be questioned. On the other hand, herbicides can, by killing weeds, reduce the availability of weed seeds for birds.

Researchers at the University of Leeds paired thirty-two farms with conventional farms and concluded that the small benefits to birds, bees and butterflies and increases in biodiversity did not compensate for the lower yields on the organic farms (Benton, 2010).The organic farms in the study produced less than half the yield of their conventional counterparts and only showed a 12% increase in biodiversity.

Other organic advocates assert that organic farming is kinder on the environment and that farm animals are more humanely treated on organic than on traditional farms. Is there any basis for accepting the idea that organic farmers have a monopoly on better animal welfare? Conventional farmers know that the productivity and health of livestock is very dependent on how they are treated.

Taste

It used to be said that organic food tasted better than conventional but blind tests have shown that people are not able to distinguish organic food on taste (Fillion and Arazi, 2002 and Basker, 2009). It has been realized that fresh, locally grown vegetables, whether organic or not, are likely to taste better than vegetables that have been in the distribution chain for some time.

A recent sting by a couple of Dutchmen at a food meeting in Houten revealed how even food experts can be hoodwinked into believing that food they were given was very tasty simply by telling them it was organic when in fact it was food from a high street fast food chain (Houten, 2014).

Carbon footprint, soil loss, sustainability, local markets and raw materials

The supporters of organic farming say it is superior to conventional in having less of an impact on global warming and therefore climate change. The Rodale Institute believes that organic farming can provide a solution to climate change (LaSalle and Hepperly, 2008). Rodale states that extrapolation of the data from its farming system and pasture trials shows that more than 100% of current annual CO_2 emissions could be sequestered by a global switch to organic practices (Anonymous, 2014).

There was a call, at its annual conference (Soil Association, 2013), that more attention should be given to the massive loss of soil and to soil degradation in the UK that could be halted by farming organically.

Organic movements are giving particular attention to promoting the idea that farming organically is more sustainable than conventional farming. It is said that consumers are motivated by the environmental, ethical practices and sustainability aspects of organic farming. However, a 2014 consumer research study by the European Food Information Council (EUFIC) found that organic-associated eco-labelling claims linked to sustainability concepts are rarely translated into purchases and that sustainability labelling claims do not play a major role in consumers' food choices (Grunert *et al.*, 2014). This study agreed with others in indicating that health-related concerns linked to pesticides, hormones, antibiotics and GM crops are the main drivers of the organic consumer purchasing behaviour.

In 2007 in the UK two shipments of organic carrots went rotten on their way from farm to retailer via cleaning and packaging plants because they had not been treated in the field with the appropriate pesticide. This was followed by increased promotion of local sales of organic produce linked to eating products in-season and reducing the number of miles the food travels and thus its carbon footprint. Local sales of vegetables in the UK in box schemes have proved popular.

Little attention is paid, at least publicly, to the problems that mankind will have to face, namely of finding how to maintain soil fertility and food production when oil and fertilizer raw materials become very expensive and especially when eventually they run out.

Perceptions, labelling and promotion of organic food in USA

Buyers of organic food in the USA do so to avoid toxic pesticides, synthetic hormones, antibiotic residues and GM organisms (Herther, 2011). Many Americans are said to believe that consumption of conventional genetically engineered food is the primary cause of deteriorating public health and childhood disease. The Organic Consumers Association (OCA, 2014) writes that the consumption of conventionally grown food has spawned an epidemic of obesity, diabetes, cancer, antibiotic-resistant infections, behavioural disorders, learning disabilities, autism, Alzheimer's and heart disease. Organic dairy products and grass-fed meat are believed to contain more vitamins, important trace minerals, essential omega-3 fats and cancer-fighting antioxidants than conventional chemical food, according to the OCA.

Public perceptions about organic food have inadvertently been affected in the USA by the legislation (in 2000) on the labelling of organic products. Despite attempts by the USDA to clarify the situation the American public has been led to believe that a product bearing the USDA organic seal implies that the product is healthier, safer and more nutritious. The organic movement has not been slow to capitalize on the situation. Lupp (2009) has shown that the health and environmental claims that are frequently included on labels of products carrying the USDA organic seal are false or misleading. Pig and poultry products in the supermarkets in the USA are sometimes labelled as being hormone-free, when in fact all poultry and pigs are not allowed to be treated with hormones; such innuendos create needless fear and mistrust for consumers purchasing meat. Hormone supplements may be used on calves (e.g. by placing a slow release tablet under the skin of the ear); provided evidence is available it may be possible to say on the label "no hormones administered". Similarly, evidence has to be presented before a meat or poultry product can be labelled "no antibiotics added".

More than half of shoppers in the USA say they believe that labelling something as organic is an excuse to charge more, and more than one-third say they believe organic is just marketing jargon with no real value or definition, according to a recent survey of 2,002 adults (Mintel, 2015). Studies on who is buying organic food in the USA have shown that 20–30% of organic consumers are dedicated to organic food and are likely to be motivated by considerations about their own and the Earth's health (Hartman

Group, 2008, and Natural Marketing Institute, 2008b).

The American consumer of organic food thinks that the terms "natural" and "organic" mean essentially the same thing, namely the absence of pesticides, growth hormones, antibiotics and GM crops (Hartman Group, 2010).

Organic companies in the USA promote organic food alleging health benefits associated with the absence of genetically modified products, hormones, antibiotics and pesticides. These messages are further promoted by advocacy groups that regularly amplify the health risk allegations linked to conventional foods and contrast them to the goodness and ethics of organic production. The voluminous literature generated between 1988 and 2014 advocating the safety aspects of avoiding conventional food and GM crops has been reviewed by Chassy *et al.* (2014) in a wide-ranging study, commissioned by Academics Review, a non-profit group of independent agricultural experts. It was concluded that the perceived safety concerns about pesticides, hormones, antibiotics and GM crops are the critical components driving sales of organic food and that consumers have been misled about comparative studies on product food safety and health attributes of organic food. The review also revealed extensive evidence of widespread, collaborative industry marketing activities as a primary cause for these misperceptions. Other factors, such as sustainability, environmental benefits and even organic certification, do not motivate the average

consumer to purchase organic products in the absence of health risk claims.

Chassy *et al.* (2014) pointed out that the anti-pesticide and anti-GM crops advocacy groups promoting organic food are likely to have a combined annual budget exceeding $2.5 billion and that the organic industry is one of the major donors to these groups. Insofar as the organic movement predated the development of GM crops, could it be that the initiation and responsibility for driving the anti-genetically modified movement lies with organic advocates?

Chassy *et al.* (2014) concluded that the American taxpayer is funding a national organic programme that is misleading consumers into spending billions of dollars on organic products based on false health and quality claims. The US government agencies, including the US Food and Drug Administration, the Federal Trade Commission and the US Department of Agriculture, are all criticized for not enforcing truthful, non-misleading claims and for not protecting the consumer against abuses.

Forrer *et al.* (2000) were perceptive many years ago when they pointed out that if consumers became satisfied with the safety of conventional food and if the furore over genetic modification died down the potential for organic food would be limited.

Promotion of organic food was sullied in 2015 by a consortium of sellers of organic food in the USA who produced YouTube videos that grossly misrepresented conventional farming and used children to do so under the title "New

Macdonald". Unfortunately this type of negative promotion is not new in the USA (Savage, 2015).

Some organic farmers in the USA have protested about a new grading system for produce and flowers introduced by the supermarket chain Whole Foods (2015) that they say devalues the organic label of the USDA. Whole Foods has introduced a rating system called "responsibly grown", which is based on the answers farmers give about how they protect the soil and wildlife on their farms, whether they limit their use of pesticides, how they conserve energy and irrigation water and how they treat their workers (Whole Foods, 2015). A farm's produce is graded as unrated, good, better or best and labelled with brightly coloured stickers. Whole Foods has traditionally sold organic food in its stores, which will now be displayed along food labelled "responsibly grown". All certified organic produce will be rated good at least.

Reasons for farming organically

It would appear that farmers are motivated to farm organically either because they have a fundamental belief in the system or because they see it as a way to increase profit.

The creation of improved social systems and communities was in the minds of many of the early supporters of organic farming and the notion is still alive. For example, the mission of the Chicago-based Angelic Organics

Learning Center (2013) is to empower people to create sustainable communities of soils, plants, animals and people through educational, creative, and experimental programmes. The Learning Center is the educational partner of Angelic Organics, a vibrant biodynamic community-supported farm. There is a desire to live in harmony with nature, causing minimum damage to the environment. Such farming can be sufficient and appropriate for subsistence farming but its ability to feed huge non-rural populations limits its value to society as a whole.

The profit motive could be strong in the growing US organic market, leading to farmers converting to organic. A meta study by Crowder and Reganold (2015) in the USA covering fifty-five crops in fourteen countries demonstrated that organic agriculture can earn farmers more (22 to 35%) than conventional farming largely as a result of premium prices; this was despite lower yields of organic crops.

The rapid recent expansion of the US organic market has been facilitated by the conversion of intensive multi-animal farms. In 2014 Cornucopia complained to USDA, on the basis of aerial photographs, that fourteen huge, certified organic dairy and chicken farms were flouting the regulations (Cornucopia, 2014). Few if any animals were seen outside. The Cornucopia Institute is committed to protecting the integrity of organic farming and produce.

Further evidence of the power of the organic label is provided by the marketing in the USA of hydroponically grown vegetables, which is permitted by the USDA despite the advice to the contrary by the National Organic Standards Board. Many countries prohibit organic hydroponic production in their own markets, including Mexico, Canada, Japan, New Zealand, and most European countries. As a result organic hydroponic producers often grow exclusively for foreign markets. At present most of the hydroponic organic produce sold in the USA is grown in Mexico, Canada, or Holland (Cornucopia, 2015). The nutrient solutions normally used in hydroponic systems are soluble refined salts (man-made), which in other circumstances would be anathema to organic proponents.

The goal of developing sustainable food production techniques is attractive to organic farmers, especially in view of the progress they have made without using fertilizers and pesticides. Work on integrated pest management is likely to benefit all types of farming. Breeding of crops with improved attributes for low-fertility conditions may well be hampered by the organic movements objecting to genetic engineering.

Detecting organic fraud

In recent decades organic products have created a valuable niche market, allowing farmers and processors to earn premium prices. "Certified organic" has become a marketing tool. Farmers in developing countries have also benefited from their organic produce fetching higher prices in export markets.

It was perhaps inevitable that the financial rewards of an organic label would lead some farmers and processors to supply conventional products fraudulently labelled as organic. This has occurred with milk products in USA, eggs in Germany and fake organic certificates in Italy (Entine, 2014).

Certification procedures are seemingly complied with but then the farmer continues to use fertilizers and pesticides. It is possible for very low levels of conventional pesticides to be found in organic crop samples, as occurred in 2012 when the USDA's National Organic Program detected pesticides in 43% of samples. It was presumed that accidental drift was the reason. However, conventional pesticides are sometimes found in organic crops at the same levels as in conventional crops, which suggests misuse.

There is sufficient concern in the organic community for the factors associated with non-compliance with EU regulations to be assessed; Gambelli *et al.* (2014) carried out a literature survey. Non-compliance was found to be most likely with livestock products and on the larger arable farms. Farmer attitudes were judged to be important but difficult to identify.

There are concerns that standards and independent third-party verification may be insufficient to prevent fraud. The search has begun for a reliable method that would enable organic food to be distinguished from food produced by conventional farming. It is difficult to envisage from an understanding of plant physiology and mineral nutrition how an organic crop could be fundamentally and consistently different in composition from a conventional product. The absence of any pesticide residues or of genetically modified material could not be used as proof of the provenance of organic produce because conventional farmers may sometimes not use pesticides.

A test that at least could accurately identify organic food for use in cases of suspected fraud rather than as a prerequisite for labelling produce as organic would be very useful.

The power of Inductively Coupled Plasma Optical Emission Spectroscopy to determine simultaneously many elements in samples was employed by Laursen *et al.* (2011) when developing a multi-element fingerprinting technique for authenticating organic wheat, barley, faba bean, and potato samples. Using fourteen elements, combined with chemometrics (a statistical procedure) they were able to discriminate between organic and conventional crops; the fourteen elements were the essential macronutrients Mg, P, S, K, and Ca) and the micronutrients B, Mn, Fe, Cu, Zn, and Mo as well as non-essential elements Na, Sr, and Ba. The discrimination power was further enhanced by analysis of more elements (Br, Ce, Cd, Cl, Cr, Co, Ga, Ge, I, Ni, Rb and W). It was concluded that multi-element fingerprinting with chemometrics has the potential to authenticate organic crops.

The finding that organic manures have a different isotopic nitrogen composition

from synthetic fertilizers is the basis of a new technique that was evaluated in the UK (Bateman *et al.*, 2005, and Bateman and Kelly, 2006). Nitrogen has two stable isotopes, ^{14}N (normally 99.63% of total) and ^{15}N (0.37%). ^{14}N and ^{15}N are lost as ammonia from manure at sufficiently different rates so that the ratio between the two stable isotopes will be altered and will be subsequently mirrored in the crop. It was hoped that this would provide a consistent and measurable indication as to whether fertilizers based on N fixation by man had been used or not. However, it was found for all crop types analysed that there is overlap between the values of the organically and conventionally grown crops. There appear to be so many complicating variables that render the method inaccurate. The hope had been to build up a database for different crops and products.

A new technique based on ^1H-NMR (proton nuclear magnetic resonance spectroscopy), otherwise described as linear discriminant analysis, has been proposed by researchers in Germany (Hohmann *et al.*, 2014). The technique, which is already in use for authenticating organically produced honey and olive oil, was successful in differentiating between tomatoes grown with or without fertilizers in greenhouses or outdoors. However, there was sufficient data overlap that, combined with unexpected differences between individual producers, means refinements are required.

The possibility of tagging a liquid such as olive oil with magnetic recoverable nanoparticles (Fe_2O_3) containing a short length of DNA as a tracer is under investigation in Switzerland (Puddu *et al.*, 2014). The particles would be added at the place of origin of the oil and later extracted by magnetic separation enabling the producer to be identified and so provide a label that cannot be removed. It is difficult to see the organic movement accepting this technique because it involves DNA and nanoparticles.

It is revealing that no distinguishing differences between the composition of organic and conventional crops have so far been discovered. Whilst some differences may be found in experiments it would appear that organic crops are not fundamentally different from conventional.

An iPhone application is currently available named Lapka (2012), a personal organic monitor consisting of a stainless steel probe that, when pierced through raw fruits and vegetables measures electrical conductivity. The conductivity is said to correlate with the concentration of nitrate ions from nitrogen-based fertilizers and so can indicate whether the product is organic or not. There is no explanation as to how the probe knows the provenance of the NO_3 anions. Organically grown crops often contain less but the same NO_3 anions.

End product testing of organic products is a controversial topic. The Soil Association is of the view that it is essential for the maintenance of the integrity of organic farming that the definition of organic food is based on the production process rather than the end product. It is not clear why the definition of a process is incompatible with the testing of an end product. In contrast, the Organic Trade Association in the USA (OTA, 2003) approves of testing for pesticide residues and feels it has a place in organic certification. The USDA, like the Soil Association, views organic certification as being process and not product-based that depends on the certifying agents attesting to the reliability of organic farmers and processors to follow a set of production standards.

The USDA sanctions the testing of organic products for prohibited residues (such as pesticides, synthetic fertilizers or antibiotics) or excluded substances (e.g. genetically engineered organisms) only under two circumstances. First, if a certifying agent believes that the farmer is intentionally using prohibited substances or practices. Second is to assess the compliance with the requirement for certifying agents to test 5% of their certified operations each year.

Most of the organic bodies, including IFOAM, support the USDA's lenient testing protocols and oppose more frequent mandatory testing of organic products for prohibited and excluded substances. Organic agriculture is basically a trust-based system, which is to be contrasted with the regular testing of conventionally grown products for pesticide residues.

A simple and acceptable method for authenticating organic crops remains a Holy Grail.

The attitude to testing of organic food contrasts with that of a few years ago in the UK with the attitudes to food irradiation. One of the main objections to irradiating food was that it was felt necessary to have a foolproof test to allow irradiated produce to be identified. Organic movements across the world prohibit food irradiation, as in the UK. However, irradiation is currently permitted in the UK on seven categories of food including fruit, vegetables, herbs and spices, fish and shellfish and poultry, and in the EU on only one food group, namely dried aromatic herbs and spices.

Summary

1. The term organic farming was coined in 1940 by Lord Northbourne, a follower of the biodynamic movement.

2. Organic farming is essentially farming according to the rules and regulations of the certifying organizations and of national or regional regulations, as for example in the USA and the EU.

3. The Romans farmed without the use of fertilizers and pesticides and appreciated the value of manure and legumes as do many peasant farmers, particularly in Africa, today. Some may argue, looking back with the current organic rules in mind, that they were farming organically. Their low productivity suggests that there would seem to be little merit in considering their methods.

4. There are basically two organic farming groups, firstly the older biodynamic farmers and secondly the mainstream organic farmers. The two groups share many features, notably their reliance on farmyard manure and composts and the use of leguminous crops for their nitrogen-fixing ability.

5. Biodynamic farming is distinguished from mainstream organic farming firstly by the use of nine preparations, which are applied in homeopathic quantities, and secondly by their planning of farming operations according to the lunar and zodiacal calendars. Biodynamic farms should ideally be mixed and self-contained.

6. Mainstream organic farming is essentially a collection of positive and negative aspects. The positive aspects of the use of manure, legumes and crop rotations by organic farmers are practices not only used for centuries but also by conventional farmers.

7. The prohibitions of the use of soluble inorganic fertilizers, of man-made pesticides, antibiotics and of GM crops together constitute the main negatives by which mainstream organic farming is identified.

8. The promotion of organic food to the public mainly rests on claims of health improvements, superior composition, avoidance of pesticides and GM crops and, to a lesser extent, on animal welfare.

9. Some advocates of organic farming, especially in the USA, see it as the solution to most of society's ills, including global warming. Others feel it is inevitable that sustainable food production will only be achieved by farming organically.

10. A fundamental dislike of what is called the industrialization of farming and of the involvement of big agrochemical companies motivates others, including NGOs, to support organic farming as being a better way of life.

11. Initially most organic farmers around the world were probably motivated by their belief in the inherent benefits of

organic farming and not by potential profits.

12. The rapid growth in the US organic market and the fact that much organic farming in developing countries is geared to exports suggests that the profit motive is now a major factor.

13. The public's perception of organic farming is the key to organic food marketing success. Surveys indicate that the public have accepted the notion that organic food is healthier, tastes better and contains fewer chemicals. The public also believe that organic farming is better for livestock and for wildlife.

14. False labelling of organic produce has stimulated research into developing a test to permit conventionally grown produce to be distinguished from organic. A foolproof test is awaited.

15. Research has not revealed any compositional feature of organic crops that could be used to differentiate them from conventional crops. This suggests that there are no grounds for thinking that organic crops are fundamentally different from other crops.

References

ASA (2000) Advertising Standards Authority Organic food complaint upheld, Adjudications 12 July 2000. Available at www.gene.ch/gentech/2000/Jul/msg00047.html

Ames, B.N. and Gold, L.S. (1997) The causes and prevention of cancer: gaining perspective. *Environmental Heath Perspective Supplements* 105, 4. 864–873.

Angelic Organics Learning Center (2013) www.learngrowconnect.org

Anonymous (2014) Regenerative Organic Agriculture and Climate Change A Down-to-Earth Solution to Global Warming. Available at http://rodaleinstitute.org/assets/RegenOrgAgriculture AndClimateChange_20141001.pdf

Balfour, E.B. (1943, revised 1949) *The Living Soil*. Faber and Faber, London.

Balfour, E.B. (1975) *The Living Soil and the Haughley Experiment*. Faber and Faber.

Baranski, M., S´rednicka-Tober D., Volakakis, N., Seal, C., Sanderson, R., Stewart G.B., Benbrook, C., Biavati, B., Markellou, E., Giotis, C.,Gromadzka-Ostrowska, J., Rembiałkowska, E., Skwarło-Son´ta, K., Tahvonen, R., Janovska, D., Niggli, U., Nicot, P. and Leifert, C. (2014) Higher antioxidant and lower cadmium concentrations and lower incidence of pesticide residues in organically grown crops: a systematic literature review and meta-analyses. *British Journal of Nutrition* 112, 5, 794–811.

Basker, D. (2009) Comparison of taste quality between organically and conventionally grown fruits and vegetables *American Journal of Alternative Agriculture*, 7, 3,129–136. doi: 10.1017/S0889189300004641.

Bateman, A. and Kelly, S. (2006) Discriminating between organically and conventionally grown crops using stable isotope and multi-element analysis. UK Food Standards Agency Food Standards Agency Project Number: Q01076 UEA Contract Number: RO2330.

Bateman, A., Kelly, S. and Jickells, T. (2005) Nitrogen isotope relationships between crops and fertilizer: Implications for using nitrogen isotope analysis as an indicator of agricultural regime, *Journal of Agricultural and Food Chemistry*, 53, 5760–5765.

BBC (2009) Farming Today, 9 February.

BBC (2014) Countryfile, 20 July.

Benton, T. (2010) Organic farming shows limited benefit to wildlife. University of Leeds. Science Daily. *Ecology Letters.* Available at www.sciencedaily.com/releases/2010/05/100505102553.htm

Bruce, T. (2013) 'Corpus analysis of the word organic', MA Assignment, University of Birmingham, England.

Chassy, B., Tribe, D., Brookes, G. and Kershen, D. (2014) Organic marketing report. *Academics review.* Available at http://academicsreview.org/wp-content/uploads/2014/04/Academics-Review_Organic-Marketing-Report1.pdf

Codex Alimentarius Commission (1999/2001) Guidelines for the production, processing, labelling and marketing of organically produced foods CAC/GL 32-1999/Rev 1, 2001, Rome.

Conford, P. (2001) *The Origins of the Organic Movement.* Floris Books, Edinburgh.

Cornucopia (2014) Think your milk and eggs are organic? These aerial farm photos will make you think again. Available at www.washingtonpost.com/blogs/wonkblog/wp/2014/12/11/think-your-milk-and-chicken-are-organic-these-aerial-farm-photos-will-make-you-think-again/

Cornucopia (2015) See no evil. Organic Industry Bigwigs Dispute What Cornucopia's Aerial Photos Reveal. The Cultivator. Available at www.cornucopia.org/wp-content/uploads/2015/03/Spring-2015-Cultivator-for-website.pdf

Crowder, D.W. and Reganold, J.P. (2015) Financial competitiveness of organic agriculture on a global scale. *Proceedings of the National Academy of Sciences* 112, 24, 7611–7616. doi: 10.1073/pnas.1423674112.

Demeter (2014) Available at www.demeterbta.com/www.biodynamic.org.uk

Dangour A.D., Dodhia, S.K., Hayter, A., Lock. K. and Uauy, R (2009) Nutritional quality of organic foods: a systematic review. *American Journal of Clinical Nutrition.* 90, 3, 680–685. doi: 10.3945/ajcn.2009.28041.

Duchy Home Farm (2014) Available at http://duchyofcornwall.org/home-farm.html. www.princeofwales.gov.uk

Edmondson, J.L., Davies, Z.D., Gaston, K.J. and Leake, J.R. (2014) Urban cultivation in allotments maintains soil qualities adversely affected by conventional agriculture. *Journal of Applied Ecology* 51, 4, 880–889. doi: 10.1111/1365-2664.12254.

Entine, J. (2014) Consumers willing to pay premium for organics, but benefits mostly psychological. Available at www.geneticliteracyproject.org/2014/06/17/consumers-willing-to-pay-premium-for-organics-but-benefits-mostly-psychological/#.U6Gq3_ldWSo

EU (2007) Council Regulation (EC) No. 834/2007.

EU (2014) What is organic farming? Available at http://ec.europa.eu/agriculture/organic/organic-farming/what-is-organic-farming/index_en.htm

FAO (1999) Position paper on Organic Agriculture. Available at www.fao.org/docrep/meeting/X0075e.htm

FAO (2013) Organic Agriculture: African Experiences in Resilience and Sustainability, Eds. Auerbach, R., Rundgren, G. and El-Hage Scialabba, N. Available at www.fao.org/docrep/018/i3294e/i3294e.pdf

Fillion, L., and Arazi, S. (2002) Does organic food taste better? A claim substantiation approach *Nutrition and Food Science*, 32, 4, 153–157. doi: 10.1108/00346650210436262.

Forrer, G., Avery, A., and Carlisle, J. (2000) Marketing and the organic food industry: A history of food fears, market manipulation and misleading consumers. Center for Global Food Issues. Washington D.C.

FoE (2004) Friends of the Earth. Food and farming. Available at www.foe.co.uk/sites/default/files/downloads/food_farming.pdf

FSA (2006) Available at www.defra.gov.uk/farm/organic/consumers/#21

FSE (2003) Farm Scale Evaluation GM crops. Available at http://webarchive.nationalarchives.gov.uk/20080306073937/http://www.defra.gov.uk/environment/gm/fse/results/fse-summary-05.pdf

Gold, L.S., Slone, T.H., Manley, N.B. and Ames, B.N (2002) Misconceptions About the Causes of Cancer 1415–1460. In: D. Paustenbach, D. (ed.) *Human and Environmental Risk Assessment: Theory and Practice* New York: John Wiley & Sons.

Gambelli, D., Solfanelli, F. and Zanoli, R. (2014) Risk assessment in EU organic certification system: a systematic literature review 725–728. In: Rahmann, G. and Aksoy U. (Eds.) *Proceedings of the 4th ISOFAR Scientific Conference. Building Organic Bridges*, Istanbul, Turkey.

Greenpeace (2010). The Case for Organic Agriculture. Available at www.greenpeace.org/usa/en/campaigns/genetic-engineering/our-vision/organic/

Grunert, K.G., Hieke, S., Wills, J. (2014) Sustainability labels on food products: Consumer motivation understanding and use. *Journal of Food Policy*. 44, 177–189. Available at http://dx.doi.org/10.1016/j.foodpol.2013.12.001

Hartman Group. (2008) The many faces of organic 2008. Bellevue. WA. Available at www.hartman-group.com/hartbeat/organics-today-who-buying-and-what-next

Hartman Group. (2010) Beyond organic and natural. Available at http://store.hartman-group.com/beyond-organic-and-natural

Herther, K. (2011) MamboTrack Health and Natural Consumer Outlook Survey Report & Press Release. *Mambo Sprouts Marketing*. 29 December 2011.

Hoefkens C., Verbeke, W., Aertsens, J., Mondelaers, K. and van Camp, J. (2009) The nutritional and toxicological value of organic vegetables: Consumer perception versus scientific evidence, *British Food Journal* 111, 10, 1062–1077.

Hohmann, M., Christoph, N. Wachter, H. and Holzgrabe, U. (2014) ^{1}H NMR Profiling as an Approach To Differentiate Conventionally and Organically Grown Tomatoes *Journal of Agricultural and Food Chemistry 62*, 33), 8530–8540. dx.doi.org/10.1021/jf502113r.

Houten (2014) Serving McDonalds to food experts? Available at https://www.youtube.com/results?search_query=organic+food+houten

IFOAM (2013) *Annual report.* Available at www.ifoam.bio/sites/default/files/annual_report_2013_web.pdf

ISOFAR (2015) Report from 3rd African Organic Conference in Nigeria, October 2015 www.isofar.org/isofar/index.php/2-uncategorised/76-report-from-3rd-african-organic-conference-in-nigeria-october-2015

Jahanban, L. and Davari, M. (2014) Organic Agriculture and Nanotechnology 679-682. In: Rahmann, G. and Aksoy, U. (Eds.) *Proceedings of the 4th ISOFAR Scientific Conference. Building Organic Bridges*, Istanbul, Turkey.

Lapka (2012) Lapka for iPhone: five sensors to measure the world, inspired by NASA and Yves Saint Laurent. Available at www.theverge.com/2012/12/11/3726638/lapka-iphone-sensors-yves-saint-laurent-meets-nasa

LaSalle, T.J. and Hepperly, P. (2008) Regenerative Organic Farming: A Solution to Global Warming. Available at https://grist.files.wordpress.com/2009/06/rodale_research_paper-07_30_08.pdf

Laursen, K.H., Schjoerring, J.K., Olesen, J.E., Askegaard, M., Halekoh, U. and Husted, S. (2011) Multielemental fingerprinting as a tool for authentication of organic wheat, barley, faba bean, and potato. *Journal of Agricultural and Food Chemistry* 59, 4385–4396. doi.org/10.1021/jf104928r

Lean, G. (2009) *Daily Telegraph*, 7 August.

Lucas, C. (2008) BBC Farming Today, 5 November.

Lupp, B. (2009), The New Organic – It's Time for a National Green Certification Program. Michigan State University College of Law. Available at http://digitalcommons.law.msu.edu/king/132

McKenzie, A.J. and Whittingham, M.J (2010) Birds select conventional over organic wheat when given free choice. *Journal of the Science of Food and Agriculture* 90, 861–869.

McKevitt, S., and Ryan, T. (2013) *Project Sunshine: How science can use the sun to fuel and feed the world.* Icon Books Ltd, London.

Mielke, H.W., Anderson, J.C., Berry, K. J., Mielke, P.W., Chaney, R.L. and Leech, M. (1983) Lead concentrations in inner-city soils as a factor in the child lead problem. *American Journal of Public Health* 73, 12, 1366–1369. Available at http://dx.doi.org/10.2105/AJPH.73.12.1366

Mintel (2015) Shoppers question validity of organic labels. Available at https://www.geneticliteracyproject.org/2015/05/18/shoppers-question-validity-of-organic-labels

Moore-Colyer, R.J. (2001) Back To Basics: Rolf Gardiner, H.J. Massingham and "A Kinship in Husbandry". *Rural History* 12, 1, 85–108. doi: 10.1017/S0956793300002284.

Natural Marketing Institute. (2008b) Connecting values with consumers. *LOHAS Journal* 19–22. Available at www.lohas.com/sites/default/files/consval_sm.pdf

Northbourne, Lord. (1940) *Look to the Land*, J.M. Dent, London.

Oates, L., Cohen, M. and Braun L. (2014) The Health and Wellness Effects of Organic Diets 9–12 In: Rahmann, G. and Aksoy U. (Eds.) *Proceedings of the 4th ISOFAR Scientific Conference. Building Organic Bridges*, Istanbul, Turkey.

OCA (2014) Organic Consumers Association. Available at www.organicconsumers.org

OTA (2003) Organic Trade Association position on pesticide residue testing. Available at https://www.ota.com/search/site/pesticides

Padel, S. (2011) The UK Market for organic food. BioFach Congress 2011. Available at www.organicresearchcentre.com

Paull, J. (2014) Lord Northbourne, the man who invented organic farming, a biography. *Journal of Organic Systems*, 9, 1, 31–53.

PAN (2015) Pesticide Action Network. Available at www.panna.org and www.pan-uk.org/

Pfeiffer, E. (1938, 1947) *Biodynamic farming and gardening* (Anthroposophic Press (1938) Faber and Faber (revised edition 1947).

Pretty, J., Noble, A.D., Bossio, D., Dixon, J., Hine, R.E., Penning de Vries, F.W.T. and Morison, J.I.L. (2006) Resource-Conserving Agriculture Increases Yields in Developing Countries. *Environmental Science and Technology* 40, 4, 1114–1119. doi: 10.1021/es051670d.

Pretty, J., Toulmin, C. and Williams, S. (2011) Sustainable intensification in

African agriculture. *International Journal of Agricultural Sustainability* 9, 1, 5–24.

Puddu, M., Paunescu, D., Stark, W.J. and Grass, R.M. (2014) Magnetically Recoverable, Thermostable, Hydrophobic DNA/Silica Encapsulates and Their Application as Invisible Oil Tags *American Chemical Society Nano, 8,* 3, 2677–2685. doi: 10.1021/nn4063853.

Purves, D. (1985) *Trace element contamination of the environment*. Elsevier Science.

Reynolds, L. (2013) Organic Farming Movement Marginal but Growing Worldwide. Available at www.ipsnews. net/2013/01/op-ed-organic-farming-movement-marginal-but-growing-worldwide

Rodale News (2014) Available at www.rodalenews.com/ organic-farming-climate-change

Savage, S. (2015) Only organic, Genetic Literacy Project. Available at http:// geneticliteracyproject.org/2015/03/03/ only-organic-stonyfields-gary-hirshberg-promote-hate-speech-against-american-farming

Smallwood, M. (2014) Give Me Spots on My Apples. Available at www. mariasfarmcountrykitchen.com/ give-me-spots-on-my-apples

Smith-Spangler, C., Brandeau, M.L., Hunter, G.E., Bavinger, C., Pearson, M., Eschbach, P.J., Vandana Sundaram, Hau Liu, Schirmer, P., Stave, C., Olkin, I. and Bravata, D.M. (2012) Are organic foods safer or healthier than conventional alternatives? A systematic review. *Annals of Internal Medicine American College of Physicians.*157, 5, 348–366. Available at http://annals.org/article. aspx?articleid=1355685

Soil Association (2009) Soil Association website, 4 November 2009.

Soil Association (2012) The Road to 2020. Towards healthy, humane and sustainable food, farming and land use. Available at https://www.soilassociation.org/ media/6484/road-to-2020.pdf

Soil Association (2013) Soil Association 'means nothing to most people. Available at http://www.foodmanufacture. co.uk/Ingredients/Soil-Association-means-nothing-to-most-people-Monty-Don?utm_source=copyright&utm_ medium=OnSite&utm_ campaign=copyright

Soil Association (2014) Stop genetic modification. Available at www. soilassociation.org/gm

Soil Association (2014a) Available at www.soilassociation.org/trade/ marketingsupport/advertisingclaims

UNEP-UNCTAD *(*2008) Organic Agriculture and Food Security in Africa, United Nations, New York and Geneva. Available at www.*unep.ch*/etb/ publications/insideCBTF_OA_*2008*.pdf

USDA (2009) *National Organic Program (NOP)* Available at www.ams.usda.gov. AMSv1.0/NOP. Accessed November 2009.

Watson, G. (2014) Is organic food better for you? Nutrition, anti-oxidants and new research. Available at www.huffingtonpost. co.uk/guy-watson/organic-food-better-for-you_b_5578419.html

Whole Foods (2015) Available at www.wholefoodsmarket.com/ responsibly-grown/what-we-measure

Willer, H. and Lernoud, J. (2016) (Eds) The world of organic statistics and emerging trends 2014. FiBL-IFOAM. Available at https://shop.fibl.org/ fileadmin/documents/shop/1698-organic-world-2016.pdf

CHAPTER 5

Biodynamic Agriculture – Key Aspects and Development

"Extraordinary claims require extraordinary evidence" Carl Sagan

Rudolf Steiner (1861–1925) originated but did not provide the name biodynamic agriculture during his agricultural course of eight lectures (notes in Appendix) in 1924 (Steiner, 1924). The leading players in the development of his ideas were the Koliskos and Pfeiffer. Born in Vienna, both Lili Kolisko (1889–1976) and her husband Eugen (1893–1939) were medical doctors who believed in homoeopathy (Young, 2010). Eugen and Lili had initially worked on homeopathic remedies for foot-and-mouth disease. Pfeiffer (1899–1961) was an electrical engineer (initially employed by Steiner to install stage lighting) who studied inorganic chemistry after joining Steiner. Steiner's ideas were also studied by anthroposophists at the Goetheanum in Dornach, Switzerland.

The essential features of biodynamic agriculture remain as stated by Steiner:

i.   The farm is treated as a unique living organism.

ii.   The fertility and health of the soil is maintained by replacing nutrients removed in crops and livestock by the application of manure and compost enhanced by special preparations.

iii. The preparations, which are herbal, are applied at homeopathic rates.

iv.   Man-made fertilizers are not used.

v.   The timing of many agricultural operations is dictated by the phases of the Moon and the planets.

Dislike of modern agriculture

Similar to the supporters of mainstream organic agriculture, the proponents of biodynamic agriculture were motivated by their dislike of the industrialization of agriculture and by the reports of soil erosion in the USA. Biodynamic agriculture favours what is conceived as natural and generally takes a strong anti-science stance.

Fertilizers were first disliked by anthroposophists, and as a result by bio-dynamic farmers. Kolisko and Kolisko (1978) wrote at length about how farmers damage their eyes and lungs at the same time as they are poisoning the soil when using man-made fertilizers. However, the main objections would appear to be related to the disruption of soil micro-flora and micro-fauna by inorganic salts and the importance of not working against nature.

There seems to have been a reaction in the first quarter of the 20th century by anthroposophists in Europe to the studies on the mineral nutrition of plants by Liebig and to the activities of the fertilizer industry, particularly the fixation of nitrogen by the Haber–Bosch process. This was at a time when it was fully accepted that plants obtained carbon from the atmosphere and all the mineral elements from the soil. The key part played by the sun in light-driven photo-synthesis and in warming the Earth was appreciated but that was the extent of the cosmic influences. The anthroposophists believed that other cosmic forces were involved. Not surprisingly, the anthroposophists faced hostility from the German fertilizer industry.

Investigation of Steiner's ideas by the Koliskos

Steiner entrusted Lili Kolisko with studying all his suggestions and this she started in 1924 at the Goetheanum, where she worked with her husband until 1939 when they moved to England. The exact part played by Eugen Kolisko is not known as he died in 1939 just before they finished the book *Agriculture of Tomorrow*, which was subsequently published in 1946 under their joint names and which incorporated Eugen's work.

The Koliskos were probably responsible for introducing Steiner to medical homoeopathy, which led to the agricultural use of homoeopathy and the extension of the idea that substances can became more potent when considerably diluted (potentization). The Koliskos did the pioneering work on the homeopathic preparations. Knowledge that the auxins, the plant growth hormones, are highly active at very low concentrations may have stimulated their work on homeopathic dilutions.

In view of the central role of the Koliskos in developing Steiner's ideas it is curious that they are not mentioned by Pfeiffer (1935, 1938 and 1947). At the same time the Koliskos do not mention Pfeifer (Kolisko and Kolisko, 1946). Conford (2001), in his detailed thesis on the origins of the organic movement, does not discuss the work of the Koliskos. Likewise, Paull (2011), when detailing the early development of biodynamic agriculture, mentions Lili Kolisko but does not quote her. What was it about their work that was responsible for them being ignored?

The Koliskos had been very affected by the soil erosion occurring in North America and the book *Rape of the Earth* (Jacks and Whyte, 1939). They looked upon their work as contributing to regenerating agriculture and human health. Similar views were later expressed by Balfour (1943).

Lili and Eugen Kolisko developed a chromatographic method of visualising the quality of food, called capillary dynamolysis. This is a qualitative procedure that involves dipping filter paper first in the test solution say of a plant juice or extract, followed, after drying, by dipping the paper in a metal salt. The patterns seen on the filter paper when the salt (commonly 1% $AgNO_3$) reacts with the test solution were thought to indicate the hidden formative forces and the inner qualities of substances in the test sample. Kolisko and Kolisko (1978) presented many illustrations of capillary dynamolysis on a range of organic materials, from which they concluded that farmers and gardeners who are still using man-made fertilizers do not realize that they are poisoning mankind. Work at the University of Kassel is said to have scientifically validated the procedure of capillary dynamolysis.

Image-forming methods described as bio-crystallization, capillary dynamolysis and circular chromatography were used by Fritz *et al.* (2014) to compare wine produced from grapes grown conventionally, organically and biodynamically. The images were compared with a catalogue of reference images. The results were interpreted as indicating higher product quality of biodynamic wine.

The nine biodynamic preparations

Eight of the original nine preparations (preparation 508 is not mentioned by all Biodynamic Associations) are mandatory for biodynamic agriculture. They are all used homoeopathically and in accordance with the biodynamic calendars.

Steiner passed on directions for making and using the preparations to Pfeiffer at the Goetheanum and in his lectures. This indicates that he was satisfied that work at the Goetheanum, possibly by Lili Kolisko, had already shown that the preparations induced the right kind of reaction in compost as indicated by her capillary dynamolysis procedure. Steiner was, however, very secretive about the preparations and did not want information on them passed on to people other than anthroposophists.

Pfeiffer (1938) followed Seiner's instructions about secrecy and only gave details of the preparations to restricted audiences, pointing out that the practical instructions were not intended for the general public. Pfeiffer (1947) did not describe how to make the preparations and he gave only limited information on their use when presenting comparative yield data.

The secrecy surrounding the nine preparations was broken with the writings of Kolisko and Kolisko (1978), on which the following notes on the rationale for and use of the nine biodynamic preparations are largely based. Some Biodynamic Associations are still somewhat reluctant to provide details. However, at a biodynamic conference in the USA in 2014 delegates were given details on the making and use of the biodynamic preparations (BDA USA, 2013).

Procedures for making the preparations differ but an essential feature for six of them is that they involve burying the material in the soil in different receptacles such as cow horns, stag's bladders, cow intestine, ox mesentery and animal skulls. All the preparations are known by their number not by a description indicating their composition. Varying rates of use are mentioned in the literature. Here the rates given by Demeter (2014) are quoted.

Cow horn preparations for application to soil (500) and to crops (501)

Steiner believed that in the living animal, cow horns (not bull horns) prevented the life forces streaming out from an animal and that horns could similarly confine and concentrate the cosmic forces in manure needed to enliven the soil (Kolisko and Kolisko, 1978). Pfeiffer (1947) did not mention the use of cow horns.

Cow horn manure – Preparation 500

Preparation 500 is made from cow manure that is first fermented in a horn that has been buried (pointed end up) in the soil for six months over the winter before being used to make a soil spray to stimulate root growth and humus formation. Burying over winter ensures that the manure is exposed to forces that pass into the soil. The contents of one horn are mixed in a bucket of rainwater. After dilution (including alternate clockwise and anti-clockwise mixing for one hour) the liquid is sprinkled on the soil in the spring. The contents of one horn (about 300–500 g) is sufficient for 1 ha (Demeter, 2014).

The Biodynamic Agricultural Association in UK recommends that preparation 500 is sprayed towards evening directly on the soil prior to sowing or planting. It is understood to encourage healthy root growth, vitalize the soil and help the plant find what it needs from the soil.

Cow horn silica – Preparation 501

Steiner believed that some humus-rich soils were short of silica and he suggested the homeopathic use of a special silica preparation (501). Preparation 501 is made from powdered quartz (packed inside a cow horn and buried in the soil for six months over spring and summer). It is applied as a foliar spray to stimulate growth. One horn can produce 300–500 g of 501, enough, after dilution, for 25 ha (Demeter, 2014). Preparation 501 is applied as a fine mist directly on to the crop early in the morning. It helps to stabilize plant metabolism and enhance the qualitative development of the crop. As with preparation 500, the horn silica preparation 501 had to be carefully stirred in water when potentizing by a specified procedure for at least one hour.

Dry herbal preparations for addition to manure and compost heaps

Steiner suggested enlivening manure and compost heaps by adding homeopathic quantities of different plant materials. As little as 2 g of each preparation is used on 10 tons manure or compost. Pfeiffer (1947) reported that the preparations increase the soil bacterial content by ten times and promote earthworms; he also found that earthworms preferred soil to which preparations 500 or 507 had been applied to soil given man-made fertilizers or urine. Aracaria advise when adding more than

one preparation to a heap that they are kept apart. Demeter recommends that 1–2 cm³ each of the manure/compost preparations (503–508) is applied to 10 m³ of deep litter manure/slurry or compost.

Preparation 502. yarrow, flowers (Achillea millefolium)

Yarrow was used in the Middle Ages (and reputedly by Achilles) to treat internal and external bleeding as well as kidney and bladder problems. So it is not surprising that Steiner came up with the idea of burying yarrow flowers in the fresh bladder of a stag over winter. Why stag bladders were selected is not explained. After burying 10 g of the flowers is extracted in 100 ml water and diluted before being added to manure or compost. The contents of one bladder may be sufficient for 250 ha. Kolisko reported that pig and fox bladders were not as suitable as that from a stag. Steiner thought that the sulphur content of yarrow was important.

Preparation 503. chamomile, flowers (Chamomilla officinalis)

Steiner suggested chamomile for its effects on calcium. Chamomile tea has been known for a few hundred years to settle the gut and probably because of this the homeopathic preparation of dried flowers are wrapped in sausages made from ox intestines before burying over winter. Just 1–2 g of preparation 503 is sufficient for 2–3 cubic metres of manure or compost.

Preparation 504. stinging nettle (Urtica dioica)

Concoctions made from nettles were much appreciated in the Middle Ages for the treatment of many diseases and conditions (Culpeper, 1653), from the clearing of the lungs to treatment of mouth sores and jaundice. Tightly packed whole shoots collected at the start of flowering are buried in the soil over the summer and winter. Kolisko used 1 g of the composted nettles in 10 ml water, which is then diluted further before use.

Preparation 505. oak bark (Quercus robur)

Steiner, who abhorred the use of all inorganic chemicals including lime, looked upon the tannin- and calcium-rich oak bark as an acceptable way of adding calcium to the soil. He believed that homeopathic amounts of calcium in oak bark would be sufficient. It is likely that he was also aware of the medicinal uses of concoctions made from oak bark (an astringent and for controlling haemorrhages (Hill, 1812). Steiner also looked upon oak bark as a general remedy for plant diseases. Small pieces of oak bark are buried in the skull of a domestic animal in a damp area. Lili Kolisko experimented with several skulls, including sheep, pigs and cow. She concluded that it was very important to use a fresh skull. Burying the bark in an earthenware pot was not successful as the bark did not decompose. The contents of one skull were sufficient for 250 ha.

Preparation 506. dandelion flowers (Taraxacum officinale)

According to Steiner, dandelions have the capacity to regulate the relation between silicic acid and potassium. Adding the dandelion preparation to manure or compost will pass on this capacity to the crops. Faded dried flowers are packed into the mesentery of an ox and buried over winter for use in the spring. Kolisko used 1–2 g of preparation 506 on a manure heap of 1.5–2.3 m³. The medicinal uses of boiled dandelion roots (for "promoting urine flow and bringing away gravel" (Hill, 1812) were known and commented on by Kolisko but not by Steiner.

Liquid compost preparations Preparation 507. valerian flowers (Valeriana officinalis)

Steiner thought that adding valerian to manure or compost heaps would help the crops in connection with phosphorus (Kolisko and Kolisko, 1978). The flowers were soaked in rainwater for a few days, after which one part of the solution was added to nine parts rainwater to make the first potency. Kolisko sprinkled the liquid formulation at the eighth potency (which effectively means sprinkling 1 ml in 100,000 litre rainwater) on the manure heap.

Preparation 508. horsetail (Equisetum arvense)

Concoctions made from horsetail were known to be effective in stopping bleeding both internal and external, and as a remedy for kidney disease. The silica content of horsetail, otherwise called the silica plant, was the attraction for the workers at the Goetheanum. Steiner considered horsetail as a cure of plant diseases. He suggested that a strong infusion of horsetail is diluted and applied in homeopathic quantities to diseased areas in fields. One part of horsetail is boiled for at least one hour in ten parts water to make preparation 508. Kolisko and Kolisko (1978) reported excellent results in treating all kinds of plant diseases, including fungi and mildew, and also pests with the fifth potency. This amounts to 1 ml of preparation 508 in 100 litre water. According to Kolisko, if a plant is diseased it means that it has lost contact with the universe.

Aracaria Biodynamic Farm recommends boiling 100 g dried horsetail in 2 litre water and sprinkling the liquid formulation on soil around plants, preferably at full or new moon. The British Biodynamic Association has excluded the horsetail preparation 508 (BDA, UK). The Biodynamic Association, which can supply all the other preparations, gives precise information on storage in order to preserve their all-important radiating power (BDA, UK, 2016).

Barrel preparation

This preparation, which was originally developed by Maria Thun, has been known by a number of different names, including "Cow Pat Pit" (Aracaria in Australia) or even the "Maria Thun" but is now known as the "Barrel Preparation". Barrel Preparation is made from matured cow manure, to which is added small amounts of finely ground basalt and ground egg shells as well as several biodynamic preparations. The Biodynamic Association in the UK (BDA UK, 2016) markets a barrel preparation but without information on how it is made. Aracaria says its "Cow Pat Pit" compost contains preparations 500 plus 502–507 and that it enhances fungal and bacterial growth in the soil. Research carried out after the Chernobyl disaster showed that the barrel preparation helped to reduce the effects of radioactive fallout (BDA UK, 2016).

The barrel preparation is dissolved in water (two tablespoons in 5 litre) and sprayed over the soil or compost. Alternatively, it can be used in solid form by sprinkling small amounts on to composts, manures, slurries or planting mixes. It is also recommended when composting domestic garden and kitchen waste.

Sources of preparations

Many Demeter farmers and gardeners produce their own preparations, sometimes with their Biodynamic advisor. However, in the case of shortages there are several commercial suppliers.

Testing of preparations

Kolisko and Kolisko (1978) experimented with potencies of the various preparations up to the sixtieth potency. Each dilution, i.e. moving to the next potency, involves diluting ten times, e.g. one part is added to nine parts rainwater from which one part is taken and again added to nine parts rainwater. At the tenth potency 1 g or 1 ml of a preparation is effectively diluted into a volume similar to that of four Olympic swimming pools (each 2,500,000 litre).

Many studies on the effects of the various preparations on plant growth and composition have been carried out (Biodynamic-Research.net). In many cases the variable results have made interpretation difficult. But clear-cut effects have been found, for example Reeve *et al.* (2010) used grape pomace (the remains after pressing, consisting of skins, pulp, seeds, and stems of the fruit) composted with manure with or without the combined six biodynamic compost preparations (502 to 507) The growth of wheat seedlings was increased by up to 60% when treated with water extracts of the finished compost and it was concluded that compost extracts can be used as fertilizer substitutes. Applications of unspecified biodynamic preparations were shown, by a pictomorphological method, to affect the growth of winter wheat (FiBL, 2014).

Heimler (2012) compared the growth and composition of two lettuce cultivars grown under conventional, organic or biodynamic farming. The yield of the conventional crops exceeded those of the organic systems but had lower concentrations of polyphenols than crops grown organically or biodynamically.

Promotion and extension of Steiner's ideas by Pfeiffer

Pfeiffer is often quoted to have turned Steiner's ideas into practice and that his book *Biodynamic Farming and Gardening* (Pfeiffer, 1938) played a major role in developing the biological dynamic methods – ultimately called biodynamic agriculture by Pfeiffer. The impact of the book was maximized by it being published simultaneously in English, German, Dutch, French, and Italian. Together with Pfeiffer's articles in magazines and his lectures, the book undoubtedly played a large part in spreading the biodynamic message across the world.

The book was based on the work at the Goetheanum from 1920 and Pfeiffer's experiences in the Netherlands where, from 1928, he converted a conventional farm to a biodynamic one. The Koliskos are not mentioned by Pfeiffer (1938, 1947) which is curious because they had all worked with Steiner at the Goetheanum from around 1920. The revised edition of his classic book (Pfeiffer, 1947) has been consulted here; it is essential reading for understanding biodynamic agriculture.

Pfeiffer was saddened that the wisdom of bygone farmers had been replaced by agricultural science and hand cultivation by machine, which he felt was inferior to manual labour. He claimed that crop yields had not increased over the previous thirty years despite a three-fold increase in nitrogen use. Pfeiffer criticized the use of man-made fertilizers to increase yields as it would lead to an increase in the requirement for water and thereby lower the water table, which he said was happening in Central Europe.

Pfeiffer was very influenced by the much-publicized soil erosion in the American prairies and assessed that about a third of the cultivated area was becoming useless – something many current American farmers would dispute. He foresaw the day when grain farmers in both USA and Canada would be living from hand to mouth. It would not be many years after Pfeiffer's prediction that grain exports from North America would be preventing famine in India and the USSR.

Pfeiffer appreciated the value of earthworms in improving both soil tilth and the humus content of soils, and was aware that sprays of copper sulphate killed earthworms. Direct drilling of crops, which involves the use of herbicides to kill all weeds, is known to favour earthworm populations, which can be devastated by mechanical cultivation. However, all organic movements prohibit the use of herbicides.

Although not present at Steiner's course, Pfeiffer had worked with him and had learnt from him his ideas, notably about the homeopathic preparations 500 and 501. Steiner had told Pfeiffer to make the benefits of these agricultural preparations known to as many anthroposophists as possible and that experiments could come later. Steiner clearly had absolute confidence in his spiritually derived ideas. An inspired Pfeiffer lectured widely, spreading Steiner's ideas to anthroposophists in the USA and Europe.

Dynamic forces

Pfeiffer explained how organisms are influenced by more than their chemistry, namely by dynamic forces (Pfeiffer, 1947). He referred to the action of auxins, hormones, enzymes and vitamins at very low concentrations – or dilutions as he called them. He cited the production of ethylene by apples and the effect it can have on neighbouring plants as a dynamic force.

Pfeiffer cited the fungicidal and bactericidal properties of a poison (amenonin) extracted from *Ranunculus acris* (tall buttercup) as evidence of plants exhibiting dynamic forces. Pfeiffer reported that he had demonstrated the benefits of the biodynamic preparations on the growth of roots by countless experiments but did not present evidence. He believed that the herbal preparations are active because in their preparation the growth-promoting substances they contain are set free and become active. Certain plants were called dynamic. Boas, a professor in Munich, was quoted as having demonstrated the efficacy of a highly dilute chamomile extract stimulating yeast at 1 in 8,000,000. Pfeiffer extrapolated from this and thought that powerful active substances must occur in manure in minute quantities. According to Boas (1932, 1937) a new approach called dynamic botany was required.

Pfeiffer followed Steiner's injunction to demonstrate how to convert farms to biodynamic and only later conduct experiments. He observed what happened on farms after conversion to biodynamic methods, whereas the Koliskos carried out experiments on plants grown under controlled conditions in pots.

Influence abroad

In 1939 Pfeiffer organized Britain's first biodynamic conference (at Betteshanger in Kent) on the estate of Lord Northbourne, a keen supporter of the organic movement in the UK. Pfeiffer's Betteshanger Conference is seen as a link between biodynamic agriculture and mainstream organic farming because it led, in 1940, to Lord Northbourne publishing his manifesto on organic farming *Look to the Land*, in which he coined the term "organic farming".

Pfeiffer lectured on biodynamic farming in the USA in the 1930s and in 1938 he emigrated to Pennsylvania, where he created a model biodynamic training farm and started the Biodynamic Farming and Gardening Association. He met J.I. Rodale, founder of *Organic Gardening and Farming* magazine, and of the organic movement in the USA. This relationship gave biodynamics a firm base in the American organic movement.

Although Pfeiffer did not visit Australia he had a major influence on the development of biodynamic agriculture there by his contributions to the country's first organic farming journal, the *Organic Farming Digest*.

Pfeiffer was an articulate and authoritative evangelist, becoming the leading advocate of biodynamics in the English-speaking world, especially in the important early years (Paull, 2009). In many ways he paralleled Lady Eve Balfour, who promoted her ideas to a large audience by lecturing widely and via her book (Balfour, 1943).

Pfeiffer (1950) looked to a future where it might be possible to spray highly diluted cow urine instead of poisonous chemicals. He reported experiments where the effects of dilutions of 1:100 billion could be detected and that the optimum effect was at a dilution of 1:2 billion. He was of the opinion that agricultural practices would have to be changed to accommodate the new knowledge. If only food production was that simple.

Handling of manure and compost

Maximizing the value of manure is a key element of organic farming. The precise details given by Pfeiffer (1947) about the making of heaps of manure and compost clearly indicates the importance he attached to manure and to its nitrogen content as a means of feeding the soil not the crop.

Pfeiffer was an advocate of mixed farming, saying that livestock play an essential part on a biodynamic farm. This means that there needs to be a balance between grass, cattle and arable crops. A typical biodynamic farm would consist of 70–75% crops and 25–30% pastures with cattle fed indoors for half the year. Pfeiffer believed that a farm without cattle was one-sided and contrary to nature. Likewise, a cattle range was one-sided.

The manure is rotted in covered heaps with access to worms from the soil beneath and is left to mature for two months before being used. The various homeopathic preparations that are added in separate holes in the top of the heap hasten rotting and promote the proper kind of fermentation. Pfeiffer (1947) states, without presenting any data, that the preparations increase the bacterial content of the manure heap ten times.

One of the main objectives of biodynamic proponents is to educate farmers how best to utilize manure. For example, the importance of ploughing in the manure quickly after distribution is stressed. Consideration is given to the source of the manure in terms of the animal and its diet and also to its moisture content.

Pfeiffer (1947) makes a single reference to Howard's Indore composting process, saying that it is reminiscent of that of the biodynamic method except that preparations were not used.

Conversion to a biodynamic farm

Pfeiffer details how to convert to a biodynamic farm, involving many aspects including the setting up a rotation, improving the quality of the manure, feeding with home-grown feed and the production of seed for future crops. The whole process is likely to take several years. His practical approach to promote biodynamic agriculture by demonstration was similar to that of the Soil Association, which some years later after the failure of the Haughley experiment concentrated on demonstrating how to farm organically.

Benefits of farming biodynamically

At the Goetheanum Pfeiffer tested, together with associates, dilutions of the preparations 500–507 mainly by bathing wheat seeds in the various dilutions and then measuring the lengths of roots and of the first and second leaves (Pfeiffer, 1947). It seems likely that this work was actually carried out by the Koliskos. Pfeiffer says that preparation 500 was particularly effective in promoting root growth, whereas preparations 500–507 improved overall growth. Nodulation in beans was greater in soil to which preparation 500 had been applied than in a soil that had been given man-made fertilizers, a result that could have been due to the negative effect of nitrate in the fertilizer and not to the positive effect of homeopathic amounts of preparation 501. Soaking seed in dilutions of preparation 500 for up to sixty minutes stimulated radish growth and tests with other preparations influenced their flavour. Pfeiffer found that preparation 501 counteracted disturbances to the geotrophic and phototrophic responses of linseed, wheat, oats, barley and sunflower seedlings.

Pfeiffer attempted to demonstrate beneficial effects of biodynamic methods on crops in field trials lasting up to four years but without success. He ascribed the failures, including negative effects, to the short period of conversion to biodynamic farming. As a result he depended on comparing productivity on farms before and after conversion to prove his point.

Balfour wrote an introduction to Pfeiffer (1947) in which she cautioned that the extent to which the reader will be able to accept his deductions will depend on whether the ecological argument has been accepted; this was not the best indication that Pfeiffer had presented her with clear evidence of benefits of biodynamic farming.

Studies showing the value of the preparations on the efficacy of manure are listed by Demeter Research. Much of the work was carried out in Germany.

Carpenter-Boggs *et al.* (2000) found that organically and biodynamically managed soils had a similar microbial status and were more biotically active than soils receiving inorganic fertilizers. Use of manure enhanced soil biological activity, but additional use of the biodynamic preparations did not significantly affect the soil biotic parameters. Similar results have been reported by Reganold (1995), which suggests that the benefits of biodynamic farming to soil quality are primarily due to the use of manure and composts.

There have been no field studies simply comparing using manure with and without preparations that would permit a meaningful assessment of the efficacy of the different preparations (Vogt, 2007).

Health

Pfeiffer compared the growth of mice fed wheat grown biodynamically with those fed wheat grown with kainit and basic slag without attempting to equate the nutrients supplied by the two methods. The biodynamically fed mice were stronger but smaller. Egg production was increased in a similar experiment with chickens fed biodynamic food. Trials with turkeys showed that health was enhanced when the birds were fed biodynamically produced grain or green food. Pfeiffer concluded that human health could be similarly improved by biodynamic agriculture. He had arrived at the same conclusion reached by Balfour (1943) but from a different direction, namely beliefs dating back to Steiner and a few experiments. Steiner believed that the use of synthetic nitrogenous fertilizers resulted in nutritionally inferior food and that it was the life forces in it that were responsible for the superiority of that which was grown according to his methods. He quoted the opinions of several doctors who believed that consuming biodynamically grown food improved health by increasing vitality and reducing allergic reactions.

Control of weeds, insects, disease and mice

For weed control Steiner advised the spreading over weedy areas an ash made from the burnt seeds of the problem weeds. In this way "something (is put) into the soil the weeds do not like" wrote Kolisko and Kolisko (1978). Kolisko mentions only one weed control experiment in which it took four years for thistles to be controlled by the application of burnt ash of thistle seeds. There is no record about the source of the seeds – if they were collected from the weedy area the act of collection may have been the control mechanism. Pfeiffer (1947) did not examine Steiner's weed control method. Current advice for controlling weeds includes mechanical cultivation and the use of rotations. Flaming is permitted but is not liked and steaming the soil is prohibited.

Neither Kolisko and Kolisko (1978) nor Pfeiffer (1947) followed up Steiner's recommendations for controlling insects in a manner analogous to the control of weeds, namely applying the burnt ash of the relevant insect to the soil. Kolisko commented that the ash works by counteracting the life process of the specific insects.

Pfeiffer stressed that disease resistance in crops was conferred by farming biodynamically. Currently diseases are mainly limited by the selection of appropriate varieties and by rotational cropping. Use of copper compounds and of sulphur has to be approved by the certifying biodynamic authority.

Steiner's strange advice about controlling field mice by scattering the ash from burning mouse skins was also not followed up by Kolisko or by Pfeiffer.

Influence of the Moon on plant growth

The Romans thought that the Moon could influence plant growth. Pliny (23 AD) gave precise instructions about harvesting, pruning and manuring in relation to the phases of the Moon in his *History of Nature*.

Steiner stressed the importance of cosmic forces emanating from the planets and the Moon and he gave instructions for planting, cultivating and harvesting based on the cycles of the sun, moon, planets and stars. The influence of the Moon on the tides had been appreciated for many years, so it was not surprising that Steiner concluded that the Moon would affect water in the soil. Steiner is quoted by Kolisko (1936) to have said that the Moon has a "colossal effect on the soil when the forces come on days of the full moon" and "that these forces spread out into the plants provided there has been rain".

Following Steiner's suggestions the Koliskos studied the influence of the Moon on the growth of a wide range of plants for fifteen years (Kolisko and Kolisko, 1978). They reported that sowing two days before the full moon resulted in better growth of several plants than sowing two days before the new moon; crops included maize, cabbage, savoy cabbage, tomatoes, gherkins, radishes, beetroot, carrot, kohlrabi, peas and beans. The results are presented as photographs. Gardeners were said to know that certain plants develop many leaves when sown during the waxing of the Moon, whereas when they are sown with the waning moon strong roots are produced (Kolisko, 1936).

Pfeiffer (1947) only mentioned the Moon once, saying that in the past farmers sowed according to the phases of the Moon and that now such practices are called nonsensical. Nevertheless, the biodynamic calendars remain an integral part of biodynamic farming.

Biodynamic sowing and planting calendars

Maria Thun, who studied astrology at the Goetheanum, is credited with being the originator of the annual biodynamic calendars. Different calendars are required to associate phases of the Moon with the local crops and their temperature and water requirements during the growing season.

Working on her farm in Germany, Thun (2000) discovered in the 1950s that vegetables planted when the Moon was in different constellations had different forms and sizes. She concluded that root crops (including onions and leeks) do best if sown when the Moon is passing in front of the zodiacal constellations associated with the Earth element; leafy crops do best when the Moon is associated with the water element; flowering plants do best associated with the air element, and fruits do better with the fire element. The twelve zodiacal constellations were classified according to four elements, namely earth, water, air and fire. In 1963 Thun presented the first of her annual lunar calendars to help gardeners and farmers decide when to sow and harvest crops based on the waxing and waning of the Moon, and the position of the planets and constellations. The calendar was subsequently translated into twenty-seven languages.

Biodynamic sowing and planting calendars have been produced for many years by Maria and Matthias Thun. They are based on more than forty years of ongoing research into the influences of the Moon, planets and constellations on plant growth. The Biodynamic Agricultural Associations detail how each month the Moon moves through all twelve constellations of the zodiac. Since ancient times the twelve zodiac constellations have been associated with each of the four elements. Three constellations are connected to each element and each element is associated with a different plant type and tissue:

Earth element: (Taurus, Virgo and Capricorn) with root crops and growth;
Water element: (Cancer, Scorpio and Pisces) with leafy crops;
Air element: (Gemini, Libra and Aquarius) with flowers;
Fire element: (Aries, Leo and Sagittarius) with fruit crops.

So when sowing, cultivating and harvesting carrots, earth/root days should be chosen (i.e. when the Moon is in Taurus, Virgo or Capricorn), for lettuce, water/leaf days and for apples, fire/fruit days.

The taste of wine (category fire/fruit) is thought to vary with the lunar calendar. Despite being dismissed as pseudo-science, some supermarkets have arranged wine-tasting sessions around good and bad days as dictated by the lunar calendar (BBC, 2009).

In the 1970s and 1980s there were many PhD dissertations in Germany, Switzerland and Austria on the biodynamic preparations and on the effects of the Moon/zodiac on crops. The Institute for Biodynamic Research (IBDF, Darmstadt, Germany) has a website, Biodynamic-Research.net. It lists the many pot and field trials testing the effects of biodynamic preparations on the yield and quality of different plant species, on soil properties, on organic fertilizers, compost and on the relationship between lunar rhythms and crop growth. It also lists the experiments that showed no effects of the preparations.

Different lunar calendars are required for different regions but the principles remain the same. For example, in Australia Aracaria produces a lunar planting and activities calendar based on the belief that the Moon exerts an influence on very small bodies of liquid such as plant sap. When water is rising during the waxing moon seedlings are thought to be able to take up liquids more easily than those sown or planted in the waning phase.

In the USA Llewellyn's *Moon Sign Book* (2014), produced since 1905, has given dates and times for various gardening activities, including when (and what) to plant, cultivate, graft, trim and/or harvest based on the Moon's relation to where constellations are in the sky. In the 2014 edition it was reported that germination of beets and carrots is better when the Moon is in Scorpio than in Sagittarius, whereas tomato seeds germinate better when the Moon is in Cancer rather than in Leo.

Biodynamic research

Biodynamic farming was compared (possibly uniquely) with mainstream organic and conventional farming in a trial (the DOK trial) started in 1978 by the Research Institute of Organic Agriculture (FiBL) in Switzerland (FiBL, 2000, 2015). Similar amounts of farmyard manure were applied in all treatments (based on similar numbers of livestock units). Composted farmyard manure with preparations was used for the biodynamic crops and rotted farmyard manure for the organic crops. Preparations 500 and 501 were sprayed several times each year. NPK fertilizers were applied as a supplement for the conventionally grown crops. Overall organic yields were about 80% of conventional farming and there would appear to have been no differences between the two organic treatments. Soil structure, microbial and earthworm activity were enhanced by both organic treatments, with the biodynamic treatment being superior to mainstream organic. Of particular interest was the finding that crops in the organically and biodynamically managed plots were more strongly colonized by mycorrhiza (30% more) and had more mycorrhizal fungal species. Colonization was suppressed by the fertilizers and/or pesticides used on the conventionally grown crops.

There are currently forty-five Demeter/Biodynamic organizations across the world, most of which are in Europe (twenty-seven), with seven in South

America, six in Asia, two in Africa and North America and one in Oceania. Biodynamic research is likewise concentrated in Europe (seven institutes) and only one in each of Brazil, Egypt and the USA. The international contact for Biodynamic research programmes is the Agricultural Section at the Goetheanum (Dornach, Switzerland) founded by Steiner (Demeter Research, Anon). The Institute for Biodynamic Research in Germany (Darmstadt) was founded in 1950, becoming the first research institution of organic farming.

Qualitative studies involving picture-forming methods such as copper chloride crystallization and paper chromatography are believed to be as valuable as nutrient analysis in assessing the benefits of biodynamic farming and the use of the preparations. Recent benefits of biodynamic preparations have been reported (Demeter Research, Anon), including increased levels of soil microorganisms (Ngosong *et al.*, 2010).

Genetic engineering and nanotechnology

Genetically modified organisms, or products derived from them, are not allowed in biodynamic agriculture. Non-organic products that are brought into the farm must be verified to be free of genetically modified organisms.

Demeter International adopts the precautionary principle with regard to the implementation of nanotechnology, and therefore excludes it from all usage

in biodynamic agriculture, and from all Demeter-certified products.

Influence of biodynamic agriculture on the origin of the mainstream organic movement

Biodynamic agriculture developed many years before Balfour started the Haughley experiment that led to the formation of the Soil Association, so the question as to whether Balfour was influenced by Steiner's ideas remains. Balfour, like Steiner, did not approve of man-made fertilizers and like him wanted farms to be mixed and self-contained, primarily to optimize the use of farmyard manure. Balfour only mentioned Steiner when referring to Pfeiffer adding organic extracts to compost heaps in a footnote in her book, *The Living Soil* (Balfour, 1943). According to her biographer, she met Pfeiffer in 1951 when she was lecturing in the USA. Her introduction in the revised edition of Pfeiffer (1947) supports his ecological approach to farming and his dismissal of the way agricultural scientists work. The homeopathic use of herbal additives to manure and compost and farming according to the planets was not followed up in the Haughley experiment, which would indicate that either Balfour was unaware of the details of the Steiner preparations or did not think they and the influence of cosmic forces were worth further attention. Nevertheless, some of the people who later helped to form the Soil Association were supporters of Steiner and of anthroposophy (Conford, 2001). In 1928 an Anthroposophical World Conference was held in London. So Steiner's anthroposophy may have indirectly influenced the Soil Association.

Swedish assessment

Kirchmann (1994) in Sweden commented that many of Steiner's statements are not provable because scientifically clear hypotheses based on them cannot be made and tested. Kirchmann concluded that Steiner's instructions are occult and dogmatic and cannot contribute to the development of sustainable agriculture.

Summary

1. The origin of biodynamic agriculture was entirely the brainchild of Rudolf Steiner and the agricultural course of eight lectures he gave in 1924 when he presented his idiosyncratic views on the perils of modern agriculture.
2. The industrialization of agriculture and especially the use of man-made fertilizers, which were thought to poison the soil by killing microorganisms, were disliked.
3. The development of biodynamic agriculture depended not only on the activities of the Koliskos, to whom Steiner entrusted the responsibility of investigating his ideas, but also to other anthroposophists at the Goetheanum and particularly on Pfeiffer.
4. There are strong elements of spirituality in biodynamic farming. Farmers need to be in tune with cosmic forces as well as nature.
5. The two key aspects of biodynamic agriculture are maximising the benefits of manure and the homoeopathic use of various, mainly herbal, preparations.
6. Homoeopathy was extended from medicine to agriculture, probably under the influence of the Koliskos.
7. The herbal preparations have mainly been studied by testing extracts on seedling growth and by Kolisko with her capillary dynamolysis procedure.
8. Particular attention is given to two preparations made a) from manure buried in cow horns and b) finely ground silica similarly buried. The first preparation is, after much dilution, applied to the soil and the second to growing crops.
9. Pfeiffer not only coined the name biodynamic agriculture but he also played the major part in promoting it in his books (in several languages) and lectures. In many ways Pfeiffer and Balfour followed similar promotional paths.
10. Pfeiffer emigrated to the USA, where he influenced Rodale, whose research activities continue today.
11. Belief in the effects of lunar and cosmic forces on plant growth was a driving force that is still evident in the use of biodynamic calendars.
12. Clear benefits of biodynamic (and of mainstream organic) farming on mycorrhizal colonization have been found on a range of crops.
13. The benefits of biodynamic agriculture have been assessed by comparing farm productivity before and after conversion to biodynamic agriculture.
14. Some animal experiments have indicated beneficial effects of biodynamic food on health.

15. Biodynamic agriculture could have had an indirect effect on the development of mainstream organic agriculture via supporters of Steiner helping to form the Soil Association.

References

Aracaria. Biodynamic Farm, Mullumbimby. Available at www.aracaria.com.au

Balfour, E.B. (1943) *The Living Soil – Evidence of the Importance to Human Health of Soil Vitality, with Special Reference to National Planning.* Faber and Faber, London.

BBC (2009) Can lunar cycles affect the taste of wine? BBC News Magazine. Available at http://news.bbc.co.uk/1/hi/magazine/8008167.stm

BDA, UK (2016) Guidelines 2 Biodynamic compost preparations. Available at https://www.biodynamic.org.uk/discover/#bd-preps

BDA, USA (2013) Biodynamic compost preparations. Available at https://www.biodynamics.com/content/biodynamic-compost-preparations https://www.biodynamics.com/track/2014/biodynamic-preparations

Boas, F. (1932) Untersuchungen fur uine dynamische Grunland Biologie. *Praktische Blatter fur Pflanzenbau* 9, 173.

Boas, F. (1937) *Dynamische Botanik.* J.F. Lehman, Germany.

Carpenter-Boggs, L., Kennedy, A.C. and. Reganold, J.P. (2000) Organic and Biodynamic Management: Effects on Soil Biology. *Soil Science Society of America Journal* 64, 1651–1659.

Conford, P. (2001) *The origins of the organic movement.* Floris Book, Edinburgh.

Culpeper, N. (1653) *The Complete Herbal.* Foulsham, London.

Demeter (2014) Production standards Demeter International. Available at www.demeter.net/sites/default/files/di_production_stds_demeter_biodynamic_15-e.pdf

Demeter Research (Anon) Available at www.demeter.net/biodynamic-research.

FiBL (2000) FiBL Dossier. Results from a 21 year old field trial. Available at https://www.fibl.org/fileadmin/documents/shop/1090-doc.pdf

FiBL (2014) Frick trial on preparations and soil. Available at www.fibl.org/en/switzerland/research/soil-sciences/bw-projekte/frick-trial-on-preparations.html

FiBL (2015) DOK-Trial, The world's most significant long-term field trial comparing organic and conventional cropping systems. Available at www.fibl.org/en/switzerland/research/soil-sciences/bw-projekte/dok-trial.html

Fritz, J., Athmann, M., Meissner, G. and Köpke, U. (2014) Quality Assessment of integrated, Organic and Biodynamic Wine using Image forming Methods, 497–499. In: Rahmann, G. and Aksoy U. (Eds.) *Proceedings of the 4th ISOFAR Scientific Conference. Building Organic Bridges,* Istanbul, Turkey.

Heimler, D., Vignolini, P., Arfaioli P., Isolani, L. and Romani, A. (2012) Conventional, organic and biodynamic farming: differences in polyphenol content and antioxidant activity of Batavia lettuce. *Journal of the Science of Food and Agriculture.* 92, 3, 551–556. Doi: 10.1002/jsfa.4605.

Hill, Sir J. (1812) *The Family Herbal. English Plants Remarkable for their Virtues.* George Virtue, London.

Jacks, G.V. and Whyte, R.O. (1939) *Rape of the Earth*. Faber and Faber, London.

Kirchmann, H. (1994) Biological dynamic farming – an occult form of alternative agriculture. Journal of Agricultural and Environmental Ethics 7, 2,173–187. doi: 10.1007/BF02349036. Available at http://link.springer.com/article/10.1007%2FBF02349036#page-2

Kolisko, E. and Kolisko, L. *Agriculture of Tomorrow* (1978) Kolisko Archive Publications, Bournemouth, England (first edition 1946).

Kolisko, L. (1936) *Moon and Plant Growth*. Kolisko Archive Publications, Bournemouth, England.

Llewellyn (2014) Llewellyn's Moon Sign Book: Conscious Living by the Cycles of the Moon. Available at http://readynutrition.com/resources/does-science-back-up-planting-by-the-cycles-of-the-moon_17082014

Maria Thun (2000) *Gardening for Life – The Biodynamic Way: A Practical Introduction to a New Art of Gardening, Sowing, Planting, Harvesting*. Hawthorn Press. Gloucester, Great Britain.

Ngosong, C., Jarosch, M., Raupp, J., Neumann, E. and Ruess, L. (2010) The impact of farming practices on soil microorganisms and Arbuscular mycorrhiza fungi: Crop type versus long-term mineral and organic fertilization. *Applied Soil Ecology* 46, 1, 134–142. doi: 10.1016/j.apsoil.2010.07.004. Available at www.researchgate.net/publication/223700365_The_impact_of_farming_practice_on_soil_microorganisms_and_arbuscular_mycorrhizal_fungi_Crop_type_versus_long-term_mineral_and_organic_fertilization

Northbourne, Lord (1940) *Look to the Land*. J.M. Dent, London.

Paull. J. (2009) How Dr. Ehrenfried Pfeiffer contributed to organic agriculture in Australia. *Journal of Bio-Dynamics Tasmania*, 96: 21–27.

Paull, J. (2011) Biodynamic agriculture: the journey from Koberwitz to the world, 1924–1938. *Journal of Organic Systems* 6, 1, 2011.

Pfeiffer, E. (1935) *Short Practical Instructions for the Use of the Biological-Dynamic Methods of Agriculture*. Rudolf Steiner Publishing Co., London.

Pfeiffer, E. (1938) *Biodynamic Farming and Gardening*. New York, Anthroposophic Press.

Pfeiffer, E. (1947) *Soil Fertility, Renewal and Preservation. Biodynamic Farming and Gardening* (original edition 1938). Faber and Faber. Available at www.soilandhealth.org

Pfeiffer, E. (1950) Vitamins, growth substances, hormones to the soil. *Farm and Garden Digest (incorporating Organic Farming Digest)*, 2, 7, 11–14.

Reeve, J.R., Carpenter-Boggs, L., Reganold, J.P., York, A.L. and Brinton, W.F. (2010) Influence of biodynamic preparations on compost development and resultant compost extracts on wheat seedling growth. *Bioresource Technology* 101, 14, 5658–5666.

Reganold, J.P. (1995) Soil quality and profitability of biodynamic and conventional farming systems: A review. *American Journal of Alternative Agriculture* 10, 1, 36–45.

Steiner, R. (1924) *Agricultural Course of Eight Lectures by Rudolf Steiner*, English translation by George Adams, preface by E. Pfeiffer, Bio-dynamic Agricultural Association, Rudolf Steiner house, London, NW1 (revised edition 1958). Available at http://wn.rsarchive.org/Lectures/GA327/English/BDA1958/Ag1958_index.

html Link to lectures: www.rsarchive.org/
    Lectures/index.php?q=A
Vogt, G. (2007) The origins of organic
    farming 9–29. In: Lockeretz, W. (Ed.)
    Organic Farming, An International History.
    CAB International, Oxford.

Young, S. (2010) Eugen and Lili Kolisko and
    Agrohomoeopathy. *Homoeopathic Journal* 3,
    12. Available at http://sueyounghistories.
    com/archives/2009/04/18/eugen-and-li
    li-kolisko-and-homeopathy

CHAPTER 6

Origin and Development of Mainstream Organic Movements

"Great oaks from little acorns grow" a 14th century proverb

The development of the mainstream organic movement in the UK and in many other countries can be traced back to Howard and particularly to the pivotal role played by Lady Eve Balfour and how she gathered around her a diverse body, mainly of men. Some of them were motivated by health concerns and others by a desire to turn the clock back to feudal times consisting of peasants and landowners. However, consideration by Conford (2001) of the activities and histories of this disparate body shows that in many ways Balfour was preaching to the converted when she wrote *The Living Soil*. Already in the 1920s and 1930s a few farmers were following the biodynamic methods of Steiner and others were composting according to Howard's Indore process (Howard, 1931 and 1940). Also in the group around Balfour were several doctors and social health workers concerned about the relationship between diet and health. The religious and spiritualist beliefs of some of Balfour's supporters also played a part.

Balfour's big contribution was to weld various medical and agricultural ideas with the objective of improving health and living conditions across the country. Apart from the obvious effects on the agricultural scene in the UK, Balfour also influenced what were probably nascent organic movements in other countries, notably in the English-speaking world, by her books and tours. It is instructive to understand how she became motivated and had such a wide influence. Michael Brander, Balfour's official biographer who was related to her by marriage, presents a fascinating history of her life (Brander, 2003), which is the basis of much of the following account.

Eve Balfour (1898–1990)

Eve Balfour was one of six children in the family of Gerald Balfour. Her uncle A.J. Balfour was Prime Minister (1902–1905) and the first Earl of Balfour. When he died the title passed to Eve's father, resulting in her acceding to the courtesy title Lady Evelyn Balfour.

Balfour and her sister, Mary had, as young girls, dreamed one day of farming and with this in mind Balfour enrolled at the age of 17 at Reading University on a two-year course to study agriculture. At the conclusion of her practical year, at the end of the First World War, Balfour took over a 50-acre derelict farm near Newport for the Monmouth War Agricultural Committee. Brander (2003) records that when she requested ammonium sulphate she was told by the committee to use basic slag instead because it was cheaper! So at this time she had

no inhibitions about man-made fertilizers. In 1919 she bought, with Mary, a rundown farm in Suffolk at Haughley with financial assistance from her father that involved the setting up of a trust for overall control. Eve, Mary and two friends (Beb Hearnden and Derry Hawker), who worked together, had problems in making ends meet for several years and to supplement their income they set up a group to play dance music and jazz and perform plays at nearby venues.

On the farm Eve was the driving force, both mentally and physically. Ploughing with horses on the flat Suffolk land was easy after the hills in Monmouth. They built up a successful dairy and retailing business on what was a mixed farm but in common with most farmers at the time found it difficult to make ends meet. Farming in the UK had experienced a boom immediately after the war but had quickly suffered, in the early 1920s, from cheap imports and the withdrawal of guaranteed cereal prices in 1921.

Other sources of income were required that resulted in Eve, in collaboration with her friend Beb Hearnden, deciding in 1925 to write a detective novel. Eventually three books, *The Paper Chase* (1926), *The Enterprising Burglar* (1928) and *Anything Might Happen* (1933), were published under the pseudonym Hearnden Balfour. It is likely that her experience in writing and involvement with publishers paid dividends when writing *The Living Soil*. It also introduced her to writing for farming magazines.

Balfour had her first taste of what she called a manurial experiment in 1928 when she found that milk production was boosted when cows were fed on grass recently treated with calcium nitrate. Their success with milk production and retailing kept the farm viable.

Signs of political activity were apparent in 1929 when Balfour contributed to the protests against wealthy brewers who were limiting the home-grown barley market by importing cheap poor quality varieties to the detriment of the nutritional value of the beer. Then during 1931–1933 she became heavily involved with organising and taking part in the resistance to the payment of the anachronistic tithes with the Suffolk Tithe-payers Association. In 1933 she was charged with unlawful assembly following one protest (Brander, 2003) but the case was dismissed. Finally, in 1935 Balfour was called before a Royal Commission on tithes. Balfour's appetite for taking up a cause was now whetted.

In 1930 the neighbouring farm (Walnut Tree) went bankrupt and was eventually bought by Alice Debenham. Debenham, who was crippled with arthritis, became a close friend of Balfour, who then managed Walnut Tree Farm for her. So began a relationship that led to the Haughley experiment.

It would appear that her ideas about the link between the soil, the environment and health were triggered by reading *Famine in England* by Viscount Lymington in 1938, and were then consolidated by discovering the work of Howard and McCarrison via the book *Wheel of Health* by Wrench (1938). In 1938 she met Howard, who by this time had influenced many farmers through his lectures and articles. Balfour was clearly attracted to Howard's ideas about humus and health, and it must have been at this time that she adopted his belief in health-promoting mycorrhizae and the harmful effects of man-made fertilizers on mycorrhizae. Balfour's dislike of man-made fertilizers was very clear at meetings with Rothamsted scientists, when she was vociferous in saying that fertilizers destroyed mycorrhizae (Cooke, 2010, Hertfordshire, Private communication). Neither Balfour nor Howard before her appeared to show any concern about the long-term availability of fertilizers.

Balfour wanted to test whether the well-being of crops, animals and man was linked to the soil. This was a topic that she discussed with Debenham and together they looked for funding for a farm-scale project. The Haughley Research Trust, controlled by the East Suffolk County Council, was formed (1940) and a subsidiary farming company, Haughley Research Farms, was created. Debenham had donated her farm to the trust and Balfour's farm was leased to the new company.

It is chastening to realize that Balfour was still relying on horse power when petrol rationing was instituted at the start of the Second World War. Ferguson tractors had become available in the inter-war years. Her minor involvement (support) at this time with the local anti-sparrow club may surprise many in the light of the modern emphasis placed on preservation of birds by organic farmers.

Haughley Research Farms (1940)

The two adjacent farms of Balfour and Debenham were combined into a single farm (216 acres) to permit the comparison in three sections of:

i.   Organic farming using compost, manure, livestock but no man-made fertilizers.
ii.   Mixed farming using livestock, man-made fertilizers and pesticides as required.

Both the organic and the mixed farm sections operated a rotation of temporary pasture alternating with arable crops and both carried a herd of dairy cows, a flock of poultry and a small flock of sheep. Livestock were fed exclusively on the produce of its own section; replacements were home bred and cereal and pulse crops were raised from home-grown seed.

iii. A stockless farm growing arable crops in rotation and using man-made fertilizers.

Howard was a consultant regarding composting. Balfour (1940) wrote a fifty-three-page monograph entitled *The Haughley Research Farms, Proposal for an Experiment on Soil Fertility in Relation to Health* to help raise funds; it became the basis of *The Living Soil* (1943). In it she presented the key aspects of her thinking, namely:

i.   The deterioration in public health in UK.
ii.   The McCarrison experiments in India.
iii. The experience of Howard with composting and the relationship between mycorrhiza, humus and disease.
iv.   The connection between whole diets and ancient farming systems in the East with organic manures.
v.   The belief that the health of man, livestock, and crops is one indivisible whole.
vi.   The need to demonstrate that the link between mycorrhiza and health is by the absorption and digestion of the fungus by crop roots.

Balfour had now taken a huge step forward from a protesting farmer to a thinker on a global scale. She thought that agricultural science was moving in the wrong direction. For example, in the preface to Pfeiffer's book (1947) on biodynamic farming she wrote about the evil that science had brought to agriculture and that she did not believe it was possible to study biology by fragmenting the subject and studying it in small plots.

The fact that she said that it would not be possible to start the Haughley experiment until the land had first been cleared of chemical traces would seem to suggest that a position had been taken in advance of any results. Balfour gave talks, e.g. to Women's Institutes in early 1940, on the subject of health and soil fertility, indicating that she was convinced about her interpretation of the work of Howard and McCarrison and of Wrench's observations.

The Living Soil (Balfour, 1943)

The idea that led to the writing of *The Living Soil* began with Alice Debenham's sister contacting Faber and Faber in the UK, suggesting that it reprinted the 1940 monograph. In due course this led in 1942 to Balfour establishing direct contact with the publisher in London. In an early letter to the company she wrote that she felt very strongly about a proper attitude to the soil was of fundamental importance to the future of mankind (Brander, 2003). During 1942 Balfour converted the monograph into a 250-page book that was eventually read by a very wide audience in many countries. As Brander points out, Balfour gave little or no basis for her assertions and that in many ways *The Living Soil* was a testament of faith rather than a scientific treatise. An early alternative title for the book, *Health, Soil and Citizen*, also indicates her thinking.

Balfour put together and developed the ideas and beliefs from a number of sources that are discussed elsewhere. Of primary importance to her was the medical testament produced in 1939 by a group of doctors in Cheshire, whose ideas were mainly based on the work in India by McCarrison on nutrition and on the findings and claims about the merits of composting by Howard. Balfour was influenced by the book *Famine in England* (1938) by the right-wing Viscount Lymington, but judging from her two books (Balfour, 1943 and 1975) she was not influenced by the basically political publication *New English Weekly*, which according to Conford (2001) provided many supporters of the organic movement with a mouthpiece. She enthused over the 1938 book of Wrench, *Wheel of Health*, which extended the McCarrison work to several other communities.

Balfour (1943) focused on:

i.   Soils need to be biologically alive for fertility to be maintained.
ii.   The addition of organic material as manure/compost is required for fertility.
iii. The best diets are based on fresh food grown naturally with recycled organic wastes.
iv.   The need for an ecological approach to farming whereby all organisms in the environment are considered.
v.   The notion that health is linked via crops to the soil.

Providing the basis for her beliefs about health, let alone evidence of the mechanisms involved, was to be a constant thorn for Balfour. The same problem remains today when many have been led to assume, largely as a result of repeated publicity (not of evidence), that organic food has health-promoting qualities. It is possible that Balfour's belief in mycorrhiza wavered with time as studies relating to them hardly featured in the Haughley experiment. The concept of health-promoting factors that originate in the soil reaching man is seldom referred to today and instead great reliance is placed on compositional comparisons of conventional and organic food with the differences being ascribed health promoting qualities (Chapter 14). Likewise, the idea that food produced in a closed system reliant on recycling is particularly healthy has few adherents. It was not surprising that the allusion to the importance of *The Living Soil* for national planning (i.e. with regard to the nation's health) that appeared in the title of the many 1943 editions was dropped by Balfour (1975) as well as by the Soil Association's (2006) version of Balfour's classic (1943).

The publication of the first edition of *The Living Soil* at the end of 1943 was so successful that by 1945 the fifth edition was being planned. During this period Balfour had received about 500 letters of support (Brander, 2003). It was a suggestion by Charles Murray that Balfour should set up a clearing house for information on her ideas that was probably the seed that grew into the Soil Association. Eleven editions of *The Living Soil* (1943) were ultimately published. Balfour's popularity grew and by middle of 1944 she had given a number of broadcast talks on compost and farming.

Balfour was not the first person to talk about organic ideas, nor was she the first to see the need to demonstrate the basic truth behind the various propositions. However, she was the right person at the right time and had the drive and ability to see it through. In this respect she deserves much credit for being the energising force of the Soil Association and for stimulating organic movements in several countries.

Soil Association (1946)

The many letters that Balfour received from readers of *The Living Soil* prompted her, in May 1945, to reply to more than 125 people of repute. She wanted to unite a diverse group that included organic farmers, doctors, agricultural journalists and right-wing reformers in order to establish the common ground, which she saw as:

i.   The concept of the soil as a living entity.

ii.   Recognition that man has to live within Nature's laws if he is not to self-destruct.

iii.   A desire to investigate Nature's laws.

iv.   A determination to resist those who wish to discard Nature's laws.

v.   The need to disseminate proven knowledge about the laws of Nature.

Eventually in June 1945, sixty founder members of what was to become the Soil Association met. A Founders Committee was elected and given the job of producing a constitution that encompassed their diverse views, a difficult task that took more than a year (Brander, 2003). The founders of the pioneer Peckham Health Centre were particularly helpful, namely Scott Williamson and Innes Pearse. Conford (2001) reveals that both Sir Albert Howard and his wife, Louise Howard, were involved with the constitution but resigned from the Founders Committee before the inaugural meeting as they objected to the decision to make the advisory panel subject to the council. Howard had long objected to experimentation and favoured demonstration and dissemination of his ideas on compost, so he was unlikely to favour the plans for experiments designed to demonstrate the benefits of organic farming. The Howards never became members and Howard's journal, *Soil and Health*, became a rival to the Soil Association's *Mother Earth* from 1946 to 1949.

Howard understood that nutrients removed by crops and in animal products needed to be replenished by composted inputs. Apparently this was not universally accepted within the organic movement, which at the time thought soils had the ability to release all the required nutrients from their mineral reserves (Soil Association, 2006).

The Soil Association has had a love/hate relation with scientific methodology but the very existence of the Haughley farm experiment clearly indicates that Balfour wanted evidence to prove it was on the right track. The sciences of chemistry and physics were seen as being completely inappropriate for studying soils and crops, especially when the subjects were fragmented and investigated piece by piece. A biological, or as it is now called an ecological, approach was called for. In 1999 a senior executive of the Soil Association giving evidence to a House of Lords Select Committee on Organic Farming and the European Union (House of Lords, 1999) showed a strong preference for holistic science and deplored the obsession with reductionist science.

The Soil Association was finally registered followed by, in November 1946, the inaugural meeting, when Balfour was appointed president. It was a post she relinquished within a year to become organising secretary. Lord Teviot took over as president. Biographical data (Conford, 2001) on several people and groups who were involved with the early organic movement are detailed in the Appendix.

Of twenty-four founder and early members of the Soil Association identified by Conford (2001), six were doctors, six were farmers using compost made according to Howard, three were farming according to the principles of Steiner, four were writers/journalists, two wished for a return to old rural lifestyles, two had extreme right-wing tendencies and one was a scientist.

The messages from Steiner's lectures in Poland and especially from the writings and lectures of Howard had been taken on board several years before Balfour emerged as the figurehead in the UK. It has been suggested (Gill, 2011) that Balfour's beliefs in spiritualism were possibly acceptable to the anthroposophists of the Steiner school, who were accordingly more likely to be attracted to her views and become early members of the Soil Association. Gill concluded that the New Age/spiritualist religious thinking of Balfour and associates might also have contributed to their ideas being rejected by the scientific community.

Reaction to The Living Soil and the Soil Association

It was inevitable that orthodox agricultural scientists brought up to observe the scientific method of hypothesis followed by experiments to check and double check the hypothesis before formulating a theory or explanation for a phenomenon would not approve of the untested ideas of Steiner or beliefs such as the ability of humus to prevent disease in man. The ideas expressed in *The Living Soil* and later by supporters of organic farming were dismissed as being purveyors of nothing more than muck and mystery, a description first coined by Howard to undermine the credibility of Steiner's ideas (Gill, 2011).

In his articles and lectures Howard does not reveal why he was opposed to the way agricultural experiments were carried out other than to say that it was not possible to study a problem by breaking it down into its constituents and then let specialists study each part separately. Could it have been that Howard's practical background in farming, despite his science degree, led him to distrust specialists who by definition could not see the whole picture that he was able to see? Balfour was trained as an agricultural scientist and must have been exposed to the ways of science at Reading.

Strong opposition also came from the fertilizer industry which was in its infancy at the time. Such a reaction was not surprising in view of Howard's description of man-made fertilizers as devil's dust (Rodale, 1947).

Haughley Experiment (1945–1969) – Results

The objectives of the Haughley experiment were to investigate and compare on a farm scale the cumulative effects of the three contrasted types of soil management on:

i.   Soil fertility.
ii. The nutritional quality of the produce.
iii. The health and productivity of farm animals fed exclusively on the feed produced on the farming unit.

The Soil Association took over and financed the experiment from 1947. After experiencing financial problems, including changed ownership of the farm, it decided in 1969 to stop the experiment and to keep the organic section as a demonstration farm for the benefit of members. The apparent loss of soil fertility in the closed system organic farm and the concern about the low productivity of organic crops hastened the termination of the experiment. It was later appreciated that the decline in fertility should have been expected in view of the fact that there were no inputs of nutrients in the closed organic section to replace those removed in the produce. The finding that the decline in soil fertility had taken about thirty years to become manifest was taken as valuable information, so making a virtue out of what was an inadequate experimental design.

Balfour had been very busy lecturing and in publicising *The Living Soil* on overseas tours, so she had been unable for several years before the Haughley experiment was terminated to supply the requested corrections and additions to the early editions (Brander, 2003). However, the opportunity was afforded after 1969 to add the results of the experiment as a new section in *The Living Soil*, which was eventually published in 1975. The long delay before publication suggests that Balfour may not have been delighted with the results of the Haughley experiment. Accounts of the results are available from Balfour (1975), from her presentation to IFOAM (International Federation of Organic Agriculture Movements) (Balfour, 1977) and from the summary given in the Soil Association's (2006) edition of Balfour's *The Living Soil*. The following review of the results is based on Balfour (1975 and 1977). Much of Balfour (1975) was written as a narrative intended for a largely lay readership and not as a scientific paper. In her 1977 conference paper at IFOAM Balfour outlined the contribution the experiment had made towards the recognition of the importance of ecological awareness in agriculture. She also talked about the best philosophical and pragmatic approaches when developing a sustainable agriculture.

The Haughley experiment was the first ecologically designed agricultural research project on a full farm scale. Balfour saw it as means of filling the evidence gap on which the claims for the benefits of organic husbandry were based. It was decided that the only way to achieve this was to study the cycling of nutrients on a whole scale, under contrasting systems on the same soil and under the same management. Such a comparison of organic farming with conventional, mixed farming had never been done before.

Balfour did not present any yield data in her discussion paper at IFOAM in 1977. However, Stanhill (1990) published ten years of data showing that there was gradual decline in organic yields in the organic closed cycle section and that the organic yields were lower than nearby conventional farms in the county.

Soil fertility

In her summary of the more important findings Balfour emphasized the value of the thousands of soil analyses that had involved monthly sampling of every field for ten years. It was found that the N, P and K contents varied considerably during the season, with maximal levels occurring in the growing season. Secondary and trace elements were also determined.

The fluctuations of available soil P were far more marked on the organic section than on the mixed farm section, particularly on humus-rich soils. It was concluded that this was evidence of the importance of biological activity in ensuring an adequate supply of nutrients. Crop P analysis revealed no comparable variation and there were no parallel measurements or observations on crop growth.

The nutrient status of the soil in the

organic section remained as high as in the mixed farm section for most of the experiment, a feature that was thought to indicate how little of the minerals applied as fertilizers are recovered in crops; without comparing the quantities of nutrients supplied in manure and compost with that supplied in the man-made fertilizers it is difficult to accept this indication. Lucerne on the organic section was more resistant to weevils than those on the mixed section in two years.

Crop yields

The generally greater yields of all crops except peas on the mixed farm were evident (Table 6.1).

Balfour (1975) reported there was less insect damage on crops in the organic section and that the livestock lived longer. No more details were given on what to Howard would have been a very important aspect.

Livestock

There was considerable variability in milk production. For twenty years milk production from cows on the organic farm was higher by about 15% when measured on the basis of the amount of concentrates fed, which in the case of the organic section were made on farm. This was despite the poorer organic grass growth than on the pastures on the mixed farm; there was no explanation for this. The behaviour of the herds and flocks differed between sections, with those on the organic section being noticeably more contented.

More laying hens were kept on the mixed farm because of the better cereal yields. Fleeces from sheep on the mixed farm were heavier and of better quality than from the organic farm in three out of four years (1965–1968). It is to be noted that initially there were very few animals on each section, namely fifteen cows, thirteen breeding ewes and 250

Table 6.1 Average yields (over varying number of years preceding 1968) of crops on the mixed section expressed as percentage of yield on the organic section (Balfour, 1975)

	Years	Average Yields Mixed/Organic %
Wheat	16	122
Oats	7	136
Barley	12	132
Beans	16	114
Peas	16	87
Kale	13	159
Mangolds	6	162
Lucerne silage	5	142
Lucerne hay	6	122
Temporary pasture hay	13	130

laying hens. Soil fertility was closely studied but relatively little was reported about crop and livestock productivity. There were no consistent differences in the composition of crops or livestock products from the different sections. The absence of any detailed mention of crop and animal diseases is disappointing in view of these aspects being fundamental to the project. Balfour had been convinced that mycorrhizae were important and had postulated that they were capable of supplying, directly to the roots, some beneficial compounds equivalent to vitamins for animals. This was another aspect that was left for future research that Balfour envisaged would be carried out, but not by the Soil Association.

The Soil Association concentrated, after the demise of the Haughley experiment, on demonstrating and publicising organic methodology. It had become evident there was a need for a system that would show that food had been produced according to the Soil Association standards and so began the certification process in 1967, probably the first set of organic standards in the world.

Rumen digestion study on fistulated heifers at Huntingdon

Two heifers were used to study and compare the effects of organic with mixed section produce from Haughley on the bacterial populations and fatty acids in the rumen of cows fitted with a fistula. However, no results were reported by Balfour (1975).

Organic associations in English-speaking countries

It is likely that language played a large part in the dissemination of organic ideas from Europe to the rest of the world. The message of Howard and the books and lectures of Balfour were easily spread to English-speaking countries, which contrasted with the spread of the ideas of Steiner that initially were treated as secrets for the converted anthroposophists. The fact that Howard corresponded with people all over the world (Rodale, 1947) indicates that he planted his ideas in several countries, including Australia, Canada, India, New Zealand and Rhodesia, and some of those in Central America.

Australia

Three organizations in Australia can claim to be the world's first advocates of organic farming. Two of them predated the Soil Association in the UK, namely the Australian Organic Farming and Gardening Society (founded in 1944) and the Victorian Compost Society (founded in 1945). The Living Soil Association of Tasmania (1946), founded by Shoobridge, followed in 1946 (Paull, 2011). A key feature of the Tasmanian group was the recycling of all vegetable and animal wastes in order to follow Nature's laws. The interest in Indore composting in Australia suggests a strong Howard influence.

Steiner's influence had reached Australia in the late 1920s with the formation, in 1929, of the Agricultural Experimental Circle of the Anthroposophical Society. There may not have been many biodynamic farms in Australia when Balfour visited but she was impressed by their demonstrations of organic farming and on one farm of the growing of all the herbs required for the various biodynamic preparations.

Balfour's success with *The Living Soil* led to a year-long overseas lecture tour to New Zealand and Australia (1958–1959), where she found more than 100 members of the Soil Association (UK) and recruited 100 more (Paull, 2011). A second wave of organic groups formed after Balfour's visit included the Soil Association of South Australia and the Organic Gardening and Farming Society of Tasmania. In 2011 there were 362 members of the Soil Association (UK) in Australia, which leads the world with 12 million certified organic agricultural hectares accounting for 32% of the world's total organic area (Willer & Kilcher, 2011). Care should be taken when interpreting this huge organic area, which mostly includes low rainfall areas where crops/pastures do not respond to fertilizers; this permits the farmers to benefit from being certified organic.

Canada

The Canadian organic agriculture movement can be traced to the 1950s, in part inspired by visits to Canada of foreign experts, such as Pfeiffer, and through the distribution of literature from Europe, the UK and the USA. The first formal organization, the Canadian Organic Soil Association (later renamed the Land Fellowship), was founded (1953) by Christopher Chapman, a film maker who produced two films of direct relevance to organic agriculture, *Understanding the Living Soil* and *A Sense of Humus*.

New Zealand

It is possible that Lord Bledisloe, who was an early supporter of the organic movement, may have influenced the development of organic agriculture in New Zealand as a result of his serving as governor (1930–1935).

The development of organic farming in New Zealand owes much to the activities of Dr Guy Chapman, a dentist who was convinced about the relationship between diet (fresh food and vitamins) and dental caries and particularly to the way crops benefit from compost. He helped to found the Humic Compost Society in 1941 and ran demonstrations on how to build a compost heap (Murphy, 2011). The society eventually became the Soil Association of New Zealand in 1970, retaining its motto "healthy soil, healthy plants, healthy people". Growing concerns about the environment led to the organic association campaigning against the addition of fluoride to drinking water and for a nuclear-free New Zealand.

In 1982 the New Zealand Biological Producers Council was formed comprised of members of the Soil Association, the Biodynamic Farmers and Gardeners Association and the Henry Doubleday Association. The Biogro symbol used to identify organic produce was registered. Eventually, in 1987 the Ministry of Agriculture took steps to help promote organically grown food. Current campaigns include opposition to genetically modified crops, food additives and the use of pesticides (Murphy, 2011; Organic NZ, 2014). During her antipodean trip Balfour visited New Zealand, where she was horrified to find that some authorities were urging that topdressings of superphosphate mixed with DDT be made compulsory throughout New Zealand (Balfour, 1959a).

USA

Aldo Leopold (1887–1948), the ecologist and environmentalist, is credited with being at the forefront of the organic movement in the USA. He was primarily concerned with the conservation of wildlife, with biological diversity and living within the bounds of nature, which were the subjects of his widely read and influential book, *A Sand County Almanac*. He wanted people to live in harmony with the Earth and to focus on how food crops are grown rather than on the end products.

Rodale, J.I. (1898–1971)

Several people were probably involved with the propagation of organic ideas in the USA but Jerome I. Rodale was the most well-known advocate of organic farming there. He had been aware of the work of McCarrison and of the lectures of Steiner but it was his learning about the ideas of Howard in the first edition of *An Agricultural Testament* (Howard, 1940) that inspired him. In his effusive memoriam to Howard, Rodale (1947) commented that Howard, who he had not met, had been a prolific letter writer and had corresponded with dozens of Americans. Howard's work was clearly Rodale's inspiration.

Rodale was a prosperous businessman, owning an electric wiring factory, who understood that eating organically grown foods was the way to a healthy life. He established the Rodale Organic Gardening Experimental Farm in 1940 and published the *Organic Farming and Gardening* magazine from 1942, in which he expressed his opposition to pesticides and to man-made fertilizers. This magazine and its successor became the Bible of the developing organic movement in the USA. Rodale composted organic wastes according to Howard's Indore process but found that the required two turnings of the compost could be replaced by using earthworms, an improvement approved by Howard.

Rodale (1945) wrote *Pay Dirt*, to which Howard contributed an introduction. Although Rodale realized that the assertions linking humus with animal health were based on observations and were not supported by experimental evidence, his book closely followed Howard's claims of the importance of composting, conserving and feeding the soil in order to work within the laws of nature. The Rodale Institute, founded in 1947, continues to carry out research on organic agriculture, advocating policies that support farmers and to educate people about how organic is the safest, healthiest option for both people and the planet (Rodale, 2011).

Farming systems trial

In 1981 the Rodale Institute started the Farming Systems Trial (FST), America's longest-running, side-by-side comparison of conventional and organic farming of maize and soybean. The details and results of this trial, which is still running, are given in Chapter 13. A no-tilling treatment was introduced in 2008 to compare with traditional tillage; this has required Rodale to develop an innovative no-till roller/crimper as an alternative to the use of the banned total herbicides. Genetically modified maize and soybeans were introduced to the FST in 2008 in view of the fact that at the time 94% of all soybeans and 72% of all maize grown in the USA were genetically modified.

Rodale associates many of the benefits of organic farming with increases in soil organic matter, the conservation of soil water especially in drought years, and of soil N during high rainfall conditions. Other benefits include the reduction of soil erosion linked to the crop rotations to cover crops and the biological control of pests following enhanced biodiversity.

Organic associations in other countries

Early developments in organic farming in Europe were largely confined to the German-speaking nations, namely Germany, Austria and parts of Switzerland. This was a direct result of the influence of Steiner but there was activity elsewhere. For example, Pfeiffer created a biodynamic farm in the Netherlands in the late 1920s.

France

Interest in organic farming in France received a boost following the foundation in 1972 of the International Federation of Organic Agriculture Movements (IFOAM) in Versailles. SYNABIO, the National Union of Organic Companies, was created in 1976 to serve the organic farming sector. The first official recognition of organic farming in France was the Agricultural Reform Law of 1980, when a national committee was made responsible for the organization and specifications of French organic farming. In 2001, the Agence BIO, a French public interest group, was set up to develop and promote organic production. Since 2007, INAO, the French National Institute for Origin and Quality, has been in charge of the following up of the implementation of organic farming.

Germany

Steiner's lectures in 1924 led to the development of the organic movement in

Germany and research from the 1930s at the Goetheanum in Switzerland by the anthroposophists maintained the Steiner impetus. In 1971 Bioland, Germany's largest organic producer organization, was founded. The Association of Organic Processors, Wholesalers and Retailers was set up in 1983, with many of its members being organic farming pioneers.

Italy

Possibly because of limited contact with developments in the rest of Europe, organic agriculture only got under way in Italy in the 1960s and 1970s; self-regulated standards were set up in the 1980s. Since then there has been a massive increase in the organically farmed area.

Japan

In the 1930s, Masanobu Fukuoka (1913–2008), a microbiologist working in soil science and plant pathology, had doubts about modern agricultural methods. In 1937, he quit his job as a research scientist, returned to his family's farm and devoted the next sixty years to developing a radical no-till, no fertilizer, organic system based on mulch and legumes for growing grain and many other crops, now known as natural or ecological farming. His books, *One Straw Revolution* and *Natural Farming*, have been widely read. The Japan Organic Agriculture Association, consisting of producers (currently about 600) and consumers, was formed in 1971

to develop and expand the organic agriculture movement (JOAA, 1993).

Netherlands

Pfeiffer brought the ideas of Steiner to the Netherlands in the 1920s. Support for the organic movement was apparent in 1976 when the independent, evidence-based Louis Bolk Institute was founded to link social issues with research on sustainable agriculture, as well as linking organic farming with nutrition and health.

Spain

As in Italy, organic farming came late to Spain, probably due to the language barrier and limited contact with the rest of Europe. The first Biocultura Organic Fair took place in Madrid in 1984, followed by national organic farming legislation in 1989.

Switzerland and the DOK trial

In 1946 the Muller family of farmers founded the Co-operative for Cultivation and Utilization, as well as the journal *Culture and Politics*, to propagate a natural and sustainable approach to farming and soil fertility that included using rock dusts.

Research Institutes of Organic Agriculture (FiBL) were set up in Switzerland in 1973, in Germany in 2001 and in Austria in 2004 to carry out research and consult on organic agriculture. FiBL is also involved with research projects

Table 6.2 Public research expenditure (Euros) on organic and conventional agriculture in 2012 (Rahmann and Aksoy, 2014)

	Organic agriculture Million	Conventional Agriculture Billion
Germany	87	4
Europe Excluding Germany	124	
Rest of world	36	
Global		40

and initiatives in Europe, Asia, Latin America and Africa.

Following concerns about nitrate contents of vegetables and pesticide residues FiBL set up, in 1978, a long-term experiment (known as the DOK trial) near Basel, Switzerland, to compare biodynamic (D), organic (O) and conventional (K for the German "konventionell") production of the arable crops wheat, potatoes, maize, soya and grass-clover leys on the same site. Results are discussed in Chapter 13.

International research

Although much less is spent on organic agriculture research than on conventional (Table 6.2), there is certainly no lack of activity and enthusiasm. Several hundred papers were presented at the four conferences organized by ISOFAR (International Society of Organic Agriculture Research) in Australia (2005), Italy (2008), South Korea (2011) and Turkey (2014); the proceedings cover a total of more than 3,500 pages.

Conferences devoted to organic agriculture in Africa (AOC, African Organic Conference) were held in Uganda (2009), Zambia (2012) and in Nigeria (2015) with the theme of achieving social and economic development through ecological and organic agricultural alternatives.

Summary

1. Balfour's background as an agricultural student and a farmer enabled her to see the relevance of the ideas of Howard, McCarrison and Wrench to the agricultural scene in England. Her early writing experience would have been of benefit when writing *The Living Soil.*

2. Balfour and her farmer neighbour, Alice Debenham, saw the need to test her ideas on a farm scale and so began the Haughley experiment. A monograph written to elicit funds for the experiment was later expanded into Balfour's classic book, *The Living Soil.*

3. The 1943 edition of this book attracted much attention among a disparate group of people, from doctors to farmers and to social reformers. These were mostly people who were already convinced of the need to change the way society and particularly agriculture was organized.

So in many ways Balfour was preaching to the converted.

4. The fact that several early members of the Soil Association were already farming organically indicates that the ideas of both Steiner and Howard had been heeded in Europe before Balfour emerged as the figurehead in the UK.

5. In 1945 Balfour arranged a meeting of people who had approved of *The Living Soil* and from this group there eventually emerged the embryonic Soil Association, formed in 1946.

6. Although Howard was involved in the early days of the Soil Association, he did not become a member. Nevertheless, he had a major influence in many countries by his writings and lectures.

7. The Haughley experiment which lasted twenty-five years produced some very interesting results, notably regarding the variation in soil nutrient contents throughout the year. However, there was no evidence that suggested that organic farming led to better animal health.

8. Yields of all crops except peas were lower in the organic section. Organic farming led to soil fertility declining and the experiment was eventually converted to a demonstration farm.

9. It is now accepted that the design of the Haughley experiment was inadequate in that there was no provision for the nutrients removed in the organic products to be replaced by inputs. It was inevitable that soil fertility and stock-carrying capacity fell over time.

10. It is likely that neither the Haughley Research Trust nor the Soil Association would have become a reality without the vision, drive and determination of Balfour.

11. Although the Haughley experiment did not provide the anticipated results, the development of mainstream organic farming did not appear to be hindered.

12. The spread of ideas on organic farming around the world depended on the language and the activities of the prime movers Howard, Balfour and Pfeiffer. The self-imposed secrecy did not help the early biodynamic movement.

13. Since 2005 international conferences on organic agriculture have attracted much attention.

References

Balfour, E.B. (1940) The Haughley Research Farms. Proposal for an Experiment in Soil Fertility in Its Relationship to Health. Haughley, Suffolk: Haughley Research Farms. Available at https://www.cabdirect.org/cabdirect/abstract/19412200222

Balfour, E.B. (1943) *The Living Soil – Evidence of the Importance to Human Health of Soil Vitality, with Special Reference to National Planning.* Faber and Faber, London.

Balfour, E.B. (1959a) Diary of the 1959 New Zealand Tour, Part I. *Mother Earth*, 1, 2, 183.

Balfour, E.B. (1975) *The Living Soil and the Haughley Experiment.* Faber and Faber, London.

Balfour, E.B. (1977) Towards a Sustainable Agriculture, The Living Soil. Address given to an IFOAM conference in Switzerland in 1977. Available at http://journeytoforever.org/farm_library/balfour_sustag.html

Brander, M. (2003) *Eve Balfour, Founder of the Soil Association and Voice of the*

Organic Movement. The Gleneil Press, East Lothian.

Conford, P. (2001) *The Origins of the Organic Movement*. Floris Books, Edinburgh.

Gill, E. (2011) *The early history of the organic food and farming movement in Britain, specifically the career of Soil Association founder, Lady Eve Balfour*. PhD thesis University of Wales, Aberystwyth.

House of Lords (1999) Organic Farming and the European Union European Communities – Sixteenth Report. Available at www.publications. parliament.uk/pa/ld199899/ldselect/ldeucom/93/9301.htm

Howard, A. and Wad, Y.D. (1931) *The Waste Products of Agriculture – Their Utilization as Humus*. Oxford University Press, London.

Howard, A. (1940) An Agricultural Testament, Appendix C. The Manufacture of Humus from the Wastes of the Town and the Village. Oxford University Press, London.

JOAA (1993) TEIKEI system the producer-consumer co-partnership and the movement of the Japan Organic Agriculture Association. First IFOAM Asian Conference Saitama, Japan. Available at www.joaa.net/english/teikei.htm

Lord Lymington (1938) *Famine in England*. The Right Book Club, Soho London.

Murphy, J. (2011) Compost club grew naturally. *Otago Daily Times*, 8 July 2011. Available at www.odt.co.nz/lifestyle/home-garden/168148/compost-club-grew-naturally

Organic NZ (2014) Available at http://organicnz.org.nz

Paull, J. (2011) The Soil Association and Australia: From Mother Earth to Eve Balfour. *Mother Earth* 4, 13–17.

Pfeiffer, E. (1947) *Soil Fertility, Renewal and Preservation. Biodynamic Farming and Gardening (original edition 1938)*. Faber and Faber. Available at www.soilandhealth.org

Rahmann, G. and Aksoy, U. (2014) Building Organic Bridges with Science. Foreword. In: Rahmann, G. and Aksoy U. (Eds.) *Proceedings of the 4th ISOFAR Scientific Conference. Building Organic Bridges*, Istanbul, Turkey.

Rodale, J.I. (1945) *Pay Dirt, Farming and Gardening with Composts* Kessinger publishing, Montana.

Rodale, J.I. and Howard, A. (1947) Sir Albert Howard – In memoriam. *Organic Gardening* 2, 6. Available at www.soilandhealth.org

Rodale (2011) Farming Systems, Celebrating 30 years. Available at http://rodaleinstitute.org/assets/FSTbooklet.pdf, www.rodaleinstitue.org

Soil Association (2006) Reprint with additions of Balfour (1943). Soil Association.

Stanhill, G. (1990) The comparative productivity of organic agriculture. *Journal Agriculture Ecosystems and Environment* 30, 1–2, 1–26. doi: 10.1016/0167-8809(90)90179-H.

Willer, H. and Kilcher, L. (2011) The World of Organic Agriculture: Statistics and Emerging Trends 2011. IFOAM and FiBL.

Wrench. G.T. (1938) *Wheel of Health. A Study of the Hunza People and Keys to Health*. Daniel Company Ltd, London.

CHAPTER 7

Organic Areas, Crops and Markets across the World

"There are lies, damned lies and statistics" Mark Twain

Most of the data for mainstream organic areas and crops are for 2012 and 2014 taken from the very detailed annual reports by FiBL (2015) and by Lernoud and Willer (2014, 2016). IFOAM (International Federation of Organic Agricultural Movements) receive data from certifying organizations in over 170 countries. Data on biodynamic areas and crops are from Demeter (2015, 2016).

About 2.3 million farmers/producers (1.9 million in 2012) out of a total of about 570 million farmers practised some form of certified/accredited organic farming across the world in 2014.

Areas of organic crops

Although biodynamic agriculture started several years before mainstream organic farming it was not destined to become the main organic movement. Globally there are 164,323 ha farmed biodynamically compared with more than 43.7 million ha of mainstream organic farming. Biodynamic agriculture has hardly moved out of its birthplace, with Europe accounting for 87% of the global biodynamic area. Most biodynamic farmers (4,956) are in Europe, notably Germany, France and Italy. Table 7.1 (Demeter, 2015).

About 1% of agricultural land in the world was farmed organically in 2014.

The statistics of the mainstream organic area are distorted by the huge organic areas in Australia, Argentina and to a lesser extent Uruguay that together contribute nearly half (21.5 million ha) of the total organic area of 43.7 million ha worldwide in 2014. Europe is the principal player in the mainstream organic movement, followed by Latin America. There has been considerable growth in organic farming since 2000, when globally there were just over 17 million ha.

The inclusion of what are called certified "Wild collection and bee-keeping areas" by IFOAM could easily appear to inflate the overall area and apparent importance of organic agriculture. IFOAM reports that globally there were about 37.5 million ha of such areas in 2014. The main activities are bee-keeping, collection of berries and wild medicinal plants. Collection of berries in Finland, accounts for 18% of global wild areas and in Zambia bee-keeping accounts for 24%. The 1 million organic beehives represent about 1% of global beehives. According to IFOAM, a total of 69 million ha were considered organic in 2012 (certified organic areas plus wild areas plus aquaculture and some forests). The wild collection areas are not included in Tables 7.1 and 7.2.

It is not surprising, in view of the fact that organic agriculture was a European construct, that more agricultural land than elsewhere has been devoted to organic farming there, notably in Spain, Italy, France, Germany, Poland, Austria, the UK, Sweden, Czech Republic and Ukraine. Swiss consumers, particularly German-speaking, have the highest per capita spending on organic food

Table 7.1 Biodynamic areas in 2014 and mainstream organic areas in 2014 according to region (Demeter and IFOAM)

	Biodynamic ha	Mainstream Organic million ha	% of all Agricultural land
Africa	3,653	1.263	0.1
Asia	5,929	3.567	0.3
Europe	142,313	11.625	2.4
Latin America*	6,935	6.785	1.1
North America	4,848	3.082	0.8
Oceania	1,003	17.342	4.1
World	164,323**	43.664	1.0

* Central, South America and Caribbean.
** 4,956 farmers.
In 2015 the global biodynamic area was 170,833 ha farmed by 5,091 farmers (Demeter, 2016).

worldwide, with 46% buying organic products every week (SwissInfo, 2016). The most popular organic products in Switzerland are eggs, fresh bread and vegetables.

Denmark has announced ambitious plans to double the area of organic crops by 2020 (Jorgensen, 2015). There will be increased subsidies to promote organic farming, organic exports and greater consumption of organic food in the public arena, e.g. schools and armed forces. It is not clear where the nitrogen required by Danish crops, animals and humans will come from after the use of man-made N fertilizer is curtailed. The alternatives would seem to be either devoting more land for nitrogen-fixing cover crops or the importing of N-rich animal feeding stuffs.

Main organic crops and areas

Details of the types of organic crops grown in each region are reported by Lernoud and Willer (2014 and 2016) from which the following summaries have been derived. Grassland, namely permanent pasture plus range land, accounts for 63% of the organic area and arable crops 18%. The main permanent crops are fruit, nut and beverage crops, especially coffee and cocoa, which comprise about 8% and 2.5% respectively of the total organic area of these crops.

The dominant position of grassland farming in Europe and North America reflects the influence of the local market for organic livestock products in these areas. Much of the meat and wool from the grazing lands in Argentina and Uruguay will be destined for export markets. The apparent emphasis on permanent organic crops in Africa is in many cases linked to exported coffee, tea and cocoa.

In Australia there are extensive areas with low and erratic rainfall where wheat will only respond to fertilizers in years of above average rain; applying fertilizers under such circumstances would be a waste of time. Until there was good evidence that wool fibre diameter is a genetically inherited trait, sheep farmers in Australia firmly believed that pastures of native grasses that virtually starve the

Table 7.2 Organic crop areas as % of total organic area according to region (Lernoud and Willer, 2014)

	Arable Crops	Permanent Crops	Other Crops	Permanent Grassland	No Details
Africa	15	47	2	3	33
Asia	40	18	2	1	39
Europe	42	10	3	44	1
Latin America	3	12	–	68	17
North America	38	4	2	38	18
Oceania	0.5	0.5	–	97	2
World	18	7	2	63	10

sheep was the main way to get high-priced super fine wool (Possingham, Adelaide, 2015, personal communication). For this reason many growers refused to apply phosphate to pastures even in moderate and higher rainfall areas. They thus became organic growers by default and then benefited from higher prices.

Only the main organic crops and main countries are included in the following summary Tables 7.3 to 7.8 (Lernoud and Willer, 2014 and 2016). For regional totals see Table 7.1.

Africa

Total organic 1,263,000 ha (3% world organic).

Permanent crops: coffee: 223,000 ha, olives: 125,000 ha, nuts: 38,000 ha, cocoa: 38,000 ha and tropical fruit: 17,000 ha.

Permanent grassland: 71,000 ha.

Arable crops: oilseeds: 124,000 ha, cotton: 32,000 ha, medicinal: 20,000 ha, cereals: 6,800 ha and vegetables: 6,000 ha.

Uganda has the largest organic area in Africa, followed by Tanzania, Ethiopia and Tunisia. Uganda also has the largest number of organic farmers – 189,000 in 2012. The Ugandans may be benefiting from their naturally fertile soils; the stories of sprouting fence posts are perhaps not all apocryphal.

There were about 580,000 organic producers in Africa in 2012 with an average farm size of < 2 ha. Most African organic crops, namely coffee, olives, cocoa, oilseeds and cotton, are exported.

Asia

Total organic 3,567,000 ha (8% world organic).

Table 7.3 Africa: Biodynamic areas and farmers (in 2015) and mainstream organic areas (in countries with more than 30,000 ha) in 2014

	Biodynamic ha	Number of Farmers	Mainstream million ha	% of all Agricultural land
Dem. Rep. Congo	–	–	0.09	0.4
Egypt	2,438	133	0.08	2.3
Ethiopia	28	1	0.16	0.5
Ghana	81	1	0.03	0.1
Madagascar	–	–	0.03	0.1
Morocco	120	1	–	–
South Africa	123	3	0.04	<0.1
Sudan	–	–	0.13	0.1
Tanzania	–	–	0.19	0.5
Tunisia	780	127	0.14	1.4
Uganda	116	1	0.24	1.7

Arable crops: cereals: 755,000 ha, oilseeds: 443,000 ha, cotton: 172,000 ha, green fodder: 94,000 ha, pulses and beans: 19,000 ha, vegetables: 14,000 ha, aromatic plants: 12,000 ha and sugarcane: 9,000 ha.

Permanent grassland: 28,000 ha.

Permanent crops: coffee: 113,000 ha, tea: 58,000 ha, tropical fruit: 53,000 ha, nuts: 47,000 ha and coconut: 30,000 ha.

The leading countries are China and India. There are 685,000 organic producers in Asia (600,000 in India) with an average farm size under 5 ha. In Asia most organic produce is consumed locally.

Europe

Total organic 11,625,000 ha (27% of world organic).

Arable crops: green fodder crops:

2,041,000 ha, cereals: 1,911,000 ha, pulses and beans: 299,000 ha, oilseed crops: 245,000 ha and vegetables: 131,000 ha.

Permanent grassland: 4,800,000 ha.

Permanent crops: olives: 492,000 ha, grapes: 266,000 ha, nuts: 180,000 ha, temperate fruit: 127,000 ha and citrus: 38,000 ha.

Spain has the largest organic area in Europe followed by Italy, France, Germany and Poland.

There are 320,000 organic producers in Europe with an average farm size of about 35 ha.

The detailed data for the twenty-seven member states of the European Union (European Commission, 2013) are in line with Table 7.5.

Limited data is available on organic livestock. In the EU organic cattle plus sheep and goats comprise about 3% of the herd. The largest producers of organic cattle are Austria, France, and UK, each with about

Table 7.4   Asia: Biodynamic areas and farmers (in 2015) and mainstream organic areas (in countries with more than 30,000 ha) in 2014

	Biodynamic ha	Number of Farmers	Mainstream million ha	% of all Agricultural land
China	41	3	1.92	0.4
India	4,958	403	0.72	0.4
Indonesia	–	–	0.11	0.2
Israel	85	2	<0.01	1.4
Kazakhstan	–	–	0.29	0.1
Nepal	117	2	0.01	0.2
Sri Lanka	912	513	0.06	2.3
Thailand	–	–	0.04	0.2
Turkey	1,066	168	0.49	2.0

350,000 head of certified animals. The largest producers of organic pigs in the EU are Germany, Denmark and France, each with more than 165,000 head. The UK dominates the organic sheep market with 1.2 million head. France is the leading EU member in the organic poultry sector with more than 10.9 million birds (European Commission, 2013).

Most organic food produced in Europe is consumed locally and is supplemented by imports.

Table 7.5 Europe: Biodynamic areas and farmers (in 2015) and mainstream organic areas (in countries with more than 30,000 ha) in 2014

	Biodynamic ha	Number of Farmers	Mainstream million ha	% of all Agricultural land
Austria	5,657	184	0.53	19.4
Belgium	57	9	0.07	4.9
Bulgaria	152	1	0.07	2.4
Croatia	54	1	0.05	3.8
Czech Republic	5,720	6	0.47	11.1
Denmark	2,279	32	0.16	6.3
Estonia	–	–	0.16	16.2
Finland	400	18	0.21	9.4
France	9,873	439	1.13	4.1
Germany	72,588	1,476	1.05	6.2
Greece	295	20	0.26	3.1
Hungary	5,723	15	0.12	2.7
Ireland	23	3	0.05	1.3
Italy	11,524	388	1.39	10.8
Latvia	–	–	0.20	11.2
Lithuania	518	7	0.16	5.7
Luxembourg	544	9	–	–
Netherlands	5,948	134	0.05	2.5
Norway	532	22	0.05	4.6
Poland	3,520	17	0.66	4.3
Portugal	382	6	0.21	6.3
Romania	146	1	0.29	2.1
Russian Fed.	–	–	0.24	0.1
Sicily	583	22	–	–
Slovakia	161	1	0.18	9.5
Slovenia	141	22	0.04	5.9
Spain	4,465	75	1.71	6.9
Sweden	808	15	0.50	16.4
Switzerland	4,196	250	0.13	12.7
Ukraine	–	–	0.40	1.0
United Kingdom	4,958	95	0.52	3.0

The area farmed organically as well as the number of organic farmers and livestock are both continuing to decline in the UK (Defra, 2015) from a peak in 2008: this is despite retail sales of organic food bouncing back in 2014 and 2015 after the 2008–2009 financial crash (Soil Association, 2013, 2015). The reasons may be that imported organic produce is contributing significantly to the increased sales and that the high cost of organic animal feed (i.e. non GM) is reducing the profitability of organic farming.

Although organic farming is very well established in Germany with more than 16,500 farmers in 2010 there was a significant number (more than 1,250) of farmers who reverted to conventional farming between 2007 and 2010 (Heinze and Vogel, 2014); over the same period there was a net increase of just under 2,700 organic farmers. No single reason has been identified for the reversions but it is thought that the lower productivity of organic farms is an important factor.

The Russian president, Vladimir Putin, told the Russian parliament in 2015 that his country is bidding to become the world's biggest supplier of organic food and could become the largest supplier of healthy, ecologically clean, high-quality food (FW, 2015); this follows an increase of mainstream organic farming in Russia from 150,000 ha in 2012 to 245,000 ha (sixty-eight producers) in 2014.

Over the years the Finnish government has set several goals for the development of the organic food production but none has been reached. In a study of the reasons Nuutila and Kurppa (2014) call for government intervention in the form of more effective taxation and subsidies.

North America

Total organic 3,082,000 ha (7% world organic).

Arable crops: cereals: 557,000 ha, green fodder: 430,000 ha, oilseeds: 124,000 ha, vegetables, 64,000 ha and pulses and beans: 49,000 ha.

Permanent grassland: 1,284,000 ha.

Permanent crops: temperate fruit: 19,000 ha, grapes: 16,000 ha, nuts: 9,000 ha, berries: 9,000 ha and citrus: 7,000 ha.

There are 16,000 organic producers in North America, with an average farm size of about 180 ha.

Certified organic crops covered roughly 0.6% of American cropland in 2014. Only a small percentage of the main American field crops are grown under certified organic farming systems, namely maize (0.3%), soybeans (0.2%), and wheat (0.6%). On the other hand, organic vegetables amounted to 6% of the total American market and organic fruits and nuts to 4%. American farmers have been attracted to organic farming by the lure of higher price premiums.

Latin America

Total organic 6,785,000 ha (15% world organic).

Arable crops: cereals: 123,000 ha,

Table 7.6 North America: Biodynamic areas and farmers (in 2015) and mainstream organic areas in 2014

	Biodynamic ha	Number of Farmers	Mainstream million ha	% of all Agricultural land
Canada	653	3	0.90	1.3
USA	4,195	107	2.18	0.6

sugarcane: 61,000 ha, vegetables: 52,000 ha, oilseeds: 46,000 ha, medicinal crops: 13,000 ha and industrial crops: 11,000 ha.

Permanent grassland: 4,546,000 ha.

Permanent crops: coffee: 408,000 ha, cocoa: 206,000 ha, fruit: 129,000 ha, coconut: 14,000 ha and grapes: 11,000 ha.

The leading countries are Argentina (mostly in the Patagonian steppes), Uruguay, Brazil and Mexico. The leading vegetable grower was Mexico with 46,000 ha. The wild collection areas cover a further 2,875,000 ha, mainly in Brazil, Bolivia, Argentina and Peru.

In most Latin American countries organic farm sizes are <5 ha, which contrasts markedly with the extremely large sizes of organic farms in Argentina (more than 2,500 ha per farm), Uruguay (1,500 ha) and the Falkland Islands (50,000 ha). In southern Patagonia organic sheep estancias may cover many thousands of ha. There are slightly more than 300,000 organic farmers in Latin America with more than 169,000 in Mexico.

Table 7.7 South and Central America and Caribbean: Biodynamic areas and farmers (in 2015) and mainstream organic areas (in countries with more than 30,000 ha) in 2014

	Biodynamic ha	Number of Farmers	Mainstream million ha	% of all Agricultural land
Argentina	1,047	22	3.06	2.2
Bolivia	–	–	0.11	0.3
Brazil	2,833	51	0.71	0.3
Chile	1,296	18	–	–
Colombia	–	–	0.03	0.1
Costa Rica	11	1	–	–
Dominican Rep.	404	4	0.17	8.5
Ecuador	30	77	0.05	0.6
Falkland Islands	–	–	0.40	36.3
Mexico	277	3	0.50	2.3
Nicaragua	–	–	0.03	0.7
Paraguay	794	2	0.05	0.3
Peru	80	2	0.26	1.2
Uruguay	–	–	1.31	8.8

Oceania

Total organic 17,342.000 ha (40% world
  organic)
Permanent grassland: 16,728,000 ha.
Permanent crops: coffee: 18,000 ha,
  coconut: 12,000 ha and nuts:
  9,000 ha.

The organic area in Oceania is dominated by the Australian area of 17,150,000 ha. Australian organic farms have an average size of just under 6,200 ha – a marked contrast with most of the world. Apart from New Zealand (110 ha per farm) and Samoa (45 ha), organic farm sizes in the rest of Oceania are <5 ha. In 2005 about 97% of the certified organic area was extensively grazed (Wynen and Fritz, 2007). There were more than 14,500 organic farmers in Oceania in 2012, mostly in Papua New Guinea (9,185) and only 2,129 in Australia.

About half the income of Australian organic farmers is derived from livestock products with beef accounting for 20%, wool milk and eggs 15%, sheep and lamb 5% and poultry 8%. There is a massive beef industry over much of the northern parts of the continent, which is based on open free-range grazing. The cattle are mostly the Brahman breed from India.

Fruit, vegetables and grains accounted for 40% of the organic market. The low and erratic rainfall in many of the wheat growing areas and on land grazed by beef cattle and sheep renders fertilizer applications unprofitable. Many of the farmers of such areas are thus not organic farmers by choice but by circumstance. Similarly there are impoverished peasant farmers in Africa and Asia who currently do not have a choice about using fertilizers and pesticides and who some would like to be considered organic even though they are not certified.

Biodynamic crops and farms

A total of 164,000 ha were farmed biodynamically by fewer than 5,000 farmers in 2015. The leading countries were Germany with 44% of the global total area, followed by France (6%), Italy (7%), the Netherlands (4%) and then by Austria, Czech Republic, Hungary, India, Spain and the UK, each with about 3%.

Incomplete (about 70%) data on crops grown on individual biodynamic farms is available for the global biodynamic

Table 7.8   Oceania: Biodynamic areas and farmers (in 2015) and mainstream organic areas (in countries with more than 30,000 ha) in 2014

	Biodynamic ha	Number of Farmers	Mainstream million ha	% of all Agricultural land
Australia			17.15	4.1
New Zealand	1,003	23	0.11	0.9
Samoa			0.04	14.3

Table 7.9 Approximate global biodynamic crop areas (2015)

	ha
Grassland and fodder crops	35,800
Cereals	20,780
Fruit	18,345
Legumes	11,140
Vegetables	5,590
Other crops + fallow	11,980
Nuts	380
Oil crops	1,464

area in forty-nine countries (Table 7.9) (Demeter, 2014).

A wide but not complete range of crops is grown biodynamically, from bananas in the Dominican Republic to tea plantations in India and herbs and spices in Egypt (Demeter, 2014).

The average size of a biodynamic farm in 2015 was 33 ha. The basic requirement for livestock to be integrated into arable holdings recognizes the central role animals play in farming biodynamically. Market gardens, orchards and other perennial crops use composts to replace manure. Although animals are not an absolute requirement, each biodynamic farm aims to be a unique self-contained unit.

Biodynamic livestock numbers totalling more than 270,000 (on the 70% global area for which data is available) are dominated by poultry (202,000); cattle and sheep account for 44,000 and 10,000 respectively (Demeter, 2014). It would have been expected that livestock, particularly cattle, would figure in all countries practicing biodynamic farming but there are many countries where this does not appear to be the case. Composting and the use of the homeopathic preparations will be a problem in these countries in the absence of accessible animal manure. Mixed farms dominate in Europe with livestock taking a prominent position, notably in Germany with 31,000 cattle on biodynamic farms.

Mainstream organic markets

(Values are US dollars)

The global market (retail level) for organic food and drink was estimated at $80 billion in 2014 ($29 billion in 2004) (Lernoud and Willer, 2014). Sales of organic products were concentrated in two regions – North America with 43% of the global market, and Europe with 44% (Lernoud and Willer, 2016 and Rahmann and Aksoy, 2014). The European market was worth $35 billion in 2014 with Germany accounting for $10.5 billion and France $6.4 billion. Sales in the non-EU countries in Europe were about $8 billion (Kahl *et al.*, 2012). The combined local food and drink markets in Africa, Asia, Latin America and Oceania are very small, totalling about $6.5 billion; some exports may be included in this total.

The global market for alternative and complementary medicine is of the same order as that for organic produce. In 2011 it was estimated to be about $60 billion (*The Economist*, 2011); the UK market is thought to be $2.4 billion.

USA

The North American organic retail sales market was valued at about $36 billion in 2014 (Table 7.10). In the USA organic sales accounted for nearly 5% of total food sales; the organic food market has grown by between 5–10% per annum over the last six years. There were 14,540 certified organic farms in the USA, making monitoring difficult (Statistic Brain, 2013 and Wall Street Journal, 2014). The market is unique in that there are several huge organic food companies. Legislation that permitted organic processors to label their products with a stamp of the US Department of Agriculture has inadvertently promoted organic sales as the customer has been led to believe that the USDA approves the quality of the products bearing its stamp; this is not the case. It is possible that the use of the USDA organic label may now be challenged on constitutional grounds; this follows a 2015 US Supreme Court case, Reed vs. the Town of Gilbert, when the imposition of rules governing signage were deemed unconstitutional. The legality of special labelling to identify foods produced by a particular process (e.g. organic) unrelated to the health or safety of the protected product may now be called into question. The USDA label is based on a production process and it is not related to the quality of the produce.

A report commissioned by Academics Review, an independent non-profit organization in the USA (Chassy *et al.*, 2014), was far from complimentary when discussing the ways in which the organic movement in the USA had exploited the food safety and health concerns of American consumers in their promotion. Marketing campaigns are mainly based on the fear factors at all levels, from labelling to advertising. It was noted that anti-GMO and anti-pesticide advocacy groups promoting organic alternatives have combined annual budgets exceeding $2.5 billion and that the organic industry is one of the major donors.

According to the Natural Marketing Institute, there is an additional US market worth $117 billion of natural food products, which includes some organic and health foods together with supplements and personal and home-care products (NMI, 2010). Natural products are certified by the Natural Marketing Institute but they are not legally defined and although they may generally be minimally processed, there are no requirements to provide proof. In a survey on the understanding of food labels in the USA it was found that many believed natural food has the same benefits as organic food but at a lower price CRNRC (2016). The survey also revealed that more people purchase natural foods than organic foods – 73% versus 58%. The organic lobby is calling on the Food and Drug Administration to ban the use of the term "natural" on food packaging.

Organic fruits and vegetables dominate the American organic market, with a 43% share, followed by dairy products (15%), packaged foods (10%), beverages (including soymilk (10%), breads/grains (10%) meat, fish, and poultry

(3%) (ERS, 2013). Non-food agricultural products account for a further $2.2 billion, comprising, for example, laundry detergents using organic coconut oil and aloe vera. Organic bananas, coffee, olive oil and mangoes were the major organic imports in 2012 (ERS, 2014).

The American consumer is prepared to pay a premium for organic products, which are now available in about 75% of conventional grocery stores. In 2012 wholesale prices of most conventionally grown vegetables were 30–65% of the prices of organically grown vegetables (artichoke 65%, cabbage 42%, carrots 40%, cauliflower 45%, lettuce 51%, onions 36%, spinach 37%, Swiss chard 55%, potatoes 30%, tomatoes 45% and sweet potatoes 60%). In 2010 comparable figures for fruit were apples Braeburn 64%, apples Gala 52%, avocados 70%, bananas 65%, oranges 50%, pears d'Anjou 67%, raspberries 76% and strawberries 58% (ERS, 2014).

The USA can claim to have the world's first fully organic city – Maharishi Vedic City, Iowa, founded in 2001. The tiny settlement of 1,300 firstly banned the sale of all non–organic food and in 2005 it voted to ban the use of all synthetic pesticides and fertilizers within city limits (MDC, 2013).

The success of the organic market is resulting in large farms becoming organic producers in the USA (Cornucopia, 2015). There still are thousands of small organic farms, but in terms of what they produce and what consumers buy they represent an increasingly small fraction of the total organic market. Cornucopia,

which is a supporter of organic farming, has admitted that when people buy organic produce they are buying the story behind the label. Aerial photos by Cornucopia have revealed several large-scale organic operations including a poultry farm that is licensed for more than 1 million birds. Large organic feedlots have been found where the requirements to provide grazing for animals or even to give them time outside are not met. Inadequate monitoring by certifying agents is blamed.

The coexistence of organic farming with conventional, particularly those growing GM crops, was the subject of a report by the USDA (2016). Some (1%) organic farmers reported losses (totalling $6 million over four years) incurred by them as a result of the contamination of their products with GM. Organic livestock farmers suffer from the high prices of organic soybean and maize, which are two to three times those of conventional crops; organic soybean and maize only account for 0.3% and 0.2% respectively of the total area of these crops in the USA compared with more than 90% that are GM. Whether such production costs should be compensated is debatable in view of the fact that the organic movement sets its own rules, which means the high costs can be seen as self–inflicted. The USDA has not set a threshold for allowable levels of GM materials in organic products. The local market of organic animal feed in the USA is dependent on imports of maize from Romania and soybeans from India. The price premium of organic maize has not been sufficient to offset the lower yields and the three-year transition period required before becoming a certified organic farmer (OTA, 2015). The USDA (2016) report fails to mention the benefits that organic growers obtain from the superior suppression of weeds and pests on neighbouring GM fields.

The many agricultural practices followed by organic farmers in the USA were surveyed in 2014 (USDA, 2016a). It was reported that of a total of 14,093 organic farmers, 50–70% used compost/manure and maintained buffer strips and 30–50% practised biological pest control, selected pest-resistant varieties and practised reduced tillage.

Europe

In Europe sales of organic food in Germany and France exceeded sales in the UK in 2013 (Table 7.10). In contrast, sales of Fairtrade food in the UK equalled the combined sales in the rest of Europe. Whereas sales of organic food fell in the UK during the recession (2008) sales in Austria, Denmark, France, Germany and Switzerland continued to grow. In the UK this was attributed to the passive role of successive UK governments in supporting organic food and farming. Fairtrade standards are set by Fairtrade International Labelling Organization (FLO), which prohibits GM crops. Fairtrade criteria require sustainable farming techniques and some will be organic farmers who benefit from higher prices. Fairtrade Premiums (additional to Fairtrade Price) are often used in the community to train producers in organic and sustainable techniques such as composting and recycling.

Functional foods are now competing with organic foods in Europe (Kahl *et al.*, 2012). The UK market, which

Table 7.10 Sales (billion $) of Fairtrade, functional food and organic produce (Lernoud and Willer, 2014a and 2016, Kahl *et al.* 2012 and Leatherhead Food, 2014)

	Fairtrade 2013	Functional 2011	Organic 2014
USA	1.68	16.4	35.9
Germany	0.81	3.38	10.5
France	0.44	2.01	6.8
China	–	–	5.0
Canada	0.21	–	2.96[*]
UK	2.55	3.70	2.57[*]
Italy	0.09	–	2.53[*]
Switzerland	0.44	–	2.09[*]
Austria	0.16	–	1.32[*]
Sweden	0.29	–	1.27[*]
Japan	–	17.3	–
Australia	–	0.9	–

[*] Data for 2012

is the largest in Europe followed by Germany and France, exceeds the organic market (Table 7.10). The term functional food, which was first used in Japan in the 1980s, covers a very broad range of products that deliver additional benefits over and above their basic nutritional value. Foods or drinks containing probiotics, prebiotics, or fortified foods (e.g. folic acid fortified bread or breakfast cereals) are classified as functional foods. Globally the biggest markets for functional food are the USA, Japan, the UK, Germany and Australia, in that order.

There are clear differences in preferences for speciality foods in Europe. Whereas in most of Europe and particularly in Germany and France the organic market is dominant, in the UK the organic market lags behind that of the functional and Fairtrade produce.

Little information is available on organic products imported into Europe, which include coffee from Brazil, kiwi fruit from New Zealand, rice from Thailand, bananas from Costa Rica, cocoa from Peru and pineapples from Uganda (EU, 2015). Likewise, little is known about the recent imports of organic produce into the UK. In 1998 a House of Lords Select Committee on Organic Farming and the European Union reported that 80% of organic fruit and vegetables consumed in the UK is imported (House of Lords, 1998).

Asia, Latin America and Africa

Many Asian and Latin American, but only a few African, countries are exporters of organic produce. Local sales of organic food are limited but there will consumption by some self-sufficient farmers. Brazil is an exception in having a thriving local market with a very wide range of organic products, including cosmetic items, supplied by more than forty companies (Organics Brazil, 2014). Local promotion is linked to developing products that are distinctive of the six regional zones from the Amazon to the Cerrado and the Pampas.

In India organic food sales are growing on the back of rising living standards and mistrust of food quality. Premiums for organic food of 10–20% seem to be accepted. The major organic food-producing states in India include Maharashtra, Andhra Pradesh, Karnataka and Uttar Pradesh (TechSci Research, 2014). In Kerala organic fruit and vegetables are actively promoted alongside organic tea, spices and fragrances. There were aspirations for the entire state to be organic by the end of 2016.

The situation in Tanzania and in neighbouring countries in East Africa indicates how local organic sales can be developed on the back of the existing export markets, mainly in Europe, which often involves overseas companies. Mella *et al.* (2007) see a huge potential for promoting the organic brand for crops and livestock products from smallholders in Tanzania, who use few pesticides and fertilizers. The exported organic coffee, cashew nuts and honey is mostly produced by smallholders. The local market is not well developed, largely because of the low level of awareness about organic products and the cost. Nuts (cashew and macadamia) and vegetables are the main organic exports from Kenya. The high costs of certification required for both the export and the local market are a restraint. Subsidies provided to conventional farmers by the government for agro-chemicals and pesticides indicate the official position and are seen as an impediment to organic agriculture.

The costs and problems of converting to organic agriculture have been strong disincentives preventing small farmers in many countries from becoming certified producers, which historically involved the participation of the organic authorities in the importing countries. There are promising signs of help in the form of regional certification schemes that are being developed with the encouragement of the UN, e.g. UNEP (2007).

Oceania

Organic export sales from Australia amount to about 20% total value of the organic market. The domestic market was dominated by dairy and meat products. Most of the organically certified products from the Pacific islands are for export to Australia and New Zealand, notably coconut products and tropical fruit. The exports to Japan from Oceania are growing.

Summary

1. After nearly 100 years organic agriculture has only made a minor impact on agriculture across the world. About 1% of global agricultural land is farmed organically.

2. Globally there are about 43.7 million ha of mainstream organic farms and 164,000 ha of biodynamic farms. Some form of organic agriculture was practised by 2.3 million people around the world in 2014 out of a total of about 570 million farmers.

3. The organic farming area has more than doubled since 2000 when there were about 17 million ha worldwide.

4. The statistics are somewhat distorted by the inclusion as organic of more than 21 million ha of extensive rangeland and pampas in Australia, Argentina and Uruguay with 39%, 7% and 3% respectively of the world's organic area. The low and erratic rainfall in the grazing and wheat areas in Australia mean that application of fertilizers are usually and unpredictably a waste. The farmers became organic by circumstance rather than by choice.

5. Oceania accounts for 40% of the global mainstream organic area, Europe with 27%, Latin America with 15%, Asia with 8%, North America with 7% and Africa with 3%.

6. IFOAM, the leading international association of organic movements, is the prime source of information on organic farming. Apart from data on organically certified crops, IFOAM also collects information on the areas where wild berries and fruits are collected and beekeeping is carried out. Such areas, which are estimated at 37.6 million ha, are not considered to be relevant here.

7. Some including the IFOAM consider that many peasant/subsistence farmers who have no opportunity to

use man-made fertilizers and pesticides should be thought of as organic farmers.

8. The character of organic farming differs markedly between regions depending on whether the crops are grown for local consumption or for export.

9. In Europe and North America, with a combined 34% of the world's organic area, production is mostly for local consumption: much organic land is grassland devoted to livestock production. In the rest of the world, with the exception of Australia and Argentina, organic grassland plays a minor part.

10. Permanent crops, mostly organic beverage crops and fruit, figure mainly in Latin America and Africa with a total of 18% of the world's organic area. Many of these crops are for export. East African countries are prominent exporters.

11. In Asia, with a total 8% of the global organic area, arable crops are prominent and are mostly consumed locally.

12. Organic agriculture is mainly carried out on small farms of <2 ha in Africa, and of <5 ha in Asia, in Latin America and in Oceania excluding Australia, where the average organic farm size is 6,200 ha.

13. Organic farm size in Europe is 35 ha and in the USA 180 ha.

14. The global biodynamic organic farm size is 33 ha. As biodynamic agriculture is based on mixed farming and recycling of manure, it does not at present lend itself to the small peasant farms in Africa, Asia, Latin America and Oceania.

15. The global value of the organic market of $80 billion is dominated by consumption in the USA with 43% and in Europe with 44%.

16. Organic food is suffering from competition from Fairtrade and functional food. In the UK the functional food and Fairtrade markets together exceed the organic market.

17. The American market has seen considerable growth since 2000. The granting of permission to certified organic producers to add a USDA stamp on the label has inadvertently and misguidedly given the impression that the USDA approves of the quality of organic food.

18. Outside Europe and North America the global sales of organic products are valued at about $6 billion, mostly exports. Local sales of organic food are believed to be limited except in Brazil, India and China.

References

Chassy, B., Tribe, D., Brookes, G., Kershen, D. and Schroeder, J. (2014) Organic Marketing Report Why Consumers Pay More for Organic Foods? Fear Sells and Marketers Know it. *Academics Review.* Available at http://academicsreview.org/wp-content/uploads/2014/04/Academics-Review_Organic-Marketing-Report1.pdf

CRNRC (2016) Consumer Reports National Research Center Survey Research Report. Food Labels Survey. Available at www.consumerreports.org/content/dam/cro/news_articles/health/PDFs/ConsumerReports-Food-Labels-Survey-April-2016.pdf

Cornucopia (2015) Cornucopia Institute: Organic farms industrializing as sustainability benefits of GMO crops rise. Available at http://geneticliteracyproject.org/2015/01/12/cornucopia-institute-organic-farms-industrializing-as-sustainability-benefits-of-gmo-crops-rise

Defra (2015) Organic farming statistics 2014. Available at https://www.gov.uk/government/uploads/system/uploads/attachment_data/file/444287/organics-statsnotice-23jun15b.pdf

Demeter (2015) Demeter statistics. Was available at http://database.demeter.net

Demeter (2016) Certified Demeter operations in member countries of Demeter-International (2/2016) Available at www.demeter.net/statistics

ERS (2013) Growth Patterns in the U.S. Organic Industry. Available at www.ers.usda.gov/amber-waves/2013-october/growth-patterns-in-the-us-organic-industry.aspx#.U_NY9fldWSq

ERS (2014) Available at www.ers.usda.gov/topics/natural-resources-environment/organic-agriculture/organic-trade.aspx

European Commission (2013) Facts and figures on organic agriculture in the European Union. Available at http://ec.europa.eu/agriculture/markets-and-prices/more-reports/pdf/organic-2013_en.pdf

EU (2015) Agriculture and rural development – organic farming. Available at http://ec.europa.eu/agriculture/organic/index_en

FW (2015) *Farmers' Weekly*, 5 December 2015.

FiBL (2015) Data on organic agriculture 2005-2013. The Organic-World.net website maintained by the Research Institute of Organic Agriculture (FiBL), Frick, Switzerland. Available at www.organic-world.net/statistics

Heinze, S. and Vogel, A. (2014) Reversion from organic to conventional agriculture in Germany. 347–348. In: Rahmann, G. and Aksoy U. (Eds.) *Proceedings of the 4th ISOFAR Scientific Conference. Building Organic Bridges*, Istanbul, Turkey.

House of Lords (1998) Select Committee on European Communities, Sixteenth Report. Available at www.publications.parliament.uk/pa/ld199899/ldselect/ldeucom/93/9302.htm#a5

Jorgensen, F. (2015) Organic Action Plan for Denmark. Working together for more organics. Available at http://en.fvm.dk/fileadmin/user_upload/FVM.dk/Dokumenter/Landbrug/Indsatser/Oekologi/7348_FVM_OEkologiplanDanmark_A5_PIXI_English_Web.pdf

Kahl, J., Załęcka, A., Ploeger, A. Bügel, S. Huber, M. (2012) Functional food and organic food are competing rather than supporting concepts in Europe. *Agriculture*, 2, 4, 316–324. doi: 10.3390/agriculture2040316

Leatherhead Food (2014) Future directions for the global functional food market. Available at https://www.leatherheadfood.com/publication/future-directions for-the-global-functional-foods-market

Lernoud, J. and Willer, H. (Eds) (2014) The world of organic statistics and emerging trends 2014. FiBL-IFOAM Report Research Institute of Organic Agric. (FiBL), Frick and International Federation of Organic Agricultural Movements, Bonn. Revised version 24 February 2014.

Lernoud, J. and Willer, H. (2014a) Organic Agriculture Worldwide. Available at http://orgprints.org/28216/19/lernoud-willer-2015-02-11-vss.pdf

Lernoud, J. and Willer, H. (Eds) (2016) The world of organic statistics and emerging trends 2016. FiBL-IFOAM Report. Available at https://shop.fibl.org/fileadmin/documents/shop/1698-organic-world-2016.pdf

MDC (2013). Available at www.maharishivediccity-iowa.gov

Mella, E.E., Kulindwa, K., Shechambo, E. and Mesaki, S. (2007) The integrated assessment of organic agriculture in Tanzania. Policy options for promoting production and trading opportunities for organic agriculture. http://hdl.handle.net/20.500.11810/2792

NMI (2010) Natural Marketing Institute. LOHAS Market-Size Data Released: A $290 Billion Opportunity. www.newhope.com/trends/lohas-market-size-data-released-290-billion-opportunity

Nuutila, J.T. and Kurppa, S. (2014) The Finnish Organic Food Chain – An Activity Theory Approach 235–238. In: Rahmann, G. and Aksoy U. (Eds) *Proceedings of the 4th ISOFAR Scientific Conference Building Organic Bridges*, Istanbul, Turkey.

Organics Brazil (2014) Brazilian organic and sustainable producers. Available at www.organicsbrasil.org/upload/tiny_mce/catalogo/catalogo-en.pdf

OTA (2015) US Forced to import corn as shoppers Demand organic food. Available at www.bloomberg.com/news/articles/2015-04-15/romanian-corn-imports-to-u-s-surge-as-shoppers-demand-organic

Rahmann, G. and Aksoy, U. (2014) Building organic bridges with science, Foreword In: *Proceedings of the 4th ISOFAR Scientific Conference. Building Organic Bridges*, Istanbul, Turkey.

Soil Association (2013) Market report. www.naturalproducts.co.uk/soil-association-2013-organic-market-report-reveals-significant-areas-of-growth-in-organic-sales

Soil Association (2015) Organic Market Report 2016 Available at https://www.soilassociation.org/certification/farming/organic-market-report-2016

Statistic Brain (2013) Statistic Brain Research Institute Available at www.statisticbrain.com/organic-food-statistics

SwissInfo. (2016) Swiss are top purchasers of organic products worldwide. Available at www.swissinfo.ch/eng/society/buying-organic_swiss-are-top-purchasers-of-organic-products-worldwide/42074012

TechSci Research (2014) Available at www.techsciresearch.com

The Economist (2011) There is no alternative. Virtually all alternative medicine is bunk; but the placebo effect is rather interesting, 19 May. Available at www.economist.com/node/18712290

UNEP (2007) Aid for Trade Case Story. The East African Organic Products Standard. Submission by the United Nations Environment Programme. Available at www.oecd.org/aidfortrade/47719232.pdf

USDA (2016) Greene, C., Wechsler, S.J., Adalja, A. and Hanson, J. Economic Issues in the Coexistence of Organic, Genetically Engineered (GE), and Non-GE Crops, EIB-149, U.S. Department of Agriculture, Economic Research Service.

USDA (2016a) Organic Survey (2014) Volume 3, Special studies, part 4. Available at www.agcensus.usda.gov/Publications/2012/Online_Resources/Organics/ORGANICS.pdf

Wall Street Journal (2014) Organic-Farming Boom Stretches Certification System. Available at www.wsj.com/articles/organic-farming-boom-stretches-certification-system-1418147586

Wynen, E. and Fritz, S. (2007) NASAA and organic agriculture in Australia 225–242. In: Lockeretz, W. (Ed.) *Organic Farming, an International History* CABI, Oxford.

CHAPTER 8

Regulation and Certification of Organic Farming

"If you have ten thousand regulations you destroy all respect for the law"
Winston Churchill

The regulation and certification of organic farming is usually carried out at the local level against the background of national and regional agreements. It is the local organic associations who monitor the certification process. In most cases the rules and regulations cover many pages of instructions, to which the reader is referred. Only the key aspects of the regulations will be considered here.

Features and comparisons of biodynamic and mainstream organic farming are discussed in Chapters 2, 6 and 7.

Biodynamic farming

The production standards that producers in each member country need to comply with before they can use the Demeter and Biodynamic trademarks are issued by Demeter International (Demeter, 2014). The trademarks are intended to assure consumers that foods and beverages have been grown and produced according to the strict Demeter standards. The Demeter brand is closely controlled and there is a comprehensive verification process.

The Demeter symbol was introduced, and the first standards for Demeter quality control of biodynamic farmers were formulated, in 1928. In the USA a Demeter Biodynamic Trade Association was founded in 1938 to spread information about biodynamic methods through educational activities, research and demonstrations.

Demeter International was set up in 1997 by the coming together of nineteen independent Demeter organizations from Austria, Australia, Brazil, Canada, Denmark, Egypt, Finland, France, Germany, Great Britain, Ireland, Italy, Luxembourg, Netherlands, New Zealand, Norway, Sweden, Switzerland and the USA. Demeter International essentially coordinates the individual certification organizations and currently represents about 5,000 biodynamic farms. Its tasks also include advancing the public understanding and acceptance of the biodynamic methods.

In 1998 the Organic Federation of Australia was established to promote both biodynamic and mainstream organic farming. The International Biodynamic Association (IBDA) was founded in 2002 to protect the trademarks "Demeter" and "Biodynamic" all over the world. The International Biodynamic Council (IBDC) was created in 2012 consisting of Demeter International and IBDA. There are Biodynamic Associations in more than thirty countries.

The biodynamic movement continues to be active, for example in 2014 there was an international conference on anthroposophic nutrition at the Goetheanum in Dornach (Switzerland) where the Anthroposophical Society was founded in 1913.

Apart from complying with the production standards of the Biodynamic Association, it is also necessary in Europe for certified biodynamic farms to comply with EC regulations 834/2007 and 889/2008 governing organic agriculture. In the USA the provisions of the Organic Foods Production Act must also be met.

The Demeter standards exceed regional and government mandated organic regulations regarding nutrition, pest and weed control insofar as they also require specific features relating to the use of homeopathic preparations (Demeter, 2014). The Demeter Standards refer to their "Star Calendar", an annual sowing calendar about the timing of farming operations in relation to the zodiacal calendar. It is recommended that the cosmic rhythms should be considered when deciding the best timing of farming operations for both crops and livestock.

The most important prerequisite for conversion to biodynamic farming is an active interest in the laws of nature and especially for the farmer to embrace a holistic view of the natural world, which goes beyond the knowledge gained purely from natural science. The farmer should also be spiritually in tune with the forces of nature.

Biodynamic farms should be self-contained. A biodynamic farm is seen as a unit that should produce all the inputs needed. If inputs must be brought in, they should come from biodynamic or organic farms.

In order to be certified it is mandatory for arable farms to carry ruminants to provide the necessary manure (maximum 2.0 livestock units/ha where one unit is equal to the amount of manure produced in a year by one mature dairy cow). One manure unit corresponds to 80 kg N and 70 kg P_2O_5. Cow horn manure, the horn silica and the compost preparations must be applied at least once every year at the appropriate time, and plant growth stage, for an "in conversion to Demeter" or Demeter certificate to be issued.

The permitted maximum application of N on arable farms is 112 kg/ha derived from manure or compost produced on the farm. This is based on the amount of manure produced by the animals that the farm could support from its own grass and fodder crops. Crop yields will inevitably be restricted. All organic manures and composts have to be treated with the various compost preparations (see Chapter 5).

Synthetic nitrogen sources and Chilean sodium nitrate are totally prohibited. Phosphorus and potassium fertilizers are only permitted if a need has been established and with the approval of the certifying body. There is no mention in Demeter UK standards (Demeter, 2014) about the sprinkling of burnt weed seed ash on the soil for weed control or of the use of insect ash for pest control.

The fungicidal use of copper salts is not permitted on vegetables but they can be used on perennial crops (maximum 3 kg/ha Cu). No genetically engineered crops or inputs containing them are allowed. Seed and planting material must be Demeter-certified if available and if not, certified organic quality is allowed. Treated seeds are strictly forbidden.

Development of mainstream organic standards

Credit for setting up the world's first organic farming association in 1944 goes to the Australian Organic Farming and Gardening Society. Early contributors to its journal included several organic notables such as Howard (their patron), Balfour, Pfeiffer and Rodale (Paull, 2008) indicating that the organic message was being successfully spread around the world. They were concerned with promoting conservation of the natural environment and with improving soil health. They disliked the use of fertilizers and sprays. However, there was no talk, at least in the early days, of formulating regulations covering the practices of which they approved.

United Kingdom

Apart from Demeter (in 1928) there was no formal definition or recognition of organic farming until 1967, when the Soil Association created the world's first set of guidelines for creating and sustaining a living soil under the title *"Standards for organically grown food"*. The four pages of guidelines were published in its magazine, *Mother Earth* (Soil Association, 1967). Farmers were invited to sign a declaration of intent to farm according to the standards; the Soil Association in return agreed to include the farmer in its publication, *Wholefood Finder*.

In 1973 the Soil Association took the next step and formed the Soil Association Organic Marketing Company Ltd, later to become the Soil Association Certification Ltd in order to concentrate on certification. Through the 1970s and early 1980s the inspection element was informal and cursory, but this gradually changed as organic farming became more established. Today the current organic standards of the Soil Association are covered by many sections (totalling more than 800 pp, including some repetition) dealing with all aspects including farming and growing (crops and livestock 248 pp), abattoirs and slaughtering (91 pp), livestock markets (33 pp), hides, leathers and skins (24 pp), aquaculture (56 pp), seaweed (31 pp), food and drink (87 pp), textiles (45 pp), health and beauty product (96 pp), forestry and timber (120 pp) and ethical trade standards (18 pp) (Soil Association, 2016). The growth and extension of standards from the four pages in 1967 to more than 800 indicates the detailed complexity of the certification process.

The Soil Association standards predated those of IFOAM, which was founded in 1972 by Balfour and representatives of organic associations in Australia, South Africa, Sweden and Switzerland. IFOAM published its standards in 1980, which were the first to be internationally agreed.

In 1987 the UK government formed UKROFS (UK Register of Organic Food Standards, 2001) and later ACOS (Advisory Committee on Organic Standards) to draw up minimum UK organic standards, to register the organic certifiers including their inspectors, and to certify those wishing to bypass the private bodies. Details of current regulations for the UK can be found in the compendium for organic food production, which is based on, and complies with, EEC directive 2092/91 (Defra, 2016).

Europe

The EU took an interest in mainstream organic farming in 1991 (directive EEC 2092/91) and in livestock in 1999 (EU 1804/99) following the standards developed by IFOAM in 1980. The European Commission subsequently drafted new regulations on organic production and labelling, namely the European Organic Action Plan in 2007 and 2008 (834/2007 and 889/2008) (EU, 2015).

The EU Commission continues to be interested in organic farming, probably under the influence of France, Germany and Italy with their relatively large organic areas. It is likely that political pressure will lead to a tightening of the rules. For example, organic farmers are currently (2015) allowed to feed their animals 5% of conventionally produced feed and to use non-organic seeds under some conditions but they would have to go 100% organic under new regulations being discussed (Rabesandratana, 2014). Consideration and discussion of equivalency arrangements continues to exercise European bureaucrats (EU, 2015a). Organic reform plans are currently being debated in the EU (Brussels Briefing, 2015). There is disagreement about whether or not to accept the presence of pesticides in some organic products. Those who feel this should not be a problem say that the reforms should concentrate on the organic production process and not on the end product; this attitude is in line with many in the EU regarding GM crops. No agreement had been reached on the EU organic farming regulations as of December 2016.

GM crops and synthetic pesticides could in time be tolerated in organic food under new EU rules following the voting by MEPs not to strip organic agricultural products of their certification if they are contaminated with GMOs or pesticides, as the majority of cases of contamination are accidental (Euractiv, 2016).

Several countries followed the EU's 1991 lead and set up national standards, including exporting countries. This led to a proliferation of national organic laws that prompted the WHO and FAO to attempt a harmonization with the publishing in 1999 of their *Guidelines for the production, processing, labelling and marketing of organically produced foods* in the Codex Alimentarius; the EU regulations were used as the starting point. Standards were published in the USA in 2002, Japan in 2001 and Australia in 1987.

USA

The USDA started a National Organic Program in 2002 to support organic farmers and processors. The differing standards that had emerged by the late 1990s across many states and private certification organizations were harmonized. Certification is mandatory for operations when organic sales exceed \$5,000. The regulations of the National Organic Program complied with the Organic Foods Production Act of 1990.

Some regulations appear to be inherently difficult to monitor, such as those relating to manure application and the use of sodium nitrate. The instructions for manure are precise – it must be composted unless it is to be used on a crop not intended for human consumption or is incorporated into the soil not less than 120 days before harvest of a product whose edible portion does not have direct contact with the soil (CFR, 2015). Sodium nitrate is only allowed in the USA if it is to supply no more than 20% of the crops N requirement (CFR, 2015). Sodium nitrate is not allowed to be used by organic farmers in Europe and according to the Codex Alimentarius (WHO/FAO, 2013) and the IFOAM because it was similar to synthetic N fertilizers and was unrenewable.

Products certified to contain more than 95% organically grown produce can display the label "USDA-Organic". Products containing a minimum of 70% organic materials can be labelled "made with organic ingredients" and even products with less than 70% can list individual organic ingredients on the label but the USDA seal cannot be used.

Organic farmers should use organic seed and according to the USDA's

National Organic Program this is a regulatory requirement subject to availability. However, the shortage of organic seed in the USA (Organic Seed Alliance, 2016) has resulted in considerable use of conventional seed. Organic farmers consider that the commercial/conventional varieties are not suitable as they have been bred to respond to high fertilizer inputs: organic growers want varieties that respond well to the slow release of nutrients from organic soil amendments.

As of 2015 there is a problem in the USA as to whether crops grown without soil by hydroponics can be classified as organic (Dixon, 2015). According to the USDA, they can be classified as organic but the National Organic Standards Board objects.

Mainstream organic farming regulations

There are a huge number of rules and regulations governing organic farming and products from the local to the regional and international levels. The IFOAM standards and the Codex Alimentarius Guidelines for Organic foods of WHO/FAO (2013) set the principles and the basic international organic reference standards.

The regulations in the EU countries (EU 2007), the USA (USDA, 2015) and Japan (Japan, 2007, 2014), the most important markets for organic food, are detailed; compliance with certification for imports into these countries was historically managed by bodies such as the Soil Association (UK), Ceres (USA), EcoCert (France and Japan), IMO (Germany) and Bio Suisse (Switzerland).

Although Codex Alimentarius lists the permitted substances it advises that the regulations of the local certification body or authority should prevail. The Codex Alimentarius is careful to prohibit the use of factory farming sources for most products, including manure.

As of 2012 some eighty-eight countries had appropriate legislation for organic agriculture (Rahmann and Aksoy, 2014). These are countries where standards have been officially endorsed by IFOAM as being in line with their common objectives and requirements or organic standards based on health, ecology, fairness and animal care principles.

Many local organic movements have their own private organic standards.

Worldwide there were 576 certification bodies in 2012, mostly located in the European Union (208), the United States (57), South Korea (33), Japan (33), China (27), Canada (21) and India (17). Asia now has more organic certification bodies than Europe (IFOAM, 2013).

This plethora of regulations and certifiers has necessitated the setting up of many country-to-country agreements in order to facilitate trade. There is growing awareness of the need for schemes that guarantee organic products are what they say they are.

The co-operation between neighbouring countries to produce regional organic standards has been a welcome development in East Africa, in South-east Asia and in Oceania.

As organic farming had its roots at the farm gate level it was inevitable that local rules and regulations were developed to take account of local circumstances. As time went by new regulations were added to maintain the integrity of the basic principles and which at the same time were attractive to consumers supporting a natural way of living. The prohibition of man-made fertilizers was followed by prohibition of synthetic pesticides, of irradiating food and most recently of genetically modified (GM) crops and, at least in the UK, of nano-materials. The regulations are now very detailed and wide ranging. It becomes more desirable than ever to develop a test that would prove a product labelled "organic" was in fact produced according to the rules.

There are many prohibitions that can be bypassed by the organic farmer provided he has obtained prior permission from the certifying body, e.g. in the UK man-made compound fertilizers, potassium sulphate, all the micronutrients, liming agents and even certain proprietary nitrogenous fertilizers can be used.

A requirement for sulphur by forage legumes grown organically was anticipated following the clean air acts (UK) and the reduction of sulphur deposits from the atmosphere. The application $MgSO_4$ and $CaSO_4$ as sources of sulphur (80 kg S/ha) was successfully tested on lucerne and clover-grass mixtures in trials by Becker *et al.* (2014) in Germany. The prevention of sulphur deficiency on organic farms is permitted when the need is determined.

The organic farmer can, under prescribed circumstances, thus take advantage of the practices of conventional farmers when organic farming is found to be wanting. Prohibited antibiotics and vaccines may be used provided a vet has prescribed them.

IFOAM norms for organic agricultural production

Insofar as IFOAM has had a major impact on many organic regulations around the world it is instructive to outline some of the important aspects of its standards that are relevant to comparisons of organic with conventional farming. Full details can be found in IFOAM, 2014.

IFOAM developed detailed standards for organic agriculture covering 134 pp (IFOAM, 2014) of regulations under the headings:

i. Organic agriculture is long term, ecological and system based.

ii. Building up soil fertility is long term and biologically based.

iii. Inputs of synthetic and of potentially harmful chemicals are to be avoided.

iv. Pollution and degradation of the environment should be minimized.

v. Specified unnatural practices are excluded.

vi. Animals to be treated responsibly.

vii. Natural health of animals to be promoted.

viii. Organic integrity should be maintained throughout the supply chain.

ix.   Workers in organic agriculture should be treated fairly.

There is much to commend the IFOAM standards to conventional farmers with the exception of items iii and v, for which local standards are likely to vary between countries and certifiers. Synthetic fertilizers or materials such as rock phosphate made soluble by chemical means are prohibited, as is the natural product Chilean nitrate.

Responsible conventional farmers have farmed and are likely to continue to farm in compliance with many of the IFOAM regulations but at the same time they benefit from the use of synthetic herbicides, insecticides, fungicides, antibiotics and feed additives. Conventional farmers can benefit from GM crops and from embryo transfer in animal breeding practices that are denied to organic farmers. Organic farmers are permitted to use artificial insemination to aid livestock improvement, which is perceived as a natural process. The regulations cover all aspects of farming and go into great detail. For example, under the umbrella of treating animals responsibly organic beekeepers are told not to deliberately kill bees during harvesting and that they should not clip the wings of queen bees (done to deter swarming).

The use of natural processes and products are central to the regulations of organic farming. So it is difficult to reconcile the acceptance by the organic movements of crop varieties developed by using radiation to induce random beneficial mutations in a hit and miss

procedure that can cause many unknown mutations at the same time. Site-directed mutagenesis, a more precise procedure, is now being used. Initially (after the Second World War) X-rays were used to induce mutations in seed but were superseded by exposure to gamma rays from ^{60}Co.

More than 3,000 varieties of the main arable crops have been produced by such unnatural mutagenesis; this includes the barley (Golden Promise) used to make organic beer. In contrast to the acceptance of mutagenesis, the precise and relatively minor changes accompanying genetic engineering are prohibited by all organic movements. It is illogical to prohibit unnatural techniques selectively – if genetic engineering is prohibited then so should mutagenesisbe.

By their position paper on GM crops IFOAM (2002) presented its absolute opposition to genetic engineering, saying that it presents unprecedented dangers to the biosphere and poses risks for organic producers. It felt that the practices of genetic engineering are incompatible with the principles of sustainable agriculture and pose unacceptable threats to human health.

IFOAM (2016) has issued a consultation document in which it takes a new conciliatory look at its opposition to genetic engineering that allows for the use of GM and the new breeding techniques provided they are used responsibly and are based on clear evidence of benefits.

The wide-ranging regulations are a problem for certifiers who have to carry out inspections (Schmid, 2007), as well as for the producers and processors to comply with. It is possible that the addition of new rules may be to the detriment of organic movements. The downfall of some ancient civilizations has been attributed to the development of too much bureaucracy and too many inhibiting and self-defeating rules. Maybe there is a lesson here for the organic movements.

Highly restrictive rules, such as self-imposed zero tolerance to GM contamination, can lead to confrontation between conventional and organic farmers. In 2014 in Western Australia an organic farmer unsuccessfully sued a neighbour for contaminating his farm with sheaves of oilseed rape blown in from the neighbour's GM crop (Paull, 2014). Organic crops in Australia are regulated by the National Association for Sustainable Agriculture, which has a zero tolerance policy for the presence of genetically modified material. The self-imposed zero tolerance was found to be not acceptable by the court as a basis for a claim. In contrast, in the USA the USDA does not automatically decertify organic producers if gene flow is found to have occurred from a neighbour's GM crop and in the European Union products can still be labelled organic provided the contamination does not exceed 0.9%.

Manure and fertilizers

Application of manures and slurries top the list of permitted soil conditioners but they should not be the only source of nitrogen. Several mineral substances are also permitted but with the exception of trace elements (for which both nitrate and chloride forms are prohibited) they should not be produced by chemical means. So limestone, kieserite, Epsom salt, potassium sulphate and potassium chloride (obtained by physical processes), rock phosphate, sodium chloride, sulphur and biodynamic preparations are permitted.

Notable by their absence are any man-made fertilizers. It is difficult to understand why fertilizers such as potassium chloride and potassium sulphate made by refining processes are prohibited when the same chemicals are allowed provided they are physically prepared from rich ores. The objection, which is based on the way the material is produced not the nature of the end product, is consistent with other objections such as those to GM crops; it is the technique that is seen as unnatural, not the nature of the end product.

Pesticides and additives

Organic farmers are permitted to use several (about forty) crop protectants and growth regulators, including pyrethrum, vinegar, calcium chloride, copper salts (sulphate and oxychloride), paraffin, lime sulphur, sodium bicarbonate, sulphur, carbon dioxide, ethyl alcohol, iron phosphates and soft soap. Historically rotenone was permitted because it is a natural product but its use has been banned in some countries.

Fungal preparations (e.g. spinosad), bacterial preparations (e.g. *Bacillus thuringiensis*) and the release of parasites, predators and sterilised insects are also permitted. It might have been expected that the harmful effects of copper salts on soil microbes would have precluded their use. The inclusion of *Bacillus thuringiensis* is curious insofar as organic farmers are not allowed to grow crops such as Bt cotton and Bt maize, which have the ability to produce the Bt toxin internally following breeding using gene transfer.

There is a long list of about sixty permitted additives for post-harvest handling and processing, including lactic acid, ascorbic acid, ammonium phosphate, mono calcium phosphate, potassium and ammonium carbonates, potassium chloride, potassium nitrate, sulphuric acid, ammonium sulphate and ethylene. Substances of certified organic origin must be used if commercially available. It is difficult to understand the logic of permitting the use of refined man-made chemicals, including soluble nitrogen and phosphorus compounds, in one part of the organic system when in another part they are prohibited on principle because they are not natural or sustainable.

Organic equivalence arrangements

These are several arrangements between countries whereby certified organic produce in one country is accepted in others without the need for further local certification and paperwork. They now exist, with minor restrictions (e.g. the exclusion of antibiotics), between the EU and the USA (2012), between the EU and Japan, Switzerland, USA and Canada (2011), and between USA and Japan and Korea (2014). Further equivalence arrangements are being prepared between the USA and Costa Rica and Mexico. These equivalence arrangements are being set up in the interests of facilitating trade despite differences between countries in defining what constitutes an organic product.

Following a UNEP-UNCTAD initiative the East African Organic Products Standards (EAS 456:2007 and ICS 67.020) was launched in 2007 to harmonize organic standards in East Africa and so promote both local consumption and the export market (UNEP, 2007). A similar harmonization, the Pacific Organic Standard, was adopted in 2008 by ten Pacific Island countries, Australia and New Zealand after a project financed by the International Fund for Agricultural Development (IFAD) and implemented by IFOAM; it became the third regional standard after those in Europe and East Africa.

Trade of organic produce is considerably facilitated by countries agreeing equivalence status. For example, UK importers do not need authorization for imports from eleven non-EU countries that have equivalency arrangements, including Argentina, Australia, Canada, Costa Rica, India, Israel, Japan, New Zealand, Switzerland, Tunisia and USA.

Non-food crops and products

A wide range of non-food organic products are available (Appendix) which seemingly benefit from being labelled organic.

Certified organic cotton is grown in several countries, particularly India. It is said organic cotton benefits cotton producers and the environment in developing countries by avoiding the harmful effects of toxic pesticides and the associated health problems and deaths common in non-organic cotton production. Consumers are said to benefit from garments that are manufactured/processed without the use of tens of thousands of acutely toxic chemicals classified as hazardous (Soil Association, 2013).

The benefits to the farmer and the environment can also be conferred by Bt cotton, which is very widely grown in India. Pesticide residues as well as chemicals (toxic dyestuffs) used in processing cotton may occur in conventional cotton. It is said that these residues may cause allergies, skin rashes and respiratory problems.

Organic and natural cosmetics are certified according to the standard COSMOS developed in Europe in 2010 by BDIH (Germany), Cosmebio & Ecocert (France), ICEA (Italy) and Soil Association (UK) (COSMOS, 2014). In common with other organic standards, GM crops, irradiation and nanomaterials are prohibited.

In the USA organic alcohol, wine and beers are certified by the USDA and in Australia certified biodynamic and organic wines are both produced.

Efforts directed towards minimizing illegal logging have led to the involvement of Woodmark, the Soil Association's certification scheme for forest management and assessment. Woodmark is now a Monitoring Organization for the European Union Timber Regulation (EUTR No: 995/2010) on illegal logging. Forest management certificates have been issued in more than thirty countries and currently certify more than 14,000,000 ha of forest.

Woodmark also assesses whether forests are complying with the regulations of the Forest Stewardship Council, an international not-for-profit organization established in 1993 to promote responsible management of the world's forests. Its main tools for achieving this are standard setting, certification and labelling of forest products. There would appear to be no precise regulations regarding the use of fertilizers and pesticides in forests.

Summary

1. The first organic standards were those set for biodynamic farmers in 1928. Demeter International and the International Biodynamic Association, founded in 2002 operate worldwide; they monitor the certification of biodynamic farmers across the world.

2. Biodynamic farmers not only comply with national and regional regulations governing organic farming but they also have to agree to use the homeopathic

preparations regularly. They also have to give due consideration to cosmic rhythms when deciding the optimal timing of farming operations.

3. The first standards for mainstream organic agriculture were created in 1967 by the Soil Association in the UK, which was followed in 1973 by the formation of a Soil Association organization to certify organic farmers.

4. The formation of IFOAM in 1972 and its standards in 1980 stimulated the development of organic standards across Europe, culminating in directives (EEC 2092/1991 and 1804/99), and led to similar work in the USA, Japan and Australia.

5. The proliferation of national organic rules prompted the FAO/WHO to attempt harmonization with the publishing of the Codex Alimentarius Commission but local certifying bodies have the final say.

6. In the USA the involvement of the USDA and the granting of permission for organic growers and processors to add a USDA stamp to organic labels have inadvertently promoted organic sales by giving the impression that the USDA favours and approves the quality of organic produce.

7. Despite the harmonization of standards, especially by IFOAM, certification of organic farms and processors is still carried out by a large number of local organizations. There were 576 certifying bodies around the world in 2012, monitoring 2.3 million organic farmers. Government regulations are in place in about 100 countries.

8. Monitoring of organic farms and processors by the certifying bodies is difficult due to the large numbers involved;

in Africa and Asia the problem is compounded by the small size of organic farms.

9. Conventional farmers are very likely to comply with many of the practices of organic agriculture regarding soil fertility, the environment and animal welfare but they do not adhere to the prohibitions of the use of man-made fertilizers and pesticides on crops.

10. It is difficult to understand the reasons for some of the organic standards, for example it appears illogical to prohibit, on the basis of how they are produced, the use of chemicals such as ammonium sulphate and phosphates on crops but to allow them for post-harvest and processing activities. Likewise, the permission to use potassium salts obtained by physical means but prohibited when they are refined chemically to produce identical salts seems illogical.

11. The use of highly toxic copper salts by organic farms when more effective and less environmentally damaging man-made alternatives are available is surprising.

12. The use of natural processes and products are central to the regulations of organic farming. Hence it is difficult to reconcile the acceptance by the organic movements of crop varieties developed by mutagenesis when at the same time banning varieties developed using genetic engineering. Mutagenesis typically involves exposing seeds or plant material with X-rays or massive doses of gamma rays from ^{60}Co.

13. Producers and processors in exporting countries have to comply with the regulations in the USA, Europe and Japan – the major importing countries; this must pose monitoring problems in

the exporting country, especially when very large numbers of small farmers may be involved.

14. There are reassuring signs of regional harmonization of regulations in East African countries and in Oceania, which should facilitate both local and export trade.

15. The setting up of organic equivalence arrangements between countries whereby certified organic produce in one country is accepted in others without the need for further local certification will also ease trade. Arrangements involving North America, Europe, Japan and Korea are in place.

16. As the rules and regulations for organic farming are self-imposed it does not seem acceptable for them to be used to influence the practices of neighbouring conventional farms. Organic farmers who wish to conform to a zero policy with regard to the presence of genetically modified plant material in their crop cannot expect compensation for minimal contamination. The attitudes in Europe and USA, where specified levels of contamination are allowed, seem preferable.

References

BDA (2012) Demeter and organic production standards. *Biodynamic Association*. Available at http://bdcertification.org.uk/index.php/bd-certification, ww.demeter-usa.org/downloads/Demeter-Farm-Standard.pdf

Becker K., Riffel, A., Fischinge, S. and Leithold, G. (2014) Benefit of sulfate fertilization in Alfalfa and clover grass mixtures in organic 535–537 In: Rahmann, G. and Aksoy U. (Eds.) *Proceedings of the 4th ISOFAR Scientific Conference. Building Organic Bridges*, Istanbul, Turkey.

Briefing (2015) Brussels Briefing on Agriculture: All you need to know for June 2015. Available at. www.vieuws.eu/food-agriculture/brussels-briefing-on-agriculture-all-you-need-to-know-for-june-2015

CFR (2015) 7-205.203-Soil fertility and crop nutrient management practice standard. Electronic code of federal regulations. Available at www.ecfr.gov/cgi-bin/text-idx ?rgn=div8&node=7:3.1.1.9.32.3.354.4

COSMOS (2014) COSMOS-standard AISBL Technical guide. Available at www.cosmos-standard.org

Defra (2016) Organic farming: how to get certification and apply for funding. Available at https://www.gov.uk/guidance/organic-farming-how-to-get-certification-and-apply-for-funding and https://www.gov.uk/organic-certification-and-standards

Demeter (2014) Production standards for the use of Demeter, biodynamic and related trademarks. Available at www.demeter.net/certification/standards

Dixon, L. (2015) Cornucopia Institute. Is Hydroponics Organic? USDA's Organic Program Allows Soil-less Practice Over NOSB's Objections The Cultivator. Available at www.cornucopia.org/wp-content/uploads/2015/03/Spring-2015-Cultivator-for-website.pdf

EU (2007) Council Regulation (EC) No 834/2007. Available at http://eur-lex.europa.eu/legal-content/EN/TXT/?uri=celex%3A32007R0834

EU (2015) Agriculture and rural development - organic farming. Available at http://ec.europa.eu/agriculture/organic/eu-policy/eu-legislation/index_en.html

EU (2015a) EU Legislation Review and Organics Trade regime: State of play Biofach. Available at http://ow.bionext.nl/sites/www.bionext.nl/files/biologica/banners/biofach_2015_-_congress_reports_-_wageningen_ur.pdf

Euractiv (2016) GMOs and pesticides could be tolerated in organic food under new EU rules. Available at http://www.euractiv eu-rules/.com/section/agriculture-food/news/gmos-and-pesticides-could-be-tolerated-in-organic-food-under-new-eu-rules

IFOAM (2002) Position on Genetic Engineering and Genetically Modified Organisms. Available at www.ifoam.bio/en/position-genetic-engineering-and-genetically-modified-organisms

IFOAM (2013) The World of Organic Agriculture Statistics and Emerging Trends 2013. FiBL and IFOAM. Available at https://www.fibl.org/fileadmin/documents/shop/1606-organic-world-2013.pdf

IFOAM (2014) IFOAM norms for organic production and processing (Version 2014). Available at http://infohub.ifoam.bio/sites/default/files/ifoam_norms_version_july_2014.pdf

IFOAM (2016) Public Consultation on the position of IFOAM – Organics International on Genetic Engineering and Genetically Modified Organisms. Available at www.ifoam.bio/en/news/2016/02/26/public-consultation-position-ifoam-organics-international-genetic-engineering-and

Japan (2007) Overview of the Organic Japanese Agricultural Standard System. Available at www.maff.go.jp/e/jas/specific/pdf/org01.pdf

Japan (2014) Organic foods. Available at www.maff.go.jp/e/jas/specific/organic.html

Organic Seed Alliance (2016) State of Organic Seed, 2016 Available at www.seedalliance.org

Paull, J. (2008) The Lost History of Organic Farming in Australia. *Journal of Organic Systems* 3, 2, 2–17 Available at http://orgprints.org/15089/1/15089new.pdf

Paull, J. (2014) GMO agriculture versus organic agriculture – Genetic trespass, a case study 17–20. In: Rahmann, G. and Aksoy, U. (Eds.) *Proceedings of the 4th ISOFAR Scientific Conference. Building Organic Bridges, at the Organic World Congress 2014*, Istanbul, Turkey.

Rabesandratana, T. (2014) Organic Farming Overhaul in Europe May Boost Research, Science Insider. Available at http://news.sciencemag.org/europe/2014/03/organic-farming-overhaul-europe-may-boost-research

Rahmann, G. and Aksoy, U. (2014) Building organic bridges with science, Foreword In: *Proceedings of the 4th ISOFAR Scientific Conference. Building Organic Bridges*, Istanbul, Turkey.

Schmid, O. (2007) Development of standards for organic farming 152–174. In: Lockeretz, W. (Ed.) Organic farming, an international history. CABI, Oxford.

Soil Association (1967) First standards for organically grown food. Mother Earth (1966–1967) 8.537–540.

Soil Association (2013) Organic cotton. Available at https://www.soilassociation.org/organic-living/health-beauty-fashion-textiles/organic-fashion-textiles/organic-cotton

Soil Association (2016) Soil Association organic standards. Available at https://www.soilassociation.org/what-we-do/organic-standards/soil-association-organic-standards

UKROFS (2001) UKROFS Standards for organic food production. Available at http://adlib.everysite.co.uk/adlib/defra/content.aspx?doc=4344&id=4346

UNEP (2007) Aid for Trade Case Story. The East African Organic Products Standard. Submission by the United Nations Environment Programme Available at www.oecd.org/aidfortrade/47719232.pdf

USDA (2015) Organic agriculture. Available at www.usda.gov/wps/portal/usda/usdahome?contentidonly=true&contentid=organic-agriculture.html

WHO/FAO (2013) Codex Alimentarius, International food standards. Guidelines for the production, processing, labelling and marketing of organically produced foods (GL32-1999). Available at www.codexalimentarius.org/standards/list-of-standards

Willer, H. and Lernoud, J. (Eds) (2014) The world of organic statistics and emerging trends 2013. FiBl-IFOAM Report. Research Institute of Organic Agric. (FiBL), Frick and Internat. Federation of Organic Agricultural Movements (IFOAM). Bonn. Revised version 24 February 2014.

World of Organic Agriculture Statistics and Emerging Trends 2013. Available at www.organic-world.net/yearbook-2013.html

CHAPTER 9

Organic Farming and Health – the Pioneers

"Healthy citizens are the greatest asset any country could have"
Winston Churchill 1943

The link between organic farming and of natural lifestyles with human health is fundamental to the organic movements. In this chapter the activities of the pioneers who played a big part in focusing attention on the relationship between farming organically and health will be discussed. In the next chapter the health studies that shed light on health in relation to organic food will be considered.

Soil, organic farming methods, diet and health

Three basic questions need to be examined. Firstly, are there any identifiable factors that link soils to crops to food, to diet and on to health other than the fact we are what we eat and most of what we eat ultimately comes from plants growing in soil? Secondly, are there fundamental aspects of the composition of organic food that have meaningful implications for human health? Thirdly, do some of the practices of organic farming lead to better human health?

What is the evidence that prompted the early advocates of organic farming to take a special interest in this topic? The interest continues to today with claims of the health-promoting chemicals sometimes found in greater quantities in organic food, e.g. Baranski *et al.* (2014), which will be considered in Chapter 14.

It is important to understand the basis for Balfour (1943) subtitling her influential initial editions of *Living Soil* with "evidence of the importance to human health of soil vitality, with special reference to national planning". *She* stressed the link from soil to health; how she came to this position is discussed in Chapter 6.

IFOAM (2006), in its leaflet Organic Agriculture and Human Health give its official approved definition that: "Organic agriculture is a production system that sustains the health of soils, ecosystems and people" and "organically-raised animals have better overall health and reduced risk of contracting or carrying diseases". It is substantiated largely by the correlative study by Swansom (2014), in which the rise in glyphosate use in the USA is linked, by means of significant correlation coefficients, to several diseases and health conditions including hypertension, stroke, diabetes, obesity, lipoprotein metabolism disorder, Alzheimer's, senile dementia, Parkinson's, multiple sclerosis, autism, inflammatory bowel disease, intestinal infections, end stage renal disease, acute kidney failure, cancers of the thyroid, liver, bladder, pancreas, kidney and myeloid leukaemia. Similar significant correlations were found between the percentage of genetically modified maize and soya planted in the USA and the same diseases; this is not surprising as most of the glyphosate is used on these crops. The correlation coefficients mostly exceed R=0.9, a value that would indicate that the link between the disease incidence and the use of glyphosate is such that 81% of the variation in one is associated with variation in the other. Swansom (2014) do not mention any connection with organic farming.

When considering diseases it is necessary to keep in mind the distinction between infectious diseases caused by a pathogen and deficiency diseases caused by insufficient intake of essential nutrients. Diseases associated with inherited conditions are not relevant here; likewise the effects of exposure to toxic pollutants. Disease and its counterpart good health can also be related to lifestyles; smoking, consumption of alcohol and lack of exercise can all be important factors, similarly dietary choices. The toxicity of pesticides will be discussed in Chapter 15.

Discussion of deficiency diseases caused by dietary inadequacies may be relevant with regard to organic food insofar as the higher prices of organic fruit and vegetables can limit their consumption by organic devotees. The early claims of the health–promoting properties of organic food were largely based on infectious diseases on twelve oxen in India and on the association of dental caries with health, diet and how the food crops were grown. In the 21st century cancer, cardiovascular diseases and the health of children are the major health concerns.

The essential nature of carbohydrates, proteins, fats, and some minerals was appreciated in the 19th century – dietary fibre that we now consider as beneficial is not essential. The discovery of vitamins in the first thirty years of the 20th century could have indicated that more was waiting to be learnt about human nutrition. It was against this background that much of the early work discussed here was carried out. The story of the connection between how crops are grown, diet and animal and human health, which marks the beginning of the organic movement in the UK, starts with Sir Albert Howard and Sir Robert McCarrison both working in India in the period 1905–1930s. They were the people who unknowingly laid the scientific framework for mainstream organic agriculture. Howard made the connection with regard to animals (by observations) and McCarrison with humans (by rat experiments).

Pioneers

Howard (1873–1947)

Howard studied agriculture at Cambridge before working as a mycologist in the West Indies (1899) and then at Wye College in England before going to India in 1905 as a government botanist at Pusa. He believed in learning from peasant farmers, who were not using inorganic fertilizers or pesticides. Pesticides then available elsewhere included arsenicals, petroleum oils, nicotine, pyrethrum, rotenone and sulphur, but they may not have been available in India. He found the crops were remarkably free of diseases and insect attack (Howard, 1943). Howard decided to follow the peasants' example and not use any insecticides or fungicides with the result that after five years he reported that insect and fungal attacks were negligible. A key element in the success was his replacing susceptible varieties with disease-resistant ones. He concluded that insects and fungi only attack unsuitable crops or varieties that are imperfectly grown.

He then extended his observations to animals, namely six pairs of strong, healthy oxen that were provided with suitable housing and with fresh green fodder, silage, and grain, all produced from fertile land. He watched how these well-fed oxen reacted to diseases such as rinderpest, septicaemia and foot-and-mouth disease, which were common in the area. Howard believed the epidemics were the result of inadequate feeding. None of Howard's oxen was segregated; none was inoculated and they frequently came into contact with diseased stock. Although his oxen rubbed noses with animals suffering from foot-and mouth, no infection occurred. The well-fed animals were able to withstand diseases provided they were well fed, as did suitable varieties of crops when properly grown. No connection could be made between composting and health of either crops or oxen as these observations were made many years before Howard worked on the Indore composting process. However, it has subsequently been implied that the oxen were fed compost-grown feed.

Howard (1943) said it was important for him to see if he could achieve the same results with crops at different places and times. This he did during the next twenty-one years at three Indian centres: Pusa (1910–24), Quetta (summers of 1910–18), and Indore (1924–31). Again no experiments were carried out. He reported that there were only a few outbreaks of fungal disease and insect attack at these three places, which he put down mainly to his regular use of farmyard manure. His confidence in the importance of the way he grew crops was so great that he wanted to import cotton bollworms and boll weevils from the USA to test his ideas on cotton: his confidence was not shared with the authorities, which was perhaps fortunate.

The closest he came to providing confirmation about the health-promoting effects of well-grown crops on animals was when he reported that the freedom from disease observed at Pusa was again experienced on oxen at Quetta and Indore (Howard, 1943). This is the anecdotal evidence of a link between composting and animal health that all dates back to Howard's observations on the twelve oxen at Pusa (1910–1924).

The wasteful use of dried manure as fuel led Howard to experiment with ways of extending the manure that was available. The Indore composting process (Howard, 1931 and 1940) made use of all the by-products of agriculture that can be converted into humus; this includes all available vegetable matter, including the soiled bedding from cattle sheds, all unconsumed crop residues, fallen leaves, farmyard manure, green manures and weeds, together with the urine impregnated earth from the floor of the cattle shed and any available wood ashes.

He was following the example of Chinese farmers. To reduce acidity earth, wood ash or chalk were sprinkled on the heaps made up of successive layers of plant material and manure. The result was the Indore process of composting, first tested in the 1920s and which in a few years was said to have more than doubled the yields of crops that were immune from disease. Conversion to humus was complete in Indian conditions in about six weeks when the compost was ready to use.

After 1931 the Indore process was taken up in a number of countries, especially on plantation crops such as coffee, tea and rubber but also on sugar cane, sisal, maize, cotton and tobacco. In all these crops the conversion of vegetable and animal wastes into humus was said to have been followed by a definite improvement in the health of the crops and of the livestock but there appear to have been no comparative studies and no data was presented by Howard. With new rural housing schemes in the UK in mind, Howard proposed that organic wastes and human excreta should be collected nightly in schemes involving composting toilets; Balfour took up this idea in *The Living Soil* (Balfour, 1943).

Howard was one of the first to draw attention to mycorrhiza in/on roots, following the publication by Rayner (1927). Current thinking about the association of mycorrhizal fungi and roots is discussed in Chapter 20. During 1938 Rayner found mycorrhizae on a large number of samples Howard had provided from crops, including, rubber, coffee, cacao, leguminous shade trees, green manure crops, coconuts, tung, cardamoms, vine, banana, cotton, sugar cane, hops, strawberries, bulbs, grasses and clovers. The common finding of mycorrhizae provided Howard with an explanation for the benefits of manure on both the growth and the health of crops. He associated the absence of mycorrhiza with diminished resistance to disease. Unfortunately Howard did not carry out any experiments or make observations that permitted him to say, for example, that diseased crops lacked mycorrhizae.

So it was that Howard extolled the value of mycorrhizae for conferring disease resistance on crops provided the fungi were kept supplied with fresh, properly made compost. Furthermore, the mycorrhizae were believed – but not demonstrated – to be inactive in soils given fertilizers. Initially Howard was not opposed to the use of fertilizers but by about 1940 he was strongly anti-fertilizer: by then he believed that fertilizers were directly responsible for the increased occurrence, since the mid-19th

century, of both crop and animal diseases. At public meetings Howard condemned the Rothamsted Experimental Station and its fertilizer experiments. The fact that experiments on wheat at Rothamsted had shown that soil fertility and high yields had been maintained for nearly 100 years by inorganic fertilizers alone was ignored by Howard.

Howard was a very practical man, a generalist and by all accounts a very forceful character, who was vociferously opposed to research conducted by teams of specialists, each working on a fragment of the whole problem. According to Howard, the agricultural investigator must be a farmer as well as a scientist. Howard writes of his forty years' experience as a farmer, not as a researcher. It is curious in view of his disapproval of experiments to note that he commented that supporters of the bio-dynamic methods of Steiner had not provided any practical examples which demonstrated the value of their theories.

Howard's six main conclusions (Howard, 1943) are revealing and deserve to be recorded in view of the pre-eminent positon he occupies at the birth of mainstream organic agriculture:

i.   Nature has provided in the forest an example that can be safely copied in transforming wastes into humus by the Indore composting process.

ii.   The slow poisoning of the life of the soil by man–made fertilizers is one of the greatest calamities that has befallen agriculture and mankind. The responsibility for this disaster must be shared equally by the disciples of Liebig and by the current economic system.

iii. The situation can only be saved by convincing people of the dangers of modern agriculture. The connection between a fertile soil and healthy crops, healthy animals and, last but not least, healthy human beings must be made widely known.

iv. As many communities as possible who can produce their own crops and food must be persuaded to feed themselves and to demonstrate the benefits of fresh food raised on fertile soil.

v. Evidence is accumulating that such healthy produce is an important factor for the well-being of mankind.

vi. The approach to the problems of farming must be made from the field, not from the laboratory. The discovery of the things that matter is three-quarters of the battle. In this the observant farmers, who have spent their lives in close contact with Nature, can be of the greatest help to the investigator.

Howard was not alone in his work in India or afterwards in England. His wife, Gabrielle (1876–1930), a plant physiologist, accompanied him to India. They worked together in India from 1905 until her death, after which Howard felt he could no longer work in India. In 1931 Howard married Gabrielle's sister, Louise (1880–1969), who had worked at the International Labour Office in Geneva. She devoted herself to his work, became an active supporter of the Soil Association and a close friend of Eve Balfour.

Howard was far from complimentary about agriculture in Europe and North America. He did not favour monoculture or the increased use of mechanization. He saw the use of man–made fertilizers as being nothing more than the commercial activities of the chemical industry wanting to make a profit from the nitrogen-fixing factories that had been developed in the First World War to manufacture explosives. He drew a connection between the increased use of chemical fertilizers and machines with his assertions about the reduction in soil fertility and the increases in crop diseases.

Howard (1943) concluded his "Agricultural Testament" with the forecast that at least half the illnesses of mankind will disappear once our food is grown on fertile soils and consumed fresh.

Sykes (1888–1965)

Friend Sykes in England was an early disciple of Howard who was admired by Balfour (1943). Sykes farmed, using only Indore compost as a source of nutrients and humus, on an infertile down land soil in Wiltshire (Sykes, 1946). He managed, four years after converting to using compost, to carry a herd of 250 head (starting with a farm that could only carry fifty head) without the aid of imported feed.

Temporary pastures, grown in rotation with arable crops, were intensively grazed. Man-made fertilizers were not used. Howard was very impressed with Sykes's success and said that his farm was a better demonstration of organic farming than the Haughley experiment (Conford, 2001). It was Howard, a hard man to refuse according to Sykes, who persuaded him to write his book, *Humus and the Farmer* (Sykes, 1946). Howard appears to have been dominant and forthright in his ways, which probably explains how he came to influence people who were probably reluctant to question him.

McCarrison (1878–1960)

McCarrison qualified as a medical practitioner at Queen's University, Belfast, and then entered the Indian medical service in 1901 when 23 years old. McCarrison was a prolific writer and lecturer (McCarrison, 1927, 1936). Early work on goitre, stimulated by the discovery of vitamins by Hopkins in 1912, led McCarrison into the field of deficiency diseases associated with inadequate diets. The First World War intervened and his research only continued after 1918. It was the UK government's concern about what could be done concerning the prevalence of diseases among Indian troops in the First World War that brought McCarrison into research contact with the fighting races of India, including the Punjabis, Brahmins, Gurkhas, Pathans, and Sikhs. The disbanded Hunzas were not in his first studies but it was not long before McCarrison, working in nearby Gilgit, became aware of the fine physique and longevity of the Hunzas who failed to get goitre and were generally free from disease.

The Hunzas (in 1911 amounting to 14,000) lived in an isolated valley in the Hindu Kush and are of particular interest because of the excellence of their diet and the fact that food production was reliant on the deposition of silt from the melting glaciers and their conservation and recycling of all vegetable wastes. Human excreta was added to composts, a custom similar to that of the Tibetans and Chinese. Their diet consisted of whole-wheat bread, barley, millet, pulses and a wide variety of vegetables. Fruit was eaten in large quantities. They consumed milk, buttermilk, clarified butter, and curd cheese and occasionally meat but rarely fish or game. Although Muslims, they drank wine. Most modern nutritionists would approve of the Hunza diet and there would be no reason to look for further explanations for their good health and longevity. Their isolation would also have protected them from infectious diseases prevalent in the Indian sub-continent.

In 1927 McCarrison was appointed Director of Nutrition Research in India, where he alone worked on human nutrition. He now had a free hand in his new laboratory at Coonoor, in the Nilgiris in Madras, where he carried out his ground-breaking research on the diets of many Indian races, whose health and physique varied widely. He conferred the exceptional health of groups such as the Hunzas, Sikhs and Pathans to rats by feeding them a very similar diet (fruit was omitted from the rat diet). The health-promoting diets given to the rats consisted of chapattis, or wholemeal bread, flour, with fresh butter, sprouted pulses, fresh raw carrots and fresh raw cabbage, unboiled whole milk, a small ration of meat with bones once a week, and an abundance of water. When given such a diet the rats did not suffer from the diseases found when they were fed inadequate diets.

When McCarrison gave his rats the diet of the poorer peoples of Bengal and of Madras, consisting of rice, pulses, vegetables, condiments and perhaps a little milk, they did not fare well. These rats suffered from many diseases, including lung diseases: pneumonia, broncho-pneumonia, bronchiectasis, pyothorax, pleurisy and haemothorax, in other words the rats got diseases of most organs and tissues in the body. McCarrrison also studied a diet of the poorer classes of England consisting of white bread, margarine, sweetened tea, boiled vegetables, tinned meats and jam. On the English diet, not only were the rats poorly grown but they developed a nervous condition – a neurasthenia.

McCarrison (1927) was aware that badly fed monkeys, pigeons, mice and guinea pigs were more susceptible to infectious diseases. McCarrison believed, based on the Hunzas, that the way the crops were grown was also important and that unless the food came from crops grown in a fertile soil, it did not possess health-promoting qualities. No evidence was presented to support

this contention: this was not an area of McCarrison's expertize.

In 1936 McCarrison gave a series of lectures in London (McCarrison, 1936), but it was not until 1938, when Wrench published *Wheel of Health*, that the significance of the McCarrison research was fully appreciated.

McCarrison's emphasis on experimental evidence is to be contrasted with the observational approach of Howard.

Wrench (1877–1949)

Wrench was a doctor who had studied the health of several disparate communities in different countries and was able to relate his own observations about the Hunzas alongside those of McCarrison.

Wrench extended the McCarrison findings to several, mostly isolated, communities such as the Eskimos, Icelanders, Faroe Islanders and North American Indians in his book, *Wheel of Health* (Wrench, 1938). Wrench highlighted the fact that these communities did not use inorganic fertilizers. He pointed out that these peoples with their varied diets lived close to nature and, most importantly, also depended on the recycling of wastes. Wrench thought that their good health depended on them living in harmony with their environment and the way crops were grown using compost. He did not associate their health with the quality of their diets. He was clearly convinced that Howard was on the right track. No evidence was supplied by Wrench to support the thesis that there is a connection between the way food is grown and health. The good health of the communities might simply have been due to isolation protecting them from imported infectious diseases and to their consumption of little sugar. The restricted diets and the reliance, by some communities, on meat suggest that diet may not have been the major factor contributing to their health. No data was presented linking health and longevity.

Wrench cited the situation in Denmark where, following the blockade of their ports in the First World War, there was a major change in the diet from one containing much bacon and ham to one based on whole wheat bread and vegetables. The result was a massive reduction in the death rate attributed to reduced consumption of meat and also of alcohol. However, the way the wheat and vegetables were grown was not thought to be relevant.

Wrench (1946) presented striking examples of the detrimental effects of civilizations on their environment. He cited many cases where, due basically to ignorance, the essential features of sound agronomy had not been complied with; these included civilizations and communities in China, Mesopotamia, the Roman Empire, in Islamic Spain, in England, in Africa since the coming of the Europeans, in Egypt, in India, in the Dutch Empire, in the British Dominions and colonies, in the USSR and in the USA. Wrench concluded that agronomists, politicians and all who are concerned with the future of western civilization should recognize the natural laws linking the management of soils to their survival.

Price (1870–1948)

Price was an American dentist known primarily for his theories on the relationship between nutrition, dental health, and physical health. He travelled widely and detailed his findings in a book, *Nutrition and Physical Degeneration* (Price, 1939). The peoples and countries he visited over nine years in the 1930s included isolated and modern Swiss, Scottish Islanders, Eskimos, North American Indians, Melanesians, African tribes, Australian aborigines, Torres Strait Islanders, New Zealand Maori and Peruvian Indians.

Price was motivated by the strong correlation he had observed in America between dental health, physical health and susceptibility to disease. Having heard rumours of native cultures that were living free of disease he thought he would try to find out if they really were healthy and what they were doing to keep themselves that way. He was specifically looking for groups of healthy people who had not been affected by western civilization. Such groups could still be found at that time, which makes his study valuable historically. Although he came across sick communities he did not appear to be interested in them.

In the USA, Price had found that dental caries was associated with malformed dental arches. He accordingly measured the incidence of dental caries and of abnormal facial structure. He found less than 1% of tooth decay in all the peoples he visited; this was despite the fact that none of the healthy people Price examined practised any sort of dental hygiene. At the same time he found no obvious signs of many of the diseases that afflicted the USA such as cancer, heart disease, diabetes, haemorrhoids and multiple sclerosis.

Price took note of diets as he suspected the key to good health and sound teeth was good food but he found only a few common dietary factors. Each group had its own particular diet but all of them provided at least four times the amounts of water-soluble vitamins and minerals (especially calcium) and at least ten times the amounts of fat-soluble vitamins as the average diets in USA. Also, all cultures consumed insects and fermented food, e.g. yoghurt, beer and fermented fish. One feature stood out, namely that many diets were rich in fat.

The foods eaten were natural, unprocessed and did not contain preservatives, additives, colourings or added sugar. When available, natural sweets such as honey and maple syrup were eaten in moderation. White flour and canned foods were not consumed and their milk products were not pasteurized or low in fat. Their animals and crops were raised on pesticide-free soil and were not given growth hormones or antibiotics. In short, these people always ate what we could today call an organic diet and it led Price to warn about pesticides, herbicides, preservatives, colourings, refined sugars and vegetable oils.

Price concluded that aspects of a modern Western diet, and particularly white flour, sugar, and modern processed and vegetable fats, cause nutritional deficiencies that are responsible for many dental and health problems.

It is somewhat curious that little attention was paid by Price to the lack of sugar in the diet in relation to the low incidence of caries he found. The absence of caries would suggest that the small amounts of sugars in fruit had no adverse effects. Maybe there was little emphasis on the relation between sugar and dental caries at the start of the 20th century despite the association being known since the writings of Fauchard (1678–1761), the father of modern dentistry. He had written in 1728 that the cause of dental caries was sugar, and that people should limit its consumption. Courses in dentistry had been established at Guy's Hospital in London and in the USA before the end of the 19th century.

The main elements of the diets of the healthy communities quoted by McCarrison, Wrench and Price had few if any common factors. The diets of the Sikhs, Eskimos, Icelanders, Faroe Islanders and North American Indians bore little resemblance to that of the Hunzas with their fruit and vegetables and little meat. The diet of the people on Tristan da Cunha with their potatoes, seabird eggs, fish, and cabbage and the Eskimos eating flesh, liver blubber and fish could not be more different from the Hunza diet. It was not surprising therefore that health was related to the fact that the peoples were believed to be living in harmony with nature and had not been contaminated by Western civilization.

There is no doubt that the work and writings of McCarrison, Wrench, Price and Balfour who all linked good health with diet had a major influence on the early organic movement. The supposed good health and the farming practices of the Hunzas were a prime example of them. However, the health of the Hunzas may not have been as good as has been portrayed. John Clark lived with the Hunza for almost two years, in 1950 and 1951, and described his experiences (Clark, 1956). He found that that they suffered from many diseases and medical conditions, and he learnt that the Hunza diet was not basically vegetarian as many had understood. Most previous visitors to the secluded Hunza valley had gone in the summer (the only possible time) and had seen that freshly grown fruit and vegetables dominated the diet. The visitors did not see the dwindling food stores leading to starvation in the winter and to the reliance on the few animals that the Hunzas could spare for food. The lack of pastures meant that few animals could be kept.

John Boyd-Orr (1880–1971)

Boyd-Orr was an eminent agricultural scientist who worked for many years at the Rowett Research Institute in Aberdeen. He received the Nobel Peace Prize for his scientific research into nutrition and was the first Director-General of the United Nations Food and Agriculture Organization. He wrote *Food, Health and Income* (Boyd Orr, 1936) which drew attention to the fact that a large proportion of the British population did not eat enough and were thus deficient in many key nutrients. Boyd-Orr's work was instrumental in ensuring that the British government considered nutrition when designing its war-time civil food rationing programme and by so doing prevented malnutrition.

Boyd-Orr appreciated that diet could have a profound influence on health and physique and that much ill-health could be attributed directly to dietary inadequacies. He also suspected that the susceptibility to infectious diseases, such as tuberculosis and other pulmonary and intestinal disorders in young children could be associated with faulty diets. Boyd-Orr referred to the work of McCarrison but not to Howard. No link was made to the involvement of humus or mycorrhiza and health. He approved of man-made fertilizers.

The Medical Testament of the Cheshire doctors (1939)

Thirty-one members of the Cheshire Medical Panel Committee presented the Cheshire Medical Testament on human nutrition and its relation to agriculture at a meeting of 600 representatives consisting of the county's local health authority, farmers and the medical profession in spring 1939. The doctors were concerned that they were preoccupied with curing sickness and that too little was being done to prevent diseases. They were motivated by the increasing incidence in England of rickets, anaemia, bad teeth and constipation. Many dentists were involved so inevitably dental health was seen as a major concern.

The doctors extolled the virtues of natural living styles of the communities such as those described by Wrench and Price. The excellent dental health of the Tristan da Cunha islanders was noted, which indicated their familiarity to Wrench (1938 and 1946). As was the case with Price (1939), the Cheshire doctors did not consider the relevance of sugar in relation to dental caries. Nutritional and health assessments of the Tristan islanders were made again after they were brought to England following the 1961 volcanic eruption (Taylor *et al.*, 1966). Between 1932 and 1955 there had been a marked deterioration in dental health after their basic diet had been supplemented with flour, sugar, jam and small quantities of biscuits, rice, canned fruit and sweets. Their natural lifestyle had not saved their teeth and health.

Drinking water on Tristan da Cunha was found to contain a low concentration of fluorine (0.1 ppm), which had been sufficient to ensure strong teeth in the absence of sugar in the diet. The Islanders suffered from anaemia and associated roundworms but there was very little cardiovascular disease.

McCarrison's work (1927, 1936) was quoted at length without advocating a particular diet. The doctors would seem to have accepted Howard's (1943) contention that it was the way compost was made and used that was responsible for the health and productivity of crops, animals and man. The extrapolation of the observation on twelve oxen to human health was cemented in the Testament.

As a result the doctors were prepared to follow Howard and to speak of the evils of man-made fertilizers.

However, the doctors baulked at accepting Howard's suggestions about the sanitary disposal of municipal and village waste and of the planning of new rural housing estates to facilitate the collection and use of night soil from compost-style toilets.

McCarrison and Howard both supported the Cheshire Medical Testament. At the presentation McCarrison highlighted the four chief faults in the diets of the majority of the people of the UK, namely i) the use of denatured wheat flour, ii) excessive use of carbohydrate foods, iii) insufficient consumption of fresh green vegetables and iv) insufficient drinking of safe milk. He commented on the excessive consumption of meat and other animal foods by some in the community. Howard contended that maintaining soil fertility must be the cornerstone of public health in the future, and that agriculture must be given its place as the foundation of preventive medicine. Medical men as well as nutrition experts, he said, were realizing that a fertile and productive soil was an asset that should be carefully looked after in the nation's interest. His words were soon to be echoed by Balfour.

Other speakers at the presentation of the Cheshire Medical Testament included Picton, Wrench, Rodale and Yellowlees. They all went on to influence the organic movement. Picton (1942) used the Testament as background for his writings on wartime food; Rodale became the leader of the mainstream organic movement in the USA, and Yellowlees (1985) forty years later lectured on the organic farming in relation to health in the Scottish Highlands.

Huntingdon feeding experiment

From 1947 to 1965 the Soil Association ran the Haughley farm experiment (Chapter 6). In 1963 laboratory rats and guinea pigs at the Huntingdon Laboratories were fed, from weaning to maturity, diets based on the produce from the organic and mixed sections in the Haughley experiment. The object was to see whether there were any growth differences that might indicate nutritional benefits of organic produce. The work of McCarrison was the motivation for this study.

Balfour (1975) reported that the results of the feeding studies had been inconclusive because the animals were fed an unnatural diet. It was concluded it would be necessary to overcome what were called technical difficulties before starting an ambitious ten-generation experiment on rats and guinea pigs; there was no follow-up.

Two fistulated dairy heifers at Huntingdon were fed produce from the Haughley experiment with the object of comparing organic with mixed section feed on digestion, the production of fatty acids and on the bacterial fauna. At the time of Balfour (1975) it had not been possible to interpret the results.

Submission to UK Parliament

In 2001 the Soil Association made many all-embracing statements about the health benefits of organic food in a memorandum submitted to the Select Committee on Agriculture of the UK parliament (Soil Association, 2001). The memorandum bore many similarities to Balfour's *Living Soil* (1943). It relied heavily on the work of McCarrison and Howard, and referred to the claims that compost-grown food conferred resistance to diseases such as foot-and-mouth. Mycorrhizae were seen as the route by which nutrients from the decomposition of applied organic matter reached crop roots. Healthy crops possessed mycorrhizae, whereas unhealthy crops did not. Moreover, it was stated that the use of inorganic fertilizers and pesticides inhibited the microbiological life in the soil, including mycorrhizae. This latter view was not supported by the review of Geisseler and Scow (2014), who reported that mineral fertilizer application led to a 15% increase in the microbial biomass above levels in unfertilized control treatments; this was based on a meta-analysis comprising sixty-four long-term trials from around the world. While the addition of organic matter can be expected to promote the activity of soil microorganisms, the application of inorganic fertilizers will not have detrimental effects on cultivated soils. In unmanaged ecosystems fertilizer N inputs generally decrease, but not eliminate, soil microbial biomass (Geisseler and Scow, 2014).

The principles and practices of organic farming were developed from research and observations. The organic movement concluded that good health in crops, animals and humans depends on soil organic matter providing the basis for crop nutrition via soil life. At the same time they noted that the Haughley experiment (Balfour, 1943, 1977) had not been properly replicated or developed in modern times.

Nevertheless, many links were made including:

i.   Pesticide use with cancers, infertility, early puberty, degenerative diseases, lowered immunity and mental health.
ii.   Application of agro-chemicals with micro-nutrient deficiencies, disease susceptibility and cancer.
iii. Intensive housing of livestock with food poisoning and antibiotic-resistant bacteria.
iv.   Feeding of animal concentrates with *E.coli* infections, cancer, diabetes, and heart disease.

The Soil Association wanted the UK government to understand that there is a direct link between food production methods and human health, and that the solution lay in farming organically. It concluded its memorandum with a plea for investment in research on organic food and farming.

Howard's beliefs in the power of mycorrhizae continued to hold sway with the Soil Association; the idea is that crops given man-made fertilizers absorb nutrients by simple diffusion whereas in organic crops nutrients are directly supplied via soil mycorrhizae. The metabolically controlled absorption of nutrients and the possible benefits of mycorrhizae on phosphorus uptake were not considered.

Summary

1. Howard played a major role in what we now know were the early days of the organic movement through his writings and lectures on his experiences in India. To many he is the founder of organic farming.

2. Howard's disaffection with the scientific method meant that he relied entirely on observation and not experimentation. The many anecdote-based claims in his *Agricultural Testament* were and continue to be widely accepted without question. He had a strong personality, which probably accounts for his ideas not being challenged.

3. He believed that crops would not get diseases or be attacked by insects if they were grown using manure or his Indore compost. Likewise, he believed the health of livestock depended on them being fed produce from healthy crops grown with manure or compost. Howard was not concerned with deficiency diseases but with infectious animal diseases.

4. Howard's claims of resistance to foot-and-mouth disease and other animal diseases were based on observations on twelve well-fed oxen made before he had carried out his work on compost production.

5. It has wrongly been assumed by many commentators that the oxen's observed resistance to foot-and-mouth was due to them eating compost-grown produce.

6. Howard's major contribution was his pioneering work in improving the utilization of manure and vegetable wastes by composting – the Indore process.

7. Howard was one of the first to draw attention to mycorrhiza in roots, to which he attached great importance particularly in relation to reducing crop and animal diseases. He believed, as some still do, that the fungal threads conducted health-promoting factors as well as nutrients to plant roots from decaying organic matter in the soil.

8. Howard did not accept the experimental methods of agricultural science and he strongly disapproved of man-made fertilizers.

9. McCarrison made a big contribution to the understanding of the importance of a good diet in order to prevent both infectious and deficiency diseases. This he did by his experiments with rats, to which he was able to transfer the health-promoting diets of some of the tribes in Northern India.

10. McCarrison's use and emphasis on experiments in which he established the clear link between diet and health is to be contrasted with the observational approach of Howard.

11. Also in contrast to Howard, McCarrison made only passing comment on the way the food is grown and the need for fertile soils. He did not connect soil properties with health.

12. Wrench, a London doctor, was largely responsible for publicising the work of Howard and McCarrison and at the same time adding his own similar experiences in other countries. He was convinced that what he saw as superior health in the isolated communities was because they lived close to nature and recycled their wastes.

13. The apparently good health, especially dental health, of peoples reported by Wrench (and later by Price) was linked to the manner in which the crops were grown using compost exclusively without the help of inorganic fertilizers rather than the diet itself and the absence of sugar.

14. Price, an American dentist, having established a connection between general and dental health in the USA studied dental health in several isolated communities. As with Wrench, Price did not establish a soil/health link. The connection between sugar and caries was not considered. It was the native diets Price admired.

15. The Medical Testament of the Cheshire doctors brought the work of Howard, McCarrison and Wrench to a much wider public in the UK but there seems to have been little apparent scrutiny of the evidence.

16. A link between health of the Tristan da Cunha islanders and their natural, conserving lifestyle was made by the Cheshire doctors. It was revealed in the 1960s that there had been a marked deterioration in dental health after their basic diet had been supplemented with flour, sugar, jam, biscuits, rice, canned fruit and sweets. Their natural lifestyle had not saved them and their teeth.

17. The health-promoting claims made by Howard about compost were still, after almost 100 years, influencing the organic proponents in the UK in 2001. A submission to the UK Parliament included

reference to the prevention of foot-and-mouth in animals fed on compost-grown crops.

References

Balfour, E.B. (1943, revised 1949) The *Living Soil with Evidence of the Importance to Human Health of Soil Vitality, with Special Reference to National Planning.* Faber and Faber, London.

Balfour, E.B. (1975) *The Living Soil and the Haughley Experiment.* Faber and Faber, London.

Balfour, E.B. (1977) Towards a Sustainable Agriculture, The Living Soil. Address given to an IFOAM conference in Switzerland in 1977. Available at http://journeytoforever.org/farm_library/balfour_sustag.html

Baranski, M., S´rednicka-Tober, D., Volakakis, N., Seal, C., Sanderson, R., Stewart G.B., Benbrook, C., Biavati, B., Markellou, E., Giotis, C., Gromadzka-Ostrowska, J., Rembiałkowska, E., Skwarło-Son´ta, K., Tahvonen, R., Janovska, D., Niggli, U., Nicot, P. and Leifert, C. (2014) Higher antioxidant and lower cadmium concentrations and lower incidence of pesticide residues in organically grown crops: a systematic literature review and meta-analyses. British Journal of Nutrition 112, 5, 794–811. doi: 10.1017/S0007114514001366.

Boyd Orr, J. (1936) Food, Health and Income: a Report on a Survey of Adequacy of Diet in Relation to Income. Macmillan and Co Ltd, London. Available at www.fao.org/fileadmin/templates/library/docs/613_OR7.pdf

Cheshire Medical Testament (1939). A Report of the Speeches of McCarrison, Howard, at a Meeting in support of the Medical Testament of the Local Medical and Panel Committees of Chester. Available at www.journeytoforever.org/farm_library/medtest/medtest_speeches.html

Clark, J. (1956) *H u n z a, Lost Kingdom of the Himalayas.* Funk & Wagnalls. New York. Available at http://biblelife.org/Hunza%20-%20Lost%20Kingdom%20of%20the%20Himalayas.pdf and http://biblelife.org/hunza.htm

Conford, P. (2001) *The Origins of the Organic Movement.* Floris Books.

Fauchard, P. (1728) Le chirurgien dentist: The Surgeon Dentist.

Howard, A. and Wad, Y.D. (1931) *The Waste Products of Agriculture – Their Utilization as Humus.* Oxford University Press. Available at www.journeytoforever.org/farm_library/HowardWPA/WPAtoc.html

Howard, A. (1940) *An Agricultural Testament, Appendix C The Manufacture of Humus from the Wastes of the Town and the Village.* Oxford University Press. Available at www.journeytoforever.org/farm_library/howardAT/ATapp3.html

Howard, A. (1943) *An Agricultural Testament* (First published in 1940). Oxford Univ. Press, New York and London. www.zetatalk3.com/docs/Agriculture/An_Agricultural_Testament_1943.pdf

Geisseler, D. and Scow, K.M. (2014) Long-term effects of mineral fertilizers on soil microorganisms – A review. *Soil Biology and Biochemistry* 75, 54–63. doi: 10.1016/j.soilbio.2014.03.023.

IFOAM (2006) Organic Agriculture & Human Health. Available at http://infohub.ifoam.bio/sites/default/files/page/files/oa_human_health_en.pdf

McCarrison, R. (1927) *Medical Testament. Diseases of Faulty Nutrition* Transactions of the Far Eastern Association of Tropical Medicine. Available at www.journeytoforever.org/farm_library/medtest/medtest_mccarrison3.html

McCarrison, R (1936) *Nutrition and National Health*. The Cantor Lectures The Royal Society of Arts 1936, published 1953. Faber and Faber, London. Available at https://www.seleneriverpress.com/historical/nutrition-and-national-health-the-cantor-lectures

Picton, L.J. (1942) Concerning the Cheshire Medical Testament. From *Feeding the Family in War-time, Based on the New Knowledge of Nutrition* by Grant, D. Harrap, London. Available at www.journeytoforever.org/farm_library/medtest/medtest_picton.html

Price, W.A. (1939) *Nutrition and Physical Degeneration – A Comparison of Primitive and Modern Diets and Their Effects*. Paul B. Hoeber, Harper & Brothers, New York London. Available at https://journeytoforever.org/farm_library/price/pricetoc.html

Rayner, M.C. (1927). *Mycorrhiza*. Cambridge University Press doi: 10.1111/j.1469-8137.1927.tb06704.x.

Soil Association (2001) Memorandum submitted by the Soil Association (R 10) Available at www.publications.parliament.uk/pa/cm200001/cmselect/cmagric/388/388m10.htm

Swanson, N.L., Leu, A., Abrahamson, J. and Wallet, B. (2014) Genetically engineered crops, glyphosate and the deterioration of health in the United States of America *Journal of Organic Systems*, 9, 2, 6–37. Available at www.organic-systems.org/journal/92/JOS_Volume-9_Number-2_Nov_2014-nson-et-al.pdf

Sykes, Friend (1946) *Humus and the Farmer*, Faber and Faber, London.

Taylor, E.C., Hollingsworth, D.F. and Chambers, M.A. (1966). The diet of the Tristan da Cunha Islanders. *British Journal of Nutrition* 20, 3, 393–411. doi: https://doi.org/10.1079/BJN19660042

Wrench. G.T. (1938) *Wheel of Health. A Study of the Hunza People and Keys to Health*. Daniel Company Ltd, London.

Wrench, G.T. (1946) *Reconstruction by Way of the Soil*. Faber and Faber, London.

Yellowlees, W.W. (1985) *Food & Health in the Scottish Highlands, Four Lectures from a Rural Practice*. Clunie Press Perthshire. Available at https://journeytoforever.org/farm_library/medtest/medtest_yelsl.html

CHAPTER 10

Organic Farming and Health – Studies and Concerns

"Let food be thy medicine and medicine thy food" Hippocrates

The notion that organic food is healthier than conventional food is widely believed by organic devotees. The organic movements' belief that organic food is healthier is not based on any human studies but largely on comparisons of the composition of organic and conventional crops. For example, the finding of potentially beneficial higher concentrations of antioxidants in organic than conventional crops is often quoted (see Chapter 14). These compounds may well have health benefits with regard to cancer and cardio-vascular disease and are thus worthy of attention, but supporting evidence is required. The benefits of consuming fruit and vegetables are often attributed to their antioxidant and secondary metabolite content.

In vitro studies with extracts of organic and conventionally produced strawberries on cancer cell (breast and colon) proliferation showed promising results (Olsson *et al.*, 2006), which was largely attributed to vitamin C acting together with other secondary metabolites. Falcarinol, a natural pesticide that accumulates in carrots and related plants in response to mould attack, was found in an *in vivo* experiment to inhibit the development of cancer tumours in rat colons (Brandt *et al.*, 2004).

Human health studies

Cancer: the million women study

A recent study (known as the "the million women study", started 1996–2001) in the UK involving 623,080 women aged 50 and over (Bradbury *et al.*, 2014) found that women who eat organic food grown without pesticides (and possibly of different composition) are no less likely to develop cancer than women who eat a conventional diet. The development of sixteen of the most common types of cancer was tracked over a nine-year period. A total of 30% of the women never ate organic food, 63% sometimes and 7% usually/always ate organic food.

The huge study co-ordinated by scientists from Oxford University in co-operation with Cancer Research UK is extremely robust because of its size; its findings strongly rebut the claims that organic food, and by implication the absence of pesticides, can reduce the risk of cancer. No difference in overall cancer risk was found when comparing the 180,000 women who reported never eating organically grown food with around 45,000 women who said they usually or always eat organic food. There were 53,769 cases of cancer in total. There was no suggestion of a connection between cancer and any unidentified health-promoting aspect of organic farming. A small (under 10%) significant increased risk for breast cancer in the organic eaters was found.

However, there was a 21% reduction in the risk of non–Hodgkin lymphoma in women who mostly ate organic food (Bradbury *et al.*, 2014). The analysis of individual cancers involved small numbers (under 150 cases – incidence 0.29%) of non–Hodgkin lymphoma in organic food eaters, compared with 0.38% in those who never ate organic food). It was concluded that the results could be down to chance. Not unexpectedly, organic supporters were not pleased that the decrease in non–Hodgkin lymphoma was being so readily dismissed. Occupational exposure to pesticides, including phenoxy herbicides, carbamate insecticides, organophosphorus insecticides and lindane, an organochlorine insecticide, was judged to be positively associated with the risk of non–Hodgkin lymphoma in a meta-analysis by Schinasi and Leon (2014). As previous studies have suggested a link between pesticide use and non–Hodgkin lymphoma, this is an aspect worth investigating further.

In the absence of other epidemiological studies that indicate a contrary conclusion it is clear that claims of an association between eating organic food and a reduction in cancer incidence can currently be dismissed. No studies are planned on the relationship between organic diets and other health attributes using the data from the "million women study".

These findings are particularly relevant as health concerns have frequently been identified as a key factor for choosing organic food. Although there is no evidence of a causal relationship between the finding of greater quantities of antioxidants such as vitamin C and phenolic compounds in organic food, the link is made to their possible cancer-reducing properties, e.g. Baranski *et al.* (2014). In this context it needs to be recalled that different fruit and vegetable varieties vary enormously in their vitamin and mineral content, for example the vitamin C content of apple varieties can differ by a factor of three (Trewavas, 2005). These differences are much greater than the trivial variations that are sometimes found in organic crops (see Chapter 14).

The finding, in the million women study, that low to moderate alcohol consumption was associated with an increased risk of a variety of types of cancer in women, including breast cancer, is to be contrasted with the absence of associations with organic food.

Pre-eclampsia

A potentially very significant recent finding from an epidemiological study in Norway concerns the probable beneficial involvement of the consumption of organic vegetables, but not fruit, on the condition pre-eclampsia (the "MoBa" study). Pre-eclampsia is a complication of late pregnancy characterized by high blood pressure and protein in the urine; cases can be severe, threatening both the mother and the foetus and leading to premature birth. The cause of the condition is not known, although it has been linked to a variety of risk factors including some of the risk factors for cardiovascular disease.

Pregnant women may be able to lower their risk of pre-eclampsia by 21% simply by eating more organic vegetables, according to the study conducted by researchers from the National Institute for Consumer Research and the Norwegian Institute of Public Health (Torjusen *et al.*, 2014).

The study, which is part of a project investigating whether organic food consumed during pregnancy provides health benefits to the mother and child, used data from the Norwegian Mother and Child Cohort Study. Between 2002 and 2008, the researchers followed 28,192 participants who were pregnant with their first child. Over the course of their pregnancies 5.3% of participants developed pre-eclampsia, which is in line with expectation.

Food consumption was monitored by questionnaires in the middle of their pregnancies. Overall, 40% of women sometimes ate at least one organic food, 7% often did so, and 2% mostly did so.

The researchers found that women who often or mostly ate organic vegetables were 21% less likely to develop pre-eclampsia than women who rarely or never did so. The incidence of pre-eclampsia in frequent consumers of organic vegetables was 4.2% (eighty-one cases in 1,951 women) compared to 5.4% (1,410/26,241) in those who consumed little or none. No connection was found between pre-eclampsia rates and consumption of organic foods other than vegetables. The absence of a connection with fruit, which are often sprayed with pesticides, is puzzling and suggests that pesticides as a group are unlikely to be involved. It was postulated that the finding of no beneficial effect or association with fruit is due to the fact that the fruit peel is often discarded, resulting in a reduced exposure to microbes compared with eating raw vegetables (Torjusen *et al.*, 2014).

The researchers found that women who ate organic food more often were likely to have healthier lifestyles and diets, including a higher consumption of vegetables, fruits, berries, whole grains and a lower consumption of meat, processed meat, white bread, cakes and sweets. They tended to be younger and have a lower body mass index than women who ate organic foods less frequently. They were slightly less likely to smoke and more likely to have reached a higher level of education. The effect of organic vegetable consumption on pre-eclampsia risk was independent of these confounding factors and socioeconomic and lifestyle differences. It was also independent of other pre-eclampsia risk factors, including gestational weight gain.

While eating lots of vegetables is advisable for all pregnant women, the new study suggests that consuming organic vegetables may provide small additional benefits.

The workers suggest that the association between pre-eclampsia and the consumption of organic vegetables could be that organic vegetables may alter the exposure not only to pesticides but also to secondary plant metabolites and/or

the influence of the composition of the gut microbiota. The connection with gut microbiota is interesting in view of the increasing attention being given to the effects of the microbes in the gut on various aspects of the human condition (Bosscher *et al.*, 2009). In the current context the possibility exists that the mix of gut microbes may be altered by some component in organic vegetables that then affects the immune system. It is understood that gut microbial ecology can affect not only the local gut immune system but also be implicated in immune-related disorders, such as type 2 diabetes and obesity (Cerf-Bensussan *et al.*, 2010).

Children, milk, eczema and omega-3 fatty acids

Floistrup *et al.* (2006) and Alfven *et al.* (2006) reported that eczema, allergies and asthma were generally, but not consistently, lower in children from anthroposophic families (supporters of biodynamic agriculture, see Chapter 5) than in children from other families; lifestyle factors associated with anthroposophy include reduced use of antibiotics, a longer period of breast feeding, less use of aspirin as well as consumption of organic food. Data (the PARSIFAL study) was collected from five European countries on 6,630 children aged 5 to 13 years, 4,606 from Steiner schools (anthroposophic) and 2,024 from reference schools. It was concluded that certain features of the anthroposophic lifestyle, such as restrictive use of antibiotics and antipyretics, are associated with a reduced risk of allergic diseases in children. It has been postulated that the development of a population of more beneficial intestinal microflora compared with children not living in families with anthroposophical lifestyles was responsible – a feature also suggested with respect to the risk of pre-eclampsia.

Consumption of organic dairy products by mothers (as measured at thirty-four weeks of gestation) and by infants (over 2 years after birth) was associated in the Dutch KOALA Birth Cohort Study with a slightly lower, but not statistically significant, eczema risk (Kummeling *et al.*, 2008). 2,306 infants were fed conventional food, 283 a moderately organic diet and 175 a strictly organic diet (biodynamic food); the percentages of infants suffering eczema were 32, 33 and 29% respectively. The influence of the mothers' diet could not be distinguished from that of the infants'. It is possible that the lower risk of eczema in children who used organic dairy products was actually due to the high consumption of organic dairy products by the mother, conferring protection started in the uterus and continued during lactation. Mothers with a high proportion of organic dairy intake had been shown in the KOALA study to have higher concentrations of rumenic and transvaccenic acids in their breast milk (Rist *et al.*, 2007). The risk of eczema and of allergic sensitization is lower in children with mothers with high vaccenic and rumenic acid in their breast milk (Thijs, 2011). No link was found between the consumption of organic meat, fruit, vegetables or eggs, and the proportion of organic products in the total diet with the development of eczema.

Allergic diseases during infancy were investigated in a prospective study on 330 children in Sweden from families with an anthroposophic lifestyle (who also preferred organic food) or from conventional families. Children from anthroposophic families had a significant 75% reduction in risk of sensitization during the first two years of life compared to children from conventional families (Stenius *et al.*, 2011). As with the PARSIFAL study, it was not possible to identify whether it was organic food or other lifestyle factors that accounted for the association.

In response to the indication that organic milk is more nutritious the Food Standards Agency (FSA, 2012) commented that while the nutrient profile of organic milk appears to be different from non-organic milk, care must be taken when drawing conclusions as to the nutritional significance of such differences. Dairy sources of omega-3 polyunsaturated fatty acids are not a viable alternative to eating oily fish. Milk contains the shorter chain forms of omega-3 polyunsaturated fatty acids (alpha-linolenic acid), while the forms present in oily fish are the long chain fatty acids eicosapentaenoic (EPA) and docosahexaenoic acids (DHA). Research has shown that the short chain forms found in plant and dairy sources do not appear to be as beneficial as those found in oily fish, which are known to be protective for cardiovascular disease, and which also may have beneficial effects on foetal development. Although the shorter forms can be metabolized to the longer forms, the conversion appears to be very limited in humans. It is estimated that it would be necessary to drink very large amounts of organic whole milk daily to obtain the benefits of consuming one portion of oily fish (FSA, 2012). Rosen (2010) calculated that it would be necessary to drink 25 litres of conventional milk every day in order to get the benefits of the weekly intake of 56–85 g of salmon and that converting to organic milk could reduce the daily requirement to 15 litres.

Turchini *et al.* (2012) pointed out that the publicity given to short chain omega-3 fatty acids in milk can mislead the public and have a negative effect on the ability of consumers to choose a more healthy diet containing long chain omega-3 fatty acids in oily fish.

USA Stanford University meta-analysis

A meta-analysis at Stanford University's Center for Health Policy (Smith-Spangler *et al.*, 2012) mainly involved comparisons (237) of the composition of organic and conventional foods (see Chapter 14) but it also included seventeen studies on humans. Only three studies involved the clinical outcomes of the consumption of organic food on humans; the remaining fourteen examined health markers (for example, serum lipid or vitamin levels).

No significant differences were found in trials lasting up to two years between those consuming organic with those eating conventional food when assessed by allergic outcomes (eczema, wheeze and atopic sensitization). The consumption of organic meat in the winter was a risk factor for *Campylobacter* infection. The risk of *Escherichia coli* contamination did not differ between organic and conventional produce. Bacterial contamination of retail chicken and pork was common but unrelated to farming method. However, the risk of isolating bacteria resistant to three or more antibiotics was higher in conventional than in organic meat. Two studies reported significantly lower urinary pesticide levels among children consuming organic versus conventional diets, but studies of biomarker and nutrient levels in serum, urine, breast milk, and semen in adults did not identify clinically meaningful differences. There were no long-term studies of health outcomes on people consuming organic versus conventionally produced food.

Similar findings were reported by the American paediatricians Forman and Silverstein (2012). They reviewed the advantages and disadvantages of organic food and concluded that while there is evidence that organic food is likely to contain fewer pesticide residues and can differ in composition from conventional food, there is no direct evidence of any meaningful nutritional benefits on children eating organic foods compared with those who eat conventionally grown food products. They found no support for claims of benefits of organic food to human health. At the same time they found no evidence of any detrimental or disease-promoting effects of an organic diet.

European review of health implications of organic food

A multi-author, wide-ranging review was produced, primarily for the European Parliament at the request of the Science and Technology Options Assessment Panel of the Directorate-General for Parliamentary Research Services, entitled *Human health implications of organic food and organic agriculture* EPRS (2016). Five policy options were developed, including no action, support organic agriculture by investing in research, development, innovation and implementation and support of the business environment of organic agriculture through fiscal instruments.

The overall conclusion was that there was very little evidence to support the notion that organic food is healthier than conventional. There were said to be indications that organic food may reduce the risk of allergic diseases and obesity but the evidence was not conclusive; this was the PARSIFAL study. Likewise, the evidence from animal experiments that suggested that identical feed from organic and conventional production has different impacts on early development and physiology was unclear. The negative effects of exposure during pregnancy to organophosphate insecticides

on children's cognitive development was highlighted and by implication extended to other pesticides. It is regrettable that phytoalexins were not mentioned, which meant that their existence and relevance with regard to unprotected organic crops as well as the amounts of potentially toxic compounds consumed were not discussed. The higher content of omega-3 fatty acids in organic milk, and probably also meat, compared to conventional products was not thought to be nutritionally significant in light of other dietary sources. The relevance of the less beneficial short chain omega-3 fatty acids in milk was not considered.

The tendency for lower cadmium contents of organic food than in conventional food was seen as being an issue that is very relevant to human health because the current exposure in Europe to cadmium is reported to be close to, and in some cases above, tolerable limits. It was concluded that further investigation was required. The variation in cadmium in the diet and in rock phosphate is discussed in Chapter 14.

The impression was given, in view of the absence of clear data supporting organic farming, that the main objective of the European survey was to substantiate the funding by the EU of organic research and the promotion of organic farming in Europe.

Palaeolithic diet studies

Palaeolithic diets may be thought of as being natural diets and as such are of interest. The Louis Bolk Institute in cooperation with the Universities of Wageningen, Groningen and Girona (Spain) carried out a two-week nutritional intervention study with people exhibiting at least two of the risk factors associated with cardiovascular disease and type 2 diabetes, namely abdominal obesity, high blood pressure, high glucose and abnormal lipids levels in the blood (Boers *et al.*, 2014). Thirty-two subjects (aged 45–60 years) were fed either a Palaeolithic-type diet or a healthy diet providing the same number of calories. The Palaeolithic-type diet consisted of lean meat, fish, fruit, leafy and cruciferous vegetables, root vegetables, eggs and nuts. Dairy products, cereal grains, legumes, refined fats, extra salt and sugar were excluded. There were a few similarities with the diet of the Hunzas.

The main goal was to study whether a Palaeolithic-type diet alters the risk factor values independently of weight loss. The Palaeolithic-type diet was beneficial in that it lowered the systolic blood pressure (by 9.1mm Hg), the diastolic blood pressure (by 5.2mm Hg), total cholesterol (by 0.52 mmol/l) and triglycerides (by 0.89 mmol/l) and increased HDL-cholesterol (by 0.15 mmol/l). The diet also had a positive effect on insulin resistance.

Human health concerns

Nitrates and health

Nitrogen compounds have gained bad reputations as pollutants and because of their possible involvement with climate change. Those who object to the use of man-made nitrogen fertilizers have capitalized on the bad reputation of nitrogen compounds. The finding of lower concentrations of nitrates in organic food is perceived as a plus point. Despite the pivotal role of nitrogen in crop growth and in sustaining human life there is a growing literature on the nitrogen problem, e.g. Sutton, M.A. *et al.* (2011). The occurrence of nitrogen compounds in the diet and their effects on health are discussed here.

Nitrates were the first nitrogen compounds to be considered harmful to humans, being implicated in two health aspects – initially in the 1940s (in USA) with the blue baby syndrome (methaemoglobinaemia) and then in 1956 (in UK) following the discovery of the carcinogenicity of nitrosamines and their possible generation in the gut from nitrates. Online searches for nitrates and health will bring up warnings about avoiding nitrates and raise the spectre of links between nitrates and leukaemia, ovarian, colon, rectal, bladder, pancreatic and thyroid cancer but without substantiating evidence. Two book-length reviews of the nitrogen issue (L'hirondel and L'hirondel 2001 and Addiscott, 2005) discussed the limited evidence to support such claims about dietary nitrate.

Methaemoglobinaemia

Nitrites can react with haemoglobin to form methaemoglobin and so reduce oxygen transport in the blood. Methaemoglobinaemia is of no known consequence in adults but can be fatal in babies, particularly in those under 3 months whose gut is not sufficiently acidic to prevent colonization with bacteria that reduce nitrate to nitrite. Nitrites also occur in some foods. The presence of nitrates in water used to make baby feeds was suspected as being the cause of the blue baby syndrome but it was eventually shown that bacterial contamination of the water plays a critical part (Addiscott, 2005).

Despite the fact that cases of methaemoglobinaemia were very rare – the last case in the UK was in 1972 – the EU Commission issued directives governing nitrates in drinking water and in vegetables, particularly spinach.

Cancer

Nitrates and nitrites, used for curing meat for many years, were perceived as being potentially carcinogenic if nitrosamine compounds were formed in the gut. In the 1960s it was found that the provision of sodium nitrite resulted in the formation of dimethylnitrosamine in the gut that could theoretically cause liver tumours in rats (*Lancet*, 1968) However, chronic feeding of nitrite to rats did not induce tumours. The amounts of nitrates and nitrites that can be used in meat curing are governed by regulations in many countries.

After it had been found that dietary nitrate could be reduced to nitrite in saliva the spotlight fell on nitrate as posing a potential cancer risk (Spiegelhalder *et al.*, 1976).

N-nitroso compounds have repeatedly been shown to be carcinogenic in animals. However, reviews such as that by the FAO/WHO Expert Committee on Food Additives (JEFCA, 2002) of the epidemiological and toxicological studies have failed to establish a definite link between nitrate intake and risk of humans developing cancer. There is no suggestion that consuming large amounts of nitrates from leafy vegetables is associated with an increased risk of cancer. In fact, epidemiological studies have shown that fruit and vegetables can protect against certain cancers (Steinmetz and Potter, 1996) (see Appendix for Chapter 14).

Hypertension and cardiovascular disease

Trials (e.g. Appel *et al.*, 1997) that showed diets rich in fruit and vegetables can reduce blood pressure reinforced the recommendations for the consumption of more fruit and vegetables with regard to cardiovascular health. It has been suggested that antioxidants in the diet are involved. However, a detailed review (Hollman *et al.*, 2011) failed to link the anti-oxidant effects of polyphenols with cardiovascular health. Moreover, concentrations of polyphenols in blood are low compared with other antioxidants due to them being metabolized following ingestion.

So the search was on to find a common factor in fruit and vegetables that accounted for their ability to lower blood pressure and benefit cardiovascular health. Several workers have drawn attention to the potential benefits of nitrates on human health via conversion to nitrites, including McKnight *et al.* (1999), Benjamin (2000 and 2008) and Gladwin *et al.* (2006). Lundberg *et al.* (2011) in Sweden reviewed the now extensive literature, commenting that the amounts of nitrate and nitrite needed for the beneficial effects on the cardiovascular system are readily achieved via our everyday diet.

Nitric oxide (NO) and nitrites are known to be active in several ways linked to cardiovascular conditions, for example by relaxing blood vessels thereby reducing blood pressure and by reducing the tendency for platelets to come together and form clots.

Webb *et al.* (2008), starting from the hypothesis that nitrates in vegetables are the source of protective nitrites and nitric oxide in humans, demonstrated a pathway for the effect. They showed that the beneficial effects of nitrates, supplied in beetroot juice (500 ml) containing nitrate at around 700 mg/l), in reducing blood pressure lasted for up to twenty-four hours. The mechanism appeared to be firstly the reduction of nitrate to nitrite (by bacteria present on the tongue after absorption in the gut and mouth) followed by the reduction of the nitrites to nitric oxide in the stomach. It now seems likely that it is the action of nitrites/NO as vasodilators and as inhibitors of platelet aggregation that are responsible for the beneficial effects of nitrates. The workers suggest that high dietary nitrate intakes

can play an important role in delaying the development of and in the treatment of hypertension. Long-term clinical trials are ideally required as Lidder and Webb (2013) conclude in their detailed review. However, there seems little doubt that a diet rich in leafy vegetables would assist in maintaining healthy vascular systems and that vegetables should take pride of place ahead of the more acceptable fruit. The importance of vegetables for cardiovascular health is discussed in Chapter 14 and the Appendix. The possibility of adverse effects of low blood pressure will need to be considered.

Studies with overweight subjects confirmed the blood pressure reducing ability of nitrate-rich beetroot juice but at the same time showed that regular consumption was required to maintain the effect (Jajja *et al.*, 2014). The relatively high nitrate content of green leafy vegetables, especially spinach, lettuce and rocket, would indicate that they are likely to be the most powerful protectors.

Toxicology of nitrates

The toxicology of nitrates needs to be considered in view of the continuing impression that nitrates are harmful.

Nitrates are rapidly absorbed in the upper gastrointestinal tract in humans with maximum concentrations in blood being achieved after about ninety minutes. Removal from the body is also rapid, with half being excreted in about five to seven hours.

The WHO (1995a) considers that the

oral lethal dose of nitrate in adult humans is probably around 20 g nitrate, which equates with sixty portions (of 80g) of the most nitrate-rich spinach.

There was no evidence that nitrates are carcinogenic in mice or rats when administered in the feed or drinking water (Grosse *et al.*, 2006). The European Food Safety Authority Contamination Panel agreed with these conclusions, noting that available evidence does not support the contention that nitrate intake from drinking water or diet is associated with increased cancer risk in humans (EFSA, 2008). Nitrates also get a clean bill of health with regard to reproductive and developmental toxicity.

Dietary nitrates and nitrites

About 85% of dietary nitrate is derived from vegetables that vary very widely in their nitrate content (Table 10.1), with most of the remainder from drinking water (Knight *et al.*, 1987). Fruit contribute very little to dietary nitrate.

The virtues of the relatively low nitrate contents of organic crops have been praised but the fact that several factors can influence nitrate contents makes it unlikely that a particular farming method will have an overriding and regular effect. Nitrogen fertilization and light intensity at harvest were identified as the major factors

Table 10.1 Nitrate concentrations and contents of 80 g portions of vegetables and fruit. Vegetable data derived and calculated from Lidder and Webb (2013) and fruit data from Food standards, Australia, New Zealand (2010)

	NO_3 mg/kg	mg NO_3 in 80g portion
Vegetables		
Spinach	965–4259	77–340
Lettuce	970–2782	78–220
Radish	1060–2600	85–208
Beetroot	644–1800	51–145
Chinese cabbage	1040–1859	83–150
Cabbage	333–725	27–58
Carrot	121–316	10–25
Potato	81–713	6–57
Onion	23–235	2–20
Tomato	27–170	2–14
Fruit		
Strawberries	70–170	5–14
Banana	45–140	4–11
Apple	<5–40	<1–3
Grapes	<5–38	<1–3
Orange	<4–15	<1–1
Mango	<5–9	<1–1
Pineapple	<5–9	<1–1

influencing nitrate levels in vegetables (Cantliffe, 1973); a similar association was found to be critical in determining nitrate levels in spinach by Schuphan *et al.* (1967).

A further illustration of the variability in results of studies comparing conventional with organic farming was evident from the Californian study by Muramoto (1999) on spinach and lettuce, two nitrate accumulators. It was found that the nitrate contents of conventionally grown spinach exceeded those of organic crops, whereas for lettuce the contents were similar.

The notion and assumption that nitrates are harmful has been perpetuated by organic supporters referring to the formation of potentially carcinogenic nitrosamines and the finding that the nitrate content of organically grown crops can be significantly lower than in conventionally grown products.

The meta–study (Baranski *et al.*, 2014) extolling the virtues of organic farming on the health-promoting chemicals in crops, commented on the 30% lower nitrate and the 87% lower nitrite contents in organic crops than in their conventionally grown counterparts without identifying the crops. Regardless of whether nitrate and nitrite intakes are physiologically relevant, it is clear that the variation between organic and conventional foods are very small compared with the very wide variation in nitrate contents of different vegetables (Table 10.1). Stockdale (2008) reached the same conclusion. Reducing the dietary intake of nitrate can be more easily achieved by avoiding leafy vegetables rather than by consuming organic vegetables.

Acceptable daily intake of nitrate

The Acceptable Daily Intake (ADI) set by the European Food Safety Authority for nitrate is 3.7 mg/kg equating to 260 mg/day for a 70 kg adult. The World Health Organization (WHO) first set an upper limit for nitrate in food in 1962 derived from studies showing that daily intakes of about 500 mg of sodium nitrate per kg body weight had been found to be completely harmless to rats and dogs in two-year studies. This value was divided by 100 to give an ADI for humans of 5 mg sodium nitrate/kg (3.7 mg nitrate/kg), a figure that was used by the Joint FAO/WHO Expert Committee on Food Additives (JECFA, 2002).

Consumption of a few vegetable portions (Table 10.1) can considerably exceed the ADI as do the amounts of nitrates shown to be beneficial and those in the DASH diets (US diets rich in nitrates designed to reduce blood pressure). Hord (2011) is not alone in calling for a review of the ADI and the regulations on nitrates in the diet, e.g. Powlson *et al.* (2008).

Nitrites

Nitrite concentrations generally tend to be very low in fresh undamaged vegetables, e.g. a maximum of 36 mg/kg NO_2 in spinach and 2 mg/kg in lettuce. However, levels can increase rapidly if the vegetables are pureed and stored at room temperature (Food standards, Australia, New Zealand, 2010).

EU and nitrates

Despite the wide acceptance of the health benefits of eating vegetables, the lack of evidence of harmful effects of dietary nitrates and the finding that nitrates can reduce the risk of cardiovascular disease, the European Commission requested a scientific opinion from the Panel on Contaminants in the Food Chain on nitrates in vegetables. A massive study followed involving nearly 42,000 nitrate analyses in ninety-two vegetable varieties from twenty member states. The nineteen scientists who wrote the EFSA report (2008) appeared somewhat reluctant to support the notion of possible benefits of nitrate intakes in vegetables when they concluded that: "Despite being a major source of nitrate, increased consumption of vegetables is widely recommended because of their generally agreed beneficial effects for health," and: "Overall, the estimated exposures to nitrate from vegetables are unlikely to result in appreciable health risks" (reprinted by kind permission of EFSA).

The panel also concluded that there was no need to consider revising the ADI for nitrate because its estimates of daily consumption of nitrates did not exceed it. This was not the conclusion of Hord *et al.* (2009), who reported that nitrate intakes in normal daily intakes of foods such as spinach and leafy vegetables often exceed the ADI. Although aware of the vasodilating ability of nitrate

metabolites, the EFSA (2008) report did not consider this aspect when assessing the relevance of current nitrate intakes.

Nitrates are still seen as a threat to European water quality and thence on health by Grizzetti, writing in the detailed appraisal of the nitrogen problem (Sutton *et al.*, 2011).

Nitrate and nitrite in drinking water

The World Health Organization drinking water guideline levels are 50 mg/l for nitrate (as NO_3) and 3 mg/l for nitrite (as NO_2). The EC 1980 Drinking Water Directive (80/778/EEC) set the same limit for nitrate in potable water; in the USA the limit is 44 mg/l. The guideline values were established to protect young infants from methaemoglobinaemia. The guideline also indicates that water with a nitrate concentration of up to 100 mg/l nitrate is safe for adults and children over 3 months. The WHO has also set a provisional guideline level for nitrite in drinking water of 0.2 mg/l for long-term exposure (WHO, 2008).

Nitrite is rapidly oxidized to nitrate in water and is rarely detected in well-oxygenated or chlorinated water (NHMRC, 2004). L'hirondel *et al.* (2006) concluded that more than enough evidence had been gathered to confidently say that nitrates are not the threat they were once thought to be and that raising the drinking water limit for nitrates would not lead to any known health risks.

It has become increasingly difficult to justify the very low conservative concentrations of nitrates in drinking water and there have been calls for further research (Ward *et al.*, 2005) as well as calls for the regulatory bodies to consider all available data on the beneficial physiological roles of nitrate and nitrite. The heavy hand of bureaucracy weighed down by the precautionary principle and the reluctance to accept evidence makes reconsideration of nitrate limits unlikely. Costs will continue to be incurred for the monitoring of nitrate vulnerable zones and for water purification. The fact that drinking water only contributes about 15% nitrate intake means that it is only part of the so-called nitrate problem.

High nitrate levels can readily occur in well water in subsistence farming due to improper handling of human and animal waste (Smil, 2001).

Whether the amounts of nitrates leached to drainage and to drinking water under organic and conventional farming will differ will depend on the quantity, timing and method of application of manures and fertilizers. Clearly if less nitrogen is applied or available after mineralization the amounts of nitrate leached will be less. Conventional farming with fertilizers provides an efficient way of applying nitrogen at the preferred time and in a manner designed to maximize utilization and thereby minimize nitrate leaching. In contrast the time required for the mineralization of nitrates from manures and the decaying remains of fertility-building legume crops mean that there is more time for nitrate leaching, especially at times when the crop requirement is minimal.

Mycotoxins

Mycotoxins are a group of naturally occurring toxic chemicals produced by certain fungal moulds that are known to grow on several crops and foods including cereals, nuts, spices, dried fruits, apple juice and coffee, especially under warm and humid conditions but also in cold climates on such foodstuffs as stored cereals and nuts. The FAO estimates that about a quarter of the world's agricultural produce is contaminated with mycotoxins (Wu, 2007) and, in the last ten years, mycotoxins have accounted for 30–60% of food and feed rejections at European Union borders.

The mycotoxins of most concern from a food safety perspective include the aflatoxins, ochratoxin A, patulin and toxins produced by *Fusarium* moulds, including fumonisins, trichothecenes and zearalenone. In some cases the toxins can be fatal. Mycotoxins can cause a variety of adverse health effects in humans. Aflatoxins are the most toxic and have been shown to be genotoxic and able to cause cancer in animal species. It is possible that they may cause liver cancer in humans. As with pesticide residues a tolerable daily intake for most mycotoxins has been established, being the quantity that someone can be exposed to daily over a lifetime without it posing a significant risk to health.

Mycotoxins can be prevented by the prophylactic use of fungicides so in theory they are likely to occur more frequently on organic foods. However, at present there is no conclusive evidence to indicate that organic food is generally more prone to contamination by mycotoxins than conventional food. Olsen (2008) reviewed several studies and were unable to identify consistent significant differences in the mycotoxin levels in organic and conventional food products. Where differences were found the estimated intakes were not a cause for concern.

The European Food Safety Authority considered the available data from field studies on the influence of pesticides on mycotoxin production (EUFIC, 2013 and EFSA, 2014); it concluded there was insufficient evidence to confirm that fungicides play a prominent and consistent role in limiting or preventing the production of mycotoxins.

In Canada, Roscoe *et al.* (2008) found low levels of multiple mycotoxins in many breakfast cereals, including deoxynivalenol (40% of samples), fumonisins and ochratoxin A (each in more than 30% of samples). The regular occurrence of mycotoxins in breakfast cereals in Canada may be related to the common practice of leaving wheat swathes on the ground for a long time after harvest: there was no suggestion of a link to organic farming.

Gourama (2015), in a study of twenty-two fungi on a wide range of foods including peanuts, rice, corn, soybeans and cashew nuts, found no significant differences in the levels of fungal contamination between organic and conventional samples.

Antibiotics

Antibiotics are seldom used by organic farmers and usually only when approved by a veterinarian. There is continuing concern about the growing numbers of antibiotic-resistant bacteria that pose a serious global threat to public health. There is little doubt that restricting the use of antibiotics in agriculture is desirable. The EU banned the use of antibiotics as growth promoters in 2006. In the USA antibiotics are still used in agriculture so it is possible for resistant bacteria to enter the food chain from the use of manure from antibiotic-treated animals.

Udikovic-Kolic *et al.* (2014) in the USA surprisingly found that the application of dairy cow manure enhanced the proliferation of resident antibiotic-resistant bacteria in soil even though the cows from which the manure was derived had not been treated with antibiotics. Although there is a theoretical risk of the build-up of resistant bacteria in the soil there is no suggestion that the contamination of vegetables with manure could pose a risk for human health.

Diseases

Australia

Consumption by children of unpasteurized organic bath milk was linked to *E. coli* infection in 2014. One 3-year child died and four others, also children, suffered from a life-threatening kidney disease in Melbourne (Calligeros, 2014). Several organic milk producers in Victoria sell raw milk called bath milk, which is described as tasty, alongside other food products. The organic producers had been getting round the prohibition of raw milk sales, which had been forbidden by law in Australia since the 1940s, by labelling the product "not for human consumption". It was presented in plastic bottles alongside pasteurized milk and was satisfying a demand for raw milk. The Victorian government has now forced the producers to add a bitter gagging agent to the raw milk if they want to sell it as "bath milk". This has restricted sales but has fostered the development of a black market. The "bath milk" was initially intended for cosmetic use as a body lotion with natural disease-healing properties. It is not known how many Australians bathed in the milk and so followed Cleopatra, who according to legend, bathed in asses milk to preserve her beauty and youth.

Brazil

Puño-Sarmiento *et al.* (2014) assessed the potential for harm of organic poultry manure arising from the presence of *E. coli* in chicken litter that had not been properly composted before use. They concluded that pathogens may remain in organic manure after the heating phase of composting, which can lead to infections, a situation exacerbated by the presence of antibiotic-resistant strains of *E. coli.*

Germany

The danger of rushing to conclusions about links between the use of manure and disease was exemplified in Germany in 2011 when a serious outbreak of a foodborne illness (causing diarrhoea, kidney failure and death) occurred. The cause was a novel strain of *E. coli* that was ultimately found to be carried on fenugreek sprouts (Burger, 2012). It appeared ominous that the sprouts were grown on an organic farm in Germany. In all, 3,950 people were affected and fifty-three died, mostly in Germany. A handful of cases were reported in other countries (CDCP, 2011). It transpired that infected fenugreek seeds had been imported from Egypt and that, according to the authorities, the organic procedures on the German farm were not implicated. However, suspicions led to the trade of some organic vegetables in Europe being disrupted. This conclusion has been challenged by a detailed examination by Ammann (2016); the links between organic farming and the outbreaks were said to have been clearly documented and that despite denials from the organic industry that organic practices were to blame. Moreover, earlier outbreaks had been ignored.

UK

The use of farmyard manure always carries the risk of disease organisms being transmitted via crops, particularly vegetables, to humans. Between 70 and 100 million t of farm manures are used each year on conventional and organic crops in the UK. This manure can carry food-borne, disease-causing microorganisms such as *Campylobacter, Cryptosporidium, Escherichia coli (E. coli 0157), Giardia, Listeria* and *Salmonella* (Environment Agency UK, 2013). As very little manure and slurry is applied to vegetable and salad crops in the UK the risk of infection is more likely to be restricted to imported vegetables.

USA

The US Centers for Disease Control and Prevention reported in the late 1990s that people who eat organic and natural foods are much more likely than the rest of the population to be infected by *E. coli, 0157:H7* (Avery, 2006).

In Minnesota (Mukherjee *et al.,* 2004) found more *E. coli* on produce labelled organic than on conventional produce, including tomatoes, leafy greens, lettuce, green peppers, cabbage, cucumbers, broccoli, strawberries and apples. In certified organic produce they found that the differences in *E. coli* contamination were not as great, which suggests that the labelling of the products as organic was suspect. The percentages of *E. coli*-positive samples in conventional and organic produce averaged 1.6 and 9.7% respectively. All the certified organic farms used aged or composted animal manure. Organic lettuce had the largest contamination of *E. coli* (22.4%). In the same study *Salmonella* was only found on lettuce from organic farms. In contrast, the Stanford meta-analysis

(Smith-Spangler *et al.,* 2012) found no differences in bacterial contamination levels between organic and conventional crops.

An outbreak of hepatitis A in the USA (several states) in 2013 was linked to organically produced pomegranates in Turkey by a study combining genetic analysis of the virus with epidemiology (Craft, 2014). The tainted Turkish pomegranate seeds had somehow been contaminated by microscopic amounts of faecal matter and were included in a five-berry combination product sold as an "Organic Antioxidant Blend". The Federal officials concluded that as the fruit could have been contaminated either by unclean hands or polluted water at any point in the growing, harvesting, processing or handling there appears to be no proof that the farming system was to blame.

Some concern, possibly misguided, is being shown in the USA about the possibility that organic farming may be associated with outbreaks especially of *Salmonella* and *E. coli*; eighteen outbreaks identified over a twenty-two-year period have been responsible for 779 illnesses and three deaths. Fifteen of the eighteen outbreaks were associated with foods that were definitely or likely to be organic, according to the US Department of Agriculture. However, because of the limited data the Centers for Disease Control and Prevention were unable to make conclusions about the relative risk of foodborne illnesses according to farming system (Harvey *et al.,* 2016 and CDCP, 2016).

Human pathogens

van Bruggen *et al.* (2008), in a wide-ranging review, considered the occurrence of human pathogens in organic and conventional foods because it was thought that the use of animal manure and the non-use of antibiotics by organic farmers might be responsible for outbreaks of intestinal diseases; this was despite the fact that up to 2006 only one case of *E. coli* had been associated with organic produce (spinach) (FDA, 2006).

Foodborne diseases, notably those caused by *Campylobacter*, *Salmonella* and the pathogenic *E. coli*, are of particular concern in Europe and North America. Livestock are reservoirs and carriers of these human pathogens. Animal manure applied to vegetables is an obvious source of pathogens but there would seem to be no reason for assuming that the use of manure whether by organic or conventional farmers is likely to lead to infections. No significant differences were found in pathogen prevalence in most of the studies comparing organic and conventional farming (van Bruggen *et al.*, 2008). The only exception was organic poultry, which often had a higher incidence of *Campylobacter* than conventional poultry. On the other hand, *Salmonella* incidence was either similar or lower in organic broilers than in conventional. It is thought that where livestock have outdoor access there is a danger that contact with wild animals such as mice, rats and birds will increase the risk of infection. No significant differences were found in the occurrence of *Campylobacter*, *Salmonella* and *E. coli* between organic and conventional cows and cattle, most probably because all had access to pastures. *Salmonella* and *E. coli* are seldom found on vegetables.

Nanomaterials

In 2008 the Soil Association in the UK was the first organization to ban the use of manufactured nanoparticles under its organic standards; a precautionary approach in line with its organic principles and designed to safeguard public health. There is evidence that some nanoparticles are toxic and potentially hazardous but the subject is in its infancy. The precautionary principle, combined with the fact that nanoparticles are man-made, is a strong motivator.

Fears about the damage to health by nanoparticles have been expressed by several public interest groups in the USA, including the Center for Food Safety and Consumers Union (CFSC, 2015). The Friends of the Earth (FoE, 2011, 2014), which campaigns to keep toxic and risky technologies out of the food we eat, considers that nanoparticles in food are harmful.

The advisory organic body (National Organic Standards Board) in the USA does not approve of the use of nanoparticles in food and petitioned the USDA National Organic Program for man-made nanomaterials to be prohibited from certified organic products (Cornucopia, 2014). The FDA (Food and Drug Administration) in the USA does not require nanoparticles to be mentioned on food labels.

The dislike of nanoparticles appears to be based on the fear that when ingested they may readily move into and through the body and be harmful. Nanoparticles are found in nature, e.g. in smoke and in flour. More than 1,000 consumer products in the USA contain nanoparticles, including many foods, fruits and cosmetics. Nanoparticles of TiO_2, for example, are used to whiten mints and yoghurt. It is likely that the current widespread use of nanomaterials in food will make formulating regulations about nanoparticles in organic food difficult.

Pesticides

Synthetic modern pesticides are of particular concern to organic farming organizations, which not only ban their use but also promote the idea that their residues are all likely to be harmful to health. While it may be argued that there have been risky pesticides that we would not accept today, this is not a reason for condemning all synthetic pesticides. The Agricultural Health Study (AHS, 2011, 2013) in the USA provides valuable information on individual pesticides. Unlike the nitrate problem, which started with the infantile condition of methaemoglobinaemia, there is no basis for the concerns about pesticides in general. The complex situation relating to all kinds of pesticides including those made by plants themselves, the phytoalexins, are discussed in Chapter 15.

Neurotoxicity

Many insecticides target the nervous system of insects and it is reasonable to think, in view of the similarities in brain biochemistry, that some insecticides may also be neurotoxic to humans. Current safety testing of insecticides does not include developmental neurotoxicity.

Laboratory experimental studies suggest that many pesticides currently used in Europe – including organophosphates, carbamates, pyrethroids, ethylenebisdithiocarbamates, and chlorophenoxy herbicides – can cause neurodevelopmental toxicity (Bjorling-Poulsen *et al.*, 2008). As the adverse effects on brain development can be severe and irreversible, preventing exposure should be the priority.

Epidemiological studies on farmworkers and their families in California (the CHAMACOS study) have related maternal urinary concentrations of organophosphate metabolites in pregnancy with abnormal reflexes in neonates, adverse mental development at 2 years of age, attention problems at 3½ and 5 years, and poorer intellectual development at 7 years (EPRS, 2016).

Only two prospective studies from Europe addressing associations between urinary levels of pesticides and neurodevelopment in children from the general public have been published. Both studies are based on the PELAGIE study in France (designed to investigate mother–child effects of herbicides); results for organophosphates and pyrethroids have been presented (Viel *et al.*, 2015 and Cartier *et al.*, 2015). While no adverse effects on cognitive function in 6-year-old children were related to maternal urine concentrations during pregnancy of organophosphate or pyrethroid metabolites, the children's own urinary concentrations of pyrethroid metabolites in particular were related to impairments in verbal and memory functions. The European study (general public exposed to pesticide residues) did not confirm the results from the US (farm families possibly exposed to pesticides when applied), showing that exposure during pregnancy to organophosphate insecticides at levels found in the general population may harm brain development in the foetus. The scientific evidence is incomplete and further studies are required, especially of low-level exposures.

Research on potential effects of organic food on human health

Market forces and the convictions of organic movements that farming organically is the way to a sustainable agriculture will inevitably mean that studies aimed at demonstrating the health benefits of organic food will continue to be made. The Louis Bolk Institute is just one organization that is conducting such research (Louisbolk. org). Animal experiments in which a benefit has been observed will be followed by examination of the diets in order to identify the health-promoting factor(s).

The fundamental question is whether or not the practice of organic farming is itself responsible for any perceived benefits?

Apart from the generation of protective natural pesticides (phytoalexins) in an attacked organic crop, which is not the intention of the organic methodology, there are few reasons for thinking that the composition of an organic crop is likely to be basically different from a conventional one (see Chapter 14). There is no evidence on humans that organic food is likely to promote health.

In the memorandum submitted by the Soil Association to the Select Committee on Agriculture in the UK Parliament (Soil Association, 2001) it called for research into the health effects of intensive farming.

The status quo of organic food in relation to health and the prospects for research was reviewed by Huber *et al.* (2011) in the Netherlands (Louis Bolk Institute) and in Poland. They concluded that the variation in outcomes of comparative studies on crop composition is very high because it is affected by many factors such as fertilization, ripening stage, plant age at harvest, and weather conditions. Apart from the lower nitrate and pesticide residues contents, attention was focused on the usually higher levels of vitamin C and phenolic compounds in organic products and the higher levels of omega-3 fatty acids in milk from organically raised animals. They concluded that it is difficult to draw conclusions from compositional data about the links between health and organic food, a situation compounded by the lack of clinical evidence of the health benefits of particular components of the organic diet.

If it was the case, as thought by Howard in India in the 1920s, that cattle were capable of being protected from catching infectious diseases such as foot-and-mouth simply by eating compost-grown feed then it should have been straightforward to duplicate the results and also to examine the implications for other diseases. Small animal trials could have been used to demonstrate the basic health-promoting properties of organic food.

Health of farmers

USA

Pesticides, which are by their nature biologically highly active chemicals, can present hazards, particularly during their application. Workers may also be exposed when working in a field where pesticides have recently been applied and by breathing in pesticide drift. Organic farmers are not exposed in these ways.

Farmers in many countries have lower overall death and cancer rates than the general public. In the United States the lower death rates among farmers for heart disease and cancers of the lung, oesoph-agus, bladder and colon are thought to be due, at least in part, to lower smoking rates, as well as more physi-cally active lifestyles and dietary factors (AHS, 2011). However, compared with the general population in the USA, the

rates for certain types of cancer appear to be higher among agricultural workers, which may be related to exposures that are common in their work environment. For example, farming communities have higher rates of leukaemia, non-Hodgkin lymphoma, multiple myeloma, and soft tissue sarcoma, as well as cancers of the skin, lip, stomach, brain, and prostate. Apart from pesticides farm workers are exposed to many substances, including engine exhausts, solvents, dusts, animal viruses, fertilizers, fuels, and particular microbes that may account for some of these elevated cancer rates. However, studies on humans reported to date have not allowed researchers to sort out which of these factors may be linked to which cancer types; studies have been ongoing since 1993 on 90,000 participants on the development of cancer and other dis-eases on farmers and their wives in North Carolina and Iowa (AHS, 2011).

There are indications that some pesti-cides may be linked to specific cancers but further research is needed to confirm these findings. The situation is complicated by the finding of associations between live-stock farming and cancer. Poultry farmers had higher rates of colon cancer and non-Hodgkin lymphoma than farmers who did not raise poultry. Sheep farmers had a greater likelihood of developing multiple myeloma than farmers who did not raise sheep. Performing veterinary services on the farm was associated with a greater risk of developing Hodgkin lymphoma (AHS, 2013). It is clear that the relation-ship between the occurrence of cancer and farming are complex.

Amish

The very low rates of cancer in Amish people were mistakenly believed to be due to the fact that they were organic farmers and did not use pesticides. However, while there are a few organic Amish farmers most are not and they use pesticides, fertilizers and grow genetically modified crops (Amish America, 2016).

Agrican study in France

Agrican has monitored the health of French agricultural workers (184,000 with average age 64 years) since 2005 and has linked, by regression analysis, cancer incidence and respiratory diseases with working with thirteen crop and five livestock groups as well as with pesticide use. According to Alerte (2014), as of October 2014 nothing has been conclusively demonstrated. The study, which compares the data with that for the general public, is due to run until 2020. Compared with the general population, the life expectancy of the agricultural population is confirmed to be higher, which is partly explained by less smoking as well as a healthier lifestyle. Incidence of lung, bladder and pancreatic cancer are lower than in the general public. For most other cancers there were no differences. However, malignant melanoma (skin cancer) in women, probably because of higher exposure to UV, was greater in the agricultural population, as was multiple myeloma (an uncommon cancer that affects special white blood cells formed in the bone marrow). The use of pesticides does not appear to be a risk factor for lung cancer. Working with cattle for many years was generally beneficial, probably by boosting the immune system but working with field peas increased the risk of lung cancer. Curiously, the risk of prostate cancer was higher for growers of sunflowers, tobacco, fruits and potatoes as well as those working with horses and pigs.

Summary

1. Studies on human health and the consumption of organic food are necessarily epidemiological and so are not able to identify causal relationships.
2. No connection was found in a massive study on women linking reduced incidence of many kinds of cancer with organic food. In the absence of other epidemiological studies that indicate a contrary conclusion it is clear that claims of an association between eating organic food and a reduction in cancer incidence can be dismissed.
3. However, the indications of a lower incidence of non-Hodgkin lymphoma in organic food eaters is worthy of further study.
4. Eating organic food was associated with a slightly reduced risk of pre-eclampsia (from 5.4 to 4.2%) in pregnant mothers in a Norwegian study. Epidemiological studies of this kind are correlations and do not establish cause and effect.
5. No epidemiological studies have been reported on men.
6. The finding of differences in the composition of organic and conventional milk

has been associated with reduced incidence of eczema in children. Omega–3 fatty acids have been implicated but the relevance of the length of the fatty acid chain needs to be considered.

7. Nitrates gained a bad reputation as being harmful to human health when it was thought that high concentrations of nitrates in drinking water were responsible for methaemoglobinaemia (blue baby syndrome). The theoretical possibility that nitrates in the diet promote the generation of carcinogenic nitrosamines in the gut added to the concern about nitrates.

8. Although the last case of methaemoglobinaemia in the UK was in 1972 the government mandated the EU directive on nitrates in drinking water, which resulted in considerable arable areas being designated as nitrate-vulnerable zones. Such zones impose costs on the farmer, the water companies and the taxpayer.

9. Nitrosamines are known to be carcinogenic in animals but they have not been definitely implicated in any human cancers.

10. Nitrate ingestion is followed by the rapid conversion to nitrite and nitric oxide, which can reduce blood pressure and platelet aggregation; both effects are likely to reduce the occurrence of adverse cardiovascular events.

11. Intakes of nitrates in vegetables, especially leafy green vegetables, can easily exceed regulatory limits for dietary nitrates and so call into question the rationale behind current nitrate and nitrite levels in drinking water.

12. The acknowledged beneficial health effects of nitrates should lead to a reconsideration of regulations on the permissible amounts of nitrate in drinking water and a better understanding of the health benefits of including leafy vegetables in the diet. It may well be concluded that conventionally grown food with higher nitrate contents would be preferred to organic crops with lower nitrate contents.

13. The nature of the health benefits of fruit are assumed to be, but are not proved to be, associated with their antioxidant content.

14. Whilst in theory organic production methods might be expected to lead to health problems caused by fungal toxins and by bacterial contaminants there is scant evidence of such effects in practice.

15. The lifestyle of farmers is probably responsible for them being generally healthier and for them living longer than members of the general public.

16. Farmers also suffer less from most cancers but indications of associations between some cancers and pesticide use are being studied alongside other aspects of farming activities.

References

Addiscott, T.M. (2005) Nitrate, Agriculture and Environment, CAB International, Wallingford, UK. doi: 10.1079/9780851999135.0000.

AHS (2011) Agricultural Health Study. Available at www.cancer.gov/cancertopics/factsheet/Risk/ahs

AHS (2013) Agricultural Health Study Update. Available at http://aghealth.nih.gov/news/2013.html

Alerte (2014) Agricultural study on cancers in agriculture: nothing convincing

at the moment. Available at http://
alerte-environnement.fr/2014/03/24/
etude-agrican-sur-les-cancers-en-milieu-
agricole-rien-de-probant-pour-linstant

Alfven, T., Braun-Fahrlander, C.,
Brunekreef, B., von Mutius, E., Riedler,
J., Scheynius, A., van Hage, M., Wickman,
M., Benz, M.R., Budde, J., Michels, K.B.,
Schram, D., Ublagger, E., Waser, M. and
Pershagen, G. (2006) Allergic diseases and
atopic sensitization in children related to
farming and anthroposophic lifestyle—the
PARSIFAL study. *Allergy*, 61, 4, 414–42.

Amish America (2016) Do Amish use
pesticides? Available at http://
amishamerica.com/do-amish-use-pesticides

Ammann, K. (2016) EHEC (Escherichia coli
O104-H4) Outbreak Germany 2011 linked
to organic farming. Available at http://
www.ask-force.org/web/AF-16-EHEC-
Sproutbreak/Ammann-EHEC-Outbreak-
2011-Norhern-Germany-20160303.pdf

Appel, L.J., Moore, T.J., Obarzanek,
E., Vollmer, W.M., Svetkey, L.P., Sacks,
F.M., Bray, G.A., Vogt, T.M., Cutler, J.A.,
Windhauser, M.M., Lin, P.H. and Karanja,
N.A. (1997) Clinical trial of the effects
of dietary patterns on blood pressure.
DASH Collaborative Research Group. *The
New England Journal of Medicine* 336, 16,
1117–1124.

Avery, A. (2006) *The truth about organic foods.*
Henderson Communications, Missouri.

Baranski, M., S´rednicka-Tober D.,
Volakakis, N., Seal, C., Sanderson, R.,
Stewart G.B., Benbrook, C., Biavati, B.,
Markellou, E., Giotis, C., Gromadzka-
Ostrowska, J., Rembiałkowska, E., Skwarło-
Son´ta, K., Tahvonen, R., Janovska, D.,
Niggli, U., Nicot, P. and Leifert, C. (2014)
Higher antioxidant and lower cadmium
concentrations and lower incidence of
pesticide residues in organically grown

crops: a systematic literature review and
meta-analyses *British Journal of Nutrition*
112, 5, 794–811.

Benjamin, N. (2000) Nitrates in the human
diet-good or bad? *Annales. de Zootechnie,
INR. Sciences* 49, 3, 207–216.

Benjamin, N. (2008) Nitrates in the
human diet 263–275. In: Givens,
I., Baxter, S., Minihane, A. M. and
Shaw, E. (eds) *International Workshop
on the Effects of the Environment on
the Nutritional Quality and Safety of
Organically Produced Foods, Reading,* CABI
Publishing.

Bjorling-Poulsen, M., Andersen, H.R.
and Grandjean, P. (2008) Potential
developmental neurotoxicity of pesticides
used in Europe. *Environmental Health* 7, 50.
doi: 10.1186/1476-069X-7-50.

Boers, I., Muskiet, F., Berkelaar, E., Schut,
E., Penders, R., Hoenderdos, K., Wichers,
H. and Jong, M.C. (2014) Favourable
effects of consuming a Palaeolithic-type
diet on characteristics of the metabolic
syndrome: a randomized controlled pilot-
study. *Lipids in Health and Disease* 13,160.
doi: 10.1186/1476-511X-13-160.

Bosscher, D., Breynaert, A., Pieters, L, and
Hermans, N. (2009) Food-based strategies
to modulate the composition of the
intestinal microbiota and their associated
health effects. *Journal of Physiology and
Pharmacology*. 60, Suppl. 6, 5–11.

Bradbury, K.E., Balkwill, A., Spencer, E.A.,
Roddam, A.W., Reeves, G.K. Green,
J., Key, T.J., Beral, V. and Pirie, K.
(2014) Organic food consumption and the
incidence of cancer in a large prospective
study of women in the United Kingdom.
British Journal of Cancer 110, 2321–2326.
doi: 10.1038/bjc.2014.148.

Brandt, K., Christensen, L.P., Hansen-
Møller, J. and Hansen, S.L. (2004) Health

promoting compounds in vegetables and fruits: A systematic approach for identifying plant components with impact on human health. *Trends in Food Science and Technology* 15, 7–8, 384–393. doi. org/10.1016/j.tifs.2003.12.003.

van Bruggen, A.H.C., Franz, E. and Semenov, A.M. (2008) Human pathogens in organic and conventional foods and effects of the environment 160–189. In: Givens. I., Baxter S., Minihane, A.M. and Shaw E. (eds) *International Workshop on the Effects of the Environment on the Nutritional Quality and Safety of Organically Produced Foods, Reading,* CABI Publishing.

Burger, R. (2012) EHEC O104:H4 In Germany 2011: Large Outbreak Of Bloody Diarrhea And Haemolytic Uraemic Syndrome By Shiga Toxin–Producing *E. Coli* Via Contaminated Food. Available at https://www.ncbi.nlm.nih.gov/books/ NBK114499

Calligeros, M. (2014) Toddler dies, four children seriously ill after drinking raw cow's milk. Available at www.theage.com. au/victoria/toddler-dies-four-children-seriously-ill-after-drinking-raw-cows-milk-20141210-124lx8.html

Cantliffe, D.J. (1973) Nitrate accumulation in table beets and spinach as affected by nitrogen, phosphorous and potassium nutrition and light intensity. *Agronomy Journal* 65. 4, 563–565.

Cartier, C., Warembourg, C., Le Maner-Idrissi, G., Lacroix, A., Rouget, F., Monfort, C., Limon, G., Durand, G., Saint-Amour, D., Cordier, S. and Chevrier, C. (2015) Organophosphate Insecticide Metabolites in Prenatal and Childhood Urine Samples and Intelligence Scores at 6 Years of Age: Results from the Mother-Child PELAGIE Cohort (France). *Environmental Health Perspective.*

CDCP (2011) Centers for Disease Control and Prevention, Outbreak of Shiga toxin-producing *E. coli* O104 (STEC O104:H4) Infections Associated with Travel to Germany (Final Update). Available at www.cdc.gov/ecoli/2011/ecolio104

CDCP (2016) Foodborne Disease Outbreak Surveillance System, foodborne illness outbreaks, National Outbreak Reporting System, organic food, Sam Crowe, US Centers for Disease Control and Prevention, USDA National Organic Program.

Cerf-Bensussan, N. and Gaboriau-Routhiau, V. (2010). The immune system and the gut microbiota: friends or foes? *Nature Reviews Immunology* 10, 735–744. doi: 10.1038/ nri2850.

CFSC (2015) Nanotechnology in our food. Available at www.centerforfoodsafety.org/ issues/682/nanotechnology

Cornucopia (2014) Nanomaterials in Organic Food? The USDA Is Looking the Other Way. Available at www.cornucopia org/2014/10/nanomaterials-in-organic-food-the usda-is-looking-the-other-way

Craft, C. (2014) Disease detectives trace last year's hepatitis A outbreak to Turkish pomegranates. Sacramento Bee, Healthy choices, Sacramento CA, USA, Sacramento Bee Company. Available at www.sacbee.com/news/local/health-and-medicine/healthy-choices/article3611719. html

EFSA (2008) Nitrate in vegetables Scientific Opinion of the Panel on Contaminants in the Food chain1 (Question No EFSA-Q-2006-071). *The EFSA Journal* 689, 1–79.

EFSA (2014) Mycotoxins. Available at www.efsa.europa.eu/en/topics/topic/ mycotoxins

Environment Agency, UK (2013) Nutrients, fertilisers and manures.

Available at https://www.gov.uk/managing-nutrients-and-fertilisers

EPRS (2016) European Parliamentary Research Service. Human health implications of organic food and organic agriculture. Available at www.europarl.europa.eu/RegData/etudes/STUD/2016/581922/EPRS_STU(2016)581922_EN.pdf

EUFIC (2013) The European Food Information Council. Organic food and farming: scientific facts and consumer perceptions. Available at www.eufic.org/article/en/page/RARCHIVE/expid/Organic_food_and_farming_scientific_facts_and_consumer_perceptions

FDA (2006) FDA Statement on Foodborne E. coli O157:H7 Outbreak in Spinach. Available at www.fda.gov/NewsEvents/Newsroom/PressAnnouncements/2006/ucm108761.htm

Floistrup, H., Swartz, J., Bergstrom, A., Alm, J.S,, Scheynius, A., van Hage, M., Waser, M., Braun-Fahrländer, C., Schram-Bijkerk, D., Huber, M., Zutavern, A., von Mutius, E., Ublagger, E., Riedler, J., Michaels, K.B. and Pershagen, G. (2006) Allergic disease and sensitisation in Steiner school children. *Journal of Allergy and Clinical Immunology* 117, 1, 59–66.

FOE (2011) Bill to regulate cosmetics threatening consumers' health with potentially harmful nanoparticles. Available at www.foe.org/news/archives/2011-06-bill-introduced-to-regulate-cosmetics-threatening-co#sthash.uxK4oyUg.dpuf

FoE (2014) Friends of the Earth Tiny Ingredients, Big Risks: Nanomaterials rapidly entering food and farming. Available at http://media-alliance.org/downloads/TinyIngredients_BigRisks.pdf

Food standards, Australia, New Zealand (2010) Survey of nitrates and nitrites in food and beverages in Australia. Available at www.foodstandards.gov.au/consumer/additives/nitrate/documents/Survey%20of%20nitrates%20and%20nitrites.doc

Forman, J. and Silverstein, J. (2012) Organic Foods: Health and Environmental Advantages and Disadvantages. *Pediatrics* 130, 5, 1406-1415.

FSA (2012) Organic milk and eczema. Available at http://tna.europarchive.org/20120419000433 and www.food.gov.uk/foodindustry/farmingfood/organicfood

Gladwin, M.T., Schechter, A.N., Kim-Shapiro, D.B., Patel, R.P., Hogg, N., Shiva, S. Cannon, R.O., Kelm, M., Wink, D.A., Espey, M.G., Oldfield, E.H., Pluta, R.M., Freeman, B.A., Lancaster, J.R., Feelisch, M. and Lundberg, J.O. (2006) The emerging biology of the nitrite anion. *Nature Chemical Biology* 1, 6, 308–315.

Gourama. H. (2015) A Preliminary Mycological Evaluation of Organic and Conventional Foods. *Food Protection Trends*, 35, 5, 385–391. Available at www.foodprotection.org/files/food-protection-trends/Sep-Oct-15-gourama.pdf

Grosse, Y., Baan, R., Straif, K., Secretan, B., Ghissassi, F.E. and Cogliano, V. (2006) Carcinogenicity of nitrate, nitrite and cyanobacterial peptide toxins. *Lancet Oncology* 7, 8, 628–9.

Harvey, R.R., Zakhour, C. M. and Gould, L.H. (2016) Foodborne Disease Outbreaks Associated with Organic Foods in the United States. *Journal of Food Protection* 11, 182–2017 and 1953–1958. Available at www.ingentaconnect.com/content/iafp/jfp/2016/00000079/00000011/art00018

Hollman, P.C.H., Cassidy, A., Comte, B., Heinonen, M., Richelle, M., Richling, E.,

Serafini, M., Scalbert, A., Sies, H. and Vidry, S. (2011) The biological relevance of direct antioxidant effects of polyphenols for cardiovascular health in humans is not established. *Journal of Nutrition* 141, 5, 989–1009.

Hord, N.G., Tang, Y, and Bryan, N.S. (2009) Food sources of nitrates and nitrites: the physiologic context for potential health benefits. *American Journal of Clinical Nutrition* 90, 1, 1–10. doi: 10.3945/ajcn.2008.27131.

Hord, N.G. (2011) Dietary nitrates, nitrites, and cardiovascular disease. *Current Atherosclerosis Reports.* 13, 6, 484–92. doi: 10.1007/s11883-011-0209-9.

Huber, M., Rembiałkowska, E., Srednicka, D., Bügel, S. and van de Vijver, L.P.L. (2011) Organic food and impact on human health: assessing the status quo and prospects of research NJAS – Wageningen. *Journal of Life Sciences* 58, 3–4, 103–109. doi: 10.1016/j.njas.2011.01.004.

Jajja, A., Sutyarjoko, A., Lara, J., Rennie, K., Brandt, K., Qadir, O. and Sierv, M. (2014) Beetroot supplementation lowers daily systolic blood pressure in older, overweight subjects. *Nutrition Research* 34, 10, 868–875. doi: http://dx.doi.org/10.1016/j.nutres.2014.09.007

JECFA (2002) World Health Organization. Sixty-ninth Report of the Joint FAO/WHO Expert Committee on Food Additives Geneva: Evaluation of Certain Food Additives Louis Bolk Institute Available at http://apps.who.int/iris/bitstream/10665/44062/1/WHO_TRS_952_eng.pdf

Knight, T.M., Forman, D., Al-Dabbagh, S.A., Doll, R. (1987) Estimation of dietary intake of nitrate and nitrite in Great Britain. *Food and Chemical Toxicology*: 25, 4, 277–85.

Kummeling, I., Thijs, C., Huber, M., van de Vijver, L.P., Snijders, B.E., Penders, J., Stelma, F., van Ree, R., van den Brandt, P.A. and Dagnelie, P.C. (2008) Consumption of organic foods and risk of atopic disease during the first 2 years of life in the Netherlands. *British Journal of Nutrition* 99, 3, 598–605. doi: 10.1017/S0007114507815844.

Lancet (1968) Nitrites, nitrosamines, and cancer. *Lancet* 1968; 18, 1, 1071–1072.

L'hirondel, J. and L'hirondel, J-L. (2001) *Nitrate and Man, Toxic, Harmless or Beneficial?* Wallingford. UK: CABI Publishing.

L'hirondel J-L, Avery, A.A. and Addiscott, T. (2006) Dietary Nitrate: Where is the Risk? *Environmental Health Perspectives* 114, 8, A458–A459.

Lidder, S. and Webb, A.J. (2013) Vascular effects of dietary nitrate (as found in green leafy vegetables and beetroot) via the nitrate-nitrite-nitric oxide pathway. *British Journal of Clinical Pharmacology* 75, 3, 673–696. John Wiley & Sons. doi: 10.1111/j.1365-2125.2012.04420.x.

Lundberg, J.O., Carlstrom, M., Larsen, F.J. and Weitzberg, E. (2011) Roles of dietary inorganic nitrate in cardiovascular health and disease. *Cardiovascular Research* 89, 3, 525–532. doi: 10.1093/cvr/cvq325.

McKnight, G.M., Duncan, C.W., Leifert, C. and Golden, M.H.N. (1999) Dietary nitrate in man-friend or foe. *British Journal of Nutrition* 8, 5, 349–358.

Mukherjee, A., Speh, D., Dyck, E. and Diez-Gonzalez, F. (2004) Preharvest Evaluation of Coliforms, *Escherichia coli*, *Salmonella*, and *Escherichia coli* O157:H7 in Organic and Conventional Produce Grown by Minnesota Farmers. *Journal of Food Protection*, 67, 5, 894–900. PMID: 15151224.

Muramoto, J. (1999) Comparison of Nitrate Content in Leafy Vegetables from Organic and Conventional Farms in California. University of California. Available at www.agroecology.org/documents/Joji/leafnitrate.pdf

NHMRC (2004) National Water Quality Management Strategy. Australian Drinking Water Guidelines 6. Available at www.nhmrc.gov.au/publications/synopses/eh19syn.htm#comp

Olsen, M. (2008) Mycotoxins in organic and conventional food and effects of the environment 145–159. In Givens. I., Baxter S., Minihane, A.M. and Shaw, E. (eds) *International Workshop on the Effects of the Environment on the Nutritional Quality and Safety of Organically Produced Foods, Reading*, CABI Publishing.

Olsson, M.E., Andersson, S.C., Oredsson, S., Berglund, R.H. and Gustavsson, K.E. (2006) Antioxidant levels and inhibition of cancer cell proliferation *in vitro* by extracts from organically and conventionally cultivated strawberries. *Journal of Agricultural and Food Chemistry* 54, 4, 1248–1255. doi: 10.1021/jf0524776.

Powlson, D.S., Addiscott, T.M., Benjamin, N., Cassman, K.G., de Kok, T.M., van Grinsven, H., L'Hirondel, J.L., Avery, A.A. and van Kessel, C. (2008) When does nitrate become a risk for humans? *Journal of Environmental Quality* 37, 2, 291–295. doi: 10.2134/jeq2007.0177.

Puño-Sarmiento, J., Gazal, L.E., Medeiros, L.P., Nishio, E.K., Kobayashi, R.K.T. and Nakazato, G. (2014) Identification of Diarrheagenic *Escherichia coli* Strains from Avian Organic Fertilisers. *International Journal of Environmental Research and Public Health* 11, 9, 8924–8939. doi: 10.3390/ijerph110908924.

Rist, L., Mueller, A., Barthel. C., Snijders, B., Jansen, M., Simões-Wüst, A.P., Huber, M., Kummeling, I., von Mandach, U., Steinhart, H. and Thijs, C. (2007) Influence of organic diet on the amount of conjugated linoleic acids in breast milk of lactating women in the Netherlands. *British Journal of Nutrition* 97, 4, 735–743. doi: 10.1017/S0007114507433074.

Roscoe, V., Lombaert, G.A., Huzel, V., Neumann, G., Melietio, J., Kitchen, D., Kotello, S., Krakalovich, T., Trelka, R. and Scott, P.M. (2008) Mycotoxins in breakfast cereals from the Canadian retail market: A 3-year survey. Food Additives and Contaminants 25, 3, 347–355. doi.org/10.1080/02652030701551826.

Rosen, J.D. (2010) A Review of the Nutrition Claims Made by Proponents of Organic Food. *Comprehensive Reviews in Food Science and Food Safety* 9, I3, 270–277. doi/10.1111/j.1541-4337.2010.00108.x/full.

Rouse, I.L., Beilin, L.J., Armstrong, B.K. and Vandongen, R. (1983) Blood-pressure-lowering effect of a vegetarian diet: controlled trial in normotensive subjects. *Lancet*. 1, 8314–5, 5–10.

Schinasi, L. and Leon, M.E. (2014) Non-Hodgkin lymphoma and occupational exposure to agricultural pesticide chemical groups and active ingredients: a systematic review and meta-analysis. *International Journal of Environmental Research Public Health*. 11, 4, 4449–4527.

Schuphan, W., Bengtsson, B., Bosund, I. and Hylmo. B. (1967) Nitrate accumulation in spinach, *Plant Foods for Human Nutrition* 14, 4, 317–330. doi: 10.1007/BF02418777.

Smil, V. (2001) *Enriching the Earth: Fritz Haber, Carl Bosch and the Transformation of World Food Production*. MIT Press, Cambridge, Massachusetts.

Smith-Spangler, S., Brandeau, M.L., Hunter, G.E., Bavinger, C., Pearson, M., Eschbach, P.J., Vandana Sundaram, Hau Liu, Schirmer, P., Stave, C., Ingram Olkin and Bravata, D.M. (2012) Are organic foods safer or healthier than conventional alternatives? A systematic review. *Annals of Internal Medicine*. 157, 5, 348–366. doi: 10.7326/0003-4819-157-5-201209040-00007.

Soil Association (2001) Organic Food and Farming Report 2000. Soil Association.

Soil Association (2001a) Organic food and farming. Myth and reality. Organic vs non-organic the facts. Available at http://orgprints.org/9042/1/Myth%26Reality(237)_.pdf

Soil Association 2011) Nanotech – a risky and unpredictable technology. Available at http://organic-market.info/news-in-brief-and-reports-article/Nanotech.html

Spiegelhalder, B., Eisenbrand, G. and Preussmann, R. (1976) Influence of dietary nitrate on nitrite content of human saliva: possible relevance to *in vivo* formation of N-nitroso compounds.*Food and Cosmetics Toxicology* 14, 6, 545–548. doi: 10.1016/S0015-6264(76)80005-3.

Steinmetz, K.A. and Potter, J.D. (1996) Vegetables, fruit, and cancer prevention: a review. *Journal of the American Dietetic Association* 96, 10, 1027–1039. doi: 10.1016/S0002-8223(96)00273-8.

Stenius, F., Swartz, J., Lilja, G., Borres, M., Bottai, M., Pershagen, G., Scheynius, A. and Alm, J. (2011) Lifestyle factors and sensitization in children - the ALADDIN birth cohort. *Allergy*, 66, 10, 1330–1338. doi: 10.1111/j.1398-9995.2011.02662.x.

Stockdale, E. (2008) Impacts of environment and management on nitrate in vegetables and water 276–306. In: Givens. I., Baxter S., Minihane, A.M. and Shaw E. (eds) *International Workshop on the Effects of the Environment on the Nutritional Quality and Safety of Organically Produced Foods, Reading,* CABI Publishing.

Sutton, M.A., Howard, C.M., Erisman, J.W., Billen, B., Bleeker, A., Grennfelt, P., van Grinsven, H and Grizzetti, B. (2011) *The European Nitrogen Assessment – Sources, Effects and Policy Perspectives.* Cambridge University Press.

Thijs, C., Müller, A., Rist, L., Kummeling. I., Snijders, B.E., Huber, M., van Ree, R., Simões-Wüst, A.P., Dagnelie, P.C. and van den Brandt, P.A. (2011) Fatty acids in breast milk and development of atopic eczema and allergic sensitisation in infancy. *Allergy* 66, 1, 58–67. doi: 10.1111/j.1398-9995.2010.02445.x.

Torjusen, H., Brantsæter, A.I.., Haugen, M., Alexander, J., Bakketeig, L.S., Lieblein, G., Stigum, H., Næs, T., Swartz, J., Holmboe-Ottesen, G., Roos, G. and Meltzer, H.M. (2014) Reduced risk of pre-eclampsia with organic vegetable consumption: results from the prospective Norwegian Mother and Child Cohort Study. *British Medical Journal Open* 2014; 4:e006143. doi: 10.1136/bmjopen-2014-006143.

Trewavas, A.J. (2005) Organic Food and Farming: hidden agenda, *Association of Applied Biologists Letters*, Issue 56.

Turchini, G.M., Nichols, P.D., Barrow, C. and Sinclair, A.J. (2012) Jumping on the omega-3 bandwagon: distinguishing the role of long-chain and short-chain omega-3 fatty acids. *Critical Reviews in Food Science and Nutrition* 52, 9, 795–803. doi: 10.1080/10408398.2010.509553.

Udikovic-Kolic, N., Wichmann, F., Broderick, N.A. and Handelsman, J. (2014) Bloom of resident antibiotic-resistant bacteria in soil following manure fertilization. *Proceedings of the National Academy of Sciences USA* 111, 42, 15202–15207. www.pnas.org/content/111/42/15202

Viel, J.F., Warembourg, C., Le Maner-Idrissi, G., Lacroix, A., Limon, G., Rouget, F., Monfort, C., Durand, G., Cordier, S. and Chevrier, C. (2015) Pyrethroid insecticide exposure and cognitive developmental disabilities in children: The PELAGIE mother-child cohort. *Environment International*. 82, 69–75. http://dx.doi.org/10.1016/j.envint.2015.05.009

Ward, M.H., de Kok, T.M., Levallois, P. Brender, J., Gulis, G., Nolan, B.T. and van Derslice, J. (2005) Workgroup report: drinking-water nitrate and health – recent findings and research needs. *Environmental Health Perspectives* 113, 11, 1607–1614.

Webb, A.J., Patel, N., Loukogeorgakis, S., Okorie, M., Aboud, Z., Misra, S., Rashid, R., Miall, M., Deanfield, J., Benjamin, N., MacAllister, R., Hobbs, A.J. and Ahluwalia, A. (2008) *Hypertension* 51, 784–790. doi: 10.1161/hypertensionaha.107.103523.

WHO (1995a) Nitrate (WHO Food Additive Series). Available at www.inchem.org/documents/jecfa/jecmono/v35je14.html

WHO (2008) Guidelines for drinking water quality [electronic resource]: incorporating 1st and 2nd addenda, Vol.1, Recommendations – 3rd ed. Available at www.who.int/water_sanitation_health/dwq/fulltext.pdf

Wu, F. (2007) Measuring the economic impacts of fusarium toxins in animal feeds. *Animal Feed Science and Technology* 137, 3–4, 363–374. doi.org/10.1016/j.anifeedsci.2007.06.010.

CHAPTER 11

Soil Organic Matter, Relevance and Benefits

"If you build up the soil with organic material, the plants will do just fine"
John Harrison (clockmaker early 18th century)

Appreciation of soil organic matter

The value and importance of organic manures and composts to world food production and the maintenance of good soil conditions, particularly soil organic matter, cannot be disputed but it must be remembered that perfect disease-free crops of high nutritional quality can be grown without soil using nutrient solution culture techniques. Cucumbers, tomatoes, courgettes, spinach and bell peppers are produced without soil in many countries; most of the home-grown tomatoes in the UK are grown in rock wool and fed with nutrient solutions. Even some of the bananas from the Canary Islands and the early peaches from Italy are grown in gravel irrigated with nutrient solutions. Plants are unable to distinguish between mineral nutrients derived from the dissolving of inorganic man-made fertilizers and those derived from the microbial breakdown of organic matter. Nitrate is nitrate no matter where it comes from.

Farmers have realized for many years that there is a close relationship between the level of organic matter in the soil and its fertility and as a consequence good farmers have for many generations aimed to maintain the dark-coloured humus content of their soils. Liebig emphasized the importance of humus in the mid-19th century.

Classical experiments and SOM levels

In the long-term classical experiments on wheat at Rothamsted (Broadbalk) the annual application of farmyard manure increased the soil organic matter (SOM) whereas fertilizers did not (from 1.7%); after forty years SOM levels had been doubled. Soils given manure for 170 years now contain 2.5–3.0 times more SOM than soils given fertilizers. Despite the differences in SOM, yields of cereals and root crops were very similar on both manure and fertilizer-treated soils for at least the first 130 years in the case of wheat (Johnston *et al,.* 2009). Not surprisingly this gave rise to the belief that, provided plant nutrients were supplied as fertilizers, extra SOM was of little importance in producing maximum yields of the crop varieties then available. Lawes and Gilbert, who were responsible for starting the trials at Rothamsted, never claimed that fertilizers were better than manure but they appreciated that the annual application of 35 t/ha (as in their wheat experiments) to every arable crop in the country was impractical. However, results over the last forty years of the experiments on the heavy soils at Rothamsted and on the light soils at Woburn have shown that SOM is beneficial in ways not related to its provision

of nutrients, especially for spring crops with higher yield potential and where soil conditions need to be optimal (see Chapter 13). Full elucidation of the forces at work is awaited. In general the regular applications of manure increased the soil organic matter contents substantially in the early years but the rate of increase declined after about fifty years long before there was any sign of the superiority of manure on yield.

It took more than 130 years and the growing of high-yielding varieties for the benefits of SOM to become apparent, indicating that questions about SOM remain. What are the time scales and with what farming practices do SOM contents change? Can the various soil factors that might be connected with the "organic matter effect" be identified for a given soil and their relevance assessed? Could meaningful critical SOM levels for different soils be established or will they be too dependent on the yield potential of the crops?

A critical level of 3.4% SOM was suggested for England and Wales by Greenland *et al.* (1975) based on the observed link between poor soil physical conditions and SOM. On the other hand, Loveland and Webb (2003) only found slight evidence for critical SOM levels although they recognized there is some evidence that there might be a desirable range of SOM covering a wide range of soils.

The level of SOM depends on the inputs of organic material, its rate of decomposition and the loss of existing SOM; it will differ with soil texture and climate. With regular long-term management SOM will tend slowly to an equilibrium level under the influence of these factors. The equilibrium levels will be higher in clay than in sandy soils and under permanent pasture than continuous arable cropping.

Worldwide there has been a decline in SOM levels as a result of converting grasslands and forests to arable crops; Bot and Benites (2005) suggest that the loss of SOM may be as much as 30%. On reasonably fertile soils with reliable water supply, yields in long-term arable agricultural systems have been maintained at very high levels by applying substantial amounts of fertilizers. However, in low-input agricultural systems, such as on smallholder farms in Africa, yields generally decline rapidly as SOM levels fall and it is effectively mined for its nitrogen content. However, restoration is possible through the use of fallows, mixed farming, appropriate crop rotations and by promoting crop growth by fertilizers.

Wasteful effects of increasing soil organic matter

It may be surprising to realize that the benefits of attempting to increase SOM beyond the equilibrium level is not advisable. Both C and N are mobilized during microbial decomposition of added organic matter. At the equilibrium level of SOM for any soil, climate, and farming system, all the C and N in further additions of organic matter will tend to be lost. The loss of NO_3 by leaching over the winter poses environmental problems and is also a financial loss. When cereals were given fertilizer N at optimal rates the amounts of mineral N remaining in the soil at harvest was often only a little larger than that in soil to which no fertilizer N had been applied (Glendining and Powlson, 1995). In contrast, the large amounts of mineral N in the manure-treated soil were at risk of loss by leaching (Goulding *et al.*, 2000). It appears that fertilizer N is taken up preferentially to soil N even when there may be more soil N in manure-treated soil.

Organic matter in soils

Organic matter levels in soils are strongly influenced by the application of organic materials including crop litter and their subsequent incorporation and by crop/grass root residues. SOM can be measured by the weight loss on ignition of a dry soil at 360°C or by determining the soil organic carbon (SOM = Soil Organic Carbon x 1.72). Levels of SOM in most agricultural soils are likely to be at a low equilibrium level depending on the farming system and will be lower than the levels in the preceding natural vegetation. The SOM in both the unfertilized plots and those given NPK (since 1843) on Broadbalk continuous wheat trial have been relatively constant at 1.7%.

SOM levels vary considerably between soils and different farming systems. In a survey over nine years in England and Wales, Church and Skinner (1986) found that 82% of arable soils contained 2–8% SOM, whereas 82% of continuous grassland soils contained more than 5% SOM and 42% more than 8%.

Considerable amounts of SOM can be found in subsurface soil. For example, Ward *et al.* (2016) found, in grasslands in Great Britain, that 60% of soil C was found below 30 cm; they estimated that there was 2097 Tg C in the top 1m of grassland soils in Great Britain.

Following poor harvests in 1968 and 1969 the UK government recommended that arable soils should ideally contain more than 3% SOM (the Strutt report, MAAF, 1970).

SOM in organically and conventionally managed soils

Few differences in SOM were found between organic and conventionally managed pasture soils in a survey (n=30) in the UK by Shepherd *et al.* (2002) from which it was concluded that it was not the farming system but the amount and type of organic matter returned to the soil that is important. Only small differences in favour of organic farming were found on a few arable areas where there were stockless conventional farms.

Humus and its significance

The decomposition of plant residues, dead microbes, insects and earthworms results in the formation of many compounds with various properties that can affect both soil conditions and crop growth. The raw materials for decomposition are mainly sugars, starches and proteins, which will decompose relatively quickly compared to cellulose, hemicellulose, lignin, fats, waxes and resins.

Humus, the material that gives the brown colour to soils, can be thought of as the ultimate product of the decomposition of organic matter. It is usually the major constituent of SOM. Humus exists as negatively charged colloids whose capacity to adsorb cations exceeds that of colloidal clay minerals. Humus is relatively stable because it cannot be used by most microorganisms as an energy source. It consists of fulvic and humic acids, which are both soluble in water, and humin, which is not. The relative amounts of humic and fulvic acids in soils vary with soil type and management practices. The humus of forest soils is characterized by a high content of fulvic acids, while the humus of agricultural and grassland areas contains more humic acids (Bot and Benites, 2005).

Owing to its very high capacity to adsorb cations such as K^+, Ca^{++} and Mg^{++}, humus plays an important part in holding the essential nutrient cations in the soil and thereby minimizing leaching losses.

Sandy soils rely heavily on the high cation exchange capacity (CEC) of humus for the retention of nutrients in the topsoil. The CEC is measured in terms of the total quantity of cations that can be adsorbed expressed as milli-equivalents (meq) per 100 g soil. The influence of humus as perceived by soil colour was evident from examples of CEC for some soils in Indiana, USA, given by Mengel (1993) (Table 11.1).

A soil with a CEC of I meq/100g

Table 11.1 Range of CEC values for soil types in Indiana according to colour and texture (Mengel, 1993)

Soil Type	CEC meq/100g soil
Light coloured sands	3–5
Dark coloured sands	10–20
Light coloured loams	10–20
Dark coloured silty loams	15–25
Dark coloured silty clay loams	35–40
Organic soils	50–100

would be able to adsorb in 1 ha of topsoil (20 cm) either 600 kg of Ca^{++}, or 360 kg of Mg^{++}, or 1,170 kg of K^+ or 540 kg of NH_4^+.

As humic and fulvic acids are both water soluble there is a potential for them to be leached, but little appears to be known. While much is known about the chemical composition of humus, the full significance of the different components is yet to be established.

Soil organic N occurs mainly (> 90%) as proteins, amino acids, nucleic acids and amino sugars. Inorganic N is present as ammonium cations held in the inner layers of clay minerals and on humus; nitrate anions are not held so are freely leached.

Manure and the development of antibiotic resistance

Treating livestock with antibiotics and then spreading their manure on the soil could promote the evolution of bacteria that are resistant to the antibiotics used. However, a study in the USA

now suggests that the manure itself could be contributing to resistance, even when it comes from cows not fed antibiotics (Udikovic-Kolic *et al.*, 2014). The mechanism at work is not clear but it may be related to the promotion of bacteria in the soil that naturally carry antibiotic-resistance genes, probably as defence against the antibiotics produced by some soil fungi and bacteria. Similar results have been reported in Denmark (Aarhus, 2016).

Importance and benefits of soil organic matter

SOM can improve soil conditions, fertility and crop yields in several different ways that are appreciated by farmers and growers both organic and conventional.

Source of recycled nutrients and increased cation exchange capacity

A major function of manure and composts is the ability to supply nutrients, not only N but also K, P, S, Mg and micronutrients. At the same time, humus can hold available cations by virtue of its cation exchange capacity, a feature that is particularly valuable in light-textured soils.

Improvement of soil structure

SOM, even in small amounts, can influence the formation of soil aggregates, which in turn affect the size and distribution of stable pores that are important for the passage of air (essential for root and microbial respiration) and of water. By improving soil aeration SOM can promote root penetration. Moreover, the formation of crumbs improves permeability and the ability of soils to take up and hold water. SOM effectively works by preventing soil mineral particles sticking together as a mass. Elliot and Lynch (1984) showed that soil aggregation is enhanced primarily by the ability of polysaccharides to bind soil particles together as crumbs. The larger polysaccharide molecules may be more important in promoting aggregate stability and water infiltration than the smaller molecules. The hyphae of actinomycetes and fungi also play an important role in connecting soil particles.

Aggregate formation was found by Watts *et al.* (2001) to be dependent on microbiological activity and was impaired when activity was low, e.g. when peat rather than grass was the organic source. A strong positive relationship between SOM and aggregate size was found by Castro Filho *et al.* (1998).

The presence of SOM was also positively linked to friability by Watts and Dexter (1998), whose primary interest was in using measures of friability to help determine the optimum soil water content for tillage and as an index of soil physical quality. Whereas Watts *et al.* (2001) had found the continuous application of manure on a silty clay loam (Broadbalk) had built up a better crumb structure, Johnston *et al.* (2009) did not find similar benefits on the soil texture of a sandy loam soil at Woburn even after many years of manure application.

It would be expected that the mechanical working of soil (ploughing) would be easier on soils with more SOM and this was found to be the case on the silty clay loam soil (Rothamsted) to which massive amounts of manure had been applied for many years. However, it was also found on the same soil that the continuous annual application of inorganic fertilizers had similarly improved the ease of working by about 12% (Watts *et al.*, 2006). This was despite the fact that soil SOM content was trebled by the manure and only by 28% by the inorganic fertlizers; even small increases in SOM can have a significant effect. The improvements in soil structure of the silty clay loam soil (Broadbalk) were observed some years before any sign of the superiority of manure on yield.

Stolze *et al.* (2000) using data from EU and other European countries, presented an informative report on the effects of organic farming on the environment in order to help clarify its possible contribution to European agro-environmental policy. Surprisingly, while they praised organic farming for improving soil fertility and helping to minimize erosion they were unable to identify any benefits of SOM related to soil structure across Europe.

Water infiltration, erosion and water holding capacity

When soils are not covered by vegetation, mulches and crop residues the more the soil is exposed to the impact of raindrops and wind; it is under these conditions that by improving the formation of stable crumbs SOM is particularly valuable.

By enhancing water infiltration SOM increases water retention. The effect of SOM increasing the available water capacity (AWC) in the top 30 cm of soil was assessed in a number of experiments at Rothamsted and Woburn (Salter and Williams, 1969). Increases in AWC associated with higher SOM levels were small, ranging from 4 to 10 mm. In almost all cases soils that had received manure had significantly higher AWC than un-manured soils. The amount of available water retained in the surface foot of soil ranged from 40 mm in a sandy loam to 70 mm in a silt loam under permanent grass. Organic matter is better than clay with regard to water availability because it releases water to growing plants more easily. The effects of SOM on AWC would be expected to be greater on light-textured soils.

Traditional cultivation by plough and discs tend to increase the rate of decomposition of soil organic matter and leave the soil susceptible to wind and water erosion. In contrast, reduced or zero-tillage systems leave more surface residues and a better soil surface structure, thereby improving water infiltration. In addition, organic matter decomposes more slowly under reduced-tillage systems.

Promotion of microbial activity

Although the amounts of C held in soil microbes are normally small, less than 5% of total soil C in forest and grassland and less than 2.5% in arable soils, soil microbes play a fundamental role in the recycling of nutrients from organic materials (Kallenbach and Grandy, 2011).

It is reasonable to assume that soils with high SOM levels will contain more and possibly more diverse population of microorganisms because soil microbes depend on organic compounds for growth. But how and perhaps whether microbial numbers and variety are beneficial is still an open question. The inability of organic farming systems to provide sufficient N in the spring at the time of maximum need by arable crops suggests that microbe numbers and activity are far from sufficient at that time.

Soil microbiological activity, as measured by the microbial biomass, was doubled by manure applications when compared with either NPK or no fertilizers on the classical wheat trial (Rothamsted, 2012). Mineralization, nitrification and denitrification would all have been affected. Activity of denitrifying bacteria in generating the greenhouse gas N_2O has been found to be increased in soils containing large amounts of SOM following manure applications compared with soils given inorganic N (Goulding *et al.*, 1998). The potential for N_2O generation is dependent on soil N levels, which can often be higher when organic materials rather than inorganic are used to supply N (Rothamsted, 2012).

Effect of fertilizers on soil microorganisms

One complaint about the use of man-made fertilizers is that they adversely affect soil microbial activity. The main-tenance in the Rothamsted experiments of the same wheat yields on the fertilizer-only plots and the manure plots for about 130 years clearly indicates that the num-bers and activity of soil microbes have been of little consequence.

Geisseler and Scow (2014) carried out a meta-analysis of data from sixty-four trials around the world to assess the long-term effects of mineral ferti-lizer applications on soil microbes in the topsoil. They revealed that in arable crops mineral fertilizer application (N, P and K) generally led to an increase in microbial biomass (on average about 15% over unfertilized controls) and also of SOM but only after ten years. The crop growth improvements due to the fertilizers increase the inputs of organic material in the form of decaying roots, of root exudates and of crop residues on the soil surface, which can all promote the activity of soil microorganisms.

Most attention was focused by Geisseler and Scow (2014) on the effects of N fertilizers. However, because in most long-term trials P and K were also applied the observed effects cannot all be attributed to N applications. In very acid soils (pH<5) fertilization tended to reduce microbial biomass. The applica-tion of urea and ammonia fertilizers can temporarily increase pH, osmotic poten-tial and ammonia concentrations to levels inhibitory to soil microbes but any effects are minimal and local to the fertilizer particles. Long-term repeated mineral N applications may alter composition of the microbial community. How specific

microbial groups respond to repeated applications of fertilizers varies consid-erably and is dependent on the environ-ment and crop management.

An average reduction of about 6% in the microbial biomass following N appli-cation was revealed by the meta-analysis of Lu *et al.* (2011) based on data from 206 peer-reviewed studies; microbial activity was not eliminated by N.

Kallenbach and Grandy (2011) found in a meta-analysis of agricultural systems that man-made fertilizers increased carbon in the microbial biomass by 9% on average above levels in unfertilized controls, whereas organic amendments increased microbial biomass by 36%. They found that even small quantities of organic amendments will rapidly restore microbial biomass across a range of crops but that specific responses depend upon the type and rate of inputs as well as on soil characteristics.

Long-term (about thirty years) appli-cations of manure increased soil micro-organisms by 25% in the DOK trial in Switzerland (see Chapter 13).

Biodiversity

Organic matter plays a fundamental role in sustaining soil biodiversity by provid-ing sources of energy and nutrients for the wide range of soil organisms.

Bot and Benites (2005) provide a very useful description of the wide range of organisms that together constitute a food web in the soil and are responsible for gen-erating available nutrients from organic matter. The cast of organisms includes not only the bacteria and fungi whose beneficial effects are well recognized but also protozoa, yeasts, mites, nematodes, insects and earthworms, each of which occupy their own niche in the soil. Some free-living nematodes by grazing on bacteria and fungi may control the populations of harmful microorganisms. The effects of earthworms differ according to whether they live in the litter zone or in the body of the soil, where they improve soil structure.

This diversity of soil organisms is believed to be an important but poorly understood component of both natural and agricultural ecosystems. It is possible when the soil has received heavy applications of pesticides, fertilizers, soil fungicides or fumigants that beneficial organisms may be killed. The balance between the pathogens and beneficial organisms may also be upset, allowing the pathogens to become problems (Bot and Benites, 2005). Farming of all kinds affects soil organisms in different ways and until it is possible to ascribe benefits to diversity it is not possible to assess whether one type of farming is superior to another.

How to maintain and increase soil organic matter levels

The difficulty of increasing SOM levels can be appreciated by considering the amounts of organic material required to raise the SOM level by 1%. On the basis of the top 20 cm of soil weighing 3 million kg/ha and 10 kg organic material

generating 1 kg SOM a minimum of 300 t/ha of organic material would be needed to increase SOM by 1% without any allowance for losses over time. The application of a total of 770 t/ha of manure on the classical wheat experiment (Broadbalk) raised the SOM level by about 1.4% in the first twenty-two years from 1843 (Johnston *et al.*, 2009).

Building up SOM is thus a long-term process that can be achieved by applying manure or compost, incorporating straw, growing cover crops, including grass in the rotation and by reducing or eliminating tillage. When SOM levels are very low the best way to build up levels again is to return the land to permanent pasture.

The importance of mixed farming was demonstrated by the work of Schulz *et al.* (2014) in Germany. Traditionally organic farming has relied on livestock (usually cattle), on perennial fodder legumes in crop rotations and on farmyard manure to generate and cycle nutrients. However, in line with specialization elsewhere more organic farms are becoming stockless. In Germany about 25% of organic farms no longer keep livestock. Schulz *et al.* (2014) examined whether the supply of organic matter is sufficient in stockless farming using the results from the long-term Organic Arable Farming Experiment at Gladbacherhof started in 1998. The legumes employed were in the mixed farm lucerne (harvested) and peas; on the first stockless farm lucerne was incorporated and peas as straw, and on the other stockless farm, cash crop field beans and peas provided straw. After two six-year rotations it was only in the mixed farm treatment with fodder legumes in the rotation and the application of cattle manure that SOM levels were maintained and sometimes increased. The stockless farm with fodder legumes as green manures was barely able to maintain the SOM level, while the stockless cash crop farm where organic matter supply relied on green manure catch crops and straw showed a marked reduction in SOM. Schulz *et al.* (2014) concluded that the inclusion of fodder legumes as green manure crops is essential in stockless organic farming; otherwise it may be necessary to import manure from neighbouring farms.

Soil organic matter and EU legislation

Farmers in the European Union have to meet certain standards relating primarily to the condition of land and animal welfare if they are to receive the Basic Payment Scheme (2015). A minimum SOM level is not specified but as part of the compliance regulations farmers must maintain SOM through appropriate practices. Burning of arable stubble is not allowed except for crop health reasons. As part of the Environmental Impact regulations farmers are also not permitted to plough or cultivate species-rich and semi-natural habitats.

In Ireland farmers were required to monitor and maintain SOM levels in arable soils at a minimum of 3.4% if they were to qualify for the Single Payment Scheme, the forerunner of the Basic Payment Scheme.

Summary

1. There is no doubt that soil organic matter (SOM) is a valuable and appreciated commodity. It promotes soil fertility, improves soil structure and minimizes erosion. It also provides a favourable habitat for soil organisms and plant roots.

2. Although organic manures and composts play an important part in food production it must be remembered that perfect disease-free crops of high nutritional quality can be grown without their help or in fact without soil.

3. Application of manure will raise the SOM level far more than by fertilizers, as demonstrated over 170 years at Rothamsted following the massive annual applications when SOM levels were increased by up to three times. Despite the higher SOM levels after the use of manure there were no associated yield benefits until after 130 years.

4. Converting grassland and forests to crop production inevitably leads to loss of SOM, possibly by 30% globally. SOM levels can be maintained on good soils but in low-input agricultural systems SOM levels decline because of the removal in crops of N which is not replaced.

5. Soils would appear to have an equilibrium SOM level that may take many years to establish. Attempts to exceed the equilibrium level by applying massive amounts of organic matter can result in the pollutant and wasteful loss of much of the added N and C.

6. SOM levels vary considerably between soil types and different farming systems, as do the equilibrium SOM levels.
7. Limited survey data in the UK suggests that differences in SOM levels between organic and conventional farming are unlikely mainly because many conventional farmers use manure; any differences will be related to the amounts applied not to the system of farming.
8. Humus, the material that gives the brown colour to soils and which can be thought of as the ultimate product of the decomposition of organic matter, is of particular importance because of its high capacity to adsorb cations – the cation exchange capacity (CEC). The richer the soils are in humus (and generally darker) the higher the CEC and the capacity to hold K^+, Ca^{++} and Mg^{++}.
9. Application of manure can improve soil conditions, fertility and crop yields by supplying nutrients and by improving soil structure, especially the formation of stable crumbs; better structure is valuable in connection with water infiltration and in protecting against erosion.
10. Soil microbial activity is promoted by applications of manure but it is not known whether apart from mineralization of organic matter any significant benefits ensue and if so under what conditions.
11. The claimed detrimental effects of man-made fertilizers on soil microbial activity are not supported by the variable evidence. While there may be local inhibition of microbes close to fertilizer particles any such negative effects are likely to be offset by the beneficial effects of fertilizers on shoot and root growth leaving increased plant residues.
12. Building up SOM is a long-term process and can be achieved by applying farmyard manure or compost, incorporating chopped straw, growing cover crops and by including grass in the rotation and by reducing or eliminating tillage. Very large amounts of organic matter are required.

References

Aarhus (2016). The use of animal manure increases the soil content of antibiotic-resistant genes. Available at www.sciencedaily.com/releases/2016/03/160307113729.htm

Basic Payment Scheme (2015) Cross compliance: guidance for (2015) Available at. https://www.gov.uk/government/publications/cross-compliance-guidance-for-2015

Bot, A. and Benites, J. (2005) The importance of soil organic matter. Key to drought-resistant soil and sustained food production. *FAO Soils Bulletin 80.* Available at www.fao.org/docrep/009/a0100e/a0100e08.htm

Castro Filho, C., Muzilli, O. and Podanoschi, A.L. (1998) Estabilidade dos agregados e sua relação com o teor de carbono orgânico num Latossolo roxo distrófico, em função de sistemas de plantio, rotações de culturas e métodos de preparo das amostras. *Revista Brasileira de Ciência do Solo* 22, 527–538.

Church, R.M. and Skinner, R.J. (1986) The pH and nutrient status of agricultural soils in England and Wales 1969–1983. *Journal of Agricultural Science* 107, 1, 21–28.

Elliot, L.F. and Lynch, J.M. (1984) The effect of available carbon and nitrogen in straw on soil and ash aggregation and acetic acid production. *Plant and Soil* 78: 335–343.

Geisseler, D. and Scow, K.M. (2014) Long-term effects of mineral fertilizers on soil microorganisms – A review. *Soil Biology and Biochemistry*, 75, 54–63. Available at www.sciencedirect.com/science/article/pii/S0038071714001187

Glendining, M.J. and Powlson, D.S. (1995) The effects of long continued application of inorganic nitrogen fertilizer on soil organic nitrogen-A review 385–446. In: Lal, R. and Stewart, B.A. (eds.) *Soil Management: Experimental Basis for Sustainability and Environmental Quality*. CRC Press, Boca Raton, Florida USA.

Goulding, K.W.T., Bailey, N.J., Bradbury, N.J., Hargreaves, P., Howe, M., Murphy, D.V., Poulton, P.R. and Willison, T.W. (1998) Nitrogen deposition and its contribution to nitrogen cycling and associated soil processes. *New Phytologist* 139, 1, 49–58. doi: 10.1046/j.1469-8137.1998.00182.x) 10.1046/j.1469-8137.1998.00182.x.

Goulding, K.W.T., Poulton, P.R., Webster, C.P. and Howe, M.T. (2000). Nitrate leaching from the Broadbalk Wheat experiment, Rothamsted, UK, as influenced by fertilizer and manure inputs and the weather. *Soil Use and Management*. 16, 244–250.

Greenland, D.J., Rimmer, D., Payne, D. (1975) Determination of the structural stability class of English and Welsh soils, using a water coherence test. *Journal of Soil Science* 26, 294–303.

Johnston, A.E., Poulton, P.R. and Coleman, K. (2009) Soil Organic Matter: Its Importance in Sustainable Agriculture and Carbon Dioxide Fluxes. *Advances in Agronomy*, 101, 1–57. doi: 10.1016/S0065-2113(08)00801-8.

Kallenbach, C. and Grandy, A.S. (2011) Controls over soil microbial biomass responses to carbon amendments in agricultural systems: a meta-analysis. *Agriculture, Ecosystems and Environment*, 144, 1, 241–252. Available at http://dx.doi.org/10.1016/j.agee.2011.08.020.

Loveland, P. and Webb, J. (2003) Is there a critical level of organic matter in the agricultural soils of temperate regions?: a review. *Soil and Tillage Research* 70, 1, 1–18.

Lu, M., Yang, Y., Luo, Y., Fang, C., Zhou, X., Chen, J., Yang, X. and Li, B. (2011) Responses of ecosystem nitrogen cycle to nitrogen addition: a meta-analysis. *New Phytologist* 189, 4, 1040–1050. doi: 10.1111/j.1469-8137.2010.03563.x.

MAAF (1970) *Modern farming and the soil: report of the Agricultural Advisory Council on Soil structure and Soil Fertility*. Ministry of Agriculture Fisheries and Food, HMSO, London.

Mengel, D.B. (1993) *Agronomy Guide, Fundamentals of Soil Cation Exchange Capacity (CEC)*. Purdue University Cooperative Extension Service, West Lafayette, Indiana. https://www.extension.purdue.edu/extmedia/ay/ay-238.html

Rothamsted (2012) *Rothamsted Long Term Experiments. Guide to the Classical and Other Long-term Experiments, Datasets and Sample Archive*. (Updated reprint of 2006 edition). Available at www.rothamsted.ac.uk/sample-archive/guide-classical-and-other-long-term-experiments-datasets-and-sample-archive

Salter, P.J. and Williams, J.B. (1969) The moisture characteristics of some Rothamsted, Woburn and Saxmundham soils. The *Journal of Agricultural Science* 73, 1,155–158. doi: https://doi.org/10.1017/S0021859600024242

Schulz, F., Brock, C. and Leithold, G. (2014) Livestock in organic farming – how important is it for soil fertility management? 459–462. In: Rahmann, G. and Aksoy U. (Eds.) *Proceedings of the 4th ISOFAR Scientific Conference. Building Organic Bridges*, Istanbul, Turkey.

Shepherd, M.A., Harrison, R. and Webb, J. (2002) Managing soil organic matter-implications for soil structure on organic farms. *Soil Use and Management* 18, 284–293.

Stolze, M., Piorr, A., Häring, A. and Dabbert, S. (2000) *The Environmental Impacts of Organic Farming in Europe. Organic Farming in Europe: Economics and Policy Volume 6.* Available at http://orgprints.org/8400/1/Organic_Farming_in_Europe_Volume06_The_Environmental_Impacts_of_Organic_Farming_in_Europe.pdf

Udikovic-Kolic, N., Wichmann, F., Broderick, N.A. and Handelsman, J. (2014) Bloom of resident antibiotic-resistant bacteria in soil following manure fertilization. *Proceedings of the National Academy of Sciences October.* Available at www.pnas.org/content/111/42/15202

Ward, S.E., Smart, S.M., Quirk, H., Tallowin, J.R., Mortimer, S.R., Shiel, R.S., Wilby, A. and Bardgett, R.D. (2016) Legacy effects of grassland management on soil carbon to depth. *Global Change Biology* 22, 8, 2929–2938. doi: 10.1111/gcb.13246. Epub 2016.

Watts, C.W. and Dexter, A.R. (1998) Soil friability: Theory, management and the effects of management and organic carbon content. *European. Journal of Soil Science* 49, 73–84. doi: 10.1046/j.1365-2389.1998.00129.x.

Watts, C.W., Whalley, W.R., Longstaff, D.J., White, P.R., Brookes, P.C. and Watts, C.W., Clark, L.J., Poulton, P.R., Powlson, D.S. and Whitmore, A.P. (2006). The role of clay, organic carbon and long-term management on mouldboard plough draught measured on the Broadbalk wheat experiment at Rothamsted. *Soil Use and Management* 22, 334–341.

Whitmore, A.P. (2001) Aggregation of a soil with different cropping histories following the addition of organic materials. *Soil Use and Management* 17, 263–268. doi: 10.1111/j.1475-2743.2001.tb00036.x.

CHAPTER 12

Manures and Slurries – Composition, Production and Application

"Money is like manure, of very little use except it be spread" Francis Bacon

The composition, production, and use of manure will be considered mainly using UK data and concentrating on cattle manure. Some variation can be expected between countries and regions because of differences in animal feeding stuffs and the crops grown but the amounts should be similar and the principles applicable generally. Manure can provide massive amounts of nutrients identical with those in fertilizers, it can improve soil structure but it can also pollute. Manure is not chemical-free as is often imagined as it contains a wide array of organic chemicals such as phenols, mercaptans and organic acids. As pointed out by Trewavas (2004) in a comprehensive review, it is very likely that half the chemicals in manure would be classified as carcinogens (see Chapter 15 for toxicological assessment of carcinogenicity).

Nutrient composition of manure

Apart from N, P and K (Table 12.1) manures can also supply useful quantities of sulphur (2.4 kg SO_3/t in solid manure and 0.7 kg SO_3/m^3 in cattle slurry) but only small amounts of magnesium (<1 kg MgO/t in solid manure) (HGCA, 2014). Only 15% of total

Table 12.1 Typical dry matter and nutrient contents of different organic manure types (BSFP, 2015)

	Dry matter (%)	N kg/t	P_2O_5 kg/t	K_2O kg/t
Solid manure				
Cattle	25	6.0	3.2	8.0
Pig	25	7.0	7.0	8.0
Sheep	25	7.0	3.2	8.0
Poultry	35	19.0	14.0	9.5
Slurries		kg/m^3	kg/m^3	kg/m^3
Cattle	6	2.6	1.2	3.2
Pig	4	3.6	1.8	2.4

sulphur in solid cattle manure and 35% in cattle slurry may be available for the next crop, with the remaining organically bound sulphur becoming slowly available for later crops. Magnesium inputs from manures should be regarded as making a small contribution to the maintenance of soil reserves. Manures are likely to contain all essential mineral elements including calcium and the micronutrients boron, copper, iron, manganese, molybdenum and zinc and so help to conserve and recycle them; the quantities may be small but nevertheless significant in the long term. Although there is no evidence that manures have prevented micronutrient deficiencies it may well have been that manures saved many crops before science had learnt about all the nutrient elements that are required by plants and how they can be supplied.

Cattle manure contains relatively little (10–25% of total N) ammonium (NH_4), with old manure containing the least (Chambers *et al.*, 2001). Cattle, pig and poultry slurries vary widely containing 20–70% of total N as NH_4 (EFMA, 2004). Poultry manures are usually rich in NH_4 (40–60% of total N).

The NH_4 cations derived from manure are indistinguishable from those in man-made fertilizers. Plant roots have no mechanism for distinguishing the origin of the NH_4. Regardless of origin all the NH_4 may be absorbed, lost to the atmosphere as ammonia gas or converted to nitrate. The rest of the N in manures is organic N, which is slowly mineralized over months or years and so becomes available for uptake by crops.

There can be considerable losses of NH_3 from manure and slurries. EFMA (2004) estimates that 20–30% of the nitrogen in farmyard manure may be lost during fermentation and storage and that of the 6.4 million t N (in manure and slurries) produced each year by the 1.4 billion farm animals in the European Union (fifteen countries) only 4.1 million t N will be available for crops, a total loss of 36%.

Production and recovery of manure

Conventional and organic farms

There are about 47 million animals generating manure in the UK, of which most are grazing sheep (Table 12.2).

Manure may be deposited directly on grass (or occasionally on a crop, as by sheep folded on turnips) or in stockyards or pens where the animals are kept. It is the manure that has to be moved from stockyards and cowsheds where cattle are kept over winter that has a particular value for organic arable farmers but the direct deposition by grazing animals must not be forgotten. The amounts of dung and slurries that can be collected from livestock vary with type and size of animal, and especially with the length of the housing period (see Appendix). Dairy cows will produce in a year (per animal) a total of about 15–25 t manure and beef cattle 5–10 t.

Organic livestock will produce about 3% of the manure in the UK based on livestock numbers (Tables 12.2 and 12.3).

Table 12.2 Total numbers of livestock in UK in 2013 (Defra, 2013a)

	Total million
Dairy cows	2.22
Beef cows	2.25
Calves	5.23
Pigs, breeding	0.50
Pigs, fattening	3.89
Sheep (+ lambs)	32.85

Table 12.3 Numbers of organic livestock in UK in 2013 (Defra, 2013b)

	million	% of Total in UK
Cattle	0.28	2.9
Sheep	1.00	3.0
Pigs	0.03	0.7

In 2015, around 65% of surveyed farms used manures on at least one field, with cattle manure from beef and dairy farms supplying most of the manure. A total of 34% of cattle manure and 66% of slurries were applied to grassland. Manure for winter sown crops is mainly applied in late summer prior to drilling, whereas spring sown crops and grass are predominantly dressed in the winter/spring.

According to the annual Survey of Fertiliser Practice in the UK, the total amounts of manures and slurries, including poultry manures and bio-solids (sewage sludge), applied amounted to about 80 million t in 2015 (Table 12.4). Thus substantial amounts of N, P and particularly of K are recycled.

Greaves *et al.* (1999) estimated that the P_2O_5 content of animal manures produced annually in the UK in the 1990s exceeded 120,000 t and that up to 65% of the P in manure was organic. By 2014 production had reached more than 180,000 t P_2O_5 (Table 12.4).

The amounts of nutrients shown in Table 12.4 are to be contrasted with the use (in thousand t) of fertilizer in 2014 in the UK of 1049 of N, 196 of P_2O_5 and 272 of K_2O (FAOSTAT, 2016). Manures thus have the potential to make

Table 12.4 Total amounts of manures and slurries used and nutrients contained in UK in 2015 (BSFP, 2015)

		N	P_2O_5	K_2O
	million t or m^3		thousand t	
Cattle manure	33.1	200	105	265
Cattle slurry	38.6	100	45	125
Pig manure	1.7	12	12	14
Pig slurry	1.4	5	2	3
Poultry manure	1.6	30	22	15
Total	83.5[1]	347	186	422

[1] includes other manure, sewage sludge and non-farm organic wastes.

a much larger contribution to K and P inputs rather than to N.

Recovery of P from sewage is now possible and practised in a few countries. A population of 60 million could theoretically generate more than 80,000 t P_2O_5 per year from treated sewage and so make a major contribution. Such recoveries deserve consideration in view of the finite nature of P ores and eventually of the supply of P fertilizers.

More than 5 million t of manure would potentially be required by organic farmers to satisfy all their N requirements (Table 12.5): this would amount to about 6% of total manure being used in the UK on organic crops (3% agricultural area). However, it is likely that organic cattle only generate about 2,000 t manure based on organic cattle comprising 3% of the UK total. The apparent shortfall in N on organic arable farms in the UK is likely to be made up by green manure crops such as clovers (red or white) or lucerne, for which data is not available. There will be considerable recycling of N on grassland but any net contribution by grazing animals will depend on consumption of feeding stuffs elsewhere.

Feedlot beef and dairy units on factory farms in USA

There were more than 80 million animals, excluding poultry that amount to 1.3 billion animals, on factory farms in the USA in 2012 (Table 12.6); a factory farm is defined as an operation with more

Table 12.5 Organic crop areas and calculated requirement for manure in UK 2013

	Organic '000 ha[1]	Manure t/ha[2]	Total Manure '000 t
Arable crops	49	24	1176
Grass, Temporary	94	9	846
Grass, Permanent	370	9	3330
Vegetables	9	24	216
Total			5568

[1] Defra (2015).
[2] Based on arable crops given 144 kg/ha N and all grassland 55 kg/ha (the average rates of N on conventional farming in 2012) and manure containing 6 kg N/t but without consideration of availability.

Table 12.6   Number of animals on factory farms in the USA in 2012 (Food and Water Watch, 2015)

	million
Dairy cattle	5.6
Beef cattle	12.1
Pigs	63.2
Broiler chickens	1,050.0
Egg-laying chickens	269.3

than 500 beef cattle (feedlots only), 1,000 pigs, 500 dairy cows, 100,000 egg-laying and 500,000 broiler chickens. There are more than 28,000 factory farms in the USA.

Organic standards in the USA do not limit the maximum size of organic factory farms (about 1,500) but they do loosely specify that the animals should have access to pasture, a standard that has been difficult to monitor; modifications to the standards for poultry are imminent. Though most organic farmers will meet high standards of welfare, economic incentives for organic farming have led to a growing number of organic producers raising animals in conditions virtually indistinguishable from non-organic factory farming, as revealed by Cornucopia (2014).

The factory farms produce a total of 369 million tons of manure and just under 0.4 billion m³ of slurry, each year, more than thirteen times the sewage produced by the entire US human population. In contrast to human sewage, the manure and waste from livestock operations is untreated. Instead it is stored in manure pits or lagoons before use. There is a real risk of pollution during storage. Small, mixed farms have traditionally used manure as fertilizer. The difference with factory farms is scale. They produce so much waste in one place that is applied locally to land in quantities that exceed the soil's ability to cope, leading to leaching and run-off.

An American company suitably named Black Gold Compost has converted composted cattle manure into a dry friable product, Black Kow (0.5% N, 0.5% P_2O_5 and 0.5% K_2O) primarily for home gardeners.

Application of manure: amounts, method and timing

Beef and dairy farms generate the most manure in the UK (Table 12.4). In 2013, 53% of farms in the UK used cattle manure and 17% used cattle slurry. Pig slurry and poultry manure was used on very few farms (BSFP, 2013). Overall about 65% of farms are thought to use some manure. Not surprisingly in view of its bulk, very few farmers (about 2%) exported animal manure to local farms. About 50% of farms in the UK stored manure for three to six months; this is a satisfactory situation in view of the recommendation from the Food Standards Agency that manure should be stacked for at least eight weeks to reduce the risk of spreading disease organisms.

On soils destined for arable crops most (81%) of the cattle slurry is broadcast. Rapid incorporation of manure and slurries is ideally required to minimize loss

Table 12.7   Time of incorporation of manure and slurry in 2015 (BSFP, 2013)

	Not incorporated %	Incorporation time %	
		Within 24 hours	More than 1 day
Manure	6	41	53
Cattle Slurry	14	43	43

of NH_3 and the survey found that about 40% was incorporated within one day the period when most NH_3 is likely to be lost by volatilization (Table 12.7).

According to the work of Jokela and Meisinger (2008), slurries should be incorporated, by injectors or band spreaders, immediately after collection from storage tanks and manure from heaps within one hour if 90% of the readily available N is to be conserved; delaying the incorporation of slurry for six hours and of manure for twenty-four hours is likely to result in significant losses, possibly more than 50%, of N as NH_3.

Apart from the volatilization of ammonia, N can be lost from manure by nitrate leaching, which will vary according to application rates and timing. Losses can be minimized by delaying applications until the late winter or spring. Applications of manure during the autumn–early winter period should be avoided in order to minimize the time for winter rainfall to wash nitrate out of the soil before crops can absorb it. Chambers *et al.* (2001) reported that nitrate leaching from a free-draining arable soil can amount to 20% of the total N applied when cattle slurry was applied in September and declined to about 1%

after a January application, comparable amounts following application of manure were 7% and <1%.

It is claimed that organic farming reduces nitrate leaching but a meta-study by Kirchmann and Bergström (2001) showed that this is not the case. They were unable to find any evidence that nitrate leaching is reduced by organic farming practices if the goal is to maintain the same yield levels as in conventional farming. They concluded that reported reductions of nitrate leaching are not related to the farming system but to the amounts and timing of the N applications, which seems logical.

In the UK it is recommended via the Code of Good Agricultural Practice for the Protection of Water (Defra, 2009) that manure and slurry applications should not supply more than a total of 250 kg N/ha/year to any given field, excluding droppings from grazing animals; the application of 250 kg/ha would require such massive amounts of manures and slurries (Table 12.8) that they are unlikely to be realized by organic farmers. According to EC regulations, certified organic farmers must not apply more than 170 kg N/ha/year from manures on the farm as a whole. Conventional farmers in the UK are

Table 12.8 Manure and slurry application rates (t/ha and m^3/ha) to supply 250 kg N/ha

Manure	t/ha
Cattle	42
Pig	36
Egg-laying chickens	16
Chicken litter	8
Slurries	m^3/ha
Cattle	96
Pig	70

Farmcrapapp

A free mobile smartphone app (Farm-CrapApp, 2016) has been developed in the UK by the SWARM Knowledge Hub that is designed to raise awareness among farmers about the nutrient and economic value of manures and slurries. It is based on the UK Fertilizer Manual RB209 (ADHB, 2015). The app provides farmers with an opportunity to assess visually manures and slurries and to calculate what is being provided in terms of available N, P and K at different application rates. The required input information consists of a) type of manure (farmyard manure, cattle slurry, pig slurry or poultry litter and b) the consistency of the slurries, namely "thin soup, thick soup or porridge" with 2, 6 and 10% dry matter respectively. Poultry litter includes both layer manure and broiler litter. The records allow for the age and method of application (surface or incorporated).

advised to calculate the amounts of fertilizer N required to meet a target yield after allowing for N available from the soil; the total amount of N should not exceed the crop requirement.

More arable crops were given manure than slurries but on grassland the total area treated with slurries slightly exceeded that given manure at the rates shown in Table 12.9 (BSFP, 2015). In recent years there has been a great deal of promotional activity in the UK aimed at encouraging conventional farmers to make adjustments to fertilizer inputs where manures are used. It now appears that farmers are beginning to use less fertilizer N on arable crops when they have also applied manure (BSFP, 2013).

Table 12.9 Average rates of application of cattle manure and slurry in UK in 2015 (BSFP, 2015)

	Cattle manure t/ha	Cattle slurry m^3/ha
Winter sown	24	36
Spring sown	25	34
Grass	15	25

Sewage

Suitably treated sewage sludge with permitted amounts of heavy metals can be useful in improving organic matter levels in soils. The nutrient content is likely to be very variable but it can be a useful source of N and P (Table 12.10).

Sewage sludge can, however, contain heavy metals that may be harmful to humans and animals. Its use depends on it meeting compositional regulations; the heavy metal content of the soil must

Table 12.10 Composition of typical sewage sludge (kg/m³) and cake (kg/t) (BSFP, 2015; ADHB, 2015)

	Dry matter %	Total N	Total P₂O₅	Total K₂O
Digested liquid sewage sludge	4	2.0	3.0	0.1
Digested cake	25	11.0	18.0	0.6

also have been determined and sludge should not be spread on a soil with a pH <5.0. The heavy metals of concern are cadmium, copper, chromium, lead, mercury and zinc. Organic farmers are generally not permitted to use sewage sludge and by taking this position are not able to benefit from the recycling of the nutrients that have been consumed and passed down to the end of the food chain.

In the UK organic farmers cannot use sewage sludge, effluents and sludge-based composts (Soil Assoc., 2014); the situation is the same in the USA (USDA, 2015). However, IFOAM (2014) appears to allow for the use of human excrement, stating that it should be handled in a way that reduces the risk of transmission of pathogens and parasites. Further, it shall not be applied within six months of the harvest of crops for human consumption, especially on crops with edible portions that may have been in contact with the soil.

As much of the organic matter in sewage sludge is mineralized within a few weeks it should ideally be spread shortly before sowing the crop, which can be difficult to achieve. In Europe there is Sewage Sludge Directive 86/278/EEC, which prohibits the use of untreated sludge on agricultural land unless it is injected or incorporated into the soil. Sludge may be treated by chemicals, by heat or by long-term storage but it must not be used on fruit and vegetable crops less than ten months before harvest. Grazing animals must not be allowed access to grassland or forage crops less than three weeks after the application of sludge.

Only about 40% of sewage sludge is used in agriculture in the European Union with the remaining 60% incinerated or sent to landfill.

Balfour

Balfour (1943) in *The Living Soil* describes a way, suggested some years before by Howard, of a way human excrement could be used in the UK. On new rural housing estates involving groups of twenty or so properties each would have a detached earth closet and share a garden. The basic idea of the planner was to break away from water-borne sewage systems. There would be a communal gardener whose daily job would include the removal of the contents of the earth closets and its immediate application and covering with soil. Balfour approved of the recycling of human excrement and suggested that it would be better if the night soil was composted with other organic wastes. These ideas were later jettisoned by the Soil Association.

Urine

Following the promotion by subsidies of the use of cow urine and dung in Maharashtra, India, in 2015 there were proposals that human urine, collected from multiplex cinemas in Mumbai, could be similarly promoted for use by organic farmers (*Hindustan Times*, 2015).

Recovery of nutrients from sewage

The recovery mainly of phosphorus and a little nitrogen from sewage has been achieved by Ostara Nutrient Recovery Technologies of Vancouver following research at the University of British Columbia. Plants have been set up in Canada, USA, UK and the Netherlands using a patented process (Pearl®) in which waste water in a fluidized bed reactor is seeded with magnesium chloride to form small (1–3.5 mm) crystals of struvite ($NH_4MgPO_4 \cdot 6H_2O$). The process can recover about 85% of the P and 15% of the N in the sewage. Struvite is a serious problem at sewage plants as it can build up and clog pipes, so preventing its formation would reduce costs. The product with 5% N, 28% P_2O_5 and 10% Mg is sold as Crystal Green and is promoted as a slow release phosphate (citric soluble) for sale to farmers, golf clubs and gardeners. Planned production at a plant serving 2.4 million people in Chicago is 10–15,000 t per year i.e. about 5 kg per person and a total of 3,500 t P_2O_5 per year.

Summary

1. Manures and slurries can be valuable sources of nutrients, particularly of N, P and K, and can provide useful amounts of S. Poultry manures are rich in N.
2. The content of other mineral elements including micronutrients in manures may have been important in conserving and recycling in the past. Deficiencies may have been prevented.
3. In the UK in 2015 66% of cattle manure was applied to arable crops, whereas 65% of slurries was applied to grassland.
4. Just over 14 million animals (cattle and pigs) in the UK in 2015 produced 35 million t manure and 40 million m^3 of slurry, containing a total (including poultry manure) of 347,000 t N, 186,000 t P_2O_5 and 422,000 t K_2O.
5. Manures and slurries can make a big contribution to the supply particularly of K and P by providing more K and about the same amount of P as fertilizers in the UK.
6. Provided the N in manure and slurries is conserved the contribution to the N supply is significant when compared with 1,049,000 t N in fertilizers.
7. Calculation of the amount of manure required by organic farmers in the UK to be able to supply the average national application rate of N to arable crops and grassland indicates that organic farmers would themselves only be able to generate less than half the amount needed.
8. The concentration of livestock on large factory farms as in the USA poses big

problems regarding disposal of excreta without causing pollution.

9. Careful handling of manure and slurries is required in order to minimize losses of ammonia before and after application. Ideally slurries should be incorporated immediately and manure within one hour of spreading.

10. Applications of manure and slurries during the autumn–early winter should be avoided in order to minimize nitrate leaching; for example in the UK nitrate leaching amounting to 20% of the total N occurred when cattle slurry was applied in September and declined to about 1% after a January application; comparable amounts following application of manure were 7% and <1%.

11. Suitably treated sewage sludge with permitted amounts of heavy metals can improve soil organic matter levels and provide variable amounts of N and P.

12. Organic farmers are usually not permitted to use sewage sludge.

13. About 85% of the P in sewage and 15% of the N can be recovered from sewage and plants are operating at sewage works in Canada, USA, UK and the Netherlands.

14. The recovery of P from sewage could in theory generate more than 80,000 t P_2O_5 from a population of 60 million.

15. Such recoveries deserve consideration wherever domestic sewage is treated in view of the finite nature of P ores and eventually of the supply of P fertilizers.

References

ADHB (2015) Agriculture and horticulture development board. Fertilizer handbook RB209. Available at www.ahdb.org.uk/documents/rb209-fertiliser-manual-110412.pdf

Balfour, E.B. (1943) *The Living Soil – Evidence of the Importance to Human Health of Soil Vitality, with Special Reference to National Planning*. London, Faber and Faber.

BSFP (2013) British Survey of Fertiliser practice. Fertiliser use on farm crops for crop year 2013. Available at https://www.gov.uk/government/uploads/system/uploads/attachment_data/file/301474/fertiliseruse-report2013-08apr14.pdf

BSFP (2015) British Survey of Fertiliser practice. Fertiliser use on farm crops for crop year 2015. Available at https://www.gov.uk/government/uploads/system/uploads/attachment_data/file/516111/fertiliseruse-report2015 14apr16.pdf

Chambers, B., Nicholson, N., Smith, K., Pain, B., Cumby, T. and Scotford, I. (2001) ADAS Booklet 1.Managing Livestock Manures Making better use of livestock manures on arable land. Available at www.rothamsted.ac.uk/sites/default/files/Booklet1ManuresArableLand.pdf

Cornucopia (2014) Think your milk and eggs are organic? These aerial farm photos will make you think again. Available at www.washingtonpost.com/blogs/wonkblog/wp/2014/12/11/think-your-milk-and-chicken-are-organic-these-aerial-farm-photos-will-make-you-think-again

Defra (2009) Code of Good Agricultural Practice for the Protection of Water. Available at http://adlib.everysite.co.uk/adlib/defra/content.aspx?id=4468

Defra (2013a) Farming Statistics – Livestock Populations December 2013, United Kingdom. Available at https://www.gov.uk/government/uploads/system/uploads/

attachment_data/file/293717/structure-dec2013-uk-19mar14.pdf

Defra (2013b) Organic Statistics 2013 United Kingdom. Available at https://www.gov.uk/government/uploads/system/uploads/attachment_data/file/317366/organics-statsnotice-05jun14.pdf

Defra (2015) Organic Statistics 2014 United Kingdom. Available at https://www.gov.uk/government/uploads/system/uploads/attachment_data/file/444287/organics-statsnotice-23jun15b.pdf

FarmCrapApp (2016) Available at www.swarmhub.co.uk/index.php/img/about.php?id=3919

EFMA (2004) European Fertilizer Manufacturers' Association. Understanding nitrogen and its use in agriculture. Available at www.magia-metachemica.net/uploads/1/0/6/2/10624795/nitrogen_in_agriculture.pdf

FAOSTAT (2016) Beta, Input statistics for 2014. Available at www.fao.org/faostat/en/#home

Food and Water Watch (2015) Factory Farm Nation 2015 edition. Available at www.factoryfarmmap.org/wp-content/uploads/2015/05/FoodandWaterWatchFactoryFarmFinalReportNationMay2015.pdf

Greaves, J. Hobbs, P. Chadwick, D. and Haygarth, P. (1999) Prospects for the Recovery of Phosphorus from Animal Manures: A Review. *Environmental Technology*, 20, 7, 697–708. doi.org/10.1080/09593332008616864

HGCA (2014) Sulphur for cereals and oilseed rape. Information Sheet 28. Available at www.hgca.com/

media/357116/is28-sulphur-for-cereals-and-oilseed-rape.pdf

Hindustan Times (2015) Maharashtra government plans to encourage urine use in organic farming. *Hindustan Times*, 9 May. Available at www.hindustantimes.com/india-news/maharashtra-govt-endorses-gadkari-s-advice-plans-to-encourage-use-of-urine-in-organic-farming/article1-1345397.aspx

IFOAM (2014) IFOAM norms for organic production and processing (Version 2014). Available at http://infohub.ifoam.bio/sites/default/files/ifoam_norms_version_july_2014.pdf

Jokela, W.E. and Meisinger, J.J. (2008) Ammonia emissions from field-applied manure: management for environmental and economic benefits 199-208. In: *Proc. 2008 Wisconsin Fertilizer, Aglime, and Pest Management Conference*.

Kirchmann, H. and Bergström, L. (2001) Do organic farming practices reduce nitrate leaching? *Communications in Soil Science and Plant Analysis* 32, 7–8, 997–1028 .doi.org/10.1081/CSS-100104101.

Soil Assoc. (2014) Soil Association organic standards. Available at www.soilassociation.org/organicstandards

Trewavas, A. (2004) A critical assessment of organic farming-and-food assertions with particular respect to the UK and the potential environmental benefits of no-till agriculture. *Crop Protection*, 23, 9, 757–781. doi.org/10.1016/j.cropro.2004.01.009.

USDA (2015) Organic agriculture. Available at http://usda.mannlib.cornell.edu/usda/nass/OrganicProduction/2010s/2012/OrganicProduction-10-04-2012.pdf

Part 3

Comparison of Organic and Conventional Farming

CHAPTER 13

Comparison of Yields of Organic and Conventionally Grown Crops

"Comparison is the death of joy" Mark Twain

Comparisons of the yields and the potential of organic and conventional farming can be considered with three questions in mind. Firstly, is it possible to achieve the same or better yields by organic as by conventional farming and if so what are the implications? Secondly, are there circumstances where manure outperforms fertilizers? Thirdly, what is the practical experience? The potential of organic farming to improve soil conditions and confer environmental benefits that could be set against any yield shortfall will be considered elsewhere.

There have been many studies that have compared the yields and composition of organic and conventionally grown crops often carried out by organic proponents or in response to statements about the superiority of organic farming. Of particular interest are claims relating to the potential of organic farming to feed the world and to the health-promoting aspects of organic food (see Chapters 14 and 18).

It is of considerable interest to determine whether there are any aspects of organic farming that either promote growth or improve crop composition in ways that are entirely peculiar to it. In other words, does farming organically with its emphasis on the use of manure, compost and legumes and the absence of agrochemical inputs affect crop yields and composition in ways that cannot be replicated by conventional farming?

In organic farming nutrients are supplied mainly in manure, compost and green manure crops but where specific nutrients have been identified as deficient inorganic materials can be applied, e.g. rock phosphate to supply P and polyhalite to supply potassium, calcium, magnesium and sulphur; likewise micronutrients. But does organic agriculture provide more than nutrients? Any such benefits, if identified, could also be enjoyed on conventional mixed farms.

Beneficial long-term residual effects of

manure could be expected following the build-up of soil organic matter (SOM) and its subsequent mineralization.

Assessment of the benefits of manure other than those conferred by its nutrient content would require a precise experimental design in which the same amounts of nutrients (at least N, P and K) from both the manure and fertilizer are available at the same time for the crop, say for an autumn-sown crop, in the spring/early summer. Allowance would have to be made for the loss of nutrients from the manure by leaching during the winter after autumn application. Compliance with this requirement would be very difficult for N unless it was possible to measure, in the spring, the amounts of available manure-derived N and then apply the same amount of fertilizer N. Control of weeds, pests and diseases would need to be the same, as would the crop rotation. Such a study does not appear to have been carried out because of its complexity. Consequently, reliance has inevitably been placed on interpreting the results of experiments that come close to fulfilling the required design.

The land area and time used to build up soil fertility and so generate the nutrients on which organic crops rely are not generally considered when comparing different farming systems. The conventional farmer relies on man-made fertilizers to supply the required nutrients, whereas the organic farmer relies on farming activities elsewhere in time or space. It should not be forgotten that the all-important manure does not generate

Table 13.1 Possible benefits of fertilizers and manure/compost

	Fertilizer	Manure/compost
Source of nutrients	Yes	Yes
Support of beneficial soil microorganisms	Small[a]	Yes
Improvement of soil physical conditions	?[b]	Yes

[a] by improving root growth and producing more crop litter (Chapter 11).
[b] can be improved by deep rooting crops, e.g. lucerne.

nutrients; it effectively transfers nutrients from other fields (grazed pastures or fodder crops) on the farm or from the consumption of imported protein-rich feeding stuffs.

The benefits of ploughed-in green cover crops are achieved at the expense of a crop growing season, which should be allowed for when assessing yields of a subsequent organic crop. When, for example, legumes are grown as a green manure crop to add nitrogen to the system, the yield of a following crop must be calculated as the product of two seasons, not one. When, however, the legumes serve as food crops, e.g. beans and peas, they do not constitute a lost crop production year. The ancient practice of fallowing is a simple example of the situation. The yield of an organic crop following a fallow year is associated with two years not just one, making it necessary to halve the organic yield when considering the farming system as a whole. It would only be acceptable to ignore the extra use of land if cultivatable land was not limited.

Haughley experiment

The Haughley experiment stands out as the first experiment with the object of comparing, at the whole farm level, a closed organic farming system with a conventional one. The twenty-five-year experiment did not live up to expectations. Soil fertility on the organic farm declined, as did crop yields, and the livestock showed no basic benefits from being fed organic crops. Details and results of the Haughley experiment (started during the Second World War) are discussed in Chapter 6.

Comparison of organic and conventional yields where the nutrient inputs are known

The ways manure can benefit soils are discussed in Chapter 11. In the absence of experiments that could determine unequivocally whether manure or composts confer benefits other than as a source of nutrients it is necessary to study the experiments from which clear inferences can be drawn. The long-term classical experiments on winter wheat, spring barley and root crops at Rothamsted (silty clay loam soil on clay with flints) and at Woburn (sandy loam) are very useful in this respect insofar as much is known about the amounts of nutrients applied and sufficient time has elapsed (over 160 years) for the effects such as the build-up of soil organic matter to have become evident. The methods used for control of weeds, pests and diseases control methods have been the same for all treatments which means the Rothamsted studies cannot be seen as comparisons of organic and conventional farming practices only as a comparison of their potential.

Rothamsted

Comparison of yields of winter wheat, spring barley, and root crops given either manure or fertilizers from the 1840s and 1850s

The demonstration that manure containing on average 7 kg N/t can be as effective as fertilizers in maintaining yields of several crops has been clearly shown by the long-term classical experiments at Rothamsted. From the 1850s to the mid-1970s very similar yields of winter wheat, spring barley and of mangolds (*Beta vulgaris* var. *esculenta*) were obtained from fertilizers providing the right amounts of N, P, and K were applied, as from manure applied at 35 t/ha annually (Johnston *et al.*, 2009). Over the last forty years manure has outperformed fertilizers under some circumstances. When these experiments were started the use of manure was part of conventional farming as it is on mixed farms today.

Continuous wheat (broadbalk)

More P, K and Mg has been applied in the manure than in fertilizers (Table 13.2) (Rothamsted (2012). Fertilizer N was applied in the spring (up to 1967 as ammonium sulphate or sodium nitrate, then as calcium ammonium nitrate and latterly as ammonium nitrate). The manure and the fertilizer P, K and Mg was applied in the autumn before ploughing. Best wheat yields on Broadbalk are currently about 9 t/ha (UK average in 2014 was 8.6 t/ha) whether given fertilizer or 35 t/ha of manure. Yields of

Table 13.2 Nutrients (kg/ha/year) supplied by 35 t/ha manure and by the best fertilizer treatment, which resulted in essentially the same wheat yields since 1843 (Rothamsted, 2012)

	35 t/ha Manure	Best Fertilizer
N	240	144
P	45	35
K	350	90
Mg	25	12

wheat given no manure or fertilizer have remained at about 1 t/ha regardless of the wheat variety; new improved varieties have been used when they became available.

In the absence of estimates of the amount of the manure-applied N that was available in the spring it is only possible to speculate how it came about that the manure and fertilizers had the same effect on yield. It may be that only 60% of the N in the manure was available or that the unconsidered micronutrients in the manure were beneficial. Although considerably more K was applied in manure, sufficient K was applied in fertilizers to achieve a positive K balance so it is unlikely that the relatively large amount of K in manure was important.

Up to the 1970s the best wheat yields varied from 2.5 to 3.0 t/ha. Since the introduction in 1968 of new short-strawed varieties with improved yield potential yields doubled to about 6 t/ha in continuous wheat and to more than 9 t/ha in wheat grown after a two-year break (oats followed by maize). The extra yield of 3 t/ha wheat after a two-year break was attributed to reductions in the effects of soil-borne pests and disease, particularly take-all (*Gaeumannomyces graminis var. tritici*) (Rothamsted, 2012). There were indications in the wheat grown after the break that manure supplemented with 96 kg N/ha in the spring gave slightly superior yields to the best NPK treatment (both about 9–10 t/ha).

What are the implications for organic farming? Best wheat yields on Broadbalk are currently just over 9 t/ha whether given fertilizers or 35 t/ha of manure. This amount of manure could be obtained from 2.3 dairy cows kept indoors over six winter months. Thus if all wheat in the UK (just under 2 million ha) was to be given N only as manure at 35 t/ha with the target of achieving yields of 9 t/ha, 70 million t of manure would be required. This is more than twice the annual amount of manure currently produced by dairy cows in a year in the UK and equates with about 4.6 million dairy cows – there are currently 2.2 million; a total of about 2.3 million ha of summer lowland grazing (at 2 cows/ha) would be required, an increase of 1.2 million ha. Winter feed for the extra 2.4 million cows would also be required (BSFP, 2013). If the arable crops other than wheat (about 3 million ha) were grown organically using manure this hypothetical situation would be exacerbated and render it absolutely impossible for a manure-based organic farming systems to maintain

food production and current eating habits.

Continuous spring barley (Hoosfield)

Spring-sown barley has been grown continuously since 1852, with manure being applied before ploughing in the autumn and the N in the spring. Until the 1980s, PK with appropriate amounts of N gave yields as large as those from 35 t/ha manure. Since then yields from the long-term manure applications have exceeded those from the best N, P, K, Mg treatment (Table 13.3) (Rothamsted, 2012).

It is postulated that the long-term applications of manure may by improving soil structure have facilitated root growth in the recently germinated, spring-sown barley and so helped the crop to develop a sufficiently large root system quickly enough to acquire nutrients and water. In contrast, autumn-sown crops have a longer period in which to develop an adequate root system.

Table 13.3   Hoosfield spring barley. Yields and nutrients (kg/ha/year) supplied by manure and the best fertilizer treatment 2002–2005 (Rothamsted, 2012)

	35 t/ha Manure	Best Fertilizer
Yield t/ha	6–7	5–6
N	240	96–144
P	45	44
K	350	90–180
Mg	25	35

Continuous mangolds and sugar beet (only from 1946) followed by winter wheat, barley and potatoes (barnfield)

Declining yields and area of mangolds in the UK led to a range of arable crops being tested on the plots that for more than 100 years had compared manure and fertilizers on the silty clay loam soil. Manure at 35 t/ha gave higher yields of the continuous root and subsequent arable crops than crops given large amounts of N together with P, K and Mg. It is again thought that the superiority of manure was because the extra soil organic matter (SOM) had improved the structure of a soil known to be notoriously difficult to cultivate.

The benefits of SOM became apparent when the N supplementation of the manure was increased and it was found that yields particularly of mangolds and sugar beet, and to a lesser extent of following crops of spring wheat and spring barley, were greater on manure-treated soils that contained more SOM than on fertilizer-treated soils. Supplementing the N from the manure and from mineralization of SOM with fertilizer N ensured that N supply was not limiting. For all four crops less N was needed to achieve the optimum or near optimum yield when the crops were grown on soils with more SOM (Johnston *et al.*, 2009).

Woburn

Vegetables

The benefits of building up SOM were evident on market garden crops on the sandy loam soils at Woburn after twenty-four years of the annual application of either 37.5 or 75 t/ha manure. For example, yields of red beet were larger on soils with more SOM even when 450 kg N/ha was applied to fertilizer-only plots (Johnston and Poulton, 2005).

Spring barley, potatoes, winter wheat and winter barley

Peat was used to establish two levels of SOM but without adding significant mineral nutrients in experiments between 1973 and 1980 (Johnston and Brookes, 1979; Johnston and Poulton, 2005). Four amounts of N appropriate to the crop were tested. The yields of the potatoes and spring barley were always larger on the soil with more organic matter irrespective of the amount of N applied, but yields of winter-sown cereals were independent of SOM. This provides further support to the notion that spring-sown crops that need to develop a sufficiently large root system quickly benefit from an improved soil structure.

With trials other than those with peat it is difficult to separate the N effect of long-term applications of manure and the mineralization of SOM (see Chapter 11) from other effects. Nevertheless, there are several indications that SOM levels can play a significant part by means not related to the supply of N and that yields are often larger on soils with more organic matter compared to those on soils with less. This is particularly true for more recent crop varieties with higher yield potentials.

Magruder plots
Continuous winter wheat

A continuous dryland winter wheat fertility experiment was started in 1891 on a previously uncultivated prairie silt loam soil in Oklahoma, USA, by A.C. Magruder. It is still running today (Mullen *et al.*, 2001). The experiment was initially established to evaluate the effect of manure application on wheat yields and determine how long the soils could sustain continuous wheat production. Manure was applied from 1891 to 1967 to supply 135 kg N/ha every four years, then from 1968 to the present at 270 kg N/ha every four years. It was only in 1930 that treatments evaluating the annual application of inorganic sources of N (37 kg N/ha, of P (34 kg P_2O_5/ha) and of K (34 kg K_2O/ha) were added.

Manure application doubled the very low yields from about 600-940 kg/ha in the un-manured wheat to 1,140–2,350 kg/ha. It was appreciated from very early on that there was not enough manure for all of Oklahoma's wheat, a similar conclusion to that for wheat in the UK. The yields of wheat given fertilizer NPK have either matched or exceeded those of the manure treatment since 1930. For the first twenty years there was a clear response to P but it was only after thirty years that there was a response to N. It appeared that the N in rainfall together with that mineralized from soil organic matter was sufficient to meet crop N needs for the low-yielding crops for many years. Responses to K were only

observed after ninety years. Manure was able to perform as well as fertilizers in sustaining the climate-restricted low wheat yields over long periods.

SOM levels fell in both the manured and un-manured plots from 3.58% initially to 1.50% after 108 years. Even when manure was applied every four years SOM levels were not sustained under monoculture wheat. The efficiency of use of nitrogen (NUE) applied in the manure over 109 years (1892–2001) was assessed by Davis *et al.* (2003). The NUE (N removed in the grain, minus N removed in the grain of wheat not given manure, divided by the rate of N applied) was 32.8%; this is consistent with 33% reported by Raun and Johnson (1999) for cereal grain production worldwide. Over the last seventy-one years (1930–2001) NUE in the manure and the NPK treatments were 29% and 36% respectively.

Rodale Institute

The Rodale Institute has been in the forefront of the organic movement in the USA since 1947 as a research and advocacy organization (Chapter 6). Its mission is to improve the health and well-being of people and the planet through organic farming. It is one of very few independent organizations doing research on organic farming and is accordingly involved in the practicalities rather than the theory. Studies have included a comparison of farming systems (from 1981) and more recently on the utilization of compost (1993–2001).

Farming systems trial (FST)

Since 1981 the Rodale Institute has compared two organic farming systems with a conventional grain-based farming system in the longest comparative trial in the USA (Pimentel *et al.*, 2005). The conventional system employed extension-prescribed fertilizers and herbicides in a simple five-year crop rotation of maize, soybeans, maize, maize and soybeans – a common rotation throughout the Midwest.

One organic system is manure-based and the other is a legume-based system. Fertility of the manure-based system is provided by the periodic application of composted animal manure and by leguminous cover crops; this system represents a typical livestock operation in which a long rotation of annual feed grain crops and perennial forages are grown for animal feed. The legume-based system utilizes leguminous cover crops as its sole source of nutrients; this legume system features a mid-length rotation of annual grain cash crops and cover crops.

In each of the three systems, maize/soybean production is the main research focus. In general, the organic systems have longer, more diverse rotations with cover crops between the cash crops. The crop rotation in the organic systems includes up to seven crops in eight years (compared to only two conventional crops in two years). While this means that the conventional system produced more maize and soybeans because they occur more often in the rotation, organic systems produce a more diverse array of food and nutrients and are better positioned to produce yields, even in adverse conditions.

Rodale, in common with other organic organizations, perceives the non-use of pesticides and herbicides as major health-promoting benefits of organic farming. Composted cattle manure and leguminous cover crops are the only sources of fertility. Herbicides are replaced by mechanical cultivation and weed-suppressing crop rotations.

In the early years of the FST the conventionally grown maize yields were greater than the two organic systems but latterly there has been relatively little difference between the three systems (Table 13.4).

Between 1981 and 2001, composted manure was applied at the rate of 5.6 t/ha in two out of five years, before planting

Table 13.4 Average annual yields (kg/ha) of maize and soybeans from 1981 to 2001 (Pimentel *et al.*, 2005)

	Maize		Soybean
	1981–1985	1986–2001	1981–2000
Organic, animal manure	4222	6431	2461
Organic, legume	4743	6368	2235
Conventional fertilizer	5903	6553	2546

maize. With the additional nitrogen supplied by leguminous crops in the rotation, the average total nitrogen given (over the whole period) amounted to 40 kg N/ha/year (or 198 kg N/ha/year for any given year with a maize crop). In the legume-based system, hairy vetch cover crop was incorporated before maize planting, translating into an average of 49 kg N/ha/year (or 140 kg N/ha/year for any given year with a maize crop) (Pimentel *et al.*, 2005).

Currently, the rate of composted manure application in the organic system is calculated based on its nitrogen content with a goal of applying 90 kg N/ha. The amount of manure applied to maize in the organic-manure system is roughly equivalent to the amount of manure that would be produced by animals fed on crops from this system. The residual nitrogen available from previous compost applications or leguminous cash or cover crops is allowed for after soil testing and by calculating the application rates of the composted manure accordingly. SOM levels have steadily increased in the organic systems, from an average of 3.5% in 1981 to 4.3% after more than thirty years. In the conventional system, however, soil organic carbon has remained unchanged at 3.5%.

Analysis of twenty-two years of data from the FST by Pimentel *et al.* (2005) revealed that the two organic systems outperformed conventional farming in the drought years. For example, average maize yields (kg/ha) over five drought years (during 1988 and 1998) were 6,938 for organic animal, 7,235 for organic legume and 5,333 for the conventional system. The higher yields were attributed to the greater retention and availability of water in soils with more organic matter, an effect that was particularly valuable in drought years. Soybean yields in the conventional system were half those of the organic manure system during the extreme drought of 1999, when total rainfall during the growing season was only 224 mm compared to 500 mm in normal years.

From thirty years of data from the FST, the Rodale (2011) concluded that:

i.   Organic yields generally match conventional yields but outperform them in drought years.
ii. Organic farming systems build rather than deplete soil organic matter, making it a more sustainable system.
iii. Organic farming uses less energy and generates less greenhouse gases.

Compost utilization trial 1993–2001

Composts and composted manure are fundamental to Rodale's vision of organic farming. Rodale considers mature compost to be a valuable source of slow-release nutrients that enhance soil microbial activity and build-up of soil organic matter. Rodale recognizes that fertilizers and raw cattle manure can provide soluble nutrients for plant growth but they do not build the soil's long-term biological reserves as well as composted manure does. Soil water holding capacity is increased by composted manure.

A study carried out by Hepperly *et al.* (2009) between 1993 and 2001 compared the effects of composted manures, raw cattle manure, and fertilizers on crop performance and soil properties in a maize–vegetable–wheat rotation (Table 13.5). From 1993 to 1998, red clover (*Trifolium pratense* L.) and crimson clover (*Trifolium incarnatum*) were used as annual winter legume cover crop before planting maize. From 1999 to 2001, hairy vetch (*Vicia villosa*) was the legume green manure. In this rotation, wheat depended entirely on residual N that remained in the soil after maize and vegetable production. As different crops were grown each year it is only possible to compare the overall average yields. The amounts of composted manure and dairy manure applied were based on estimates of the proportion of N that was assumed would be available (40% from composted manure and 50% from dairy manure).

Composted and raw dairy manures produced greater maize and wheat yields than mineral fertilizers, which could be attributed to greater amounts of total N supplied by the organic sources. Composted manures (one part manure mixed with three or four parts leaf compost) were better than fertilizers and raw dairy manure in building up soil nutrient levels, in providing residual nutrients for wheat and in reducing nitrate losses. After nine years, soil C and N remained unchanged or declined slightly in the fertilizer treatment, but increased with the use of composted materials. The long-term benefits of composts were evident in the wheat yields, which reflected the residual effects of applications made to previous maize or vegetable crops.

DOK Trial FiBL

Following concerns about the nitrate contents of vegetables and about pesticide residues, the Research Institute of Organic Agriculture (FiBL) in Switzerland set up, in 1978, a long-term experiment, the DOK trial (FiBL, 2000, 2015; Mäder *et al.*, 2002, 2006). Biodynamic (D), organic (O) and conventional (K for German: "konventionell") farming systems were compared on

Table 13.5   Total N applications (kg/ha) and average annual yields (kg/ha) over nine years (1993–2001) for maize, wheat and peppers (Hepperly *et al.*, 2009)

	Maize		Wheat		Bell Peppers	
Treatment	Total N	Yield	Total N	Yield	Total N	Yield
Broiler litter/leaf compost	2220	6296	298	3166	2256	1394
Dairy manure/leaf compost	2819	7181	254	3718	2353	1641
Raw dairy manure	2314	7395	238	3728	1845	1674
Mineral fertilizers	1375	6475	84	3366	1121	1556

wheat, potatoes, maize, soya and grass-clover leys on the same site. Composted farmyard manure containing the biodynamic preparations were used in system D and rotted farmyard manure in system O. Similar amounts of farmyard manure were applied in all treatments (based on livestock units). NPK fertilizers were applied as supplements to the conventional system. Much smaller amounts of N, P and K were supplied by the organic treatments. Preparations 500 (horn manure) and 501 (horn silica) were sprayed several times each year in the biodynamic treatment.

The basic objective was to determine whether organic arable farming is feasible and sustainable given the pressures from weeds, pests and diseases and whether it would produce adequate yields. Every year, three crops of the seven-year crop rotation were grown side-by-side at two fertilization intensities each. The systems being compared in the trial only differ in terms of fertilizer usage and plant protection, while crop rotations, soil cultivation and choice of cultivars were the same across all treatments. The rotation consisted of potatoes, winter wheat in years one and two and grass/clover leys in years six and seven; various crops were grown in years three, four and five, including cabbage, beets, winter wheat, soybeans and maize. No man-made pesticides were used in the organic cropping systems.

Over thirty-five years to 2012 organic yield levels for all crops were approximately 80% of those achieved by the conventional system. Potato yields showed the biggest shortfall; yields in the organic systems were 34–42% lower than in the conventional plots, mostly due to late blight. For winter wheat, yield shortfalls were about 11–14% during 1985–2005 and 33% during 2005–2012 when the conventionally grown crops were given more fertilizers; mean wheat yields in the organic systems over twenty-eight years were 3.8 t/ha. Mean soybean yields over six cropping years during 1999–2012 were similar in the organic and conventional systems. Over the same period, silage maize yields in the organic systems were 11% lower than those in the conventional system. Yields from the organic grass-clover leys were somewhat variable, being 11–13% lower than their conventional counterparts during 1978–1998, 5% higher during 1999–2005, and about 11% lower again between 2006 and 2012.

The long-term nature of the trial has permitted studies on the impact of the different production systems on soil structure and fertility. The organically and biodynamically managed plots contained 25% more soil microorganisms and exhibited higher long-term soil fertility than the conventional plots. Crops in the organically and biodynamically managed plots were more strongly colonized by mycorrhiza (and harboured more mycorrhizal species). Both organic systems led to improved soil structure. Over the whole period to date no consistent differences were found between the two organic systems, although there were occasional differences in crop performance related to the variable nutrient composition (mainly K) of the composts and manure.

Organically grown crops used less fossil fuel energy than conventional crops; the greater use of fuel for mechanical cultivation in the organic system was more than offset by the relatively large amount of energy needed to produce fertilizers and pesticides. On a unit area basis 30–50% less energy was used in the organic systems but on a unit crop basis it was only 19% lower than that in the conventional system.

Comparison of yields of organic and conventionally grown crops

It has been widely acknowledged for some time that organic yields (based on one cropping season and without allowance for an associated fallow or cover crop year) for all crops, excluding legumes, are less than those of conventional crops, e.g. Stockdale *et al.* (2000) Table 13.6. However, the reasons for the lower yields are seldom discussed, often because the studies were not designed to address such matters.

Over the last twenty-five years there have been at least six major meta-analyses

Table 13.6   Yields of organic crops as percentage of conventional crops (Stockdale *et al.*, 2000)

	%
Arable crops	60–80
Forage crops	70–100
Fruit	50–80

comparing the yields of organic and conventional farming, by Stanhill (1990), Badgley *et al.* (2007), Seufert *et al.* (2012), de Ponti *et al.* (2012), Ponisio *et al.* (2014) and Tuomisto *et al.* (20012). It was probably inevitable that the studies would attract comment, e.g. by Goulding (2009) on the Badgley study and by Ponisio on all the previous studies.

As yields levels vary widely between trials, organic yields have usually been expressed as percentages of conventional yields when aggregating the studies in order to make the comparisons. In some cases the data is subjected to sophisticated statistical operations that can obscure and may give the impression of precision. The majority of the studies used in the meta-analyses can suffer from the defect that too little attention is paid to the way in which the organic crops obtain nutrients. An inherent problem for many meta-analyses is the subjective nature of the selection process for including studies. Some studies have been used in more than one meta-analysis.

Stanhill (1990)

Stanhill collected data on twenty-six crops including maize, wheat, barley, soybeans, potatoes, beans, carrots and tomatoes. On the basis of 205 comparisons it was concluded that organic yields averaged 91% of conventional; however, milk yields were greater on organic farms. About one third of the data, mostly from the 1970s, showed the superiority of organic farming, notably biodynamic. Data from the Haughley experiment was

1 *Stromatolites*: A very early, possibly 3.5 billion years ago, life form consisting of blue-green algae photosynthesizing in cemented sediments. (Hamelin Pool, Western Australia, 2007).

2 *Tomatoes*: Plants can be grown without soil. The healthy tomatoes are growing with their roots in rock wool and supplied with nutrient solution (Ely, Cambridgeshire, UK, 2009).

3   *Pigeon Coops*: Pigeon droppings (guano) have been prized since Roman times for promoting crop growth (Egypt, 1983).

4 *Weed-free crops*: Soya beans on left growing in a soil successfully treated with a pre-emergent herbicide (Rio Grande do Sul, Brazil, 1974).

5 *Caterpillars, cabbage white:* Damage of this severity requires treatment with an insecticide; there are no options (Hertfordshire, UK, 2015).

6 *Beans for humans*: Are beans destined to be the main source of protein?

7 *Agroponicos*: Organic agriculture Cuban style. Vegetable growing in beds of soil enriched with organic waste materials (Varadero, Cuba, 2016).

8 *Crop spraying*: Crop spraying without equipment (Syria, 1974).

9   *Multi-faceted organic market:* Spices and beverages are prominent in the market in Kerala (Cochin, India, 2014).

10   *Oxen*: Is manual ploughing with oxen a viable alternative to mechanised ploughing? (Vinales, Cuba, 2016).

also used. It was significant that Stanhill accepted that there were many difficulties in comparing yields.

Badgley et al. *(2007)*

Badgley *et al.* challenged the proposition that organic agriculture cannot contribute significantly to the global food supply because of low yields and insufficient supply of organically acceptable sources of N. Using a global dataset of 293 examples, organic yields were compared with conventional (n=160) or with low-intensity food production (n=133) by estimating the average yield ratio (organic/conventional) of different food categories for developed and developing countries. In the developed countries the organic yields were on average 91% of conventional (n=138). In the developing countries the organic yields of food crops were 173% of conventional for the same range of crops (n=128). Overall organic crop yields were 30% higher than conventional. All of the previous data given by Stanhill (1990) were used (n=205), together with data from Lampkin and Padel (1994).

Organic agriculture was very loosely defined by Badgley *et al.* and was not restricted to any particular certification criteria. Man-made fertilizers and pesticides would generally not have been used. The data for developing countries that showed organic crops giving superior yields relied heavily on data where the evidence was based on yields before and after changes in farming practices following agricultural training programmes; much less data was derived from controlled experiments and paired farm comparisons. The training programmes reported, for example, by UNEP-UNCTAD (2008) and Pretty *et al.* (2001, 2006, and 2011) in developing countries, can be very valuable at subsistence level farming in introducing improved farming methods but they are not the way that a low input farmer on a small farm can become more than a subsistence farmer. The programmes introduced several resource-conserving technologies and practices, including integrated pest and nutrient management, conservation tillage, agroforestry, mixed farming and improved varieties. It is inappropriate to include such studies when comparing organic and conventional farming. Subsistence farmers do not become organic farmers simply because they do not use fertilizers and pesticides.

Having established the organic/conventional yield ratios, Badgley *et al.* calculated the potential organic global food production by multiplying the current crop yields for the different crops by the appropriate ratio. Their estimates of global biological N fixation were somewhat optimistic; they were based on the areas of arable land on which extra leguminous green manures could be grown. They concluded that organic farming could produce enough food on a global per capita basis to sustain the current human population, and potentially an even larger population, without increasing the area of agricultural land and that leguminous cover crops could fix enough nitrogen to replace the amount of man-made fertilizer currently in use. Not surprisingly, these conclusions were challenged.

Goulding *et al.* (2009) found that the Badgley claims were based on a literature survey that was selective and on overestimates of N fixation by legumes; they estimated that yields of organic wheat were 65% of conventional compared with Badgley's 85%. Moreover, the amount of N fixation by legumes is usually too small and variable to support large and consistent wheat yields. The optimistic conclusions of Badgley were also criticized by Cassman (2007) and Hendrix (2007). Avery (2007) found that more than 100 non-organic yields were incorrectly claimed as organic and that organic yields were misreported. Some unrepresentative low non-organic yields used to compare with high organic yields were counted several times by citing different papers that referred to the same data. Favourable and unverifiable studies from biased sources were given equal weight to rigorous university studies.

Seufert et al. *(2012)*

Seufert *et al.* used sixty-six studies (thirty-four different crops) where all the organic crops were certified organic; they calculated that organic yields were on average about 75% of conventional yields. For rain-fed legumes and perennials on near neutral pH soils the organic yields were only about 5% less than conventional. Organic yields were 13% lower when best organic practices were used and 34% lower yields when the conventional and organic systems were most comparable. Some of the yield differences were explained by the differences in the amounts of N inputs. When organic systems receive more N than conventional systems, organic yields were better. In other words, some organic systems appear to be N limited, whereas conventional systems are unlikely to be. Seufert *et al.* were optimistic, saying that under certain conditions that is, with good management practices and growing conditions – some organic crop types e.g. fruit and oilseed crops can nearly match conventional yields.

Connor (2008, 2013) responded to Badgley *et al.* (2007) and to Seufert *et al.* (2012), in particular pointing out that the potential of an organic farming system cannot be assessed from comparisons of annual yields of individual crops. He was concerned that the Badgley and Seufert studies would renew false hopes for switching to organic systems. He pointed out that the source of the nutrients used in organic farming is seldom considered. The use of restorative legume cover crops and the recycling of manure containing imported nutrients are often part of organic farming and need to be considered when assessing the potential for organic farming to feed the world. Land is required somewhere to generate/provide the nutrients. When nutrients are sourced abroad, as is often the case in the UK and Europe, in the form of animal feeding stuffs the only practical solution is for them to be replaced by fertilizer inputs and by the growing of legumes in the exporting country.

de Ponti et al. (2012)

De Ponti *et al.* conducted a meta-analysis study using 362 published papers comparing organic and conventional yields. Organic yields were on average 80% of conventional yields, but the variation was substantial (Table 13.7). The yield gap differed significantly between crop groups and regions, and tended to be greater when conventional yields were high.

Tuomisto et al. (2012)

Average organic yields were found to be 75% of conventional yields over all crops in a meta-analysis of seventy-one studies covering thirteen crops, including wheat, barley, oats, oilseed rape, potato, vegetables, sugar beet, leys, olive, citrus and melons by Tuomisto *et al.* (2012) (Table 13.8). The lower organic yields were variously attributed to insufficient nutrients, weeds, diseases and pests. They estimated that the lower crop yields, the

Table 13.7   Yields of organic crops as percentage of conventional crops (de Ponti *et al.*, 2012)

	%
Cereals	79
Wheat	73
Fodder	86
Potato	74
Pulses	88
Oilseed	74

Table 13.8   Yields of organic crops as percentage of conventional crops (Tuomisto *et al.*, 2012)

	%
Winter wheat	62
Spring wheat	78
Barley	65
Potato	68
Vegetables	79
Grass leys	85

reduced animal production and the land area required for fertility-building crops in organic farming would mean that in Europe 84% more cultivatable land would be required if all farming became organic.

Ponisio et al. (2014)

Ponisio *et al.* brought together 115 studies (1,071 yield comparisons) on fifty-two crops in thirty-eight countries carried out over thirty-five years, which included much of the data already used in previous meta-analyses. Ponisio *et al.* were critical of the selection methods and statistics (basically lack of variance estimates) used in the meta-analyses of Stanhill, Badgley, Seufert and de Ponti. Ponisio *et al.* (2014) excluded comparisons of improved subsistence yields following introduction of new farming methods and they sought variance estimates from the original authors in order to improve the assessment. They found that organic yields were on average for all crops only 19% lower than conventional yields, a smaller yield gap than previous estimates (Table 13.9).

Table 13.9   Yields of organic crops as percentage of conventional crops (Ponisio *et al.*, 2014)

	%
All crops	81
Cereals	79
Legumes	85
Non-legumes	82
Fruit and Nut	94
Root and Tuber	71
Vegetables	83

It was not possible for Ponisio *et al.* (2014) to distinguish between the studies where organic crops were harvested every year from those that relied on a previous fertility-building crop and were thus dependent on more than one year's cropping. However, it was found that multi-cropping (growing several crops together in the same field) and the use of crop rotations reduced the yield gap to under 10%, which is indicative of the value of fertility-building crops. They were confident that given appropriate investment in agro-ecological research to improve organic practices the yield gap for some crops could be greatly reduced or even eliminated: this does not allow for conventional yields being increased by the introduction of new practices. The poorer performance of organic farming was thought to be partly due to the fact that no crop varieties had been bred specifically for organic farming systems. Ponisio *et al.* were able to look at the yield gap between organic and conventional at different levels of N input. When N inputs in the organic and conventional crops were similar there was a small yield gap (9%), compared with a gap of 30% when more N was supplied in conventional farms. The yield gap was intermediate at 17% when N inputs were greater in organic treatments; other factors related to the organic practices must have been limiting yield, such as pests or diseases.

Forster et al. *(2013)*

Information concerning the performance of farming organically and conventionally in tropical and subtropical regions was provided by Forster *et al.* (2013) in India as part of SysCom (FiBL, 2016). They compared the productivity of a cotton–soybean–wheat crop rotation under three farming systems, namely biodynamic, organic and conventional (urea, superphosphate and potassium chloride were applied, cotton with and without Bt). During the first crop cycle (2007–2008) the organic cotton crop yields were 29% lower than conventional and the wheat crops were 27% less. During the second cycle (2009–2010), when all yields were lower the cotton and wheat yields were similar in all farming systems but the organic soybean yields were 11% lower. The use of fewer inputs made organic soybean production – as part of cotton-based crop rotations – more profitable. There were small insignificant yield differences between crops farmed biodynamically and organically.

FiBL

The Research Institute of Organic Agriculture, Switzerland, has been involved, since 2005–2007 with long-term Farming Systems Comparison in the Tropics collectively known as SysCom in three tropical countries, Bolivia (cacao), India (cotton, wheat and soybean) and Kenya (maize) FiBL (2016). It is reported that there is a strong potential for organic farming for improving farmers' incomes but in the absence of data, for India and Bolivia, on yields and inputs, including their source, it is not possible to assess the relevance of the findings; it is possible that the yields of the conventionally grown crops were so low that any improvement would have been appreciated. Muriuki *et al.* (2012) concluded that in Kenya the results suggested that organic farming using manure may be a viable option for tropical Africa. Without any input of N into the farming system from man-made fertilizers or from legumes, manure is doing no more than recycling N and fertility will eventually decline following the mining of N in soil organic matter and the loss of reactive N into the environment.

USA: Comparison of organic and conventional yields in 2008 and 2011

The several meta-analyses of the many comparative yield studies are roughly in agreement with the differences reflecting the variability of the source data. The studies by Savage (2011 and 2014) may come to be the definitive comparative study as it is based on actual reported crop yields representing 80% of crop land in the USA in 2008 and 2011. This has been possible as a result of the publishing of the organic acreage and production data for these two years, which has allowed the average organic yields to be compared with the current and historical yields of conventional crops derived from national statistics (USDA 2015, USDA-NASS). The 2008 Organic Survey was a complete inventory of all known organic farms (14,093 covering 1.48 million ha) in the USA (USDA, 2015). Organic farming is practised on about 0.44% of US cropland.

This USDA data allows the actual organic farming and conventional farming systems to be compared for nearly the whole of the country. Yields will have been influenced by factors not experienced or practised uniformly by the two systems. For example, the extent to which irrigation was used may have differed. The non-use of genetically modified crops by organic farmers may also have been to their detriment. Nevertheless, there are no reasons for doubting that best practices were not observed on both organic and conventional farms.

Savage (2014) compared the yields on a state-by-state basis, which entailed 370 crop comparisons. In 84% of the comparisons, organic yields were lower mostly by 20–50%; in 7% of the comparisons yields were equal. The rare instances where organic yields were higher than conventional (9% of total) were almost entirely for hay and silage crops (89%) and only 10% were food crops.

The lower yields of nearly all organic crops are evident in Tables 13.10, 13.11 and 13.12; there is fairly close agreement between the 2008 and 2011 surveys.

Savage calculated that if all crops had been grown organically in 2014 at least an extra 44 million hectares would have been required to achieve the same national production; this area is equivalent to the entire parkland and wild lands in the lower forty-eight US states; the comparable area in 2008 was 49 million ha. Total farmland in USA is about 370 million ha, of which about 130 million ha consists of arable crops. The main extra required areas are shown in Table 13.13. For hay crops 2.67 million fewer ha would be needed.

Table 13.11 Organic yields of tree, nut and vine crops in 2011 as percentage of conventional crops (Savage, 2014). Data for 2008 survey are shown in parenthesis

Cranberries	33
Figs*	36
Strawberries*	39 (58)
Grapes	51 (51)
Walnuts	52
Pears*	68
Oranges	74
Peaches	75
Coffee	75
Apples *	75 (88)
Dates*	90
Raspberries	128

*Most widely grown organic tree, nut and vine crops.

Table 13.10 Organic yields of row crops in 2011 as percentage of conventional crops (Savage, 2014). Data for the 2008 survey are shown in parenthesis

Cotton	55
Flax*	57 (83)
Rice	61 (59)
Groundnuts	63
Maize grain	65 (71)
Barley	67 (70)
Soybean	69 (66)
Winter wheat	71 (60)
Sunflower	74
Dry edible beans	77 (81)
Sorghum	79 (52)
Haylage, silage*	81
Sorghum silage*	83
Oats*	83 (86)
Hay (not lucerne)	121

*Most widely grown organic row crops.

Table 13.12 Organic yields of vegetable crops in 2011 as a percentage of conventional crops (Savage, 2014) Data for 2008 survey are shown in parenthesis

Spinach *	29
Tomatoes	39 (63)
Carrots*	51
Peppers	60
Cabbage	62 (43)
Celery	68 (69)
Potatoes	70 (72)
Onions	79
Squash*	90
Peas	112
Sweet corn*	116
Sweet potato	129
Snap beans	144

*Most widely grown organic vegetable crops.

Table 13.13	Extra areas (million ha) of selected crops required if all production in the USA had been organic in 2014

Maize, grain	18.25
Soybeans	15.26
Winter wheat	5.34
Miscellaneous row crops	3.56
Cotton	3.10
Spring wheat	2.61
Rice	0.74
Sorghum	0.69
Maize silage	0.65
Barley	0.52
Grapes	0.40
Groundnuts	0.31
Almonds	0.26
Lettuce	0.24
Sunflower	0.22
Dry edible peas	0.20
Potatoes	0.18
Dry edible beans	0.17

Some implications of conversion to organic

The notion that organic farming had a huge potential was partly derived from the misinterpreted work of UNEP-UNCTAD (2008), Pretty and Hine (2000) and Pretty *et al.* (2006 and 2011), which had led to suggestions that yields might increase significantly under organic farming.

As was clearly evident from the classical long-term trials at Rothamsted and the practical experience in the USA, the requirement for extra cattle and land to sustain organic farming on a large scale cannot be met.

CWS Agriculture compared a stockless organic and a mixed conventional farm in a seven-year investigation of the technical feasibility and the economic consequences of organic farming (Leake, 1999). Organic farming was found to be as profitable as conventional farming but depended on obtaining price premiums for the produce. Yields of organic winter wheat were 60% and winter oats 75% of conventional yields. Beans were found to be a more successful fertility building break crop than peas. A complete switch to organic farming would result in a 44% drop in UK wheat production based on the CWS Agriculture results. Only two organic crops suffered severely from pests: a crop of peas to an attack of aphids and a crop of wheat to slugs. Weeds, however, caused complete crop failure on several occasions and were rated as a major problem.

The extent to which organic agriculture might be able to provide as much food as conventional farming in England and Wales was addressed by Jones and Crane (2009) in a statistical study partly sponsored by the Soil Association. They used production data from a sample of 176 farms, classified basically as organic, and then computed the national production if the whole farmed area was converted and achieved the measured organic production levels. A problem with this study was that most organic farms in the study were livestock or mixed farms and that there were very few organic arable farmers on which to base production levels (Table 13.14).

It was calculated that if all cereals in England and Wales were organic, production would fall to about 60% of

Table 13.14   Types of organic farms in England and Wales used in the study (Jones and Crane, 2009)

Livestock	74
Dairy	39
Mixed	18
General crops	15
Cereals	12
Horticulture	11
Poultry	6
Pigs	1
Total	176

conventional production – a result similar to that of Leake (1999). It was impossible to calculate the situation for oilseed rape and sugar beet as there is no domestic organic production. A wholly organic agriculture would produce forage peas/beans and potatoes equivalent to the conventional. A similar situation exists with vegetables for which national production could be maintained under an organic regime.

A basic assumption behind these calculations is that the resources used by the existing organic farmers would be available for all farmers in the hypothetical organic situation. The sources of the nutrients applied on the few organic arable crops were not identified. The alternatives are N fixation by clovers in grass leys grazed in the summer, by legume cover crops that will be ploughed in and the consumption of animal feeding stuffs; all would lead to the generation of nutrient-recycling manure. Whatever the source, the implications are that the converted arable farms would have to introduce animals and/or grow legume cover crops.

A wholly organic dairy sector could produce as much as 70% of conventional milk volumes. As livestock enterprises are common on organic farms, a wholly organic agriculture would supply 68% more beef and 55% more lamb than conventional agriculture. Such increases would have to be accompanied with more grazing in the arable regions, which must be allowed for. The extra beef and lamb would pose problems, especially at a time when the overconsumption of red meat is increasingly seen as a health risk. Many see a reduction in the consumption of livestock products as a key component of the way organic agriculture can feed the world (Chapter 19).

The pig and poultry meat sectors would be hardest hit by transition to organic production. With the end of intensive systems, production would fall to some 30% of conventional levels, although there would be some potential for expansion of production on to the extensive areas of grassland available on the envisaged organic farms. The extra grassland area involved is not quantified. For eggs a wholly organic agriculture is estimated to produce around 73% of the conventional egg numbers.

As England and Wales is not self-sufficient in food production there is no suggestion that a wholly organic agriculture could feed the current population. The achievement of self-sufficiency by organic farming would require revolutionary changes, at least in eating habits and in the organization of agriculture (Chapters 18, 19).

Nutrient sources for organic farms

N application in manure is sometimes restricted by organic regulations, e.g. in the UK organic farmers can supply manure providing up to 250 kg N/ha/ year to a given field provided that the average application over the whole farm does not exceed 170 kg N/ha/year (Soil Association, 2016). On the basis that 1 t cattle manure contains 6 kg N, an organic farmer can apply up to 40 t/ha manure to any one field, which is similar to the amounts supplied annually to wheat in the Broadbalk experiment. The EC Regulation 1804/1999 also directs that organic farms should not apply more than 170 kg/ha N overall. The question is: where does the manure and its N come from?

It is possible, in several European countries, for organic farmers to obtain manure/nutrients from conventional farms provided the amounts are limited and are not the output of factory farming; such manure must not be the basis of manuring programmes. According to European organic regulations, animals must be fed with organically produced grains and fodders (EC, 2008). However, some derogations exist that allow the use of conventionally produced animal feeding stuffs when organic supply is low (OFIS, 2013).

Kirchmann *et al.* (2008) reported that 75% of organic mixed farms in Austria and Sweden imported some fodders from conventional farms and that 27% of stockless organic farms imported manure from conventional farms. It was concluded that organic agriculture was often dependent on nutrients derived from conventional farming; clearly under such circumstances organic farming cannot be described as sustainable.

In a study in south-west France, Nesme *et al.* (2012) found that most organic farms imported organic materials and/or compost and manure, the latter from neighbouring farms or urban areas. Nowak *et al.* (2013), also in France, showed in a two-year study on sixty-three organic farms that inputs from conventional farming of N, P and K amounted on average to 87, 9 and 16 kg/ha/ year respectively or 23%, 73% and 53% of the N, P and K applied. These inputs were mostly in the form of manure and were higher in farms with low stocking rates, which are quite common in Europe; e.g. 66% of organic farms were stockless in France in 2011 (Nowak *et al.* 2013; AgenceBio, 2013). Similar results have been reported in Denmark, where in 2011 approximately 25% of nutrients given to organic crops came from manure from conventional farms (Oelofse *et al.*, 2013).

Concern is now being shown in Europe about the reliance organic farming places on conventional farming because it runs counter to the fundamental concepts of organic agriculture. The organic movement in Denmark has set a maximum of 70 kg N /ha/year that can be sourced from conventional manure and has plans to ban the use of all conventional materials by 2022 (Oelofse *et al.*, 2013). It is known that organic farmers in Austria,

France, Denmark and Sweden have indirectly been using man-made fertilizer and most likely genetically modified soybeans in animal feeding stuffs from conventional farms for some time; the situation in other European countries is not known.

Concern about the many certified organic farmers in Norway that use manure from conventional farms in Norway led Mckinnon *et al.* (2014) to consider the undesirable contaminants that might be transferred; the main suspects are heavy metals, veterinary medicines and pesticides. Although the potential contaminants have been identified, little appears to be known about the amounts, the fate and relevance of the ones that reach the food chain.

Another way by which nutrients from conventional farming benefit organic agriculture is through residual nutrient reserves in soils that were built up prior to conversion to organic. The fertility of most conventionally managed soils will have been increased by the long-term applications of inorganic P and K fertilizer. The residual P and K effects are most often not considered. Sewage sludge has a considerable potential as a source of P in particular but is currently not allowed in organic farming.

Summary

1. Comparisons of organic and conventional farming need to consider firstly whether or not the yielding potential of both systems is the same and secondly what is the current practical experience.

2. It is known that manure as well as being an important source of nutrients can improve soil structure and influence soil microorganisms; these effects may be reflected in crop performance.

3. The definitive experiment to answer the questions about yield potentials and whether there are organic farming practices that cannot be replicated by conventional farming has not been carried out, largely because of the near impossibility of matching the N supply and timing in both systems.

4. The long-term classical experiments on winter wheat, spring barley and root crops at Rothamsted have provided much information because the history of nutrient inputs is documented and because sufficient time (over 160 years) has elapsed for effects to become apparent.

5. From the 1850s to the mid-1970s the same yields of winter wheat, spring barley and mangolds were obtained with fertilizers, providing sufficient amounts of N, P, and K as with manure applied at 35 t/ha annually. This was a clear demonstration of the efficacy of both man-made fertilizers and of manure, even though considerably more nutrients were supplied in the manure than in the fertilizers.

6. However, since the 1980s evidence has accumulated showing that the massive amounts of manure applied annually for many years has not only built up soil organic matter but has also improved yields of spring-sown crops by mechanisms seemingly not linked to their nutrient content or to the nutrients released on mineralization of soil organic matter. Winter-sown cereals do not show the same effect. It is currently thought that

the improvement of soil structure by the manure has a greater effect on spring crops that need to establish an absorbing root system as quickly as possible than it has for autumn–sown crops that have a longer time to develop an effective root system. It took about 140 years and the application of just under 5,000 t/ha for this effect on spring-sown crops to become apparent.

7. It can be calculated for the UK that the conversion of all wheat farming to a manure-based system would require 2.4 million extra dairy cows and 1.2 million ha extra summer grazing – an impossible scenario even before other arable crops are also considered as well as the impossibility of consuming the extra dairy/livestock products.

8. The first whole farm comparison of organic and conventional farming set up by Balfour, the founder of the organic movement in the UK, suffered from a basic problem in that the organically farmed land was a closed system. Nutrients removed in the crop and livestock products were not replaced. Inevitably soil fertility declined. Organic systems have to import nutrients, especially N, and/or generate N by biological fixation on the farm if they are to succeed.

9. The Magruder continuous dryland winter wheat fertility experiment started in 1891 in the USA showed that the yields of the very poor crops were similarly increased by both manure and fertilizers. Mineralized N, plus that from rain, appeared to be sufficient for many years.

10. The Farm System Trial of the Rodale Institute has shown that organic yields are generally able to match conventional yields but can outperform them in drought years. The Rodale Institute values the build-up of soil organic matter and the lower environmental impact of organic farming in its quest for a more sustainable agricultural system

11. Composted manures based on dairy manure and broiler litter appeared to be more effective than fertilizers in promoting maize and wheat yields, possibly because more N was supplied.

12. The feasibility of organic farming was examined by FiBL in the DOK trial in which biodynamic and mainstream organic were compared with conventional farming. Over thirty-five years organic yield levels for all crops were approximately 80% of those produced by the conventional system. Organic potatoes showed the largest shortfall.

13. The organically and biodynamically managed plots in the DOK trial contained 25% more soil microorganisms and exhibited higher soil fertility than the conventional plots. Crops in the organically and biodynamically managed plots were more strongly colonized by mycorrhiza (and harboured more mycorrhizal fungal species). Both organic systems led to improved soil structure and generally had similar effects. The absence of differences between the effects of the mainstream organic and the biodynamic systems suggests that the biodynamic preparations had no effect.

14. A few hundred studies comparing the yields of organically and conventionally grown crops have been subjected to meta-analyses in the last twenty-five years by at least six groups.

15. With the exception of one analysis, all the studies demonstrate that organic farming is not as productive

as conventional. Organic yields vary widely according to the local circumstances. Relatively little attention has been paid to the reasons for the poorer organic yields and to the way the organic practices are managed to generate sufficient nutrients, particularly N. The yields of organic crops are often dependent on a previous fallow or cover crop year that is often ignored when comparing yields.

16. According to the meta-analyses, yields of most organic crops with the exception of legumes would appear at best to be about 80% of conventional yields. Some workers remain optimistic that organic farming can be improved and that ultimately there will only be a very small yield penalty. Provided sufficient nutrients are supplied in manure and are available for uptake at the right time there is no reason why organic yields cannot match conventional, as shown by the classical experiments at Rothamsted. The problem is that it is impossible to produce enough manure for more than a limited area.

17. A much clearer picture of actual organic and conventional yields of many crops has been provided by studies based on nearly all the organic and most conventional crops in the USA. There is good agreement with the meta-analyses in that organic yields of some crops in the USA were at best 80% of conventional in 2008 and 2011. The whole US survey, which represented about 80% of farmland, was very robust by virtue of its size.

18. For many organic crops, including maize, rice, cotton, groundnuts, grapes and several vegetable crops, yields were only 30 to 65% of conventional yields.

19. If the USA had converted to organic in 2014 at least 44 million ha more cultivatable crop land would have been required to meet food production; this would be an area as large as all the parkland and wild land in all the lower forty-eight US states.

20. The implications of conversion to organic agriculture are seldom addressed, despite the claims that organic farming is the route to sustainable food production. A study in England and Wales concluded that it would only be possible to produce around 60% of current conventional cereals production and 70% for milk if all conventional farming converted to organic. At the same time, beef and lamb production would be increased by 68% and 55% respectively, whereas pig and poultry production would fall by about 70%. Such changes with livestock products is completely out of line with the current thinking of the organic movement that meat consumption should be reduced across the world.

21. Significant numbers of organic farmers in Austria, France, Denmark, Norway and Sweden import manure from conventional farms, which is of concern because it runs counter to the fundamental principles of the organic movement. Such imports of manure to organic farms from conventional are permitted by EU regulations; however, the extent of the practice is not known. It means that indirectly organic farmers are reliant on man-made fertilizers and even of genetically modified crops. This is a clear indication that a farming system that does not permit fertilizer inputs to replace the nutrients exported in crop and livestock products is not sustainable.

References

AgenceBio (2013) Available at www. agencebio.org

Avery, A. (2007) Organic abundance report: fatally flawed. *Renewable Agriculture and Food Systems* 22, 4, 321–323. doi: 10.1017/S1742170507002189.

Badgley, C., Moghtader, J., Quintero, E, Zakem, E., Chappell, M.J., Aviles-Vazquez, K., Samulon, A. and Perfecto, I. (2007) Organic agriculture and the global food supply. *Renewable Agriculture and Food Systems* 22, 2, 86–108. doi: 10.1017/S1742170507001640.

BSFP (2013) British Survey of Fertiliser Practice. Fertiliser use on farm for the 2013 crop year. Available at https://www.gov.uk/government/uploads/system/uploads/attachment_data/file/301475/fertiliseruse-statsnotice-08apr14.pdf

Cassman, K.G. (2007) Editorial response by Kenneth Cassman: Can organic agriculture feed the world – science to the rescue? *Renewable Agriculture and Food Systems* 22, 2, 83–84.

Connor, D.J. (2008) Organic agriculture cannot feed the world. *Field Crops Research* 106, 187–190.

Connor, D.J. (2013) Organically grown crops do not a cropping system make and nor can organic agriculture nearly feed the world. *Field Crops Research* 144, 145–147. doi.org/10.1016/j.fcr.2012.12.013.

Davis, R.L., Patton, J.J., Teal, K.Y., Tang, M.T., Humphreys, J., Mosali, Girma, K., Lawles, J.W., Moges, S.M., Malapati, A., Si, J., Zhang, H., Deng, S., Johnson, G.V., Mullen, R.W. and. Raun, W.R. (2003) Nitrogen balance in the Magruder plots following 109 Years in continuous winter wheat. *Journal of Plant Nutrition* 26, 8, 1561–1580. doi: 10.1081/PLN-120022364.

EC (2008) Commission regulation (EC) No 889/2008 Rules for the implementation of Council Regulation (EC) No 834/2007 on organic production and labelling. *Official Journal of the European Union* L250.

FiBL (2000) FiBL Dossier. Results from a 21 year old field trial. Available at https://shop.fibl.org/fileadmin/documents/shop/1090-doc.pdf

FiBL (2015) DOK-Trial, The world's most significant long-term field trial comparing organic and conventional cropping systems. Available at www.fibl.org/en/switzerland/research/soil-sciences/bw-projekte/dok-trial.html

FiBL (2016) Organic equals conventional agriculture in the tropics. Available at www.systems-comparison.fibl.org/fileadmin/images/subdomains/syscom/News/SysCom_Kenya_International_media_release.pdf

Forster, D., Andres, C., Verma, R., Zundel, C., Messmer, M.M. and Mäder, P. (2013) Yield and Economic Performance of Organic and Conventional Cotton-Based Farming Systems – Results from a Field Trial in India. *PLOS ONE* 8(12): e81039. doi: 10.1371/journal.pone.0081039.

Goulding, K.W.T., Trewavas, A.J. and Giller, K.F. (2009) Can organic farming feed the world? A contribution to the debate on the ability of organic farming systems to provide sustainable supplies of food. *Proceeding 663 International Fertilizer Society, York, UK.*

Hendrix, J. (2007). Editorial response by Kenneth Cassman: Can organic agriculture feed the world – science to the rescue? *Renewable Agriculture and Food Systems* 22, 2, 83–84.

Hepperly, P., Lotter, D., Ulsh, C.Z., Seidel, R. and Reider, C. (2009) Compost, manure and synthetic fertilizer influences, crop yields, soil properties, nitrate leaching and crop nutrient content. *Compost Science and Utilization.* 17, 2, 117–126.

Johnston, A.E., and Brookes, P.C. (1979) Yields of, and P, K, Ca, Mg uptakes by crops grown in an experiment testing the effects of adding peat to a sandy loam soil at Woburn, 1963–77. *Rothamsted Experimental Station Report for 1978*, Part. 2, 83–98.

Johnston, A.E., and Poulton, P.R. (2005) Soil organic matter: Its importance in sustainable agricultural systems. *Proceedings. 565 International Fertiliser Society* York, UK.

Johnston, A.E., Poulton, P.R. and Coleman (2009) Soil Organic Matter: Its Importance in Sustainable Agriculture and Carbon Dioxide Fluxes. *Advances in Agronomy* 101. 1–57.

Jones, P. and Crane, R. (2009) CAS Report 18 Centre for Agricultural Strategy, University of Reading. Available at https://www.researchgate.net/publication/242130448_England_and_Wales_under_organic_agriculture_how_much_food_could_be_produced

Kirchmann, H., Katterer, T. and Bergstrom, L. (2008) Nutrient supply in organic agriculture – plant availability, sources and recycling in Organic Crop Production – Ambitions and Limitations 89–116. In: Kirchmann, H. and Bergstrom, L. (eds) *Organic Crop Production – Ambitions and Limitations* Springer, Dordrecht, The Netherlands. Available at http://pub.epsilon.slu.se/3510/1/Organic_Crop_Production_Chapter5_2008.pdf

Lampkin, N. and Padel, S. (1994) *The Economics of Organic Farming – An International Perspective*, CABI, Wallingford, UK.

Leake, A.R. (1999) A report of the results of CWS Agriculture's Organic Farming Experiments 1989–1996. *Royal Agricultural Society of England Journal*, 160.73-81.

Mäder, P., Fliessbach, A., Dubois, D., Gunst, L., Fried, P., Niggli, U. (2002) Soil fertility and biodiversity in organic farming. *Science* 296, 5573, 1694–7. doi: 10.1126/science.1071148.

Mäder, P., Fliessbach, A., Dubois, D., Gunst, L., Jossi, W., Widmer, F. and Gattinger, A. (2006) The DOK experiment (Switzerland). 41–58. In: Raupp, J., Pekrun, C., Oltmanns, M. and Köpke, U. (eds). *Long-term Field Experiments in Organic Farming International Society of Organic Agriculture Research (ISOFAR), Scientific Series.* Berlin, Germany.

Mckinnon, K. Serikstad, G.L. and Eggen, T. (2014) Contaminants in manure – a problem for organic farming? 903–904. In: Rahmann, G. and Aksoy U. (eds) *Proceedings of the 4th ISOFAR Scientific Conference. Building Organic Bridges*, Istanbul, Turkey.

Mullen, R.W., Freeman, K.W., Johnson , G.V. and Raun, W.R. (2001) The Magruder Plots – Long-Term Wheat Fertility Research. Better Crops. 85, 4, 6–8 Available at www.ipni.net/ppiweb/bcrops.nsf/$webindex/7FDAD6021219B07385256AEA005BFFEE/$file/01-4p06.pdf

Muriuki, A., Musyoka, M., Fliessbach, A. and Forster, D. (2012) Resilience of Organic versus Conventional Farming Systems in Tropical Africa: The Kenyan Experience 180. In: Tielkes, E. (ed.) *Tropentag 2012 – Resilience of agricultural systems against crises - Book of abstracts*, Cuvillier Verlag, Göttingen, Germany.

Nesme, T., Toublant, M., Mollier, A., Morel, C. and Pellerin, S. (2012) Assessing phosphorus management among organic farming systems: a farm input, output and budget analysis in Southwestern France. *Nutrient Cycling in Agroecosystems*, 92, 225–236, doi: 10.1007/s10705-012-9486-0.

Nowak, B., Nesme, T., David, C. and Pellerin, S. (2013) To what extent does organic farming rely on nutrient inflows from conventional farming? *Environmental Research Letters* 8, 4, 044045. doi: 10.1088/1748-9326/8/4/044045.

Oelofse, M., Jensen, L.S. and Magid, J. (2013) The implications of phasing out conventional nutrient supply in organic agriculture: Denmark as a case *Organic. Agriculture*. 3, 1, 41–55. doi: 10.1007/s13165-013-0045-z.

OFIS (2013) Organic Farming Information System – Ingredient Authorizations. Available at http://ec.europa.eu/idabc/en/document/2085/16.html#top

Pimentel, D., Hepperly, P., Hanson, J., Seidel, R. and Douds, D. (2005) Report 05-1 Organic and Conventional Farming Systems: Environmental and Economic Issues. Cornell University and Rodale Institute.

Ponisio, L.C., M'Gonigle, L.K., Mace, K.C., Palomino, J., de Valpine, P. and Kremen, C. (2014) Diversification practices reduce organic to conventional yield gap. *Proceedings of the Royal Society B* 282, 1799. doi: 10.1098/rspb.2014.1396.

de Ponti, T., Rijk, B. and van Ittersum, M.K. (2012) The crop yield gap between organic and conventional agriculture. *Agricultural Systems* 108, 1–9. doi.org/10.1016/j.agsy.2011.12.004.

Pretty, J. and Hine, R. (2001) Reducing Food Poverty with Sustainable Agriculture: A Summary of New Evidence. Final report from the SAFE World Research Project. Available at http://siteresources.worldbank.org/INTPESTMGMT/General/20380457/ReduceFoodPovertywithSustAg.pdf

Pretty, J., Noble, A.D., Bossio, D., Dixon, J., Hine, R.E., Penning de Vries, F.W.T. and Morison, J.I.L. (2006) Resource-Conserving Agriculture Increases Yields in Developing Countries. *Environmental Science and Technology* 40, 4, 1114–1119. doi: 10.1021/es051670d.

Pretty, J., Toulmin, C. and Williams, S. (2011) Sustainable intensification in African agriculture. *International Journal of Agricultural Sustainability* 9, 1, 5–24.

Raun, W.R. and Johnson, G.V. (1999) Improving nitrogen use efficiency for cereal production. *Agronomy Journal* 91, 357–363. Available at www.plantstress.com/articles/min_deficiency_m/nue.pdf

Rodale (2011) The Farming systems trial – celebrating 30 years. Available at http://rodaleinstitute.org/search/farming+systems+trial

Rothamsted (2012) Rothamsted long term experiments. Guide to the Classical and other Long-term Experiments, Datasets and Sample Archive. (Updated reprint of 2006 edition). Available at www.rothamsted.ac.uk

Savage, S.D. (2011) A comparison of 2008 organic row crop yields to historical yield trends. Available at www.scribd.com/doc/48257015/Organic-Historical-Comparison-2-5-11

Savage, S.D. (2014) The Organic Yield Gap, An independent analysis comparing the 2014 USDA Organic Survey data with USDA–NASS statistics for total crop production. Available at https://

www.scribd.com/doc/283996769/
The-Yield-Gap-For-Organic-Farming

Seufert, V., Ramankutty, N., Foley, J.A.
(2012) Comparing yields of organic and
conventional agriculture. *Nature* 485, 7397,
229–232. doi: 10.1038/nature11069.

Soil Association (2016) Soil Association
Organic Standards. Available at
https://www.soilassociation.org/
what-we-do/organic-standards/
soil-association-organic-standards

Stanhill, G. (1990) The comparative
productivity of organic agriculture.
*Agriculture Ecosystems and
Environment* 30, 1–2, 1–26.
doi: 10.1016/0167-8809(90)90179-H.

Stockdale, E.A., Lampkin, N.H., Hovi,
M., Keatinge, R., Lennartsson, E.K.M.,
Macdonald, D.W, Padel, S., Tattersall,
F.H., Wolfe, M.S. and Watson, C.A. (2000)
Agronomic and environmental implications
of organic farming systems. *Advances in
Agronomy*, 70, 261–327. doi.org/10.1016/
S0065-2113(01)70007-7.

Tuomisto, H.L., Hodge, I.D., Riordan,
P and Macdonald, D.W. (2012) Does
organic farming reduce environmental
impacts? – A meta-analysis of European
research. *Journal of Environmental
Management* 112, 309–320. doi.
org/10.1016/j.jenvman.2012.08.018.

UNEP-UNCTAD *(*2008) Organic
Agriculture and Food Security in Africa,
United Nations, New York and Geneva,
2008. Available at www.unep.ch/etb/
publications/insideCBTF_OA_2008.pdf

USDA (2015) Organic Survey (2014)
Organic Farming: Results from the 2014
Organics Survey. Available at https://
www.agcensus.usda.gov/Publications/
Organic_Survey

CHAPTER 14

Composition of Organic and Conventionally Grown Crops – Health Implications

"What is it that is not poison? All things are poisons and nothing is without poison. It is the dose only that makes a thing not a poison"

Paracelsus 16th century

Claims that can be traced back to Howard and Balfour have repeatedly been made that organic food is healthier than conventionally produced food. It is one of the main reasons people give for choosing organic, especially when buying for babies and children. In 1939 the leading Cheshire doctors (see Chapter 9) who thought manure and composts resulted in healthier crops asked Sir E.J. Russell, then Director of the Rothamsted Experimental Station, whether there was any evidence to support such a notion. He replied that having searched diligently for evidence that organic manure gives crops of better quality than inorganic fertilizers no differences had yet been found. Has the situation changed?

Several years ago the Advertising Standards authority (2000) in the UK ruled that there was no evidence to support the claim of the superiority of organic food on health grounds. Despite this it would appear that this message has not reached the general public and there seem to be many who believe they are buying more nutritious and healthier food when they buy organic.

As a result of the absence of direct evidence of any health benefits, much attention has been focused by organic supporters on the presence in organic food of compounds that are potentially beneficial and also to chemicals in conventionally grown food, notably pesticide residues, that could possibly be harmful. What is not generally appreciated is that the compounds on which the spotlight is directed are just a tiny selection of the extremely large number of organic compounds with widely varying and potentially beneficial properties found in plants. The complexity of the situation is a barrier to understanding the significance of differences in crop composition, especially of individual compounds found to be associated with different farming practices. An understanding of the diverse biochemical composition of plants is needed; it will then be easier to assess the relevance of the presence of different compounds, whether they occur naturally or whether they are induced. It may be that there are potentially beneficial compounds of which we are currently ignorant.

Plant composition – secondary metabolites and bioactive compounds

Plants contain a vast array of bioactive compounds. Those present in greatest quantity are involved with growth and development – primary metabolites, e.g. carbohydrates, amino acids, proteins and lipids; they are present in all plants. In contrast, secondary metabolites tend to be specific to groups of plants and occur in small amounts. There are various estimates of the number of bioactive compounds in plants, from more than 10,000 (Brandt and Mølgaard, 2001) to possibly as many as 100,000 (Trewavas, 2005). Only a small fraction has been identified and studied. Plants are a vast storehouse of complex molecules waiting to be identified, isolated and possibly used for the benefit of mankind. Bioactive compounds with pharmacological properties would be of particular interest, e.g. glyceollin, a soybean phytoalexin with medicinal properties including anti-cancer, anti-bacterial and anti-fungal activity (Ng *et al.*, 2011). The possible influence of farming method on their production has seldom been considered.

Of importance in relation to the concerns about synthetic pesticides is the ability of plants to generate secondary metabolites that can protect them against particular insects or pathogens. These metabolites are collectively known as phytoalexins or as natural pesticides. This complex topic has been studied for many years. Fraenkel (1959), a pioneer in the field, believed that the main function of secondary metabolites is to defend plants against herbivores. It is now appreciated that these compounds can also help plants to defend against diseases. They can also have other functions such as attracting predators of pests (War *et al.*, 2012) and even of influencing the growth of neighbouring plants of a different type, e.g. some rice varieties exude ß-tyrosine, which acts as an inhibitor of weeds and bacteria (Yan *et al.*, 2015).

The notion that plants can communicate with other plants and with organisms has gone from being viewed as a lunatic fringe idea to an accepted ecological phenomenon (Adler, 2011); at a session of the Ecological Society of America in 2010 data was presented describing how plants communicate with each other, with herbivores and with predators of those herbivores. The idea that a plant being eaten by a caterpillar can signal a neighbouring plant to increase its protective alkaloids by releasing volatile organic compounds will take time to be widely accepted. As pointed out by War *et al.* (2012), our understanding of these defensive mechanisms is still limited but could lead to improvements in pest management and to reductions in the amounts of insecticides used. Plants have had more than 300 million years to evolve defence mechanisms; man has had a generation or two to understand them.

Chemical defences used by plants are just one of a range of mechanisms alongside structural defences, as discussed in detail by Freeman and Beattie (2008). Structural defences occur not only as thorns but also at the cellular level in the form of lignified cell walls. The capacity of plants to tolerate attack is also important (Turley *et al.*, 2013).

The ability of plants to defend themselves against attack by insects and pathogens varies widely between different species but also among populations of the same species. This was evident from the work of Zust *et al.* (2012) on natural populations of *Arabidopsis thaliana* and the genetic control of the production of protective glucosinolates against aphids. Where the eating pressure from the aphids was intense it was found that genes associated with resistance to aphid attack were favoured but in areas where aphid damage was low these genes were not favoured. The interplay of environmental factors and genetic variation is likely to have resulted in a wide variation in the bioactive compounds found in different varieties.

Plants that produce bioactive

compounds seem to be the rule rather than the exception. Thus, most plants including the common food and feed plants are capable of producing small amounts of such compounds. In contrast, the typical poisonous or medicinal plants contain high concentrations of the more potent bioactive compounds (Bernhoft, 2010).

Most secondary metabolites fall roughly into three classes of compounds, namely phenolics, terpenoids and alkaloids (Anon, 2013). Several secondary metabolites have powerful physiological effects and have been used as medicines, e.g. quinine. It was only in the latter half of the 20th century that secondary metabolites were clearly recognized as having important functions in plants.

Phenolics, terpenoids and alkaloids
Phenolics

Phenolic compounds are widely distributed in plants. Plant tissues may contain up to several grams per kilogram (fresh weight), and there may be more than 4,000 of such compounds. Considerable variation in crop polyphenol concentrations (dry weight basis) and dietary intakes were reported by Khokhar and Apenten (2003), e.g. cereals 220–1,030 mg/kg, legumes 340–17,100 mg/kg, nuts 0.4–380 mg/kg and daily intakes of 1 g in the USA, 23 g in the Netherlands and 28 g in Denmark. Red wine can contain 1,000–4,000 mg/l polyphenols and tea (beverage) 750–1,050 mg/l.

There is a growing interest in the role of phenolic compounds as antioxidants, in the prevention of various degenerative diseases, such as cancer and in cardio-vascular diseases. It is not known which of the polyphenols may be significant in relation to health. The existence of many polyphenols in any one group of plants may hinder research but it may facilitate speculation about their relevance.

Phenolic compounds can be produced as defence responses by injured plant tissues. Simple phenolic compounds, such as salicylic acid in the leaves of certain plants, can provide a defence against pathogens. Baxter *et al.* (2001) found the concentration of salicylic acid to be six times higher in organic vegetable soups than in soups based on conventionally produced vegetables, which are likely to have been protected from infection by pesticides.

Other phenolic compounds, the isoflavones, are synthesized rapidly in legumes when they are attacked by bacteria or fungi; they have strong antimicrobial activity. Lignin is also valuable in plant defence: plant parts containing cells with lignified walls are much less palatable to insects and animals and are less easily digested by fungal enzymes. Several phenolic compounds are highlighted in the comparative studies discussed later, including flavonoids, flavonols, flavanones, stilbenes, anthocyanins and vitamin E.

Terpenoids

Terpenoids occur in all plants and more than 22,000 have been described (Freeman and Beattie, 2008). The simplest terpenoid is isoprene, a volatile gas emitted from leaves in large quantities during photosynthesis; it may protect cell membranes from damage caused by high temperature or light. Monoterpenoids are exemplified by the aromatic oils (such as menthol) found in the leaves of members of the mint family; these oils may have insect-repellent qualities. The pyrethroids, which are monoterpene esters found in the flowers of chrysanthemum species, have been used commercially as insecticides. They fatally affect the nervous systems of insects. Many spices, including basil, oregano, rosemary, sage, thyme and cinnamon, contain monoterpenoids that can function as insecticides.

Triterpenoids comprise the plant steroids, some of which can protect plants from insect attack. Azadirachtin extracted from the seeds of the neem tree (*Azadirachta indica*) can be used as an insecticide by acting as a deterrent. Neem oil is used by some organic farmers.

Alkaloids

Alkaloids are nitrogen-containing organic compounds that accumulate in plant leaves and fruit. More than 5,000 alkaloids have been identified in plants. Many alkaloids are extremely toxic, especially to mammals, acting as potent nerve poisons and enzyme inhibitors, e.g. ricinine, strychnine and coniline (from hemlock). Most alkaloids are bitter and so act as deterrents to animals. A curious case of deterrence is that of the Cinnabar moth caterpillars. The caterpillars can eat ragwort (*Seneccio* spp.) and groundsel (*Seneccio vulgaris*) and accumulate the contained alkaloid senecionine without suffering any ill effects. By doing so they acquire their own defensive weapon against predators (Anon, 2015); this phenomenon is exhibited by other caterpillars and other alkaloids, e.g. the moth Utetheisia and monocrotaline (Wikipedia).

Several alkaloids are used pharmaceutically, including quinine, atropine, codeine, and morphine. Other alkaloids have stimulatory properties, e.g. caffeine, cocaine, nicotine and lysergic acid diethylamide – LSD (originally derived from ergot, a fungal disease on cereal heads). Fresh green tea shoots and Arabica coffee beans can contain 3% and 1% caffeine (dry weight basis) respectively. Caffeine ingestion can increase dopamine levels in the brain and so improve alertness.

Toxic effects of bioactive compounds

Many compounds that occur naturally in plants can be toxic to humans and some of the bioactive compounds that plants produce for self-defence (phytoalexins) are also potentially harmful to humans. However, it is thought that the moderately elevated concentrations of phytoalexins in unprotected crops (e.g. organic) are not only unlikely to represent any risk to man (Brandt and Molgaard, 2006, 2007) but may be health-promoting (Brandt and Mølgaard, 2001). The potential effects of phytoalexins are generally largely ignored.

Phytoalexins have been found in toxicology tests not only to be carcinogens (probably 50–60%), but also teratogens, oestrogen mimics, sterility inducers, nerve toxins, goitrogens and chromo-some breakers (Trewavas, 2005). Clearly phytoalexins should not be ignored.

Humans may have evolved mecha-nisms to metabolise and/or excrete some of the very harmful compounds. This may even apply to the ability to deal with amygdalin, a cyanogenic glycoside found in the kernels of the stones of some stone fruit (apricots, cherries, peaches, pears, plums and prunes) that are sometimes eaten after cooking. The flesh of the fruit itself is not toxic but hydrogen cyanide is formed if the fresh kernels are broken and chewed. The same cyanogenic gly-coside toxin is found in cassava roots and fresh bamboo shoots, making cooking obligatory. Bitter almonds also contain amygdalin but refining removes it.

Several glycoalkaloids are produced naturally by potatoes, the most common being solanidine, solanine and chaco-nine, which at low levels produce desir-able flavours, but are toxic at elevated levels, causing a burning sensation in the mouth. Chaconine and solanidine are known to have teratogenic properties (Trewavas, 2005). Most of this natural toxin is in the peel, or just below the peel. Valcarcel (2014) found that carotenoids, phenolics and flavonoids were also con-centrated in the same tissues in potatoes.

Bioactive compounds in organic and conventional crops: meta-analyses

Over the years there have been very many studies of widely varying quality, many not peer reviewed and without statisti-cal analysis, comparing the composition of organically and conventionally grown crops. These studies have been aggre-gated in several meta-analyses, which have highlighted the presence of greater concentrations of a number of com-pounds in organic crops. A fundamental problem exists when carrying out meta-studies, namely that of selecting appro-priate data. Ideally only data from trials where the soils and climatic conditions were similar, the same varieties grown and which were sampled at the same time should be used (Kumpulainen, 2001).

Studies from 1997 to 2015
Woese et al. *(1997)*

Woese *et al.* in Germany reviewed more than 150 studies (1926–1994) that evalu-ated desirable and undesirable constitu-ents in organically and conventionally produced food. The studies covered cereals, potatoes, vegetables, fruit, wine, beer, bread, milk and eggs. Conventional vegetables normally had higher nitrate contents and lower dry matter contents. Otherwise no consistent differences were found. In many cases contradictory results were found. Half the studies on vegetables (lettuce, cabbage and spinach) showed no difference in vitamin C content but in the remainder there were tendencies for organic crops to contain more vitamin C.

Worthington (2001)

Worthington surveyed the existing lit-erature, comparing nutrient contents of organic and conventional vegetable and fruit crops. The nutrients included ten essential mineral elements, vitamin C and ß carotene. She reported that organic crops contained significantly more vitamin C (27%), iron (21%), magnesium (29%), and phosphorus (14%) than conventional crops and significantly less nitrate (15%).

Organic center (2005 and 2008)

The Organic Center in the USA, a non-profit organization, is a source of information on scientific research about organic food and farming. It has produced two reviews (Benbrook, 2005; Benbrook *et al.*, 2008) comparing the composition of organic and conventional crops. Much of the interest in the secondary metabolite composition of crops arises from the assumption that the health-promoting properties of fruit and vegetables are due to their antioxidant content.

Total antioxidant capacity, total phenolic content, the polyphenols quercetin and kaempferol, vitamins A, C, and E, potassium, phosphorous, nitrates and total protein were determined in 232 carefully selected matched pairs of studies (Table 14.1).

Although the meta-analysis may show that on average organic crops may have an apparently better composition and contain, for example, more phenolics and vitamin C, the effects are not consistent. This suggests factors may be involved that are operating irregularly or intermittently and which are not directly related to the organic practices. The reasons for

Table 14.1 Comparison of organic and conventional foods (Rosen, 2008 citing data by Benbrook *et al.*, 2008). Percentages are based on the total number of matched pairs of studies

	% where Organic Higher	% where Conventional Higher
Total Antioxidant capacity	88	12
Total Phenolics	71	24
Quercetin	87	7
Kaempferol	55	45
Vitamin C	63	37
ß-Carotene	50	50
Vitamin E	62	38
Nitrate	17	83
Protein	15	85

the inconsistencies need to be established before attempting to make generalizations based on the entire database.

Most of the paired results used by Benbrook were from single years; was the considerable year-to-year variation a possible confounding factor? The Organic Center accepted that the secondary metabolites can be synthesized by plants in response to external stimuli such as pest and pathogen attack as well as being influenced by soil type and conditions but this aspect was not followed up. The supposition is that crops grown organically are inherently more resistant to pathogens and pests than conventional crops and can cope well without the protection of pesticides; this explains why no consideration was given to the effects of varying pest pressure in the matched studies.

Rosen (2008) critically reviewed the work of Benbrook and came to the conclusion that a case had not been made for the nutritional superiority of organic food, and that even if there was reliable evidence the customer would not know whether it had been a good year for organic or for conventional, what the variety was and what were the soil conditions.

Mitchell et al. (2007)

Comparisons of organic and conventional tomatoes formed part of the trial called Long Term Research on Agricultural Systems, at Davis, California. The ten-year comparative trial on tomatoes, grown in a two-year rotation with leguminous winter cover crops, provided valuable data insofar as the same variety was grown in adjacent plots. Conventional plots received herbicides and other pesticides as needed, whereas organic crops received only organically approved pesticides, such as sulfur and Bt compounds. Composted poultry manure was applied initially at 45 t/ha, declining to 18 t/ha as soil organic matter levels rose. The organic crops received more N from winter cover crops, giving a total of 440 kg N/ha/ year declining to 230 kg (over the years). Conventional N inputs varied less from about 170 to 210 kg N/ha according to local best practice.

Organic and conventional yields were similar. The organic tomatoes had appreciably more of the flavonoids quercetin (79%), naringenin (31%) and kaempferol (97%) based on the ten-year mean differences than conventional tomatoes. Organic crops contained 115 mg/kg quercetin, 40 mg/kg naringenin and 63 mg/kg kaempferol (dry matter basis). The levels of the flavonoids increased over time in both systems; however, the rate of increase in the organic system was higher. The differences were attributed primarily to the contrasting soil fertility management practices. The workers concluded that it was the behaviour and quantity of N in the organic and conventional systems that had the strongest influence and they suggested over-fertilization (of N in both conventional and organic systems) might reduce the health benefits from tomatoes and should be avoided. There was no suggestion of the involvement of any pest or disease attacks.

Dangour et al. (2009, 2010)

The systematic examination of the available published literature (1958–2008) by Dangour *et al.* was designed to review the strength of the evidence that human health effects could be attributed to eating organic rather than conventional food. The reviews were commissioned by the Food Standards Agency in the UK.

In their literature assessment of the data on the compositional comparisons of organic and conventional food, Dangour *et al.* (2009) were only able to identify fifty-five studies (from 52,000 articles) that were satisfactory. It was found that conventionally produced crops had a significantly higher content of nitrogen, and organically produced crops had a significantly higher content of phosphorus and higher titratable acidity. No evidence of a difference was detected for the remaining eight of eleven crop nutrient categories analysed, namely vitamin C, phenolic compounds, Mg, K, Ca, Zn, Cu and total soluble solids.

Assessment of human health effects was a prerequisite for a study to be included by Dangour *et al.* (2010). Only twelve studies (eight human and four *in vitro* studies) met their inclusion criteria. The studies were very heterogeneous in terms of design, exposure and health outcomes. There was one study that suggested an association of reported consumption of strictly organic dairy products with a reduced risk of eczema in infants; the other studies showed no evidence of differences in nutrition-related health outcomes that result from exposure to organic or conventionally produced foodstuffs.

The fact that only twelve studies (from 98,000 articles) were judged to be acceptable for the review indicates the extremely limited nature of the evidence base on this subject.

A limitation of Dangour *et al.* (2010) was the fact that very few crops were covered by the study, namely fruit (apples, grapes, oranges, strawberries, tomatoes and two studies on wine) and carrots. The lack of health-related data involving vegetables is unfortunate in view of the importance of their consumption to human health. There are many studies comparing the composition of organic and conventional crops that did not experimentally assess health

effects. The European Food Information Council also concluded that there is no evidence suggesting significant nutritional differences between organic and conventionally produced food (EUFIC, 2013).

Rosen (2008, 2010)

Rosen presented a detailed examination of the evidence put forward by proponents of the superiority of organic food. He considered all the studies and claims up to the Quality Low Input Food project (QLIF), whose findings were later covered by Baranski *et al.* (2014). He concluded that the zeal of organic food proponents to fulfil their missions leads them to ignore the facts that many of their claims are made on the basis of research articles that were not reviewed by independent scientists and data that was not statistically significant; in doing this he felt that they stretched the truth. He particularly took objection to the criticism of conventional vegetables because some contain higher levels of nitrate than organic vegetables. Not only is there scant evidence of the deleterious effect of nitrate, there is much evidence of beneficial effects (see Chapter 10). Rosen concluded that consumers who buy organic food because they believe that it contains more healthful nutrients than conventional food are wasting their money.

Holmboe-Ottesen (2010)

Holmboe-Ottesen's review of the possible health effects of increased levels of bioactive compounds in organically grown food plants was presented at a symposium in Oslo in 2008 on several aspects of bioactive compounds (Bernhoft, 2010). She drew attention to the many factors that may affect the amounts of nutrients and bioactive compounds in crops; important factors included choice of variety, fertilizer regimen, pesticide application, soil preparation, cultivation practices (such as crop rotation), cover crops, as well as timing of harvest and storage of crop. She emphasized the importance of variety, as did the EU-funded QLIF project (Lueck *et al.*, 2008; QLIF, 2009), as the most important factor determining the content of antioxidants and nutrients in crops. Whereas conventional farmers are likely to choose varieties based primarily on high yield potential, the organic farmer may be more concerned with disease resistance properties and by so doing will grow crops that contain or can generate higher levels of protective bioactive compounds.

The QLIF (2009) project conducted on adjacent plots of organic and conventional crops on more than 3,000 ha and also found that organic food plants had increased levels of antioxidants.

Stanford University (2012)

A meta-analysis of 237 of the most relevant studies comparing organic and conventional produce was carried out at the School of Medicine, Stanford University (Smith-Spangler *et al.*, 2012). Seventeen randomized trials were included in the study on groups of people consuming organic or conventional diets. There were no significant differences between those fed organic and those consuming conventional food when assessed by allergic outcomes (eczema, wheeze and atopic sensitization, Chapter 10).

Nutrient levels, bacterial, fungal and pesticide contaminants of various products (fruits, vegetables, grains, meats, milk, poultry, and eggs) grown organically and conventionally were compared. Very few significant compositional differences were found. No consistent differences were seen in the vitamin contents and only one nutrient, namely phosphorus, was significantly higher in organic than in conventional produce; the virtual absence of phosphorus deficiency in man means that this has little clinical significance. There were no differences in the protein or fat content of organic and conventional milk, although evidence from a limited number of studies suggested that organic milk and chicken may contain significantly higher levels of omega-3 fatty acids (see Chapter 10). Statistically higher levels of total phenols were found in organic produce.

The researchers were unable to identify specific fruits or vegetables for which organically grown crops were consistently the healthier choice. The review provided little evidence that conventional foods posed greater health risks than organic products. It was found that while organic produce had a 30% lower risk of pesticide contamination, organic foods are not necessarily completely free of pesticides. Furthermore, the pesticide levels of all foods generally fell within the allowable safe limits. Two studies did find low levels of pesticide residues in the urine of children fed on organic food but the significance of these findings on health was deemed to be unclear. Additionally, organic chicken and pork appeared to reduce exposure to antibiotic-resistant bacteria.

Baranski et al. *(2014)*

Baranski *et al.* carried out a massive meta-analysis of studies comparing organically and conventionally grown crops and produce using 343 peer-reviewed publications (1992–2011) with funding mainly from the European Community and with the collaboration of the Sheepdrove Trust, an organization that promotes the development of organic and sustainable farming systems. This very valuable and unlikely-to-be repeated analysis deserves close consideration.

Studies involved comparisons of samples from matched farms (167), produce collected at retail outlets, i.e. basket studies (89), and on comparisons made in controlled experiments (208); statistical analysis by the source authors had only been carried out in a few studies, namely 61 in matched farms, 34 in basket studies and 54 in experiments.

Sophisticated statistical analysis of the data from all the comparisons (464) revealed substantial differences between the composition of organic and non-organic crops/crop-based foods (Tables 14.2 and 14.3).

Organic fruit generally contained more total antioxidants (around 20%) than

Table 14.2 Concentrations of key compounds in organic crops (fruit and vegetables) expressed as percentages of mean for compounds where mean organic values (from weighted meta-analysis) were higher than conventional (Baranski *et al.*, 2014). Values in parenthesis are from unweighted meta-analysis

	% higher than mean	Range (95 %)
Flavonols	50 (44)	28 to 72
Anthocyanins, total	44 (32)	–2 to 91[a]
Stilbenes	28 (212)	12 to 44
Flavones	26 (17)	3 to 48
Phenolic acids, total	5 (33)	3 to 6
Antioxidant activity	17 (18)	2 to 32
Carotenoids, total	17 (22)	0.4 to 34
Vitamin C	6 (29)	–3 to 15[a]

[a] negative values indicates conventional values were higher than organic.

conventional fruit: in contrast, the antioxidant content of organic and conventionally grown vegetables did not differ. Furthermore, while the overall antioxidant levels were higher in organic fruit, there were no differences in the important antioxidants, phenolic acids, flavanones, flavones and flavanols, the major contributor would appear to be the carotenoids. Fruit is not an important dietary source of carotenoids.

As discussed in detail by Del Rio *et al.* (2013), the jury is still out on the significance of the findings on polyphenols, such as the flavonoids, on human health. While

Table 14.3 Concentrations of key compounds in conventional crops expressed as percentages of mean for compounds where mean conventional values (from weighted meta-analysis) were higher than organic (Baranski *et al.*, 2014)

	% higher than mean	Range (95 %)
Vitamin E	15	49 to –19[a]
Protein, total	15	27 to 3
NO₃	30	143 to –84[a]
Fibre	8	14 to 2
Cd	48	112 to –16[a]

[a] negative values indicate organic values were higher than conventional.

there is considerable epidemiological evidence indicating that diets rich in fruit and vegetables are associated with a reduction in the risk of a number of chronic diseases, the definitive human experiments have yet to be carried out to identify the effective individual or combinations of antioxidants (see Appendix). A large number of antioxidants were included in the meta-analysis for which there were no differences in concentrations but these were not highlighted. The potentially beneficial effects of organic farming on antioxidants are ascribed to crops without distinction, which ignores the fact that the higher concentrations of antioxidants were found in fruits, not vegetables.

The wide variation exhibited and the absence of the actual concentrations of the analysed antioxidants and all other compounds mean that it is difficult to generalize about the relevance of the compositional differences. Nevertheless, the higher concentration of antioxidants seems to be a real, albeit irregular, effect. Inevitably the link was made between antioxidants and reduced risk of chronic diseases, including neurodegenerative diseases and certain cancers. Evidence to support such links is not available and until it is the best general advice is to eat the varieties of fruit and vegetable that are richer in the components considered to be health-promoting.

Fruit and vegetable varieties are known to differ markedly in their nutrient composition (Trewavas, 2004). The aggregation of massive amounts of disparate data in a meta-analysis has its attractions and seemingly provides answers that can be widely extrapolated. The variation shown across all the results was wide (Tables 14.2 and 14.3), indicating considerable inconsistency. The contribution of cultivars to the wide variation was seemingly not considered.

The possibility of an interaction between cultivar and farming system was demonstrated by Qi You *et al.* (2011) in a study of three blueberry cultivars. Two organically grown cultivars contained about 10% more phenolics and anthocyanins than conventionally grown, whereas the third contained 50% more when grown conventionally. Moreover, the variation between cultivars dwarfed any effect of farming practice. Total phenolics in blueberries varied from about 50 to 350 mg gallic acid equivalence per 100 g fresh weight and anthocyanins from 100 to 220 mg/100 g fresh weight according to cultivar.

The fact that the mix of secondary metabolites differs between species and cultivars means that the aggregation of data in a meta-analysis will obscure real effects on particular species or cultivars.

Further evidence of the difficulty of interpreting the meta-analysis data was apparent when only the replicated trials (32% of total) that were subjected to statistical analysis were considered separately. Whereas examination of all the data had indicated organic crops contained more of many potentially beneficial compounds, a somewhat different picture emerged when only the replicated studies were included.

For nearly all parameters (for which organic values were higher) the combined number of studies where conventional values were significantly higher plus those where no statistical differences were established exceeded the number where organic values were higher. The results of replicated comparative trials where statistical analysis was possible are thus at odds with the mass of data not amenable to statistical analysis. For example, while organic crops have on the basis of all the data 69% more flavonones than the mean, there were fourteen trials where conventional crops contained significantly more flavonones and eleven where there were no differences compared with twenty-four trials in which organic crops contained more. A similar situation existed with nitrate levels; while twelve conventional crops contained significantly more nitrate, there were three trials in which the organic crops contained more and seventeen where there was no significant difference. The comparable

values for cadmium were conventional 2 (more Cd), organic 1 and no significant difference 15; when all the non-replicated data is included conventional crops were judged to contain 48% more Cd than the mean values (Table 14.3).

Magnitude of organic effects

The effects of organic farming on composition of crops amount to percentage increases of 10 to 70% of individual components. In contrast, the variation in composition associated with other factors such as variety are much larger, amounting to a few hundred per cent. Likewise, the phytoalexin content can sometimes be elevated several times by pathogen or insect attack (Karban and Baldwin, 1997).

Huber et al. (2015)

Huber *et al.* in their review of recent studies also found lower nitrate levels and less pesticide residues, and usually higher levels of vitamin C and phenolic compounds, in organic plant products, as well as higher levels of omega-3 fatty acids and conjugated linoleic acid in milk from organically raised animals. However, there was much variability, probably associated with fertilizer use, ripening stage and plant age at harvest. No simple relationship between nutritional value and health effects was apparent, making it difficult to draw meaningful conclusions from compositional data about the health effects of organic foods. Some *in vitro* studies have shown higher antioxidative and antimutagenic activity in organic than in conventional food.

Huber *et al.* point to the fact that the plant world may contain as many as 75,000 or even 100,000 different compounds and that any given species may contain 7,500–10,000. Analysing hundreds or even thousands of compounds would only reach a small minority (the tip of the iceberg) of compounds that may possibly affect human health. It is likely that there will be interactions between compounds about which very little is known; the situation is very complex.

Milk and Omega-3 fatty acids

A recent meta-analysis based on 170 papers on milk (Średnicka-Tober *et al.*, 2016) concluded that organic milk, fat and dairy products are likely to contain about 56% more beneficial omega-3 fatty acids than conventional products. There may be some confusion regarding the significance of this finding relating to the chain length of the fatty acids and the fact that the fatty acid composition of milk varies with many factors including breed of cow, diet and stage of lactation (Schwendel *et al.*, 2015). Most of the omega-3 fatty acids in milk are short chain (Kliem *et al.*, 2013) with only small (often barely detectable) amounts of long chain omega-3 fatty acids, which are the omega-3 fatty acids known to be protective for cardiovascular disease. The reported 56% higher concentration of long chain omega-3 fatty acids in organic milk would, on the basis of today's average intake of all dairy products, supply an additional 8 mg long chain omega-3 fatty acids per day; a tiny amount when set against a target of 450 mg/d (Minihane *et al.*, 2008; and Givens, England, 2016, personal communication). Furthermore, switching from conventional to organic milk would increase omega-3 fatty acid (mostly short chain) intake by about 33 mg/day in the average diet; based on a daily intake of about 2.2 g of omega-3 fatty acids per day this would amount to only about a 1.5% increase, which is also negligible (Givens, 2016). As discussed by Scollan *et al.* (2008), production systems, including organic, containing a higher proportion of forage, rather than conventional concentrates will result in more omega-3 fatty acids in the milk; the benefit comes from the diet. Higher levels of vitamin E and carotenoids were also found in milk by Średnicka-Tober *et al.* (2016).

In a parallel meta-analysis on meat products (sixty-seven studies) differences in fatty acid profiles were found but the heterogeneity was high; this might have been due to differences associated with the animals and meat types represented (Średnicka-Tober *et al.*, 2016a). It is likely that the high grazing/forage-based diets of the organic farming livestock was the main reason for differences in fatty acid profiles. The relevance of the findings for human health can only be assessed when more definitive evidence is available on the health-related activity of specific fatty acids.

Iodine

It has been known for some time that organic milk usually contains less iodine than conventional milk, e.g. in Scandinavia (Rasmussen *et al.*, 2000; Dahl *et al.*, 2003), and in the UK (Food Standards Agency, 2008; Bath *et al.*, 2012). The meta-analysis (Średnicka-Tober *et al.*, 2016) indicated that conventional milk contained 74% more iodine than organic milk. It was suggested that the lower iodine levels in organic milk could be considered beneficial in order to safeguard against iodine toxicity (thyrotoxicosis).

A recent proposal by the European Food Safety Authority to reduce the iodine concentration in complete animal feed from 5 to 2 mg/kg for lactating ruminants in order to protect consumers was reported by Flachowsky *et al.* (2014). However, iodine toxicity is unlikely to occur in individuals who are on normal Western diets, particularly if they have never been iodine deficient. Thyrotoxicosis was found in the UK in towns where goitre had previously been endemic and where there had subsequently been excessive intake of iodine, including that from spring/summer milk (Barker and Phillips, 1984); the previous damage to the thyroid due to iodine deficiency had left it unable to deal with excessive intakes of iodine.

Milk produced in the winter normally has a higher iodine concentration than that produced in summer (Flachowsky *et al.*, 2014); this is because housed cows in winter obtain more iodine from the concentrates than cows outdoors in summer get when grazing. It follows that organic cows, being more dependent on forage and less on concentrates than

conventional cows (European Union, 2008), will produce milk with lower iodine contents. The greater consumption of clovers containing goitrogenic glucosides may be a contributory factor (Rasmussen *et al.*, 2000). Iodophors are permitted dairy disinfectants in both organic and conventional farming, so they are unlikely to be implicated in the differences in iodine content.

Iodine deficiency disorders are extremely serious, especially in developing countries. Iodine is most important during pregnancy and in extreme cases a shortage in the diet can result in irreversible brain damage in the foetus. Much progress has been made by the International Council for Control of Iodine Deficiency Disorders over the last forty years in preventing iodine deficiency by using iodised salt in many countries. Most of the iodine in typical Western diets comes from milk and dairy products. The replacement of conventional milk by organic milk would be likely to increase the risk of sub-optimal iodine intakes and the risk of brain impairment during pregnancy. It is understood that in the UK some organic dairy farmers are feeding iodine supplemented feed.

Mild to moderate iodine deficiency has been demonstrated in a large UK cohort of pregnant women (Bath *et al.*, 2014). The evidence of an association between low maternal iodine status in early pregnancy and poorer verbal IQ, reading accuracy and comprehension in the children (Bath *et al.*, 2013) means that any reduction in the iodine intake of females of reproductive age should be avoided; conventional milk should be preferred. The proposal to reduce the iodine concentration in animal feed seems ill-advised.

Selenium

The lower concentration of selenium in organic milk was not thought to be of significance in view of the fact that milk only provides about 12% of recommended intakes in the UK (Średnicka-Tober *et al.*, 2016).

Protein and metabolite profiles

The variation in protein profiles linked to the farming system have been revealed by proteonomic studies whereby the entire set of proteins produced or modified by an organism or system are determined. Studies on potatoes and wheat have indicated that substantial parts of the protein profiles may be affected by the farming system. For example, Rempelos *et al.* (2013) found that two proteins known to be involved in biotic stress (1, 3-ß-D-glucan glucanohydrolase, a tuber invertase inhibitor) were more abundant in the leaves of potatoes grown with compost than with mineral fertilizers. More of the proteins involved in photosynthesis were present in potato leaves grown with fertilizers than with compost. Use of pesticides had no effect on the protein profiles.

Potato crops from organic and non-organic farming systems were found to differ in 160 of the 1,100 tuber proteins studied by Lehesranta *et al.* (2007) but the significance of the differences is not known. Most of the affected proteins played parts in normal metabolism but a few were involved in defence responses, indicating that the organic potato tubers may have been under some kind of stress.

In studies on wheat flag leaves Tétard-Jones *et al.* (2013) found that the abundance of 111 proteins in organic and conventional crops differed significantly. Proteins involved in nitrogen remobilization, photosynthesis and stress response were identified. These results indicate that the effect of contrasting growing systems on nutrient utilization and wheat grain yield are likely to be complex.

Zorb *et al.* (2006) used wheat, harvested in 2003 from the controlled field trial started in 1978 at the Research Institute of Organic Agriculture (FiBL, Frick, Switzerland) to compare the metabolite profile in grain from plants grown under a biodynamic, an organic and a conventional farming system. The amounts of N, P and K provided by the three systems were about the same. Despite exclusion of fungicides from the organic production systems there was no evidence of disease attack that might otherwise have confounded the results if phytoalexins had been produced. The amounts of fifty-two different metabolites including amino acids, organic acids, sugars, sugar phosphates, sugar alcohols and nucleotides were determined. Differences (of 10–40%) were only detected in eight of the fifty-two metabolites. No differences were found in the remaining forty-four metabolites. Only a small impact of the different farming systems on the nutritional value of the wheat would be expected.

Cadmium and heavy metals

Cadmium

Higher concentrations (48%) of the potentially toxic metal Cd were found by Baranski *et al.* (2014) in conventional than in organic food when all the studies were used but in the absence of data on the actual concentrations it is not possible to assess the relevance of the Cd values and so distinguish between hazard and risk. Moreover, several studies showed no significant differences in Cd levels of organic and conventional crops. The higher Cd levels were only found in conventional cereals and not in other crops.

Eurola *et al.* (2003), one of the studies included in the meta-analysis of Baranski *et al.* (2014), found that the Cd content of oats varied between cultivars and that there were no difference between organic and conventional farming. Moreover, the Cd concentrations (0.046, 0.029, and 0.052 mg/kg dry weight over three consecutive years) were well below the maximum permitted level of 0.100 mg/kg (fresh weight) for cereals in the European Union (EC regulation 1881/2006). Zaccone *et al.* (2010), in another study included in the meta-analysis, reported higher Cd levels in pasta from ten durum wheat cultivars grown conventionally than organically (0.082 vs. 0.018 mg/kg).

Cadmium occurs naturally in all agricultural soils, in sewage sludge and to varying degrees in phosphatic fertilizers. Controversy exists regarding the potential harmful effects of cadmium in the diet (Roberts, 2014). The only known case of Cd toxicity, *itai-itai* disease, occurred on subsistence farmers in Japan growing rice on soils contaminated with industrial wastes.

The higher concentrations of cadmium in conventionally grown crops found by Baranski *et al.* (2014) were most likely due to the fact that organic farmers do not use man-made phosphatic fertilizers. Roberts (2014) concluded that the cadmium in phosphatic fertilizers does not pose a risk to human health. Nevertheless, the potential dangers of excess cadmium should not be ignored.

The cadmium content of rock phosphates varies from 1 mg Cd/kg in the igneous rocks from the Kola Peninsula to more than 100 mg/kg in the sedimentary phosphate rocks from North Africa (Kumpulainen, 2001). Mar and Okazaki (2012) found a much wider variation in cadmium contents of rock phosphate with values up to 507 mg/kg from Morocco, 199 mg/kg in phosphate from Idaho and 3 mg/kg from Florida. As organic farmers are permitted to apply rock phosphates it is advisable that they take care over its provenance if they are concerned about the cadmium content of their crops.

Cadmium is poorly absorbed in the human gut (3–5% of intake) but unfortunately it is retained for many years in the kidney and liver with a biological half-life of up to thirty years; it is an element to be avoided. Liver and kidney are likely to be very rich in cadmium, containing 0.206 mg Cd/kg compared with 0.023 mg/kg in cereals, 0.039 mg/kg in fruits and 0.067 mg/kg in vegetables (EFSA, 2009). The relevance of the 48% reduction in cadmium reported by Baranski *et al.* (2014) can only be assessed by calculating the impact on dietary cadmium intake and comparing it with the tolerable weekly intake for cadmium of 2.5 µg/kg body weight proposed by EFSA (2009). High consumption of foods rich in cadmium, e.g. by vegetarians, were estimated by EFSA to have an average weekly exposure of up to 5.4 µg/kg body weight; mean dietary weekly exposure to Cd was assessed by EFSA at 1.9–3.0 µg/kg body weight.

Cadmium has been classified as a Group 1 carcinogen by the International Agency for Research on Cancer but unless combined with a risk analysis this hazard assessment has little meaning. Smokers should be wary as cadmium absorbed in the lungs from smoking can contribute as much as dietary cadmium (WHO, 1992).

Heavy metals (C^r, Cu, Ni, Pb and Zn)

The effect of conventional and organic farming on five heavy metals in soil and semolina samples from ten different wheat cultivars was assessed by Zaccone *et al.* (2010). Significantly more heavy metals were applied in the organic farming system. However, semolina samples obtained from the cultivars grown in conventional farming system were richer in Cr (0.182 mg/kg *vs.* 0.050 mg/kg),

and Cu (0.0066 mg/kg *vs.* 0.0058 mg/kg), which suggests that some aspect of organic farming may be able to reduce the absorption and accumulation of these elements in the grain. On the other hand, semolina samples obtained from the organic farms had slightly higher concentrations of Ni (0.295 mg/kg *vs.* 0.166 mg/kg), Pb (0.094 mg/kg *vs.* 0.082 mg/kg), and Zn (0.0136 mg/kg *vs.* 0.0108 mg/kg) than those obtained from conventional farms. No generalizations are possible from such a limited study.

Relevance of the varying composition of crops associated with farming system

The conclusions of the Organic Center (Benbrook, 2005 and 2008) and by Baranski *et al.* (2014) about the composition of organic products are very similar, namely that organic farming is likely to increase the antioxidant capacity, particularly vitamin C and reduce the NO_3 and Cd contents when compared with conventional farming and as a result promote health. The reviews by Smith-Spangler *et al.* (2012) and Dangour (2009, 2010) do not support such a conclusion about health. Other workers have also reported that organic crops contain more vitamin C, antioxidants and other compounds, which is further evidence that the effects may be real but it does not mean, as is implied, that the effects of organic farming on secondary metabolites are regular, reproducible and sufficient to be beneficial to health.

Progress in the assessment of the relevance of the findings is likely to depend on learning more about the reasons for the variable effects and hopefully the identification of the particular compounds (and amounts) that are active in promoting health.

Inconsistency of organic effects

The fact that the effects of organic farming on crop composition were not universally displayed clearly indicates that there can be no guarantee that organic products will contain more of the health-promoting compounds than conventional products. It can be argued that on average an organic diet would be preferable but the finding in the meta-analyses that conventional crops can sometimes be superior means that this argument cannot be sustained.

It is widely accepted that the chemical composition of a given crop can be affected by several factors, notably soil conditions, climate, aspect, season and especially variety. Wine grapes are a prime example of how these factors can have a marked effect. Wine produced from the same variety when grown not only in different regions but also on different sides of a hill on similar soils can vary considerably in composition and character.

It is accepted that several external factors influence crop composition and most of them have much greater effects than the farming system employed (Brandt *et al.*, 2011). Varietal differences in secondary metabolite content are likely to be obscured when results are aggregated in a meta-analysis. Nevertheless, some generalizations may be validated by studies with humans. For example, a Danish study showed a higher content of flavonoids and biomarkers of antioxidant defence compounds in blood and urine of those consuming an organic diet compared with a diet produced conventionally (Grinder-Pedersen *et al.*, 2003).

The stage at which a crop is harvested can have a marked effect on the content of flavonoids and polyphenols; crops that are harvested green such as those destined for long transport are likely to contain low concentrations (Lueck *et al.*, 2008).

Vitamin C

The frequent finding of higher concentrations of vitamin C in organic produce is seen as a significant health benefit. However, these increases of possibly 30% should be viewed against the wide variation that occurs not only between different species but also between cultivars of the same species. For example, the vitamin C content of apple cultivars can vary widely, e.g. from 4 mg/100 g of flesh in McIntosh, to 7 mg in Delicious, to 11 mg in Cox's Orange and to 29 mg in Sturmer (Anon, 2015a). Apple skin and the flesh just under it is rich in vitamin C – often containing as much as 50 mg/100 g.

Fruits vary in their vitamin C content from banana with typically 9 mg/100g, grapefruit 34 mg, orange 53 mg, papaya 62 mg, kiwi 98 mg and blackcurrant

155–215 mg (Anon, 2015b). These values need to be set against the daily requirement. In the UK the RNI (Reference Nutrient Intake) is 40 mg/day, which is thought to provide a considerable safety margin (Anon, 2005); in the USA the RDA (Recommend Dietary Allowance) is 90 mg/day for adult males.

Considerable variation in the vitamin C content of kiwi fruit was reported by Ichiro *et al.* (2004). They found that the content in the fruit of different cultivars of *Actinidia deliciosa* varied from 29 to 80 mg/100 g fresh weight. In most cultivars of *A. chinensis*, vitamin C content in fruit was higher than those in *A. deliciosa*, with some cultivars containing about 200 mg/100 g. There was also a wide variation in vitamin C content of *A. arguta* fruit, with concentrations ranging from 37 to 185 mg/100 g.

The vitamin C content of vegetables vary from 15–20 mg/100 g in lettuce and rocket, to raw spinach 35 mg, raw cauliflower 50 mg, raw broccoli 110 mg and to 165 mg in raw red peppers, according to Fondation Louis Bonduelle (2015). Other reports also indicate a wide range in the vitamin C content of brassicas from 16 mg/100 g in Brussels sprouts, 19 mg in kale, 25 mg in Chinese cabbage, 50 mg in cauliflower and 75 mg in broccoli (Jahangir *et al.*, 2009).

If increasing the dietary vitamin C intake is a priority it would be advisable to select fruit and vegetables on the basis of their likely vitamin C content. The possible increases (of up to 30%) in vitamin C in organic crops are of minor significance in comparison.

Antioxidants

The main soluble antioxidants in the diet are vitamin C and glutathione. If there is sufficient vitamin C in the diet other soluble antioxidants are likely to be ineffective and accordingly irrelevant (Trewavas, 2005).

In a detailed study of phenolics in eight apple cultivars by Tsao *et al.* (2003) Red Delicious and Northern Spy were found to contain twice as much total phenolics as Empire. In a three-year study the antioxidant content (quercetin, kaempferol and vitamin C) of organic and conventional tomatoes showed considerable year-to-year variation (Chassy *et al.*, 2006). The high values found in organic tomatoes in the first year influenced the three-year average value and obscured those in the two later years when conventional tomatoes contained more antioxidants. Rosen (2008) quotes other examples of the variation of antioxidant contents with and within the season for olives, blueberries and lettuce. All these examples of irregular and inconsistent effects of farming method point to the conclusion that generalizations based on meta-analyses hide the full story and are of limited value.

The effects of the antioxidants in purees made from organic and conventional tomatoes were compared by Caris-Veyrat *et al.* (2004) after measuring the antioxidant plasma status of humans (groups of twenty females). No differences were found in the concentrations of vitamin C and lycopene (two major antioxidants) in the plasma after three weeks' dietary supplementation with 96 g/day of puree, despite the significantly higher concentration of vitamin C in the organic tomatoes.

In order to circumvent the problems of comparing different measures of antioxidant capacity the USDA developed a single method for measuring the total antioxidant capacity (ORAC oxygen radical absorbance capacity) and the total phenol contents; a database of 275 food items was built up. However, in 2012 the ORAC Database was removed by the USDA due to mounting evidence that the values indicating antioxidant capacity have no relevance to the effects of specific bioactive compounds, including polyphenols, on human health (USDA Agric. Res. Service, 2012). It highlighted the lack of evidence of a causal link between the various compounds with the postulated health conditions and the misuse of the database by the food and dietary supplement industries to promote their products and by consumers to guide their food and dietary supplement choices: a salutary lesson.

Benefits of fruit and vegetables

The observed differences in composition of organic and conventional food are thought to have health implications. The very large number of bioactive compounds precludes studying their activity individually. Investigation of the reasons for any health benefits of fruit and vegetables can provide an insight.

There is no doubt about the health-promoting capacity of fruit and vegetables, particularly on cardiovascular diseases and strokes, of fruit with cancers of the upper GI tract, and of both fruit and vegetables on colorectal cancer (see Appendix).

Polyphenols have attracted much attention. However, Hollman *et al.* (2011) in their review were unable to establish the biological relevance of any direct antioxidant effects of polyphenols on cardiovascular health in humans. Moreover, they felt that a direct antioxidant effect of polyphenols *in vivo* is questionable because concentrations in the blood are low compared with other antioxidants and because metabolism following ingestion lowers their antioxidant activity.

Investigation of the reasons for the beneficial effects of fruit and vegetables have followed two routes: firstly dietary supplementation with synthetic antioxidants and secondly prospective studies in which dietary antioxidants have been related to cardiovascular health and cancer incidence. The epidemiological studies, the supplementation trials with synthetic antioxidants and the links between dietary intakes of antioxidants and health are discussed in detail in the Appendix.

Phytoalexins and their occurrence

In humans the body's immune system recognizes an invading antigen such as a virus, bacteria or fungus and produces specific antibodies to counter it. It has been known for many years that plants react in a similar fashion when attacked by a disease (Ames *et al.*, 1990); damage caused by a feeding insect can also have the same effect. Secondary metabolites with antimicrobial activity that are induced by stress, the phytoalexins, are an important part of the plant defence system against pathogens and insects. Phytoalexins are present in all fruits and vegetables and their synthesis can be increased many times (possibly 50–100 times) by pest damage and/or disease (Karban and Baldwin, 1997).

Work at The Sainsbury Laboratory UK is opening a new door in our understanding about the way a plant cell is triggered into action to produce phytoalexins (TSL, 2016; Postma *et al.*, 2016). Bacteria and fungi can gain relatively easy access to the region between the cells, the apoplast, where they meet the outer cell membrane. The pathogen is recognized by receptors localized at this plasma membrane that then stimulate phytoalexin production within the cytoplasm.

Plants can produce many kinds of compounds, e.g. terpenoids, alkaloids, anthocyanins, phenols, and quinones, that either kill or retard the development of feeding insects. It is also thought that plants can communicate by means of volatile compounds after an insect attack and by doing so may attract natural enemies of the insect pest (War *et al.*, 2012). Wounding, touch and wind are also known to induce defence responses in plants, e.g. Chehab *et al.* (2009). Plants may also contain defence chemicals that are present before any challenge by a pathogen or pest – these are known as phytoanticipins, e.g. saponins in crops such as oats and tomatoes may confer protection against fungi. Mansfield (2000) listed many phytoanticipins in a wide variety of crop including brassicas, currants, lupins and tomato.

Phytoalexins may be selective in deterring insects, for example Hart *et al.* (1983) reported that some soybean phytoalexins deterred Mexican bean beetles but not soybean looper larvae.

The complexity of the production of phytoalexins was evident from the work of Tzin *et al.* (2015), who found a maize variety (B73) that responded positively to caterpillar feeding then became more vulnerable to aphid attack. The two phytoalexins induced in maize by specific caterpillar and aphid attacks are both derived from the same compound, a benzoazinoid (DIMBOA). In variety B73 the caterpillar response takes priority, leaving less benzoazinoid available for the production of the aphid phytoalexin. Studies on seventeen maize varieties from around the world found similar results on some but on others aphid numbers were also reduced. By determining the nature of the genetic control it should be possible to breed for varieties that are naturally more resistant to specific insects.

Phytoalexins have attracted the attention of those looking for new fungicides as well as new compounds for controlling human diseases. The agrochemical companies are interested in effective phytoalexins that could be economically synthesized or act as leads to novel compounds. The pharmaceutical industry has been interested in the antioxidant properties of phytoalexins in order to see if they could be exploited in the control of cancers and cardiovascular diseases.

Examples of phytoalexins according to plant family

Ahuja *et al.* (2012) gave many examples of phytoalexins in their study of the genetic control of the biosynthesis and regulation of camalexin and its role in plant defence in *Arabidopsis thaliana*. The production of camalexin, a fairly common phytoalexin, can be induced in *Arabidopsis* leaves by a range of plant pathogens including bacteria, fungi and viruses.

Much of the early work on phytoalexins was carried out on *Leguminosae* and *Solanaceae* with limited work on *Brassicaceae, Vitaceae, Poaceae, Malvaceae, Euphorbiaceae, Moraceae, Orchidaceae* and *Ginkgoaceae* (Jeandet *et al.*, 2013).

Brassicaceae

At least forty-four phytoalexins have been isolated from cultivated and wild brassicas, most of which are alkaloids, for example camalexin, brassinin, rutalexin, bassilexin and spirobrassinin (Ahuja *et al.*, 2012). Glucoraphanin, a glucosinolate, is a phytoalexin found in brassicas. The potential anti-cancer (breast and prostate) and cholesterol-lowering properties of glucoraphanin are currently of interest and a super broccoli containing two to three times more glucoraphanin than normal has been bred in the UK at the Institute for Food Research in Norwich; it has been on the market since 2011 (Armah *et al.*, 2015). The responsible genes were found in wild varieties of broccoli in Sicily and it took twenty-seven years of conventional breeding to develop the improved variety. Organic broccoli sprouts are sold in the USA on the basis that they contain fifty times more glucoraphanin than mature broccoli.

As allyl isothiocyanate, a phytoalexin in brassicas can harm the plant itself it is only formed after a herbivore attack; it is stored in the harmless form of a glucosinolate and when an animal chews a tissue the enzyme myrosinase is released and acts on a glucosinolate known as sinigrin to form allyl isothiocyanate (Zhang, 2010). Under normal conditions the myrosinase is physically separated from the glucosinolates. The pungency from radish and horseradish associated with allyl isothiocyanate is appreciated by humans but not by herbivores.

Allyl isothiocyanate may be an example of a compound exhibiting hormesis (see Chapter 15). It was found to be a rodent carcinogen in toxicological tests but to have anti-cancer activity in both cultured cancer cells and animal models (Zhang, 2010); it also has antimicrobial activity against a wide spectrum of pathogens.

Convolvulaceae

Sweet potatoes contain the terpenoid phytoalexins myoporone, farnesoic acid and dendrolasin (Trewavas, 2005).

Leguminosae

Plants of this family when grown under stress conditions produce phytoalexins belonging mainly to different classes of isoflavone and aglycones. In lucerne pterocarpan compounds (e.g. medicarpin) are synthesized in response to fungal or bacterial infection, whereas in pea, pisatin is the main phytoalexin. In soybean it is the glyceollins and in groundnuts the stilbene-derived compounds, including resveratrol, which are the main phytoalexins (Ahuja *et al.*, 2012). Beans contain the phenolic phaseolin and the alkaloid vicine (Trewavas, 2005).

Malvaceae

Cooper *et al.* (1996) reported on the production by resistant genotypes of cacao (*Theobroma cacao*) of four phytoalexins in response to a xylem-invading fungal pathogen (*Verticillium dahliae*). This original finding was curious in that elemental sulphur (as cyclo-octasulphur S_8) was one of the phytoalexins together with two phenolics and a triterpenoid. In cotton the terpenoid gossypol is induced by insect (particularly lepidoptera) attack.

Poaceae

Various phytoalexins can accumulate in cereals in response to pathogen attack including kauralexins and zealexins in maize, avenanthramides in oat, and the flavonoid sakuranetin in rice (Ahuja *et al.*, 2012). Phenylamides have been added by Cho and Lee (2015) to the list of phytoalexins in rice that are induced by pathogen infections and by abiotic stresses including ultraviolet radiation.

Rubiaceae

Phenolic compounds can account for at least 90% of the total antioxidant capacity of berry fruits, which are generally ten times richer in phenolics than vegetables (Hancock *et al.*, 2007). Many phenolics are known phytoalexins. Berry fruits are also usually very rich in anthocyanins.

Solanaceae

Potatoes contain solanine, chaconine and leptidine (Trewavas, 2005). Solanine is also found in tomatoes, aubergines and paprika (and deadly nightshade). Although solanine is a nasty poison in large doses the amounts normally found in food are innocuous. Moreover, it is poorly absorbed and rapidly excreted. Other plants not of the *Solonaceae* family that contain solanine include apples, cherries and okra. Pariera Dinkins *et al.* (2008) showed that glycoalkaloid concentrations in potatoes were increased more by defoliation by Colorado beetles than by hand; concentrations in the skin were raised by 58% and in the inner tissues by 48%. Capsidiol (a terpenoid) is the major phytoalexin induced in pepper, fruit and tobacco by pathogenic fungi. Scopoletin, a polyphenol, is a major phytoalexin of tobacco plants (Ahuja *et al.*, 2012). Nicotine, an alkaloid predominantly found in tobacco, also occurs in tomato, potato and aubergine; it has been used as an insecticide but is not a known phytoalexin.

Umbelliferae

Furanocoumarins (bergapten, psoralen, xanthtoxin and isopimpinellin) occur in several members of this family, notably parsnip, celery and parsley. Contact with celery is known to cause dermatitis but Beier *et al.* (2009) reported that it was only diseased celery with high levels of furanocoumarins that did so, which is further evidence of phytoalexin production. The phytoalexin content of celery was increased nearly nine times by storage for seventy minutes at minus 15°C, followed by seventy-two hours at room temperature. Dercks *et al.* (1990) found that exposure in the laboratory of celery to an acidic fog (a single four-hour exposure at pH levels associated with commercial celery production near major population centres in California) markedly stimulated the production of furanocoumarins.

Falcarinol is a phytoalexin that accumulates in carrots (to about 2 mg/kg) and other plants from the carrot family in response to a mould attack, called liquorice rot (Crosbya and Aharonson, 1967); it is a toxin that can cause dermatitis by an allergic reaction. Normal consumption of carrots does not pose a risk of any toxic allergic effect in humans. On the other hand, there is good evidence from studies on rats that falcarinol has anti-tumour properties (Kobæk-Larsen *et al.*, 2005). Such work points the way to

a better understanding of the potentially active health-promoting ingredients in fruit and vegetables.

Vitaceae

Phytoalexins of grapevines belong mainly to the stilbene family, e.g. viniferin. Variation between cultivars in stilbene concentrations at the site of infection is known to be related to their resistance to downy mildew (Ahuja *et al.*, 2012).

The presence of resveratrol (a stilbenoid) in wine has attracted attention and research on human metabolism and health following *in vitro* and *in vivo* experiments indicating beneficial effects (Smoliga *et al.*, 2011). An additional reason for the interest was the hypothesis that the health-promoting Mediterranean diet may partially be due to the modest daily consumption of red wine. Resveratrol is thought to have preventive effects with regard to cardiovascular disease, glucose intolerance and cancer. However, as of 2016, there is limited evidence to support such human health effects.

Resveratrol is produced by several plants in response to injury or attack by bacteria or fungi (Frémont, 2000). The skin of grapes, blueberries, raspberries, and mulberries are rich in resveratrol. The genes responsible for resveratrol production have recently been engineered, at the John Innes Centre UK, into a tomato that can contain as much resveratrol as fifty bottles of red wine (Yang and Butelli, 2015); a single such tomato could provide the beneficial amount of resveratrol indicated by animal experiments. One tomato also contained the amount of genistein found in 2.5 kg of tofu. Genistein, which is found in soybeans, may be involved in preventing cancers in which steroids are implicated.

Sulphur

Cooper *et al.* (1996) were the first to find, in disease-resistant genotypes of *Theobroma cacao*, that elemental sulphur could act as a phytoalexin. Nature had discovered the potential use of sulphur as a fungicide long before man. Fungitoxic sulphur has been found in the xylem of resistant lines of other species, namely tomato (Williams *et al.*, 2002), tobacco, French beans and cotton (Williams and Cooper, 2003; Cooper and Williams, 2004) when infected by vascular pathogens both fungal and bacterial.

In tomato it was found that accumulation of S in the vascular tissue was more rapid and much greater in a disease-resistant line than in a disease-susceptible one. Levels of S detected in the resistant variety were approximately 10 µg/g fresh weight excised xylem and were judged to be sufficient to inhibit spore germination of *Verticillium dahliae* (Williams and Cooper, 2003). Likewise, induced concentrations of S were sufficient in cocoa (Cooper and Williams, 2004).

The sulphur was only found and was concentrated in cells and structures in potential contact with the vascular pathogen, such as the xylem parenchyma and the cell walls of xylem vessel.

Phytoalexins in organic produce and in misshapen fruit and vegetables

Baranski *et al.* (2014) and others, e.g. Holmboe-Ottesen (2010), acknowledge that many of the antioxidants and phenolics found in higher concentrations in organic crops are known to be produced by plants in response to pests and diseases.

Organic proponents have suggested that synthetic pesticides are detrimental because they limit the production of health-promoting phytoalexins, such as the stilbenes, flavonols and flavonones (Hoffman, 2014).

It is clear from studies on many plant families that phytoalexin production is a very common phenomenon. Susceptible organic crops that are not protected by synthetic pesticides will be more likely to suffer attacks. Thus it is logical to infer that the greater amounts of phytoalexins in organic crops may be due to the occurrence and severity of a pest or disease attack. This explanation was not accepted by Brandt *et al.* (2011) and Baranski *et al.* (2014), who consider that the higher phenolic concentrations in organic crops are because mineral N fertilizers are not used. If this was the case it would be expected that because organic farmers do not use man-made nitrogenous fertilizers the finding of more phenolics would be universal, which is not the case.

Although there is evidence from Ghosh *et al.* (2008) that organically grown tea can contain more polyphenols than tea given inorganic fertilizers it is possible that this was related to the amount of N supplied and not to the source of the N. Norbæk *et al.* (2003) had found that increasing amounts of animal manure as the source of N reduced polyphenol contents in barley.

There would appear to be two explanations for the differences in composition of organic and conventional crops. The first is that the more readily available N given to conventional crops as inorganic N fertilizers results in protein metabolism and related growth being prioritised, leaving less carbohydrate for production of bioactive compounds including phytoalexins. The second is that pathogen and pest attacks stimulate the production of bioactive compounds, which will be more evident in unprotected crops. The increases in the content of bioactive compounds in organic crops shown in the meta-analyses are much smaller than the measured effects of pathogens and insects on phytoalexins, which is compatible with only a fraction of any given organic crop being attacked.

The effect of a pathogen attack will first be evident near the point of attack often as necrotic tissue. If the tissue is still expanding at the time of attack it is likely that growth will be irregular and the plant part deformed. Likewise the production of phytoalexins will be more pronounced near the point of attack but can also be evident in apparently undamaged tissue. For example, in parsnip roots fungal attack was found to increase the furocoumarin (furanocoumarin) concentration from 20 mg/kg in roots treated with fungicide to 394 mg/

kg in the damaged parts and 200 mg/kg in the visibly undamaged parts of diseased parsnips not treated with fungicide (Mongeau *et al.*, 1994; Mattsson, 2006; Ostertag *et al.*, 2002). The elevated level of furocoumarin in infected root tissue showing no damage indicates more than a localized reaction. Mongeau *et al.* (1994) reported that mice fed infected parsnip roots containing high levels of furocoumarin had enlarged livers, abnormal liver pathology and a sixteen-fold increase in hepatic cell turnover.

Similar results were reported by Ostertag *et al.* (2002); microbial infection of parsnip roots resulted in dramatic increases in furocoumarin levels from less than 2.5 mg/kg in freshly harvested parsnips to up to 566 mg/kg in roots showing mould infection after storage at room temperatures for fifty-three days. Storage of whole parsnips at 4°C resulted in an increase of furocoumarin concentrations up to about 40 mg/kg after thirty-eight days. Such findings are not compatible with the notion that the form or amount of N supply to the crop in the field is responsible for variations in phytoalexin contents. This conclusion is confirmed by the work of Schulzová *et al.* (2012), who found no significant differences between the effects of fermented pig slurry and fertilizers on the amounts of furanocoumarins in two varieties of celeriac. The average levels of furanocoumarins in the two varieties Albin and Kompakt were 2.6 mg/kg and 10.2 mg/kg respectively.

Furocoumarins are a class of phytoalexins found in beans, carrots, celery, dill, parsley, parsnips, peanuts, peas, grapefruits, lemons, lime, oranges and potatoes that can cause skin irritations and even skin cancer in some people Mattsson (2007).

The most common causes of misshapen and blemished fruit and vegetables are insect or pathogen attack; in a few cases a nutrient deficiency such as boron deficiency or physical damage, e.g. by hail, can prevent normal development. Physical damage is also known to stimulate the production of phytoalexins, as shown by the work of Green and Ryan (1972) on potatoes and Karban and Baldwin (1997) on cotton.

Clear evidence of the importance of using fungicides to reduce the production of phytoalexins was provided by the work of Mongeau *et al.* (1994) with furocoumarins in parsnips referred to above, and by the work of Magee *et al.* (2002) on resveratrol concentrations in muscadine grape varieties when attacked by mould fungi. All the five grape cultivars that had been treated with fungicides had lower resveratrol concentrations in the skins of the berries (but not in the seeds or pulp) than in the untreated ones. In all varieties the resveratrol concentration of skins from fungicide-treated vines was lower (1.33 mg/kg) than that of unsprayed vines (4.34 mg/kg). In the variety with the greatest capacity to produce resveratrol the comparable values were 2 and 9.6 mg/kg resveratrol. In two varieties the skin resveratrol concentrations were less than 1 mg/kg. Dercks *et al.* (1995) cited experimental data that wines made from fungicide-treated plants had lower concentrations of resveratrol than wine made from untreated plants.

If reducing the natural pesticide (phytoalexins) intake is desired it would be sensible and logical to avoid eating misshapen fruit and vegetables and also those showing evidence of fungal attack. It may be a cause for concern that phytoalexins and synthetic pesticides are both similarly toxic, as found in rodent to toxicology tests (see Chapter 15).

The grapefruit effect and possible parsnip effect

It has been appreciated for many years that it is inadvisable to consume grapefruit or grapefruit juice when taking certain medications (Bailey *et al.*, 2013). Bergamottin, 6',7'-dihydroxybergamottin, bergamottin, bergapten and bergaptol, are thought to be the main furanocoumarins responsible for the "grapefruit effect" (Sakamaki *et al.*, 2008); they act by inhibiting an enzyme (cytochrome P450 3A4 found in the lining of the gut and in the liver) that breaks down many drugs that are taken orally. The inhibition of this enzyme can result in far more of the active drug being present in the body than was intended. Serious side-effects may follow. About ninety drugs are now known to be affected, including many calcium blockers used to reduce blood pressure and most but not all statins; the list is probably still rising.

The effects of the furanocoumarins will be minimized if the drug is taken many hours from the consumption of the

Table 14.4   Concentration mg/l of furanocoumarins in grapefruit juice

White		Red		
Average	Maximum	Average	Maximum	Reference
7.7	16.6	5.9	12.8	Widmer and Haun, C. (2005)
6.3	10.8	4.2	6.1	Fukuda *et al.* (2000)
39	262			Chen *et al.* (2011)

grapefruit. However, it is recommended by health authorities that it is better to err on the side of caution and never have any grapefruit (or other citrus fruit containing furanocoumarins) when taking drugs known to be affected. Other citrus fruit, including Seville oranges (and accordingly marmalade), limes and pomelos, also contain furanocoumarins but they are not present in some varieties of sweet oranges.

There would appear to be considerable variation (Table 14.4) in the furanocoumarin concentrations in grapefruit juice, much of which is thought to be varietal.

The occurrence of the same inhibitor furanocoumarins not only in parsnips (especially in unprotected roots that become diseased) but also in celery, beans, carrots, dill, parsley, peanuts and peas mean that attention should be paid to the possibility that these crops may also interact with some drugs. It would be logical to think that if the crops contain as much furanocoumarins as grapefruit they could also pose a problem. Very little appears to be known about the amounts of furanocoumarins in these crops with the exception of parsnips and celery. Research on celery was stimulated by the fact that furanocoumarins can cause severe dermatitis on harvesters.

Table 14.5   Concentration mg/kg of furanocoumarins in grapefruit, parsnips and celery

	Average	Maximum	Reference
Grapefruit (n=32)	3.7		Fukuda *et al.* (2000)
Parsnips (n=110)	15.1	145	Lombaert, *et al.* (2001)
Celery (n= 114)	1.9	15.2	Lombaert, *et al.* (2001)

Table 14.6   Average consumption of furanocoumarins (mg) in grapefruit juice, grapefruit, parsnips and celery

	mg
Grapefruit juice (200 ml white)	* 1.2–3.3
Grapefruit (half)	1.8
Parsnip 80 g	1.2–11.6
Celery 80 g	0.15–1.2

* not including Chen *et al.* (2011).

Available data (Tables 14.5 and 14.6) indicate that parsnips and to a lesser extent celery can contain similar amounts of furanocoumarins as occur in grapefruit juice. Maximum values in parsnip can exceed those in grapefruit juice. This is particularly the situation with diseased parsnips and in parsnips that have been badly stored, when the furanocoumarin concentration can rise to a few hundred mg/kg (Ostertag *et al.*, 2002; Mattsson, 2006).

There would seem to be good reasons for not consuming parsnips that have not been protected by pesticides and also parsnips that have been stored at ambient temperatures just in case they have elevated levels of furanocoumarins. It is evident, however, that much needs to be learnt about the different furanocoumarins in view of the finding that they differ in potency (Ohnishi *et al.*, 2009).

Phytoalexins in the wild ancestors of crops

The main phytoalexins in cultivated varieties of potato, tomato, brassica species, parsnip, bean, lathyrus, lima bean, aconite, cassava, betel nut, mango, lettuce, cucurbits are much lower (by three to twenty times) than in their wild counterparts (Ames *et al.*, 1990).

For example, the wild potato *Solanum acaule*, the progenitor of cultivated strains of potato, has a glycoalkaloid content about three times that of cultivated strains and is accordingly more toxic. The leaves of the wild cabbage *Brassica oleracea* (the progenitor of cabbage, broccoli, and cauliflower) contain about twice as many glucosinolates as cultivated cabbage. The wild bean *Phaseolus lunatus* contains about three times as many cyanogenic glucosides as does the cultivated bean.

Trewavas (2005) writes that in breeding for better yield the phytoalexin content of several crops has been reduced two to ten-fold and in doing so has made the crops safer for humans but requiring synthetic pesticides for protection. The aim of breeding for pest resistance is to increase the phytoalexin content; such varieties may be preferred by organic farmers. It is possible that palatability has been improved as a result of the lower phytoalexin contents in the cultivated modern varieties.

Summary

1. In the absence of clinical evidence supporting the claims that organic food is healthier than conventional, attention has been focused on the bioactive compounds in crops that are thought to be beneficial and also on chemicals such as pesticide residues that might be harmful.

2. The situation is fundamentally very complex simply because of the very large number (many thousands) of bioactive organic compounds with widely varying properties found in plants. The relevance of suggestions about potentially beneficial compounds must be assessed against the background of a complex and extensive biochemistry. Plants that produce bioactive compounds/ secondary metabolites are the rule rather than the exception.

3. Most secondary metabolites fall roughly into three classes of compounds, namely phenolics, terpenoids and alkaloids.

4. Bioactive compounds are involved in many aspects of plant development, including protection against pests and pathogens, attraction of insects, signalling and influencing the growth of

neighbouring plants. Each type of plant has its own complement of bioactive compounds.

5. It is the involvement particularly of phenolic and terpenoid compounds in plant defence mechanisms that are important.

6. Plants may contain up to several grams per kilogram of phenolic compounds, and there may be a total of more than 4,000 such compounds. There are likely to be more than 20,000 terpenoids in plants.

7. Plant-derived alkaloid compounds (more than 5,000) are noted for their extreme toxicity to mammals, e.g. strychnine, and others are potent stimulants, e.g. caffeine and nicotine.

8. There have been several meta-analyses that have aggregated the data from the large number of studies comparing the composition of organically and conventionally grown crops.

9. Most attention has been focused on compounds that have antioxidant properties, such as the phenolics and terpenoids.

10. Interpretation of the meta-analyses is hampered by the facts that the data is assessed on the basis of percentage differences in composition without reference to actual amounts of the chemicals involved and the general absence of statistical analysis of the raw data as well as the way data was selected.

11. It is possible to make a few generalizations based on the findings of the meta-analyses. Organically grown crops tend to have higher concentrations of vitamin C, total antioxidant (fruit) and phenolics and lower concentrations of nitrate, protein, cadmium and synthetic pesticides. It is inferred that organic food is healthier because of these differences.

12. The penalties paid for the increased content of potentially beneficial compounds in organic crops are lower yields and, for the consumer the higher prices of organic produce, which are likely to reduce consumption.

13. The fact that the effects on composition are not consistent means that there can be no guarantee that any organic food will contain more of the health-promoting compounds than conventional food and at the same time suggests that the observed effects are the result of factors that are not always operating.

14. One meta-analysis concluded that there was no evidence of health benefits of organic food based on human and *in vitro* studies.

15. Many of the antioxidants are phytoalexins produced by plants responding to insect and pathogen attack; the higher concentrations in organic crops may be due to the fact that they were not protected by a pesticide and as a result some plants in the field were attacked and phytoalexins generated.

16. The greater concentrations of cadmium in conventional crops are very probably associated with the contaminant cadmium in phosphatic fertilizers. It is difficult to assess the relevance of the cadmium values in the meta-analyses and so distinguish between hazard and risk in the absence of data on the actual concentrations. The amounts of cadmium in crops will depend on the source of the phosphate rock.

17. Organic milk has been found to contain more fatty acids (especially omega-3) than conventional and is accordingly assessed as superior. However, as the omega-3 fatty acids in milk are mostly short chain and not the more beneficial

long chain types the claim of superiority is dubious.

18. Organic milk is likely to contain less iodine and selenium than conventional milk because conventional dairy cows will get more iodine and selenium in concentrates than organic cows get from grazing. Iodine deficiency disorders are a big problem in many countries. Irreversible foetal brain damage is a major concern.

19. In support of organic milk it was suggested that the lower iodine levels could be considered beneficial in order to safeguard against iodine toxicity (thyrotoxicosis); this was despite the fact that iodine toxicity is most unlikely to occur in individuals who are on normal Western diets and had not earlier suffered from iodine deficiency.

20. Protein profile studies have provided evidence that the farming system can affect plant metabolism.

21. The chemical composition of a given crop or food can be affected by several factors, notably soil conditions, climate, aspect, season and especially variety. The variation in vitamin C content between crops and varieties are much larger than the average variations reported in the comparative studies.

22. The significance for human health of the antioxidants and other components in the diet needs to be established before any claims about health benefits can be made linking their intake to farming method. The trail of evidence starts with fruit and vegetables and the advice to eat at least five portions (80 g) every day. This was and still is good advice.

23. There have been several huge, long-term epidemiological studies in the USA and in Europe in which very clear evidence has been gained showing positive effects of consuming fruit and vegetables on major health outcomes.

24. The messages seem clear. It is likely that fruit and vegetables are mainly beneficial by reducing deaths due to cardiovascular disease and strokes rather than by reducing cancer incidence.

25. The risk of cancers of the upper gastrointestinal tract seems to be inversely linked with fruit intake but not with vegetable intake. The risk of colorectal cancer is inversely associated with intakes of total fruit and vegetables and total fibre. There do not appear to be links to other cancers.

26. The precise reasons for the health-promoting properties of fruit and vegetables are not yet known. Much attention has been focused on the antioxidants vitamin E, ß-carotene/vitamin A, and vitamin C. The involvement of fibre should not be ignored.

27. Although correlative studies have linked vitamin C with reducing cancers of the upper gastrointestinal tract and vitamin E with inhibiting cardiovascular diseases, long-term clinical trials involving more than 100,000 people with vitamin C, E or ß-carotene supplements have so far failed to demonstrate their efficacy in reducing the risk of cardiovascular disease and cancer or in prolonging life.

28. There are indications that a combination of vitamin C, vitamin E, and ß-carotene supplements can reduce the risk of developing the advanced stage of age-related macular degeneration.

29. Overall there is not a case for healthy people taking antioxidant supplements but this does not mean eating food sources rich in antioxidants should be discouraged. It is possible that the

health benefits of food are due to the fact that they contain several antioxidants that may be working together, and not to the activity of just one or two antioxidants.

30. Prospective studies in which dietary antioxidants have been related to cardiovascular health and cancer incidence have been largely inconclusive but there is evidence of a reduction in the risk of gastric cancer.

31. All plants have the capacity to generate phytoalexins when attacked by pathogens or by insects. The finding that such compounds, which include phenolics, anthocyanins and flavonoids, are also found in higher concentrations in organic crops suggests that synthesis of these compounds may have been stimulated by attack. This phenomenon is accepted but interpreted differently by organic proponents who consider that the protective use of pesticides and use of fertilizers are detrimental by reducing the amounts of the beneficial phytoalexins.

32. The finding that phytoalexin concentrations are not increased in all organic crops/plants could be due to the fact that not all the unprotected organic crops/plants are attacked; those that are not attacked will contain smaller amounts of phytoalexins.

33. The fact that furanocoumarins in grapefruit are responsible for the serious and dangerous enhancement of the effect of many drugs – known as the grapefruit effect – raises the question as to whether furanocoumarins in other crops such as parsnips (especially organic) may also have a similar deleterious effect.

34. Diseased fruit and vegetables and those misshapen as a result of insect attack are likely to be rich in phytoalexins and should logically be discarded if reducing pesticide intake is a priority.

References

Adler, F.R. (2011) Plant signalling: the opportunities and dangers of chemical communication. *Biology Letters* 7, 2, 161–162. doi: 10.1098/rsbl.2010.0790.

Ahuja, I., Kissen, R. and. Bones, A.M. (2012) Phytoalexins in defence against pathogens *Trends in Plant Science*. 17, 2, 73–90. doi: 10.1016/j.tplants.2011.11.002.

Ames, B.N., Profet, M. and Gold, L.S. (1990) Natures Chemicals and Synthetic Chemicals: Comparative Toxicology. *Proceedings of the National Academy of Sciences USA* 87, 9, 7782–7786.

Anon (2005) Manual of Nutrition. Food Standards Agency. Tenth Edition. The Stationary Office, London.

Anon. (2013) Metabolites: Primary vs Secondary. Available at http://lifeofplant. blogspot.co.uk/2011/03/metabolites-primary-vs-secondary.html

Anon (2015) Cinnabar moth. A Nature observer's Scrapbook. Available at www. bugsandweeds.co.uk/moths%20p2.html

Anon (2015a) Natural food guide-Fruit Vitamin C Content of Apple Cultivars The Natural Food Hub. Available at www. naturalhub.com/natural_food_guide_fruit_ vitamin_c_apple.htm

Anon (2015b) Natural food-fruit vitamin C content. The Natural Food Hub. Available at www.naturalhub.com/natural_food_ guide_fruit_vitamin_c.htm

Armah, C.N., Derdemezis, C., Traka, M.H., Dainty, J.R., Doleman, J.F., Saha, S., Leung, W., Potter, J.F., Lovegrove,

J.A. and Mithen, R.F. (2015) Diet rich in high glucoraphanin broccoli reduces plasma LDL cholesterol: Evidence from randomised controlled trials. *Molecular Nutrition and Food Research* 59, 5, 918–926. doi: 10.1002/mnfr.201400863.

ASA (2000) Advertising Standards Authority Organic food complaint upheld Adjudications 12 July 2000. Available at www.gene.ch/gentech/2000/Jul/msg00047.html

Barański, M., Średnicka-Tober, D., Volakakis, N., Seal, C., Sanderson, R., Stewart, G.B.,.Benbrook, C., Biavati, B., Markellou, E., Giotis, C., Gromadzka-Ostrowska, J., Rembiałkowska, E., Skwarło-Sońta, K.,Tahvonen, R., Janovská, D., Niggli, U., Nicot, P. and Leifert, C. (2014) Higher antioxidant and lower cadmium concentrations and lower incidence of pesticide residues in organically grown crops: a systematic literature review and meta-analyses. *British Journal of Nutrition,* 112, 5, 794–811. doi: 10.1017/S0007114514001366. Supplementary material http://dx.doi.org/10.1017/S0007114514001366

Barker, D.J. and Phillips, D.I. (1984) Current incidence of thyrotoxicosis and past prevalence of goitre in 12 British towns. *Lancet*; 2, 8402, 567–570.

Bath, S.C., Button, S., and Rayman, M.P. (2012) Iodine concentration of organic and conventional milk: implications for iodine intake. *British Journal of Nutrition,* 107, 7, 935–940. doi: 10.1017/S0007114511003059.

Bath, S.C., Rayman, M.P., Steer, C.D., Golding, J., and Emmett, P. (2013) Effect of inadequate iodine status in UK pregnant women on cognitive outcomes in their children: Results from the Avon Longitudinal Study of Parents and Children. (ALSPAC). *The Lancet* 382, 9889 331–337.

Bath, S.C., Walter, A., Taylor, A., Wright, J., and Rayman, M.P. (2014) Iodine deficiency in pregnant women living in the South East of the UK: The influence of diet and nutritional supplements on iodine status. *British Journal of Nutrition* 111, 9, 1622–1631. doi: 10.1017/S0007114513004030.

Baxter, G.J., Graham, A.B., and Lawrence, J.R. (2001) Salcylic acid in soups prepared from organically and non-organically grown vegetables. *European Journal of Nutrition* 40, 6, 289–292.

Bailey, D.G., Dresser, G. and Arnold, J.M.O. (2013) Grapefruit-medication interactions: Forbidden fruit or avoidable consequences? *Canadian Medical. Association Journal* 185, 4, 309-16. doi: 10.1503/cmaj.120951.

Beier, R.C., Ivie , G.W. and Oertli, E.H. (2009) Psoralens as Phytoalexins in Food Plants of the Family Umbelliferae. Significance in Relation to Storage and Processing. *Xenobiotics in Foods and Feeds* 19, 295–310. doi: 10.1021/bk-1983-0234. ch019.

Benbrook, C.M. (2005) Elevating Antioxidant Levels in Food through Organic Farming and Food Processing. Organic Center State of Science Review. Available at https://www.organic-center.org/reportfiles/Antioxidant_SSR.pdf

Benbrook, C., Zhao Yanez, J., Davies, N. and Andrews, P. (2008) New evidence confirms the nutrition superiority of plant-based organic foods. Organic Center State of Science Review. Available at https://www.organic-center.org/reportfiles/Nutrient_Content_SSR_Executive_Summary_FINAL.pdf

Bernhoft, A. (2010) Bioactive compounds in plants – benefits and risks for man and animals 11–18. In: Bernhoft, A. (ed.) *Proceedings of symposium held at The Norwegian Academy of Science and Letters, Oslo, 2008*. Available at www.dnva.no/binfil/download.php?tid=48677

Brandt, K. and Mølgaard, J.P. (2001) Organic agriculture: does it enhance or reduce nutritional value of plant foods? *Journal of Agricultural and Food Chemistry* 81, 9, 924–931. doi: 10.1002/jsfa.903.

Brandt, K. and Mølgaard, J.P. (2006) Food Quality 305–327. In: Kristiansen, P., Taji, A., and Reganold, J. (eds.) *Organic Agriculture – a global perspective*. Victoria, Australia: CSIRO Publishing.

Brandt, K. (2007) Organic agriculture and food utilization. Issues Paper. International Conference on Organic Agriculture and Food Security, Italy: FAO Interdepartmental Working group on Organic Agriculture. Available at www.usc-canada.org/UserFiles/File/organic-agriculture-and-food-security.pdf

Brandt, K., Leifert, C., Sanderson, R. and Seal, C.J. (2011) Agroecosystem management and nutritional quality of plant foods: the case of organic fruits and vegetables. *Critical Reviews in Plant Sciences* 30, 1–2, 177–197.

Caris-Veyrat, C., Amiot, M.J., Tyssandier, V., Grasselly, D., Buret, M., Mikolajczak, M,, Guilland, J.C., Bouteloup-Demange, C. and Borel, P. (2004) Influence of organic versus conventional agricultural practice on the antioxidant microconstituent content of tomatoes and derived purees; consequences on antioxidant plasma status in humans. *Journal of Agricultural and Food Chemistry* 52, 21, 6503–6509. doi: 10.1021/jf0346861.

Chassy, A.W., Bui, L., Renaud, E.N., Van Horn, M. and Mitchell, A.E. (2006) Three-year comparison of the content of antioxidant microconstituents and several quality characteristics in organic and conventionally managed tomatoes and bell peppers. *Journal of Agricultural and Food Chemistry* 54, 21, 8244–8252. doi: 10.1021/jf060950p.

Chehab, E.W., Eich, E. and Braam, J. (2009) Thigmomorphogenesis: a complex plant response to mechano-stimulation. *Journal of Experimental Botany* 60, 1, 43–56. doi: 10.1093/jxb/ern315.

Chen, C., Cancalon, P., Haun, C. and Gmitter Jr. F. (2011) Characterization of Furanocoumarin Profile and Inheritance Toward Selection of Low Furanocoumarin Seedless Grapefruit Cultivars. *Journal of the American Society for Horticultural Science* 136, 5, 358–363. Available at http://journal.ashspublications.org/content/136/5/358.full

Cho, M.-H. and Lee, S-W. (2015) Phenolic Phytoalexins in Rice: Biological Functions and Biosynthesis. *International Journal of Molecular Science* 16 *12*, 29120-29133. doi: 10.3390/ijms161226152.

Cooper, R.M., Resende, M.L.V., Flood, J., Rowan, M.G., Beale, M.H. and Potter, U. (1996) Detection and cellular localization of elemental sulphur in disease-resistant genotypes of Theobroma cacao. *Nature* 379, 159–162. doi: 10.1038/379159a0.

Cooper, R.M. and Williams, J.S. (2004) Elemental sulphur as an induced antifungal substance in plant defence. *Journal of Experimental Botany* 55, 404, Sulphur Metabolism in Plants Special Issue, 1947–1953, doi: 10.1093/jxb/erh179.

Crosbya, D.G., Aharonson, N. (1967) The structure of carotatoxin, a natural toxicant from carrot. *Tetrahedron* 23, 1, 465–472. doi: 10.1016/S0040-4020(01)83330-5.

Dahl, L., Opsahl, J.A., Meltzer. H.M. and Julshamn, K. (2003) Iodine concentration in Norwegian milk and dairy products. *British Journal of Nutrition* 90, 3, 679–685.

Dangour, A.D., Dodhia, S.K., Hayter, A., Allen, E., Lock, K. and Uauy, R. (2009) Comparison of composition (nutrients and other substances) of organically and conventionally produced foodstuffs: a systematic review of the available literature. Report for Food Standard Agency. London. *American Journal of Clinical Nutrition* 90, 3, 680–685. doi: 10.3945/ajcn.2009.28041.

Dangour, A.D., Lock, K., Hayter, A., Aikenhead, A., Allen, E. and Uauy, R. (2010) Nutrition-related health effects of organic foods: a systematic review. *American Journal of Clinical Nutrition* 92, 1: 203–210. doi: 10.3945/ajcn.2010.29269.

Del Rio, D., Rodriguez-Mateos, A., Spencer, J.P.E, Tognolini, M., Borges, G. and Crozier A. (2013) Dietary (poly) phenolics in human health: structures, bioavailability, and evidence of protective effects against chronic diseases. *Soil and Tillage Research* 18, 14, 1818–1892. doi: 10.1089/ars.2012.4581.

Dercks, W., Trumble, J. and Winter, C. (1990). Impact of atmospheric pollution on linear furanocoumarin content in celery. *Journal of Chemical Ecology* 16, 2, 443–454. doi: 10.1007/BF01021776.

Dercks, W., Creasy, L.L. and Luczka-bayles, C.J. (1995) Stllbene phytoalexin and disease resistance in Vitis, 287–315. In Daniel, M Purkayasta, R. P. (eds.) *Handbook of Phytoalexin Metabolism and Action*. CRC Press.

EFSA (2009) Cadmium in food – Scientific opinion of the Panel on Contaminants in the Food Chain. *EFSA Journal* 980, 36–139. doi: 10.2903/j.efsa.2009.980.

EUFIC (2013) The European Food Information Council. Organic food and farming: scientific facts and consumer perceptions. Available at www.eufic.org/article/en/page/RARCHIVE/expid/Organic_food_and_farming_scientific_facts_and_consumer_perceptions

Eurola, M., Hietaniemi, V., Kontturi, M.; Tuuri, H., Pihlava, J.M., Saastamoinen, M., Rantanen, O., Kangas, A. and Niskanen, M. (2003) Cadmium contents of oats (*Avena sativa* L.) in official variety, organic cultivation, and nitrogen fertilization trials during 1997–1999. *Journal of Agricultural and Food Chemistry* 51, 9, 2608–2614. doi: 10.1021/jf020893+.

European Union (2008) Commission Regulation (EC) no. 889/2008 of 5 September 2008 laying down detailed rules for the implementation of Council Regulation (EC) no. 834/2007 on organic production and labelling of organic products with regard to organic production, labelling and control. *Official Journal of the European Union* 51, L250.

Flachowsky, G., Franke, K., Meyer, U., Leiterer, M., and Schöne, F. (2014) Influencing factors on iodine content of cow milk. *European Journal of Nutrition*, 53, 2, 351–365. doi: 10.1007/s00394-013-0597-4.

Fondation Louis Bonduelle (2015) Vitamin C. Available at www.fondation-louisbonduelle.org/france/en/know-your-vegetables/health-and-nutrients/vitamine-c-6.html#axzz3uKdeJ6OG

Food Standards Agency (2008) Retail Survey of Iodine in UK Produced Dairy Foods, Food Survey Information Sheet 02/08. London: Food Standards Agency, UK.

Fraenkel, G. (1959) Raison d'etre of secondary plant substances. *Science* 129. 3361, 1466–1470. doi: 10.1126/science.129.3361.1466.

Freeman, B.C. and Beattie, G.A. (2008) An Overview of Plant Defences against Pathogens and Herbivores. *The*

Plant Health Instructor. doi: 10.1094/ PHI-I-2008-0226-01.

Frémont, L. (2000) *Biological effects of resveratrol. Life Sciences* 66, 8, 663–673.

Fukuda, K., Guo, L., Ohashi, N., Yoshikawa, M. and Yamazoe, Y. (2000) Amounts and variation in grapefruit juice of the main components causing grapefruit-drug interaction. *Journal of Chromatography* 741,195–203.

Ghosh, B.C., Palit, S., Gupta, S.D., and Swain, D.K. (2008) Studies on tea quality grown through conventional and organic management practices: Its impact on antioxidant and antidiarrhoeal activity. *Transactions of Asabe* 51, 6, 2227–2238. doi: 10.13031/2013.25376.

Green, T.R. and Ryan, C.A. (1972) Wound-induced proteinase inhibitors in plant leaves: a possible defence mechanism against insects. *Science* 175, 4023, 776–777. doi: 10.1126/science.175.4023.776.

Grinder-Pedersen, L., Rasmussen, S.E., Bügel, S., Jørgensen, L.V., Dragsted, L.O., Gundersen, V. and Sandstrøm, B. (2003) Effect of diets based on food from conventional versus organic production on intake and excretion of flavonoids and markers of antioxidative defence in humans. *Journal of Agricultural and Food Chemistry* 51, 19, 5671–5676. doi: 10.1021/jf030217n.

Hancock, R.D., McDougall, G.J. and Stewart, D. (2007) Berry fruit as superfood: hope or hype. *Biologist* 54, 2, 73–79.

Hart, S.V., Kogan, M. and Paxton, J.D. (1983) Effect of soybean phytoalexins on the herbivorous insects Mexican bean beetle and soybean looper. *Journal of Chemical Ecology* 9, 6,657–72. doi: 10.1007/ BF00988774.

Hoffman, R. (2014) Organic produce trumps conventional – Here's why. Intelligent Medicine. Available at http://drhoffman. com/article/organic-produce-trumps-conventional-heres-why

Hollman, P.C., Cassidy, A., Comte, B., Heinonen, M., Richelle, M., Richling, E., Serafini, M., Scalbert, A., Sies, H. and Vidry, S. (2011) The biological relevance of direct antioxidant effects of polyphenols for cardiovascular health in humans is not established. *Journal of Nutrition* 141, 5, 989S–1009S. doi: 10.3945/jn.110.131490.

Holmboe-Ottesen, G. (2010) Increased levels of bioactive compounds in organically grown food plants. Possible health effects? 236–252. In: Bernhoft, A. (ed) *Bioactive Compounds in Plants – Benefits and Risks for Man and Animals*, The Norwegian Academy of Science and Letters, Oslo. Available at www.dnva.no/binfil/ download.php?tid=48677

Huber, M., Rembiałkowska, E., Srednicka, D., Bügel, S. and van de Vijver, L.P.L (2015) Organic food and impact on human health: Assessing the status quo and prospects of research. NJAS – Wageningen *Journal of Life Sciences* 58, 3–4, 103–1109. doi: 10.1016/j.njas.2011.01.004.

Ichiro, N., Yuka, Y., Miho, Y., Atsuko, S., Tetsuo, F., and Tadachika, O. (2004) Varietal Difference in Vitamin C Content in the Fruit of Kiwifruit and Other *Actinidia* Species. *Journal of Agricultural and Food Chemistry* 52, 17, 5472–5475. doi: 10.1021/jf049398z.

Jahangir, M., Hye, K.K., Young, H.C. and Verpoort, R. (2009) Health-affecting compounds in *Brassicaceae*. *Comprehensive Reviews in Food. Science and Food Safety* 8, 2, 31–43. doi: 10.1111/j.1541-4337.2008.00065.

Jeandet, P., Clement, C., Courot, E. and Cordelier, S. (2013) Modulation of Phytoalexin Biosynthesis in Engineered

Plants for Disease Resistance. *International Journal of Molecular Sciences 14*, 7, 14136–14170. doi: 10.3390/ijms140714136.

Karban, R. and Baldwin, T. (1997) *Induced responses to herbivory*. Univ. Chicago Press.

Khokhar, S. and Owusu Apenten, R.K. (2003) Anti-nutritional factors in food legumes and effects of processing 82–116. In: Squires, V.R. (ed.) *The role of food, agriculture, forestry and fisheries in human nutrition. Encyclopedia of Life Support Systems*, 82–116.

Kliem, K.E., Shingfield, K.J., Livingstone, K.M. and Givens, D.I. (2013) Seasonal variation in the fatty acid composition of milk available at retail in the United Kingdom and implications for dietary intake. *Food Chemistry* 141, 1, 274–281. doi.org/10.1016/j.foodchem.2013.02.116.

Kobæk-Larsen, M., Christensen, L.P., Vach, W., Ritskes-Hoitinga, J., and Kirsten Brandt, K. (2005) Inhibitory Effects of Feeding with Carrots or Falcarinol on Development of Azoxymethane-Induced Preneoplastic Lesions in the Rat Colon. *Journal of Agricultural and Food Chemistry* 53, 5, 1823–1827. doi: 10.1021/jf048519s.

Kumpulainen, J. (2001) Organic and conventional foodstuffs: nutritional and toxicological quality comparisons *International Fertiliser Society, Proceedings* 472.

Lehesranta, S.J., Koistinen, K.M., Massat, N., Davies, H.V., Shepherd, L.V., McNicol. J.W., Cakmak, I., Cooper, J., Lück, L., Kärenlampi, S.O. and Leifert, C. (2007) Effects of agricultural production systems and their components on protein profiles of potato tubers. *Proteomics* 7, 4, 597–604. doi: 10.1002/pmic.200600889.

Lombaert, G.A., Siemens, K.H., Pellaers, P., Mankotia, M. and Ng, W. (2001) Furanocoumarins in celery and parsnips: method and multiyear Canadian survey *Journal of AOAC International* 84, 4: 1135–43.

Lueck, L., Brandt, K., Seal, C., Rembialkovska, E., Huber, M., Butler., G., Bennett., R., Oughton., L., Nicholas, P., Kretzschmar, U., Sundrum, A. and Leifert, C. (2008) QLIF Workshop 1: Product quality in organic and low input farming systems 18–20. Paper presented at *4th QLIF Congress in the Frame of the 16th IFOAM Organic World Congress and the 2nd ISOFAR Conference*, Modena, Italy.

Magee, J.B., Smith, B.J. and Rimando, A. (2002) Resveratrol content of muscadine berries is affected by disease control spray program. *HortScience* 37, 2, 358–361. Available at http://hortsci.ashspublications.org/content/37/2/358.full.pdf

Mansfield, J.W. (2000) Antimicrobial compounds and resistance, the role of phytoalexins and phytoanticipins 325–370. In: Slusarenko, A., Fraser, R. and van Loon, L. (eds). *Mechanisms of Resistance to Plant Diseases*. Kluwer Academic Publishers.

Mar, S.S. and Okazaki, M. (2012) Investigation of Cd contents in several phosphate rocks used for the production of fertilizer. *Microchemical Journal* 104, 17–21 doi.org/10.1016/j.microc.2012.03.020.

Mattsson, J.L. (2006) Spray more for safer food. *New Zealand Geographic* 77, 12-16. Available at https://www.nzgeo.com/stories/spray-more-for-safer-food

Mattsson, J.L. (2007) Mixtures in the real world: The importance of plant self-defence toxicants, mycotoxins, and the human diet. Toxicology and Applied Pharmacology 223, 2, 125–132. doi: 10.1016/j.taap.2006.12.024.

Minihane, A.M., Givens, D.I. and Gibbs, R.A. (2008) The health benefits of n–3 fatty acids and their concentrations in organic and conventional animal-derived foods 19–49. In: Givens, D.I., Baxter, S., Minihane, A.M. and Shaw, E. (eds.) *Health benefits of organic food: effects of the environment.* CABI, Wallingford, England.

Mitchell, A.E., Hong, Y.J., Koh, E., Barrett, D.M. and Bryant, D.E. (2007) Ten-year comparison of the influence of organic and conventional crop management practices on the content of flavonoids in tomatoes. *Journal of Agricultural and Food Chemistry* 55: 6154–6159. Available at http://pubs. acs.org/doi/pdf/10.1021/jf070344%2B

Mongeau, R., Brassard, R., Cerkauskas, R., Chiba, M., Lok, E., Nera, E.A., Jee, P., McMullen, E. and Clayson, D.B. (1994) Effect of addition of dried healthy or diseased parsnip root tissue to a modified AIN-76A diet on cell proliferation and histopathology in the liver, oesophagus and forestomach of male Swiss Webster mice. *Food and Chemical Toxicology* 32, 3, 265–71.

Ng, T.B., Ye, X.J,, Wong, J.H., Fang, E.F., Chan, Y.S., Pan, W., Ye, X.Y., Sze, S.C., Zhang, K.Y., Liu, F. and Wang, H.X. (2011) Glyceollin, a soybean phytoalexin with medicinal properties. *Applied Microbiology and Biotechnology* 90, 1, 59–68. doi: 10.1007/s00253-011-3169-7.

Norbæk, R., Aaboer, D.B.F., Bleeg, I.S., Christensen, B.T., Kondo, T. and Brandt, K. (2003) Flavone C-glycoside, phenolic acid, and nitrogen contents in leaves of barley subject to organic fertilization treatments. *Journal of Agricultural and Food Chemistry* 51, 3, 809–813. doi: 10.1021/jf0258914.

Ohnishi, A., Matsuo, H., Yamada, S., Takanaga, H., Morimoto, S., Shoyama, Y., Ohtani, H, and Sawada, Y. (2009) Effect of furanocoumarin derivatives in grapefruit juice on the uptake of vinblastine by Caco-2 cells and on the activity of cytochrome P450 3A4. *British Journal of Pharmacology* 130, 6, 1369–1377. doi: 10.1038/sj.bjp.0703433.

Ostertag, E., Becker, T., Ammon, J., Bauer-Aymanns, H. and Schrenk, D. (2002) Effects of storage conditions on furocoumarin levels in intact, chopped, and homogenized parsnips. *Journal of Agricultural and Food Chemistry* 50, 9, 2565–2570.

Pariera Dinkins, C.L., Peterson, R.K., Gibson, J.E., Hu, Q. and Weaver, D.K. (2008) Glycoalkaloid responses of potato to Colorado potato beetle defoliation, *Food and Chemical Toxicology* 46, 8, 2832–2836. doi: 10.1016/j.fct.2008.05.023.

Postma, J., Liebrand, T.W., Bi, G., Evrard, A., Bye, R.R., Mbengue, M., Kuhn, H., Joosten, M.H. and Robatzek, S. (2016) Avr4 promotes Cf-4 receptor-like protein association with the BAK1/SERK3 receptor-like kinase to initiate receptor endocytosis and plant immunity. *New Phytologist.* 2010, 2,627–642. Copyright 1999–2017 John Wiley & Sons. doi: 10.1111/nph.13802.

Qi You, Baowu Wang, Feng Chen, Zhiliang Huang, Xi Wang, Pengju G. Luo (2011) Comparison of anthocyanins and phenolics in organically and conventionally grown blueberries in selected cultivars. *Food Chemistry* 125, 1, 201–208. doi: 10.1016/j. foodchem.2010.08.063.

QLIF (2009) Integrated Research Project: Advancing Organic and Low Input Food. Summary of Sub-project 2: Effect of production methods. Report 2009. Available at www.qlif.org/Library/leaflets/folder_2_small.pdf

Rasmussen, L.B., Larsen, E.H. and Ovesen, L. (2000) Iodine content in drinking water and other beverages in Denmark. *European Journal of Clinical Nutrition*. 54, 1, 57–60.

Rempelos, L., Cooper, J., Wilcockson, S., Eyre, M., Shotton, P., Volakakis, N., Orr, C.H., Leifert, C. and Gatehouse, A.M.R. (2013) Quantitative proteomics to study the response of potato to contrasting fertilization regimes. *Molecular Breeding* 31, 2, 363–378. doi: 10.1007/s11032-012-9795-7.

Roberts, T.L. (2014) Cadmium and Phosphorous Fertilizers: The Issues and the Science. *Procedia Engineering* 83, 2014, 52–59. doi: 10.1016/j.proeng.2014.09.012.

Rosen, J.D. (2008) Claims of Organic Food's Nutritional Superiority: A Critical Review. American Council on Science and Health. Available at http://acsh.org/2008/07/claims-of-organic-foods-nutritional-superiority-a-critical-review

Rosen, J.D. (2010) A Review of the Nutrition Claims Made by Proponents of Organic Food. *Comprehensive Reviews in Food Science and Food Safety* 9, 3, 270–277. doi/10.1111/j.1541-4337.2010.00108.x/full.

Sakamaki, N., Nakazato, M., Matsumoto, H., Hagino, K., Hirata, K, and Ushiyama, H. (2008) Contents of furanocoumarins in grapefruit juice and health foods. *Shokuhin Eiseigaku Zasshi*. 49, 4, 326–31. www.ncbi.nlm.nih.gov/pubmed/18787320

Schulzová, V., Babička, L. and Hajšlová, J. (2012) Furanocoumarins in celeriac from different farming systems: a 3-year study. *Journal of the Science of Food and Agriculture* 92, 14, 2849–2854. doi: 10.1002/jsfa.5629.

Schwendel, B.H., Wester, T.J., Morel, P.C.H., Tavendale, M.H., Deadman, C., Shadbolt, N.M. and Otter, D.E. (2015) Invited review: Organic and conventionally produced milk – An evaluation of factors influencing milk composition. *Journal of Dairy Science* 98, 2, 721–746. doi: 10.3168/jds.2014-8389.

Scollan, N.D., Kim, E.J., Lee, M.R.F., Whittington, F. and Richardson, R.I. (2008) Environmental impacts on n–3 content of foods from ruminant animals 5069 In: Givens. I., Baxter S., Minihane, A.M. and Shaw E. (eds) International Workshop on the Effects of the Environment on the Nutritional Quality and Safety of Organically Produced Foods, Reading, CABI Publishing.

Smith-Spangler, C., Brandeau, M.L., Hunter, G.E., Bavinger, C., Pearson, M., Eschbach, P.J., Vandana Sundaram, Hau Liu, Schirmer, P., Stave, C., Olkin, I. and Bravata, D.M. (2012) Are organic foods safer or healthier than conventional alternatives? A systematic review. *Annals of Internal Medicine, American College of Physicians*.157, 5, 348–366. http://annals.org/article.aspx?articleid=1355685

Smoliga, J.M., Baur, J.A. and Hausenblas, H.A. (2011) Resveratrol and health-a comprehensive review of human clinical trials. *Molecular Nutrition and Food Research* 55, 8, 1129–1141. doi: 10.1002/mnfr.201100143.

Średnicka-Tober, T., Barański, M., Seal, C.J., Sanderson, R., Benbrook, C., Steinshamn, H., Gromadzka-Ostrowska, J., Rembiałkowska, E., Skwarło-Sońta, K. , Eyre, M., Cozzi, G., Larsen, M.K., Jordon, T., Niggli, U., Sakowski, T., Calder, P.C., Burdge, G.C., Sotiraki, S., Stefanakis, A., Stergiadis, S., Yolcu, H., Chatzidimitriou, E., Butler, G., Stewart, G. and Leifert, C. (2016) Higher PUFA and n–3 PUFA, conjugated linoleic acid, ß-tocopherol and iron, but lower iodine and selenium concentrations in organic milk: a systematic literature review and meta- and redundancy analyses. *British Journal of Nutrition* (2016), 115, 6, 1043–1060. doi: 10.1017/S0007114516000349.

Średnicka-Tober, T., Barański, M., Seal, C.J., Sanderson, R., Benbrook, C., Steinshamn, H., Gromadzka-Ostrowska, J., Rembiałkowska, E., Skwarło-Sońta, K. , Eyre, M., Cozzi, G., Larsen, M.K., Jordon, T., Niggli, U., Sakowski, T., Calder, P.C., Burdge, G.C., Sotiraki, S., Stefanakis, A., Stergiadis, S., Yolcu, H., Chatzidimitriou, E., Butler, G., Stewart, G. and Leifert, C. (2016a) Composition differences between organic and conventional meat: a systematic literature review and meta-analysis. *British Journal of Nutrition* 115, 6, 994–1011. doi: 10.1017/S0007114515005073.

Tétard-Jones, C., Shotton, P.N., Rempelos, L., Cooper, J., Eyre, M., Orr, C.H. and Leifert, C. (2013) Quantitative proteomics to study the response of wheat to contrasting fertilization regimes. *Molecular Breeding* 31, 2, 379–393. doi: 10.1007/s11032-012-9796-6.

Trewavas, A. (2004) A critical assessment of organic farming-and-food assertions with particular respect to the UK and the potential environmental benefits of no-till agriculture. *Crop Protection* 23, 9, 757–781. doi.org/10.1016/j.cropro.2004.01.009.

Trewavas, A.J. (2005) Organic food: the hidden agenda. Association of Applied Biologists. Letters Issue 56.

Tsao, R., Yang, R., Young, J.C. and Zhu, H. (2003) Polyphenolic profiles in eight apple cultivars using high-performance liquid chromatography (HPLC). *Journal of Agricultural and Food. Chemistry.* 51, 21, 6347–6353. doi.org/10.1021/jf0346298.

TSL (2016) New insights into signalling pathway of long-established immune receptor. The Sainsbury Laboratory.

Available at www.tsl.ac.uk/news/new-insights-signalling-pathway-long-established-immune-receptor

Turley, N.E., Godfrey, R.M. and Johnson, M.T.J. (2013) Evolution of mixed strategies of plant defence against herbivores. *New Phytologist* 197, 2, 359–361. doi:10.1111/nph.12103.

Tzin, V., Lindsay, P.L., Christensen, S.A., Meihls, L.N., Blue, L.B. and Jander, G. (2015) Genetic mapping shows intraspecific variation and transgressive segregation for caterpillar-induced aphid resistance in maize. *Molecular Ecolology* 24, 22, 5739–5750. doi: 10.1111/mec.13418.

USDA Agric. Res. Service (2012) USDA database for Oxygen Radical Absorbance Capacity (ORAC) of Selected Foods, Release 2 (2010). Available at www.orac-info-portal.de/download/ORAC_R2.pdf

Valcarcel, J. (2014) Profiling phytochemical and nutritional components of potato. PhD Thesis, University College, Cork, Ireland.

War, A.R., Paulraj, M.G., Ahmad, T., Buhroo, A.A., Hussain, B., Ignacimuthu, S. and Sharma, H.C. (2012) Mechanisms of plant defence against insect herbivores. *Plant Signalling and Behaviour.* 7, 10: 1306–1320. doi: 10.4161/psb.21663.

WHO (1992) World Health Organization, Environmental Health Criteria 134, Cadmium. International Programme on Chemical Safety (IPCS) Monograph. Available at www.inchem.org/documents/ehc/ehc/ehc134.htm

Widmer, W. and Haun, C. (2005) Variation in furanocoumarin content and new furanocoumarin dimers in commercial grapefruit (*Citrus paradisi* Macf.) juices. *Journal of Food Science* 70, 4. C307–312. Copyright 1999–2017 John Wiley & Sons. doi: 10.1111/j.1365-2621.2005.tb07178.x.

Williams, J.S., Hall, S.A., Hawkesford, M.J., Beale, M.H. and Cooper, R.M. (2002) Elemental Sulfur and Thiol Accumulation in Tomato and Defence against a Fungal Vascular Pathogen. *Plant Physiology* 128, 1, 150–159.

Williams, J.S. and Cooper, R.M. (2003) Elemental sulphur is produced by diverse plant families as a component of defence against fungal and bacterial pathogens. *Physiological and Molecular Plant Pathology* 3, 1, 3–16. doi.org/10.1016/j.pmpp.2003.08.003.

Woese, K., Lange, D., Boess, C. and Bögl, K.W. (1997) A Comparison of Organically and Conventionally Grown Foods – Results of a Review of the Relevant Literature. *Journal of the Science of Food and Agriculture* 74, 3, 281–293. doi: 10.1002/(SICI)1097-0010(199707)74:3<281::AID-JSFA794>3.0.CO;2-Z.

Wojcik, M., Burzynska-Pedziwiatr, I. and Wozniak, L.A. (2010) A Review of Natural and Synthetic Antioxidants Important for Health and Longevity. *Current Medicinal Chemistry,* 2010, 17, 28, 3262–3288.

Yan, J., Aboshi, T., Teraishi, M., Strickler, S.R., Spindel, J.E., Tung, C.W., Takata, R., Matsumoto, F., Maesaka, Y., McCouch, S.R., Okumoto, Y., Mori, N. and Jander, G. (2015) The Tyrosine Aminomutase TAM1 Is Required for ß-Tyrosine Biosynthesis in Rice. *Plant Cell* 27, 4, 1265–1278. doi: 10.1105/tpc.15.00058.

Yang, Z. and Butelli, E. (2015) Scientists produce beneficial natural compounds in tomato – with potential for industrial scale up. Available at https://www.jic.ac.uk/news/2015/10/beneficial-compounds-tomato

Zaccone, C., Di Caterina, R., Rotunno, T. and Quinto, M. (2010) Soil – farming system – food – health: Effect of conventional and organic fertilizers on heavy metal (Cd, Cr, Cu, Ni, Pb, Zn) content in semolina samples. *Soil and Tillage Research* 107, 2, 97–105.

Zhang, Y. (2010) Allyl isothiocyanate as a cancer chemopreventive phytochemical. *Molecular Nutrition and Food Research* 54, 1, 127–135.

Zorb, C., Langenkamper, G., Betsche, T., Niehaus, K. and Barsch, A. (2006) Metabolite profiling of wheat grains (Triticum aestivum L.) from organic and conventional agriculture . *Journal of Agricultural and Food Chemistry* 54, 8301-8306.

Züst, T., Heichinger, C., Grossniklaus, U., Harrington, R., Kliebenstein, D. J. and Turnbull, L.A. (2012) Natural enemies drive geographic variation in plant defenses. *Science* 338, 6103, 116–119.

Pesticides – The Fear and the Problems

"The only thing we have to fear is fear itself"
Franklin D. Roosevelt – Presidential inaugural address 1932

Synthetic pesticides

Synthetic pesticides here include herbicides, insecticides and fungicides which in 2009 accounted for 46%, 25% and 26% of the global pesticide market respectively UNEP (2013). Molluscicides, acaricides and rodenticides are not considered.

Pesticides are widely used to control weeds and pests that pose risks to crops, livestock and human health. Predominantly used in agriculture, they are a highly cost-effective means of reducing crop losses and have thereby contributed to the huge increases in agricultural productivity since the Second World War. Pesticides have undoubtedly improved the lot of humanity. However, as with all technologies, they also present certain hazards. Early pesticides, especially those based on arsenic for example, were highly toxic. Modern pesticides, including most of those developed and used in the past sixty years, are generally far less toxic, but are nevertheless not entirely without hazard.

Activists continue to argue particularly in Europe that steps should be taken to reduce further the levels of consumer and environmental exposure both to pesticides and their residues in food. There is a danger that an apparent lack of understanding of risk assessment will result in legislation based on perceived hazards and, in Europe, on the precautionary principle; the latter is often invoked as a blocking tactic.

Successful pesticides are those that have the capacity to kill weeds, pathogens and insect pests without damaging the crop and disrupting the local ecology. Organic farmers disapprove of their use on principle because they are synthetic and because it is believed that their biological killing power means that even the smallest exposure to them will be harmful.

It is important to consider separately the effects of the direct exposure to pesticides when they are being applied on the farm from the intake of pesticide residues in food.

A group of Canadian Family Physicians carried out two reviews in 2004 and 2012 of world literature of studies dealing with the chronic effects of exposure to pesticides (not residues). In their first review they dealt with several health aspects including cancer, non-Hodgkin's lymphoma, leukaemia and reproductive outcomes (Sanborn *et al.*, 2004). Most of the studies compared farmers, pesticide applicators, gardeners and other occupational groups with high exposures to pesticides with the general population. Several chronic detrimental effects of pesticides were evident on groups who were occupationally exposed. It was not possible to indicate which pesticides were particularly harmful. The doctors called for a reduction in the use of pesticides and for training on their proper handling, especially in developing countries. The correct handling and application of pesticides is discussed in detail by Matthews (2016).

In their second review, Sanborn *et al.* (2012) examined the association between exposure (occupational, non-occupational and domestic) to pesticides on reproductive, respiratory, and neurobehavioral health outcomes in children and adults. The emphasis appears to have been on domestic and applicator use of pesticides rather than agricultural. Non-organochlorine pesticides were associated with deleterious reproductive outcomes. In contrast, there was little evidence indicating that the organochlorine pesticides affect reproductive outcomes. Prenatal and childhood pesticide exposure was associated with measurable deficits in child neurodevelopment across a wide age range from birth to adolescence. Sanborn *et al.* (2012) concluded that more care needs to be taken to minimize the pesticide exposures of pregnant women and children from all potential sources, whether on the farm or in the home. Maternal exposure to organophosphate and organochlorine insecticides was associated with asthma in children.

Synthetic pesticides are regularly condemned as harmful, for example the Pesticide Action Network leaflet entitled "Pesticides are a major killer" PAN UK (2013), says: "Many pesticides are highly hazardous to human health, causing severe and sometimes fatal illness among farm families and farm workers, as well as damaging wildlife and the environment, and putting toxins into the food chain," and: "The World Health Organization (WHO) estimate that over 350,000 people die every year from acute pesticide poisoning. This figure does not include deaths from cancer or other chronic diseases caused by pesticide exposure. WHO think that long-term exposure may result in upwards of 750,000 people suffering specific chronic defects and cancers each year in developing countries alone." (Permission to quote from PAN UK [2013] gratefully acknowledged).

According to the WHO (2004) and Hart and Pimentel (2002), successful suicides, using pesticides, accounted for 250,000 deaths and attempted suicides resulted in considerably more cases of temporary or permanent disability. The WHO also says that worldwide an estimated three million cases of pesticide poisoning occur every year. Intentional and unintentional pesticide poisoning has been acknowledged to be a serious problem in many agricultural communities in low- and middle-income countries. The costs of illness and injury associated with pesticide use can be a major drain on developing world economies. The United Nations Environment Programme (UNEP, 2012) estimated that between 2015 and 2020 the health costs of pesticide exposure in sub-Saharan Africa will amount to US $90 billion.

UNEP (2012) called for action to reduce the growing health and environmental harm from chemicals of all kinds; there are more than 140,000 chemicals on the global market but the health and environmental effects of only a very few of them have been evaluated. While pesticides account for a very small number of chemicals, their volume is large. In Europe a regulation (EC 1907/2006) was put in place in 2006 known as REACH that governs chemicals and their safe use. It aims to improve the protection of human health and the environment through a system of Registration, Evaluation, Authorization and Restriction of Chemicals.

PAN UK (2013) has launched a campaign to free towns of pesticides. It says that many of the pesticides used in our urban areas can cause serious illnesses such as cancer and birth defects. Hundreds of towns around the world have already gone pesticide-free (PAN UK, 2013).

The Pestizid Aktions-Netzwerk (PAN Germany, 2012) considers that all populations worldwide are affected by pesticide contamination and face the threat of chronic health disorders. Particularly at risk are people employed in agriculture because they are directly exposed to pesticides and frequently suffer from acute as well as chronic poisoning. PAN

International presented facts and figures on health hazards and the use of pesticides in suicide attempts. Its recent study reported that, of the total 1.3 billion farm workers worldwide, about 41 million are harmed by pesticides.

High exposure to organophosphate pesticides was associated with cognitive decline by Starks *et al.* (2012) in a US study on 693 licensed pesticide applicators who had reported having been heavily exposed. On seven of the nine neurobehavioral tests used to assess memory, motor speed, sustained attention, verbal learning and visual scanning and processing those who had been exposed showed no diminished ability; however, in two tests they were two to four seconds slower, which was presented as evidence of cognitive decline. The National Institute of Environmental Health Services in the USA reported that effects of early childhood exposure to certain organophosphates, which can go undetected because of absence of symptoms, can lead to lasting detrimental effects on learning, attention and behaviour (NIEHS, 2016).

In a Californian study the combined exposure to two fungicides and a herbicide, namely ziram, maneb and paraquat, increased the risk of contracting Parkinson's disease by a factor of three (Wang *et al.*, 2011); the increased risk was still substantial (80%) when maneb was omitted.

Exposure to several pesticides as a result of mishandling, improper application or by spray drift can undoubtedly have very serious health consequences.

Educational and training programmes on pesticide use will continue to be essential.

Obsolete pesticides

The difficulties of improving the handling and use of pesticides in several developing countries are exacerbated by the massive quantities of obsolete pesticides in these countries. UNEP (2013) estimated there were 537,000 t of obsolete pesticides in Africa, Asia, Eastern Europe, Latin America and the Middle East combined. Organophosphates and organometallics such as arsenic, mercury and tin-based chemicals are among the obsolete pesticides. The clean-up costs of dealing with these pesticides will be enormous. Beginning in 2005 a regional, multi-stakeholder partnership was created uniting the African Union, World Bank, FAO, CropLife and the Pesticide Action Network to co-ordinate action to address the disposal problem and other issues. Regulations governing the marketing and use of pesticides need to be tightened to prevent a repeat of this situation.

Pesticide residues

Those who believe synthetic pesticides residues in food are responsible for many serious illnesses are totally opposed to pesticides and view the presence of the merest trace of a pesticide as dangerous.

Organic food is often described as being health-promoting partially because of it is free of pesticide residues. If this was correct it would be expected that those that eat the pesticide-containing conventional food would suffer from high rates of cancer and cardiovascular diseases. Instead consumption of conventionally grown fruit and vegetables leads to significantly better health outcomes (Appendix). The notion that pesticide residues are harmful is not supported by epidemiological evidence.

It is not surprising that conventional food will contain numerically more synthetic pesticide residues than organic but without assessing whether the actual amounts are sufficient to pose a health risk such an observation has little merit.

Hidden costs of pesticides

The total costs of the use of pesticides have been estimated (based on a meta-analysis of sixty-one papers and thirty datasets) in a report in France (Bourguet and Guillemaud, 2016). The costs were set against the many benefits of pesticides killing agricultural and human pests. Of particular concern were the massive costs involved in regulation. Other costs, notably those related to the impairment of human health and to environmental damage, are normally hidden to the pesticide users and are paid for by non-users. In the USA the total costs probably reached $39.5 billion per year by the end of the 1980s, of which regulatory costs were the main contributor (possibly $22 billion).

Re-evaluation of benefit-cost ratios of pesticide use in various countries revealed that the total cost of pesticide use might have outreached its benefits by the start of the 1990s. Bourguet and Guillemaud advocate that the cost of illnesses and deaths caused by chronic exposure to pesticides need to be better evaluated; they feel that if these costs are included the benefit-cost ratio of pesticide use may have fallen below one.

The simplification of the regulatory process, e.g. by changing or possibly eliminating the rodent carcinogen testing, as discussed later, could result in huge savings.

It is to be remembered that man needs the extra food resulting from pesticide use. Weeds, pests and diseases are quoted as being responsible for 20–40% losses in global food production (e.g. Savary *et al.*, 2012). Oerke (2006) estimated the global crop losses caused by weeds, pests and disease at 26–29% for soybean, wheat and cotton, and 31, 37 and 40% for maize, rice and potatoes, respectively. Overall, weeds caused the highest potential loss (34%), with animal pests and pathogens being less important (losses of 18 and 16%). The control of weeds, whether mechanically or by herbicide, would appear to be more important than controlling pests and diseases.

Rachel Carson

Rachel Carson, with her book *Silent Spring* (Carson, 1962), was the first to draw the attention of a wide audience to the potential dangers of pesticides. She did this by using the single example of DDT, which she claimed was responsible for the deaths of raptors and which she forecast would cause cancer in humans. Carson, a marine biologist, was sadly suffering from cancer when writing *Silent Spring*. The book coincided with the beginning of the organic movement in America and it remains a constant reference for organic activists worldwide. Her message appealed to many in the public and media, who were ready to condemn the use of "chemicals"; for many years *Silent Spring* was not reviewed critically.

However, Carson never called for a ban on DDT, nor did she call for a ban on any pesticide. Contrary to popular belief, she only called for the proper use of pesticides. In view of the importance of *Silent Spring* to the organic movement it is curious that the part played by organic farmers in helping Carson with her book has only recently come to light (Paull, 2013).

Two biodynamic farmers in Long Island, New York, set Carson on the way by providing her with evidence of the harmful effects of DDT. They had compiled the evidence when they sought an injunction to stop the US Federal government spraying their property with DDT from the air and, having failed to stop the spraying, sued for damages (Paull, 2013). DDT was sprayed in a campaign to eradicate the gypsy moth, a destructive pest of hardwood trees in the eastern USA. Carson used the supplied documents as the primary input for her book but she did not acknowledge the contributions of the two biodynamic farmers. It is possible that recent court cases where the involvement of organic agriculture had been perceived to have had a negative effect on the outcome may have influenced Carson. The part played by the biodynamic farmers, who had trained at the Goetheanum, was revealed in the many letters written by Carson (Paull, 2013).

Carson's case about DDT was primarily based on aerial spraying of trees in urban areas and the deaths of birds; it had nothing to do with agricultural use of pesticides or of pesticide residues. It is difficult to understand how it came about that a book that mostly dealt with a single pesticide and a single biological class – birds – was assumed to apply to all pesticides and all organisms, and that it continues to do so. Although the validity of Carson's thesis about DDT has been questioned and, despite the many millions of deaths due to malaria and dengue fever following the banning of DDT use, there has been little dilution of the Carson effect. Malaria and the associated premature deaths of many millions of Third World children had been virtually eliminated by the use of DDT. Malaria took hold again after the DDT ban. The Western environmental activists (and their modern-day supporters) who agitated for a DDT ban, following the fuse lit by Carson, have never accepted this. It is now realized that better control of DDT use would have been a more sensible course of action.

DDT is an organo-halogen and poses a relatively low risk to humans (see HERPE values in Appendix) but was banned mainly because of its bioaccumulation and persistence in the environment. As pointed out by Trewavas (2008), there are at least 3,800 organo-halogens made naturally by marine organisms. Some of these natural organo-halogens bioaccumulate through marine food chains and have even been detected in human breast milk. Should we ban nature?

The publishing of *Silent Spring* coincided in the 1960s with the devastating foetal damage caused by the drug thalidomide given to treat morning sickness. Pesticides were perceived as another example of the inadequacy of modern science and why it should not be trusted. Thalidomide caused severe limb deformities and deaths. The legacy of thalidomide was that testing for teratogenicity was made mandatory for all agrochemicals and pharmaceuticals.

It is inevitable that organic movements will highlight any research that supports their views. So it was in 1989/1990 when attention was focused on Alar, a growth-regulating chemical widely used in conventional apple and pear orchards, after the Environmental Protection Agency in the USA had declared it a carcinogen. Coverage in the media put the topic into the public domain. The Alar cancer scare would later be exposed (Rosen, 1990; Cohen *et al.*, 1999) but not until after it had promoted the sale in the USA of natural foods. The market and the media were not interested in the amounts of Alar consumed by humans. In terms of apple juice consumption (made from Alar treated apples), humans would have to consume 500 gallons each day for seventy years to match the harmful dosage given to the mice that caused cancer.

Cocktails of pesticide residues

Some opponents of synthetic pesticides claim that residues may act synergistically and that a cocktail of pesticide residues will have greater effect than the sum of their individual effects. However, there is no evidence for such a phenomenon. Nevertheless, the European Food Safety Authority (EFSA) has developed a software tool for carrying out exposure assessments of the cumulative risks from exposure to several pesticides on particular organs or tissues (EFSA, 2014). The tool is being evaluated on groups of pesticides that may affect the thyroid and nervous systems. In the future it is planned to look at other groups of pesticides that affect different organs. One objective would seem to be assessing whether current maximum residue levels (MRL) are fit for purpose in assessing risk. Unfortunately MRLs in Europe are now used as a health-related legal tolerance level and not as a measure of the proper use of a pesticide, which is how they were derived – at least in the UK. According to EC regulation 396/2005, a MRL is the highest level of a pesticide residue that is legally tolerated in or on food or feed when pesticides are applied correctly. It is somewhat misguided and illogical to ignore the acceptable daily intake (ADI), which is the measure of the amount of a specific substance in food or drinking water that can be ingested (orally) on a daily basis over a lifetime without any appreciable health risk.

An article on the risk assessment of multiple residues has been published by the European Food Safety Authority (EFSA, 2013) that reviews the terminology, methodologies and frameworks developed by national and international agencies for the human risk assessment of combined exposure to multiple chemicals. Phytoalexins or natural pesticides are ignored by the EFSA in the article.

The large number and amounts of phytoalexins would suggest that the chances of a cocktail effect with them would dwarf those of the few and small amounts of synthetic pesticide residues. The EFSA (2014) software tool would be sorely tested when assessing the effects of the exposure to hundreds of phytoalexins on groups of organs.

Cancer

One legacy of Carson has been the preoccupation by the organic movements and supporters, the media and regulatory bodies with linking cancer to pesticide residues. What is the evidence for links with cancer?

Cancer is responsible for around 23% of deaths in the Western world (Lomborg, 2001; Ames and Gold, 1997). Doll and Peto (1981) estimated the relative importance of the various causes of avoidable cancer in America. Diet (meat and fat) and tobacco were highlighted as being associated with 35% and 30% of avoidable cancers respectively. Doll

and Peto concluded that the occurrence of pesticides as dietary contaminants seemed unimportant. These early conclusions were later confirmed by other authorities. A study by the US National Research Council (1996) on carcinogens in food found that most of the naturally occurring and synthetic chemicals in the diet are present at levels below which any adverse effects on biological activity is likely and too low to pose an appreciable cancer risk. The World Cancer Research Fund (WCRF, 1997) concluded that there was no convincing evidence that food contaminants such as pesticides increased the risk of any cancer or that there was evidence of any likely causal relationships. The Canadian Cancer Society came to the same conclusion (Ritter, 1997).

In the UK Parkin *et al.* (2011) presented somewhat similar findings to those in the USA when they estimated, from 134,000 cases, the fraction of cancers that can be attributed to past exposure, to lifestyle and environmental risk factors. Tobacco was the main risk factor accounting for 23% of cases of cancer in men and 16% women. The lack of fruit and vegetables in the diet accounted for 6% (men) and 3% (women) of cancers. Other main risk factors such as alcohol, being overweight, exposure to sun and infection, were all rated as low at <5% each.

As early as 1983 Ames drew attention to the presence in the human diet of a great variety of natural mutagens and carcinogens, as well as many natural antimutagens and anti-carcinogens including antioxidants.

Gold *et al.* (2002) presented a very useful appraisal on the subject of misconceptions about the causes of cancer. She identified nine misconceptions/untruths, namely:

i.   Incidence of cancer is soaring in the United States and Canada.
ii.   Synthetic chemicals at environmental exposure levels are an important cause of cancer.
iii.   Reducing pesticide residues is an effective way to prevent diet-related cancer.
iv.   Human exposures to potential carcinogens are primarily to synthetic chemicals.
v.   The toxicology of synthetic chemicals is different from that of natural chemicals.
vi.   Cancer risks to humans can be assessed by feeding animals with high doses.
vii.   Synthetic chemicals pose greater carcinogenic hazards than natural chemicals.
viii. Pesticides and other synthetic chemicals are disrupting hormones.
ix.   Regulation of low, hypothetical risks is effective in advancing public health.

Despite the falling incidence of colorectal cancer in Europe (Ait Ouakrim *et al.*, 2015) it is still possible to be told by the media about cancer epidemics and a possible link with pesticides. On the other hand, the recent findings that the placenta is unable to protect the foetus from carcinogens such as acrylamide, nitrosamines and dioxins will focus minds on the relevance of the maternal intake of bioactive compounds and of synthetic pesticides (Henshaw and Suk, 2015). Hopefully phytoalexins will also be considered.

Phytoalexins (natural pesticides) in the diet

The occurrence in the diet of residues of identifiable pesticides both natural and synthetic is usually ignored. What is also forgotten is that it is the quantity not the presence of any pesticide that is important.

The higher concentrations of phytoalexins that are often found in organic crops, e.g. Baranski *et al.* (2014) are compatible with their production being the result of a pest attack on the unprotected organic crops (Chapter 14). It is theoretically possible that when choosing pathogen-resistant varieties organic farmers may inadvertently have chosen those that contain more phytoalexins or phytoanticipins (natural pesticides that are active in plants before attack); little seems to be known about this.

Examples of potential toxins that may accumulate in organic crops as a response to stress-related factors include the furanocoumarins in parsnip and celery, glycoalkaloids in potatoes and glucoraphanin, and a glucosinolate in *Brassicaceae*.

In some instances the concentrations of the phytoalexins in organic crops may be increased many times but this is probably the exception. For example, glycoalkaloid contents in organic potatoes are usually higher (on average by 20%) than those in conventional crops; the concentrations are likely to be less than 200 mg/kg, which is believed to be a safe level. On the other hand, the furanocoumarin concentration in parsnip roots was increased twenty times in diseased crops compared with crops treated with fungicide (Mattsson, 2006).

Concentrations of phytoalexins in crops can vary widely. The bottom end of the scale probably starts at a few mg/kg, e.g. benzopyrene in spinach at around 7 mg/kg. Other examples are furfurol in cocoa and coffee at 55 to 255 mg/kg, caffeic acid at 50–200 mg/kg in apples, pears, plums and potatoes; at the top end of the scale are the pyrrolizidine alkaloids, which occur in thousands of plants and sometimes at concentrations greater than 10,000 mg/kg (Ames, 1990).

Only a few phytoalexins have been studied toxicologically. Gold *et al.* (2002) found that thirty-eight (53% of total tested) were judged to be carcinogens in rodents: 47% of phytoalexins were not carcinogenic (Appendix). This is similar to the carcinogenic percentage for all the chemicals (including synthetic pesticides) that have been tested. About 50 to 60% of all chemicals tested (1,275 by 1997) can be said, when supplied at very high doses (maximum tolerated dose) and/or for long enough, to be rodent carcinogens and therefore potentially hazardous to man (see Appendix). It seems reasonable to assume that about half of all the untested phytoalexins will be rodent carcinogens at high doses. In other words, natural pesticides pose the same hazards as synthetic pesticides.

However, because the exposure by humans to both natural and synthetic pesticides is very low, the risk of harm is extremely small. It is the dose that makes a chemical harmful. Examples of phytoalexins that are carcinogens include quercetin in apples, limonene in citrus and caffeic acid in coffee, carrots, lettuce, potato and celery (Trewavas, 2005).

The average daily diet is likely to contain several naturally occurring organic compounds that could concern toxicologists on grounds other than cancer; these include the teratogens chaconine and solanidine in potato, the oestrogen mimics flavonoids and isoflavones in most fruit and vegetables, the sterility inducers theobromine in chocolate, gossypol from cattle cake (via animal products), the nerve toxins tomatine in tomato, solanine in potato, curcurbitacin in courgettes and cucumber and carotoxin in carrots (Trewavas, 2005). It is thought, based on *in vitro* studies at very low concentrations, that 40% of phytoalexins are able to induce breakage of chromosomes.

Fungal attack of soybeans and French beans elicited the production of the phytoestrogens coumestrol, genistein, formononetin, and daidzein (Mattson, 2008). High levels of potato glycoalkaloids have been shown to cause birth defects and increased foetal mortality in a number of animal species (Friedman *et al.* 1992; Gaffield and Keeler, 1996; Renwick *et al.*, 1984). Based on potato glycoalkaloid responsiveness to Colorado potato beetle predation, Hlywka *et al.* (1994) concluded that when a crop is not protected from pests it may produce sufficiently large amounts of phytoalexins that may impact on food safety.

Some phytoalexins are antioxidants; others such as the indole alkaloid phytoalexins in brassicas have been found to have beneficial properties in *in vitro* experiments, including anti-carcinogenic and cardiovascular protective effects (Pedras *et al.*, 2011). It is believed that phytoalexins (stilbenes and other phenolic compounds) in groundnut have anti-diabetic, anti-cancer and vasodilatory effects (Holland and O'Keefe, 2010). The biological activities of glyceollin, a phytoalexin in soybean, include anti-proliferative and anti-tumour properties (Ng *et al.*, 2011). *In vitro* studies (Yang *et al.*, 2009) have indicated the sorghum phytoalexins 3-deoxyanthocyanins, might be useful in helping to reduce the incidence of gastrointestinal cancer.

Vladimir-Knežević *et al.* (2012) in their survey of the health effects of plant polyphenols drew attention to several studies linking increased intake of plant phenols with improved health outcomes, notably reduced blood pressure levels (Medina-Remón *et al.*, 2011) and reduced incidence of lung, prostate and breast cancer (Knekt *et al.*, 2002).

Little is known about the total quantities of phytoalexins in different diets. The very large number of phytoalexins would make it impractical to monitor the presence and the quantities of individual phytoalexins in the diet. Conventional farmers minimize the risk posed by harmful phytoalexins by applying pesticides. As pointed out by Mattsson (2008), the only way to limit our exposure to all potentially harmful chemicals both natural and synthetic in food is by using synthetic pesticides correctly. Moreover, focusing attention on the extremely small amounts of synthetic pesticides in the diet is not scientifically rational; it is very likely that we are over-testing synthetic pesticides and under-testing phytoalexins (Mattsson, 2008).

Breeding for increased resistance to pest and diseases is attractive but there is an associated risk that plants that are better able to resist insect and fungal attack may have higher levels of potentially harmful phytoalexins and phytoanticipins.

The preoccupation of many in society and especially of regulatory bodies with assessing the hazards and risks of synthetic pesticides residues has meant that no or very little attention is given to the hazards and risks of the many potentially harmful compounds in the diet such as the phytoalexins. As discussed here, consideration of phytoalexins leads to the conclusion that the obsession with the toxicity of synthetic pesticides is misdirected.

Carson was not to know that all our food contains potentially dangerous phytoalexins and secondary metabolites, and that we eat far more of these natural chemicals than we do of the man-made pesticides. Many in the organic

Figure 15.1 Representation of the amounts of natural pesticides (phytoalexins) and synthetic pesticides consumed daily in fruit and vegetables

movements still do not appear to appreciate this.

Ames and Gold (1997) reported many years ago that on average Americans ingest 5,000 to 10,000 different natural pesticides and their breakdown products each day, totalling about 1,500 mg compared with 0.1 mg of synthetic pesticides residues. Natural pesticides typically constitute 5–10% of a plant's dry weight. (Ames *et al.*, 1987).

Beier and Nigg (2001) presented data that indicated the dietary intake of natural pesticides exceeded that of synthetic by 80,000 to 100,000 times, i.e. a much greater difference than that of Ames *et al.* There are no reasons for doubting that the more conservative original data of Ames do not apply to the amounts of natural pesticides consumed in the UK, Europe and elsewhere by people who consume five helpings of conventionally grown fruit and vegetables daily. Consumers who buy expensive organic foods in order to avoid pesticide exposure are focusing their attention on <0.01% of the pesticides that they consume and ignoring the potentially harmful natural pesticides.

It is sometimes suggested that over the millennia humans have evolved the ability to consume phytoalexins without any ill effects but have not had time to do so for man-made pesticides. The fact that extremely small amounts of any phytoalexin are involved means that the pressures on reproductive performance and survival are likely to have been minimal. Furthermore, it would seem very unlikely that metabolic pathways could have evolved to handle all the possibly hundreds or thousands of theoretically toxic phytoalexins with their varied chemistry. It might have been expected that man would have evolved to be able to cope with all poisons in food, especially the most poisonous ones such as the alkaloid strychnine, but he has not.

Pesticides used by organic farmers

Pesticides are used by organic farmers, including the fungicides, copper sulphate and sulphur and the insecticides rotenone, pyrethrins, neem oil, mineral oils, Bt and Spinosad (a new organically acceptable insecticide derived from a bacterium (*Saccharopolyspora spinosa*) (Bunch *et al.*, 2014); they are generally not analysed in routine surveys of pesticide residues. Several have not been subjected to detailed toxicological studies and are believed to be associated with ill health, e.g. rotenone and the onset of Parkinson's disease, and with environmental effects, e.g. neem oil at very low concentrations is linked to the death of bumblebees.

Organic pesticides are thus not blameless and the presence of their residues means that it is not true to generalize and say that organic food is pesticide-free. Rotenone is now banned in Europe and, as of 2012, has been voluntarily cancelled by the manufacturer in the USA. Neem oil (azadirachtin) is limited to its use as a fish killer in the USA and is permitted in Europe (EU Commission, 2015). Use in the UK is no longer approved by the Soil Association following misuse. Details of pesticides and other materials permitted for use by organic farmers in North America can be found from the Canadian Organic Materials Review Institute (OMRI, 2015).

The build-up of copper in the soil as a result of using copper sulphate as a fungicide is probably the main reason (together with a concern about sustainability and the environment) why a number of winemakers, particularly in France, are dropping their organic status (Porterfield, 2016). This is not in line with the views of organic wine groups who continue to claim that organic wines contain few or no chemicals. Globally, about 1,500 to 2,000 growers produce organic wine, with about 900 in France. The new conventional growers will now have the opportunity to use more effective synthetic fungicides and to save on application costs having relinquished their organic status.

Organic researchers are trying to find alternatives to the use of copper compounds for controlling late blight on potatoes but so far without success. For example, Kühne (2014) in trials in Germany found that two new copper-free pesticides, one of microbiological origin and the other of plant origin, had little effect when compared with copper hydroxide. The effective inhibition of the growth of *Phytophthora infestans* (the cause of late blight) by volatile isothiocyanates released by various brassicas in *in vitro* experiments was not repeated in field trials in Germany when the brassicas (*Brassica juncea*, *Raphanus sativus*, *Brassica rapa*, *Vicia villosa*), which were grown as a cover crop, were shredded and incorporated in either the autumn or spring, before planting the potatoes Grabendorfer (2014).

Synthetic pesticides are sometimes found in organic crops. For example, in the 2014 survey the USDA (2016) found detectable residues of synthetic pesticides in 21% of the organic samples that were analysed; the forty different synthetic pesticide residues detected in organic food were at levels similar to those in comparable conventional food, which for these pesticides suggests

intentional use by the organic farmer and not contamination by spray drift. In both the conventional and organic samples the amounts were far too small to constitute a health/safety concern. A few of the residues were of old organochlorine or organophosphate insecticides or their metabolites, related probably to their persistence in the environment, but most were of modern pesticides used by conventional farmers.

The claims that organic food in the USA does not contain synthetic pesticide residues and that imported fruits and vegetables pose greater risks from pesticide residues than do domestic produce were discussed in detail by Winter (2012). The residues found in imported fruit and vegetables were low and were of no concern health-wise. Winter was unable to find evidence to support the contention that particular conventional fruit and vegetables should be avoided. The inclusion of a product in a list such as the "dirty dozen" (EWG, 2015) is based on the number of different pesticide residues found and not on the quantity of pesticides or on any risk assessment. In the absence of such assessment these lists have little, if any value. EWG's Dirty Dozen list of produce included apples, strawberries, grapes, celery, peaches, spinach, sweet bell peppers, imported nectarines, cucumbers, cherry tomatoes, imported snap peas and potatoes.

Similar results were found by the Canadian Food Inspection Agency in 2011– 2013 (CFIA, 2013) when more than 97% of the 19,000 samples were found to comply with the prescribed MRL levels.

Regulation of synthetic pesticides and pesticide residues

Before pesticides can be authorized for use they are tested for all possible health effects on animals. The first step is hazard identification followed by the assessment of the acceptable daily intake (ADI). The ADI is the quantity of the pesticide that can be consumed daily for life without causing any measurable effects. The ADI is used to assess the risk of ingesting the pesticide by comparing it to potential exposure.

The ADI is set after determining by long-term animal studies (e.g. rats, mice, dogs, pigs and rabbits) the lowest daily intake level that showed No Observable Effects (NOEL) in the most sensitive animal species by means of blood tests and microscopic examination of all key organs. The NOEL value, which is expressed as mg of active ingredient per kg animal body weight, is then divided by 100 to allow for humans being more sensitive than animals and for possible variation between humans. This, the ADI value is then translated to man on the basis of weight. The size of the safety factor of 100 can be appreciated by thinking about the likely safe stopping distance for a car travelling at 50 mph (80 km/h), namely 53 m, as being equivalent to the NOEL; if a safety factor of 100 was applied in order to cater for variation in people and

car performance the stopping distance would be an impossible 53 cm.

The Maximum Residue Levels (MRL) for a given pesticide, which are now legally binding in many countries, originated for example in the UK as the maximum residue levels likely to be found in a crop or food item after a given pesticide had been properly treated with the pesticide. The MRLs are not related to health risks. They are a measure of the correct, and therefore safe, use of pesticides. When an MRL is exceeded an assessment of risk is made by considering the lifetime intake in relation to the ADI. The MRLs are set to ensure consumption does not exceed the ADI even if everything consumed contains the permitted MRL. In Europe the MRL is now used as a legally binding level that should not be exceeded, which does not allow consideration of the ADI when assessing risk; this is both a confusing and unsatisfactory situation.

Pesticide residues in food are monitored by regular analysis of both domestic and imported samples by several national authorities.

UK

In the UK the Expert Committee on Pesticide Residues in Food oversees an annual programme to check food and drink for traces of pesticide residues (PRIF, 2015). More than 200,000 pesticide/commodity combinations are analysed annually; concentrations seldom exceed 1 mg/kg.

In 2014 3,615 samples of food and drink were analysed for residues of

399 pesticides. More than 56% of the samples tested did not have any detectable pesticide residues. Fewer than 2% (seventy-two) of the samples, mainly fruit and vegetables, contained a residue above the MRL (PRIF, 2015). The comparable numbers for 2012 were 3,657 samples, 393 pesticides and 63% with no pesticides and 2% with more than the MRL (PRIF, 2012).

All the samples that contain residues above the MRL are subjected to a risk assessment, particularly with babies, toddlers and the elderly in mind; this involves the Health and Safety Executive (HSE). In nearly all cases in the UK such assessments have shown that people would consume less than the ADI. All samples with residues of a pesticide not approved for use in the UK are reported to HSE for further investigation. When there are concerns, more and focused testing is carried out and if necessary the regulatory authority may take enforcement action.

Organic produce can sometimes be found to contain pesticide residues. In 2014, of 142 samples that were labelled organic, seven (four imported products) contained pesticide residues with one containing more than the MRL, and in 2012, five of the 215 organic samples contained pesticide residues. Risks to health were assessed as being most unlikely.

Europe

The European Food Safety Authority (EFSA), set up in 2002, brings together the residue data provided by member states. The European Commission became involved by introducing EC Statutory MRLs (EC Regulation 396/2005), which came into force in 2008. Everyone involved in the marketing of food and animal feed within the EU must comply.

In 2008 the EU Commission responded to a question (E-4703/08EN) about its proposed banning of pesticides and the suggestion that untreated crops would be liable to be attacked and would contain more phytoalexins that were potentially as harmful as synthetic pesticides by saying that it did not know of any scientific report to substantiate the notion of an enhanced risk to consumers if pesticide use was reduced. The EU Commission failed to appreciate the minimal risk of harm from both phytoalexins and from the synthetic pesticides it wished to ban as well as the huge difference in amounts of the two.

In 2013 the EFSA reported on the data for 80,967 samples (70% EU products) of twelve different food commodities in which 209 different pesticides were analysed (EFSA, 2015). Overall, 97.4% of the tested food samples fell within the legal limits and 54.6% of the samples contained no quantifiable residues at all. In general, products imported from outside the EU were more likely to exceed the MRL than locally produced (5.7% for imported products versus 1.4% for EU products).

Estimates of dietary exposure support the conclusion that the presence of residues found in the food products covered by the EU-coordinated monitoring programmes is unlikely to have any long-term effects on the health of consumers. The probability of being exposed to pesticide residues exceeding the toxicological threshold for short-term exposure that may lead to negative health outcomes was also judged to be low.

Nearly all the pesticide residues found in organic products were within the legal limits and only exceeded the MRL in 0.8% of the samples.

The exposure of the Belgian consumer to pesticide residues from the consumption of fruit and vegetables was studied by Claeys *et al.* (2008). They found that the exposure to most pesticides was one hundredth of the acceptable daily intake. It was only if massive amounts of fruit and vegetables were consumed that the intake of some pesticide residues could reach 9–23% of ADI.

The improvements in analytical techniques will inevitably permit smaller and smaller amounts of pesticide residues to be detected in foodstuffs and will lead to the reporting of fewer samples with no quantifiable residues – and then to possible misinterpretation. Maybe there will be a parallel increase in the understanding of the risks posed by the small amounts involved.

The precautionary attitude of legislators to limiting the exposure of humans to potential harmful substances is evident from the Drinking Water Directive in Europe (Directive 98/83/EC). According to this directive, the maximum level for an individual pesticide is 0.1 μg/l (0.1 parts per billion) after processing at the water works. This value was the limit of detection when the

directive was first introduced; it was not based on any risk to human health or to the environment.

USA

In the USA the regulation of pesticide residues in food is carried out by the US Food and Drug Administration (FDA) using the inputs of the Environmental Protection Agency (EPA). The EPA sets the tolerances for the maximum amount of pesticide residues allowed in or on a food and has been responsible (since 1991) for the Pesticide Data Program (PDP). There has been an emphasis on commodities consumed by infants. More than 99% of the products analysed (10,104) in 2013 contained residues below the EPA tolerances (PDP, 2013). More than 40% of the samples had no detectable pesticide residues and residues exceeding the tolerance were detected in only 0.23% samples (mostly imported). In 2014 when 10,619 samples were analysed 41% had no detectable residues and fewer than 0.36% contained more than the tolerance values (PDP, 2014).

Actual dietary exposures (based on market basket samples) to pesticides were compared by Winter (2015) with established toxicological criteria such as the Chronic Reference Dose using government data for 2004–2005. He found that all pesticide residues continued to be at levels far below those that could be of any health concern. Consumers should be encouraged to eat fruit, vegetables, and grains and should not fear the low levels of pesticide residues found in such foods.

International Codex MRLs

There are many national authorities that rely on Codex MRLs for pesticide residues. The FAO and WHO are together responsible for setting specific MRLs for the Codex Alimentarius Commission on Pesticide Residues. Codex MRLs are non-statutory levels. The aim is to protect the health of consumers and to ensure fair trade practices in international food trade. They are used as guidance on acceptable levels and are only relevant where they apply to commodities for which national statutory MRLs have not been set. A commercial online global MRL database has been set up (Bryant Christie, 2015).

Toxicological studies on rodents

It is remarkable that there are so few chemical-related cancers in view of the fact that at least 50% of all chemicals tested (about 1,350) had been assessed as rodent carcinogens before 2000 (Gold *et al.*, 2002). By 2011, 1,547 chemicals had been tested.

It is apparent that about 50–70% of all chemicals can cause cancer in rodents (provided sufficient is administered) and that natural and synthetic chemicals are equally carcinogenic (Table 15.1).

Since 1971, more than 900 chemicals have been evaluated by the IARC (2016); of these more than 400 have been classified as carcinogenic or probably/possibly carcinogenic to humans. The relevance of such evaluations has been questioned many years ago by Ames and co-workers (e.g. Ames *et al.*, 1990; Ames and Gold, 1997), who consider that a plausible explanation for the high frequency of cancer is that the maximum tolerated dose fed in the tests causes chronic cell killing and consequent cell replacement, a risk factor for cancer; such effects may well be limited to high doses. Ignoring this greatly

Table 15.1 Proportion of chemicals evaluated as carcinogens and mutagens (Ames and Gold, 2000 and Gold *et al.* 2008)

	Proportion	%
Chemicals tested in both rats and mice	405/688	59
Naturally occurring chemicals	98/171	57
Synthetic chemicals	307/517	59
Chemicals tested in rats and/or mice		
Chemicals in Carcinogenic Potency Database	702/1348	52
Natural pesticides	38/72	53
Mould toxins	14/23	60
Chemicals in roasted coffee	21/30	70
FDA database of drug submissions	125/282	44
Mutagens	294/393	75
Non-mutagens	217/455	48

exaggerates risks with the effect that toxicologists have identified many chemicals with effects evident at high doses that are not relevant at low doses.

With regard to pesticide residues, the amounts ingested are so small, relative to levels that have been shown to have toxicological effects, that they are toxicologically implausible as health risks. Regulatory standards have been based on a number of worst case assumptions, particularly that the relation between dose and effect is linear with no threshold (see Hormesis).

The extent to which rodent carcinogenicity tests on a chemical are currently used when evaluating acceptable daily intakes and the associated health risks may probably be far less than might be imagined (Sir Colin Berry, England, 2014, personal communication).

Adami *et al.* (2011) discuss how originally epidemiological findings on the association of a chemical with a health condition were later verified by toxicological tests in order to establish a causal link. Then with the setting up of toxicity testing guidelines, toxicology took on the primary role and was used to anticipate or predict potential adverse effects in humans. Epidemiology, in many cases, then served a secondary role to verify or negate the toxicological predictions. A process for combining the fields of toxicology and epidemiology was proposed by Adami *et al.* (2011) based on the placement of the two sets of results on a causal relationship grid.

Opposition to the use of rodent carcinogenicity studies has been mounted by People for the Ethical Treatment of Animals (PETA, 2006) in a report to the EU Commission. It concluded that the very costly studies that take up to five years produce results of dubious reliability and relevance to humans. The costs in terms of finance, skilled personnel and animal lives (typically 800 rats in a single test) and suffering vastly exceed their purported benefits. In support of its case that the tests should be abandoned for regulatory purposes, PETA cites the failure of rodent cancer tests to detect cigarette smoke, asbestos and benzene as human carcinogens. It called for the various directives relating to the "REACH" chemicals, pesticides, pharmaceuticals and cosmetics in Europe to be amended and that assessments of carcinogenic and other toxic hazards are based on a weight-of-evidence evaluation of existing data, validated by *in vitro* tests.

A motion (B8-0005/2016) was put to the European Parliament for a resolution to ban cancer tests on rodents gradually (EU, 2015a). Further action is awaited. As of May 2016 the Enquiry Unit of the European Parliament were unable to say when any decisions were likely to be made and acted upon inside and outside the Parliament; the behaviour of the EU with regard to delays of several years when acting on decisions on GM crops is not a good omen.

Alternatives to the costly two-year rodent carcinogen tests were proposed by Long (2007) based on at least two *in vitro* tests using a) Syrian hamster embryo cells and b) human lymphocytes.

Chemicals could be screened in weeks instead of years. The ability of rodent tests to predict cancer in humans may be as high as 90% but sometimes as low as 50% of the time. The Syrian hamster embryo cells test was able to identify 100% of inorganic human carcinogens and 82% of organic human carcinogens.

Toxicological assessments of coffee and wine

Consideration of the nature and quantity of chemicals consumed in coffee and in wine helps to understand the relevance of the small amounts ingested as pesticide residues. Humans consume relatively large amounts of natural chemicals every day in food and drink.

Coffee

Some of the chemicals in coffee have been classified by the International Agency for Research on Cancer (IARC, 2016) as category 2B human carcinogens, namely caffeic acid and catechol; caffeine, hydroquinone and furfural are all category 3 carcinogens for which there are indications of carcinogenicity in animals but not enough evidence to warrant classification as a human carcinogen.

There are more than 1,000 chemicals in roasted coffee, twenty-one (out of thirty tested) are rodent carcinogens, which leaves about 1,000 other chemicals in roasted coffee to be tested, of which at least half are likely to be rodent carcinogens and would worsen the apparent

situation (Nebert, 2015). These findings do not mean coffee is dangerous, but rather that the identification of a hazard needs to be combined with worst-case-scenario risk-assessment studies to assess the risk to human health.

Two cups of filter coffee will contain about 44 mg of the human carcinogens caffeic acid and catechol compared with 0.1 mg synthetic pesticide that would be consumed in five portions of fruit and vegetables each day (i.e. less than 40mg in one year). So if it is not considered risky to consume coffee there should be no concern whatsoever with consuming synthetic pesticide residues in fruit and vegetables.

There are no reasons to worry about coffee based on a US epidemiological study that followed more than 125,000 people over two decades that showed that regular coffee consumption was not associated with an increased mortality rate (Davies, 2011). There are no indications in the UK of any catastrophic effects of the national consumption of 70 million cups of coffee per day.

Coffee drinking has over the years been associated with a variety of beneficial as well as harmful health effects, including preventing type 2 diabetes and Alzheimer's and increasing the risk of melanoma and liver cancer. The differing claims may be related to the presence in coffee of a vast array of chemicals, some of which may be beneficial, for example antioxidant chlorogenic acids, and others that may be harmful (Ames and Gold, 2000).

Studies in South Korea on more than 25,000 young and middle-aged asymptomatic men and women (average age 41 years) indicated that moderate coffee consumption was associated with fewer cases of subclinical coronary atherosclerosis as measured by the occurrence of coronary artery calcification (an early indicator of coronary heart disease) (Choi *et al.*, 2015). No distinction was made between caffeinated and decaffeinated coffee, which is not commonly drunk in Korea.

In 2016 the International Agency for Cancer Research concluded that coffee was no longer classifiable as a carcinogen. This reversed its previous classification in 1991 when coffee was assessed as possibly carcinogenic to humans (Group 2B). More than 1,000 studies, including epidemiological studies, have shown that coffee drinking was not linked with the incidence of several cancers such as breast, pancreas and prostate and may even be associated with

reduced incidence of liver and uterine cancer (IARC, 2016a). For more than twenty other cancers, the evidence was inconclusive.

Wine

In studies directed to determining the best time to harvest Pinot noir grapes, Yuan and Qian (2016) measured the amounts of compounds related to aroma. Forty-nine main odour-active compounds were detected. The partial list (Table 15.2) indicates that several of the chemicals are skin irritants; others, not listed, are animal, but not human, carcinogens based on tests with high doses of pure chemical and not at the concentrations found in wine. Most of the listed skin irritants are also harmful when inhaled.

As with coffee, the concentrations of these compounds in the grapes were so very small (in µg/kg quantities) that the

Table 15.2 Behaviour of some chemicals found in Pinot noir grapes (Yuan and Qian, 2016)

	Classification	Comment **
2, 3–Butanedione	Skin irritant	Harmful if ingested or inhaled
Butanal	Skin irritant	
Butanoic acid	Skin irritant	Harmful if ingested or inhaled
Guaiacol *	Skin irritant	LD_{50} Rat oral 725 mg/kg
1–hexanal*		LD_{50} Rat oral 4890 mg/kg
β–damascenone*		Possibly effect on aroma
Benzyl alcohol		LD_{50} Rat oral 1230 mg/kg. Possible neurotoxin
p–Cresol	Skin irritant	Toxic if inhaled or ingested
m–Cresol	Skin irritant	LD_{50} Rat oral 242 mg/kg
Eugenol	Skin irritant	Harmful if ingested or inhaled Category 3 carcinogen
Benzoic acid	Skin irritant	LD_{50} Rat oral 1700 mg/kg

* compound with high flavour
** comments according to Pubchem (2016)

classifications as possible carcinogens or as skin irritants are of no consequence; there is no risk from these perceived individual hazards in wine.

Hormesis

As it is necessary to feed massive amounts (often by force feeding) of a pesticide in order to elicit effects including the induction of cancer in test animals there is a fundamental problem about how to consider toxicological data when setting the lifetime human Acceptable Daily Intakes of extremely small amounts.

It is often asserted by those who believe pesticides cause cancer that even the smallest intake of a few molecules of a pesticide poses a risk. It is assumed that the relationship between the intake and the harmful effect is linear with no threshold (LNT) below which there is no effect, see Figure 15.2. In contrast, a threshold dose-response model is generally accepted in pharmacology.

Toxicology studies seldom consider the application of small amounts

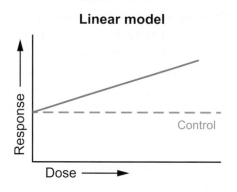

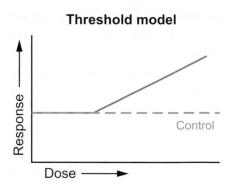

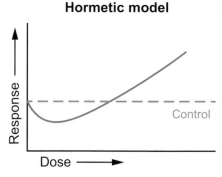

Figure 15.2   Dose response curves: possible models for relationship between dose and effect of an anticipated harmful substance. The x axis represents the dose and the y axis the response. Above control = harmful effect, below control = beneficial effect (the kind permission of Leslie Corrice (2017) is acknowledged)

well below the NOAEL. This has led those who do not distinguish between hazard and risk to accept the "Linear No Threshold" (LNT or linear) model as being applicable to pesticide residues. The effects of the massive amounts required to elicit an adverse effect are extrapolated back to zero application to estimate the effect of a small application. This has been done without actual studies on small doses; the validity of the LNT model is now doubted by many.

Both the LNT and the threshold model preclude the possibility of a hormetic effect whereby a substance that causes harm at a high dose can be beneficial at a low dose. The term hormesis was first used by Southam and Erlich (1943) to describe the effect of an oak bark compound that promoted fungal growth at low doses, but strongly inhibited it at higher doses. Until recently the hormetic model has not been widely contemplated perhaps because it seems counter-intuitive. However, based on a few thousand dose response studies, it is now understood to occur more frequently than the LNT and threshold models (Rodricks, 2003; Cohen, 2006; Calabrese, 2010).

Calabrese and Baldwin (2003) found that a stimulatory response was the most common response below the "No Observed Adverse Effect Level" (NOAEL).

It is ironic that some insecticides have been shown to exhibit hormesis in that high doses were lethal and low doses beneficial in promoting population growth (Cutler, 2013; Guedes and Cutler, 2014; Mallqui *et al.*, 2014). It has been suggested that one of the reasons for the resurgence of an insect pest after application is the hormetic effect operating on those insects that were physically protected from most of the insecticide. This effect is not to be confused with the survival and consequent multiplication of resistant insects.

Herbicide hormesis is also said to occur at sub-toxic doses of herbicides, as discussed by Belz and Duke (2014) and Belz *et al.* (2011).

Calabrese (2005) presented evidence of hormesis operating in pharmacology, for example low doses of antitumor drugs have been shown to enhance the proliferation of tumour cells. This and other findings led Calabrese to conclude that hormetic dose responses are common and are independent of the biological model, the endpoint measured, and the stressing agent; they represent a basic feature of biological responses to chemical and physical stresses.

In vivo studies with low doses of penicillin (and streptomycin) increased mortality in mice given an LD_{50} dose of deadly bacteria while preventing death at higher doses (Randall *et al.*, 1947).

It is manifestly misleading to suggest or imply that ingesting small amounts of pesticide residues is likely to be harmful. The risk of harm from small intakes of pesticides is overstated. It is possible that a pesticide found to be carcinogenic at high doses in animal studies could be beneficial when given at low doses – both to the animal and man. Pesticide residues, when present, occur at very low concentrations at which hormetic effects are theoretically possible. So while high acute and chronic exposures to the actual pesticides can be harmful, the levels of residues in food that are well below danger thresholds may even be beneficial.

It remains to be seen how data on hormesis can or will be used in risk assessment but it is difficult to envisage that those involved with traditional toxicological testing will be prepared to relinquish their entrenched position.

Hormesis should not be confused with homeopathy (which plays a major part in biodynamic agriculture). The homeopathic dilutions of an active substance are such that only the impression of the substance is normally present in the applied dilution. Hormesis requires the presence of measurable amounts of the active ingredient.

Associations or links between human health conditions and food components and exposure to chemicals are often incorrectly portrayed as indicating causal relationships. Even epidemiological studies do not provide proof, only the basis for intervention and toxicological trials.

The results of rat toxicology tests are valuable for assessing the relative toxicity hazards but linking such tests to epidemiological studies can be a problem (Adami *et al.*, 2011). The relative carcinogenicity of many compounds that man is exposed to have been assessed by relating the human lifetime intake (mg/kg body weight per day) of a compound as a percentage of the amount (also mg/kg) that caused cancer in 50% of the test animals by the end of a lifetime (Ames and Gold, 1997, Gold *et al.* (2002). The

HERP (the Human Exposure dose/Rodent Potency dose) values are presented in the Appendix.

It is revealing that the amount of the carcinogen caffeic acid in 12 g coffee is one thousandth of the amount that would cause cancer in 50% of rats when consumed for a lifetime.

Radiation hormesis

The idea that humans could benefit from low-level, whole-body radiation is not new but many still find it difficult to accept in view of the widespread fear of radiation. The concept of radiation hormesis has been accepted in several countries, for example Japan, China and France, as described by Luckey (2006). The evidence comes from several sources including the thousands of people who live in good health with constant high background radiation and the health records of people chronically exposed in Taiwan to cobalt-60 (Chen *et al.*, 2007). In Japan there are indications that low dose irradiation of the torso can be an effective treatment for non-Hodgkin's lymphoma if treated early (Sakamoto, 2004). However, most governments in the USA and Europe believe that the LNT model applies to radiation and they have enacted rules and regulations accordingly.

One result of this official position has been that little attention has been given to research into the possible benefits of low-level radiation (hormesis) in the treatment of different cancers. Another result is the massive extra costs of building nuclear power plants.

Implications

Labelling or describing some pesticides as rodent carcinogens has led to the idea that conventional farming is responsible for an epidemic of human cancer caused by the consumption of pesticide residues in food. However, there has been a steady rise in life expectancy in developed countries and cancer rates (except lung cancer, caused by smoking) are falling. Using data from the Ames laboratory, the American Cancer Society has concluded that there is no evidence that the amount of pesticide residues in conventionally grown foods causes any form of cancer.

Linear extrapolation from the maximum tolerated dose in rodents to low-level exposure in humans has led to grossly exaggerated mortality forecasts. The concept of hormesis is largely disregarded. The use of the LNT model as the basis for risk assessment and risk management now needs to be reassessed. This re-examination could profoundly alter many regulations monitored by government agencies. Studies are urgently required on the effects of low doses in animal studies. The prospects for developments in this area are probably poor in view of the hazard-aware and the risk-averse times we live in.

Carcinogenic pharmaceuticals

It is apparent that 44% of pharmaceuticals are likely to be carcinogens when tested at high doses (Table 15.1 and Contrera *et al.*, 1997). This may be a further example of hormesis whereby the effects of low doses of a pharmaceutical are the opposite of high dose effects. Regardless of the hormesis aspect, there is a clear need for a much better understanding of what rodent toxicological, and especially carcinogenicity, testing means.

The International Council for Harmonization of Technical Requirements for Pharmaceuticals for Human Use is considering the guidelines regarding the use of rodent carcinogenicity when testing pharmaceuticals (ICH, 2016). Its goal is to introduce a more comprehensive and integrated approach to address the risk of human carcinogenicity of small molecule pharmaceuticals, and to define conditions under which the two-year rat carcinogenicity studies may add value to that assessment.

It is hoped that knowledge of the pharmacological targets and pathways together with toxicological and other data will, in certain cases, provide sufficient information to anticipate the outcome of two-year rat carcinogenicity studies and help in predicting the risk of human carcinogenicity. It is postulated that with sufficient information it may be possible to assess the likely risk of human carcinogenicity without conducting an expensive two-year rat carcinogenicity study.

Pharmaceutical companies will submit carcinogenicity assessment documents for all pharmaceuticals under evaluation with ongoing or planned two-year rat carcinogenicity studies to the relevant drug regulatory agencies in their region. The documents will be compared to the outcome of the two-year rat carcinogenicity studies to evaluate the accuracy of the predictions to the actual experimental results. There appears to be a fundamental desire to continue to make use of rodent carcinogenicity testing, albeit indirectly.

Hazard and risk

Many who proselytize about the dangers of synthetic pesticide residues based on the misconception that even the smallest amounts of the chemical will be harmful also seem to be unaware of the difference between hazard and risk.

The IARC (2016) clearly states that it evaluates cancer hazards but not the risks associated with exposure and that the distinction between hazard and risk is important. A chemical is considered a cancer hazard if it is capable of causing cancer in the animal studies. Risk measures the probability that cancer will occur in humans, taking into account the level of exposure to the chemical.

The result is that cancer hazards may be identified even when risks are extremely low with the known patterns of use or exposure. Recognition of such carcinogenic hazards is important because new uses or unforeseen exposures could lead to risks that are much higher than those currently seen.

Summary

1.   Synthetic pesticides are by their very nature potentially dangerous and in developing countries pesticides have too often been used to commit suicide. There is clearly much to be done both educationally and in regulating pesticide marketing to minimize such misuse. Apart from the many suicide deaths there are numerous cases of ill health caused by unsuccessful suicide attempts.

2.   The pesticide problem in developing countries is exacerbated by the presence of large stocks of obsolete pesticides that need to be disposed of.

3.   Direct occupational exposure to some pesticides as a result of application errors or by spray drift can have serious health consequences, not only on the applicator but also on his or her family and children.

4.   The harmful effects of direct exposure to pesticides, especially when they are being applied, need to be minimized by appropriate ongoing training but they are not relevant when considering the small amounts of pesticide residues in food.

5.   It was not surprising following the media attention given to Carson's book *Silent Spring* that organic movements have misguidedly highlighted the presence of pesticide residues in conventionally grown crops as serious risks to health.

6.   If this was correct it would be expected that those who eat the synthetic pesticide-containing conventionally grown food would suffer from high rates of cancer and cardiovascular diseases. Instead, consumption of conventional fruit and vegetables leads to significantly better health outcomes.

7.   Carson's case about DDT was based on aerial spraying of trees in urban areas and had nothing to do with the agricultural

use of pesticides or of pesticide residues in food.

8. Studies on the causes of different cancers have not indicated any involvement of synthetic pesticide residues with cancer but this is unlikely to stop talk of cancer epidemics and of possible links with pesticides.

9. The actual quantity of synthetic pesticides in food and in the diet is seldom considered by the opponents of pesticides.

10. It may be said that conventional food will contain numerically more pesticide residues than organic but without assessing whether the actual amounts are sufficient to pose a health risk such an observation is of little value.

11. It is usually conveniently ignored that residues of all kinds of pesticide namely the natural (phytoalexins) and the synthetic pesticides occur in the diet.

12. The daily intake of phytoalexins in a five-a-day fruit and vegetable diet is conservatively estimated at 1,500 mg and that of synthetic pesticides at 0.1 mg. Thus it would be logical to be more concerned about the phytoalexins, about 53% of which are likely to be rodent carcinogens, if reducing intake of potentially harmful pesticides was a priority. Some phytoalexins can act as teratogens, others as oestrogen mimics and nerve toxins when applied at sufficiently high concentrations in tests on rodents.

13. The phytoalexin load in organically grown crops that are attacked by a disease or pest will be increased to more than the average 1,500 mg per day in a diet based on conventionally grown food. More information on this aspect is needed as it would seem that increases in phytoalexin concentrations in attacked crops can sometimes be very substantial.

14. The rigorous nature of toxicological testing in rats results in many compounds being rated as carcinogens and then assessed as hazards: however, the very small amounts of any given compound that man is likely to encounter in a lifetime would mean that there would be no risk or a minimal one.

15. About 50% of all chemicals tested (1,547 by 2011) have been assessed as rodent carcinogens; for naturally occurring chemicals the percentage is 57%, for all synthetic chemicals 59%, for phytoalexins 53%, for pharmaceutical drugs 44% and for compounds in roasted coffee 70%.

16. It is instructive to consider the several carcinogens in coffee. However, there are no reasons for thinking that they are of concern because only very small amounts are present. The amounts of carcinogenic natural compounds in two cups of coffee exceed the amounts of synthetic pesticide residues consumed in a year from a five-a-day fruit and vegetable diet.

17. The conclusion is clear – the intake of synthetic pesticide residues in food is most unlikely to be harmful. While it may be impossible to say that there is absolutely no risk from eating food that might contain pesticide residues, the risks to health are extremely small and are vastly outweighed by the benefits of eating a healthy balanced diet containing fruit and vegetables.

18. Organic crops can contain not only pesticides approved by the organic regulations but also in a few cases, at least in the USA, as shown by surveys, synthetic pesticides; it would appear that some organic farmers have intentionally used synthetic pesticides as the amounts

of residues were similar to those found in conventional crops. There can be no guarantee, only a probability, that organic food will be synthetic pesticide-free.

19. Regular monitoring of pesticide residues in produce is carried out in developed counties. Legally permitted maximum residue levels (MRL) are seldom exceeded; a MRL is not related to the biological effect of the pesticide but is a measure of the proper use by the farmer. No detectable residues are commonly found in about 50% of samples in Europe and the USA.

20. Acceptable daily lifetime intakes of pesticide residues based on animal feeding studies have a large safety factor of 100 times built in.

21. The fear of pesticide residues fostered by organic movements has left a legacy affecting not only the public but also the regulators and politicians. It will be difficult to allay this fear.

22. The finding that pesticide residues, wherever monitored, are at levels far below those that could be of any health concern will hopefully encourage the consumption of the health-promoting fruit and vegetables.

23. The possibility, or even probability, that some pesticide residues, in amounts well below any adverse effect level, could have beneficial health effects is unlikely to be accepted let alone understood. Nevertheless, the hormetic effect is being found in many dose response measurements. Whether or not pesticide residues can have beneficial effects needs to be studied.

24. It remains to be seen how data on hormesis can or will be used in risk assessment and it may be unlikely that those involved with the traditional interpretation of toxicological testing will be prepared to relinquish their belief that the Linear No Threshold model correctly links low doses to effect.

25. There is little doubt that the rodent toxicological tests are either intentionally or unintentionally misinterpreted, especially by those wishing to influence the public and media about the presence of synthetic pesticides in food. It is facile and wrong to imply that a chemical found to be a rodent carcinogen will be a risk to human health. If this was thought to be the case humans would become afraid to eat fruit and vegetables because of all the carcinogenic phytoalexins; this would be to their detriment.

26. Concerns are being expressed about the usefulness and validity of rodent carcinogenicity tests. A motion is currently under consideration by the European Parliament that such tests should be phased out.

27. While more attention could be paid to the biochemical and possible tissue-damaging effects of realistic exposures to pesticide residues in food, a major priority is that of getting people to understand the difference between hazard and risk.

References

Adami, H.O., Berry, Sir Colin L., Breckenridge, C.B., Smith, L.L., Swenberg, J.A., Trichopoulos, D., Weiss, N.S. and Pastoor, T.P. (2011) Toxicology and Epidemiology: Improving the Science with a Framework for Combining Toxicological and Epidemiological

Evidence to Establish Causal Inference. *Toxicological Sciences* 122, 2, 223–234. doi: 10.1093/toxsci/kfr113.

Ait Ouakrim, D., Pizot. C., Boniol. M., Malvezzi, M., Bonio, l.M., Negri, E., Bota, M., Jenkins., M.A., Bleiberg, H. and Autier, P. (2015) Trends in colorectal cancer mortality in Europe: retrospective analysis of the WHO mortality database. *British Medical Journal* 2015; 351:h4970. doi: 10.1136/bmj.h4970.

Ames, B.N. (1983) Dietary carcinogens and anti-carcinogens. Oxygen radicals and degenerative diseases. Science. 221, 4617, 1256–1264.

Ames, B.N., Magaw, R., and Gold, L.S. (1987) Ranking possible carcinogenic hazards. *Science* 236, 4799, 271–280.

Ames, B.N. (1990) Dietary pesticides (99.99% all natural). *Proceedings of the National Academy of Sciences* 87, 772–776.

Ames, B.N. and Gold, L.S. (1997) The causes and prevention of cancer: gaining perspective. *Environmental Heath Perspective Supplements* 105, Suppl, 4): 865–873.

Ames, B.N. and Gold, L.S. (1997) Environmental pollution, pesticides, and the prevention of cancer: misconceptions *FASEB Journal* 11, 13, 1041–1052.

Ames, B.N. and Gold, L.S. (2000) Paracelsus to parascience: the environmental cancer distraction. *Mutation Research* 447, 3–13. Available at http://toxnet.nlm.nih.gov/cpdb/pdfs/Paracelsus.pdf

Barański, M., Średnicka-Tober, D., Volakakis, N., Seal, C., Sanderson, R., Stewart, G.B., C., Biavati, B., Rembiałkowska, E., Markellou, E., Giotis, C., Gromadzka-Ostrowska, J., Skwarło-Sońta, K., Tahvonen, R., Janovská, D., Niggli, U., Nicot, P. and Leifert, C. (2014) Higher antioxidant and lower cadmium concentrations and lower incidence of pesticide residues in organically grown crops: a systematic literature review and meta-analyses. *British Journal of Nutrition* 112, 5, 794–811. doi: 10.1017/S0007114514001366. Supplementary material. Available at https://www.cambridge.org/core/journals/british-journal-of-nutrition/article/div-classtitlehigher-antioxidant-and-lower-cadmium-concentrations-and-lower-incidence-of-pesticide-residues-in-organically-grown-crops-a-systematic-literature-review-and-meta-analysesdiv/33F09637EAE6C4ED119E0C4BFFE2D5B1#fndtn-supplementary-materials

Beier, R.C. and Nigg, H.N. (2001) Toxicology of naturally occurring chemicals in food 1–186. In: Hui, Y.H., Smith, R. and Spoerke, D.G. (eds.) *Foodborne Disease Handbook. Second Edition, Vol. 3. Plant Toxicants,* Marcel Dekker, New York.

Belz, R.G., Cedergreen, N. and Duke, S.O. (2011) Herbicide hormesis – can it be useful in crop production? *Weed Research* 51, 4, 321–332. doi: 10.1111/j.1365-3180.2011.00862.x.

Belz, R.G. and Duke, S.O. (2014) Herbicides and hormesis. *Pest Management Science* 70, 5, 698–707. doi: 10.1002/ps.3726.

Bourguet, D. and Guillemaud, T. (2016) The hidden and external costs of pesticide use. *Sustainable Agriculture Reviews* 19, 35-120.

Bryant Christie (2015) Global MRL database. Available at https://www.globalmrl.com/home

Bunch, T.R., Bond, C., Buhl, K. and Stone, D. (2014) Spinosad General Fact Sheet; National Pesticide Information Center, Oregon State University Extension Services. Available at http://npic.orst.edu/factsheets/spinosadgen.

html and Spinosad … A New Organically Acceptable Insecticide. Available at www.2ndchance.info/fleas-spinosadGarden.pdf

Calabrese, E.J. and Baldwin, L.A. (2003) The Hormetic Dose-Response Model Is More Common than the Threshold Model in Toxicology. *Toxicological Sciences* 71, 2, 246–250. doi: 10.1093/toxsci/71.2.246.

Calabrese, E.J. (2005) Cancer biology and hormesis: human tumour cell lines commonly display hormetic (biphasic) dose responses. *Critical Reviews in Toxicology.* 35, 6, 463–582.

Calabrese, E.J. (2010) Hormesis is central to toxicology, pharmacology and risk assessment. *Human and Experimental Toxicology* 29, 4, 249–261. doi: 10.1177/0960327109363973.

Carson, R. (1962) *Silent Spring.* Boston, Massachusetts, Houghton Mifflin.

CFIA (2013) Canadian Food Inspection Agency National Chemical Residue Monitoring Program 2012–2013 Report. Available at www.inspection.gc.ca/food/chemical-residues-microbiology/chemical-residues/ncrmp-report/eng/1415838181260/1415838265896

Chen, W.L., Luan, Y.C., Shieh, M.C., Chen, S.T., Kung, H.T., Soong, K.L., Y.C., Chou, T.S, Mong, S.H., Wu, J.T., Sun, C.P., Deng, W.P., Wu, M.F. and Shen, M.L. (2007) Effects of Cobalt-60 Exposure on Health of Taiwan Residents Suggest New Approach Needed in Radiation Protection. *Dose Response.* 5, 1, 63–75. doi: 10.2203/dose-response.06-105. Chen.

Choi, Y., Chang, Y., Ryu, S., Cho, J., Rampal, S., Zhang, Y., Ahn, J., Lima, J.A.C., Shin, H. and Guallar, E. (2015) Coffee consumption and coronary artery calcium in young and middle-aged asymptomatic adults. *Heart* 101, 9, 686–691 doi: 10.1136/heartjnl-2014-306663.

Claeys, W.L., De Voghel, S., Schmit, J.F., Vromman, V. and Pussemier, L. (2008) Exposure assessment of the Belgian population to pesticide residues through fruit and vegetable consumption. *Food Additives and Contaminants: Part A* 25, 7.

Cohen, B., Carlisle, J., Fumento, M., Gough, M., Miller, H., Milloy, S., Smith, K. and Whelan, E. (1999) The Fear Profiteers, Do Socially Responsible Businesses Sow Health Scares to Reap Monetary Rewards? Available at www.hudson.org/files/publications/fear_profiteers.pdf

Cohen, E. (2006) Pesticide-mediated homeostatic modulation in arthropods. *Pesticide Biochemistry and Physiology* 85, 1, 21–27. doi.org/10.1016/j.pestbp.2005.09.002.

Contrera, J., Jacobs, A. and DeGeorge, J. (1997) Carcinogenicity testing and the evaluation of regulatory requirements for pharmaceuticals, *Regulatory Toxicology and Pharmacology* 25, 2, 130–145. doi: 10.1006/rtph.1997.1085.

Corrice, L. (2017) The Hiroshima Syndrome. Radiation: The No-Safe-Level Myth. Available at www.hiroshimasyndrome.com/radiation-the-no-safe-level-myth.html

Cutler, G.C. (2013) Insects, Insecticides and Hormesis: Evidence and Considerations for Study. *Dose Response* 11, 2, 154–77. doi: 10.2203/dose-response.12-008.Cutler.

Davies, E. (2011) Chemistry in every cup. Chemistry World May. Available at http://www.rsc.org/images/Coffee%20-%20Chemistry%20in%20Every%20Cup_tcm18-201245.pdf

Doll. R. and Peto, R. (1981) The causes of cancer: quantitative estimates of avoidable risks of cancer in the United States today.

Journal of the National Cancer Institute 66. 6.1191–1308.

EFSA (2013) International Frameworks Dealing with Human Risk Assessment of Combined Exposure to Multiple Chemicals. *EFSA Journal* 11, 7, 3313 doi: 10.2903/j.efsa.2013.3313.

EFSA (2014) Pesticides: breakthrough on cumulative risk assessment. Available at www.rivm.nl/en/Documents_and_publications/Common_and_Present/Newsmessages/2016/Breakthrough_on_cumulative_risk_assessment_exposure_to_pesticides_in_food

EFSA (2015) European Union report on pesticide residues in food. EFSA Journal 13, 3, 4038. doi: 10.2903/j.efsa.2015.4038.

EU Commission (2015) Pesticides database. Available at http://ec.europa.eu/food/plant/pesticides/eu-pesticides-database/public/?event=homepage&language=EN

EU (2015a) Motion for a resolution. Motion for a European Parliament resolution on the gradually banning of cancer tests on rodents. Available at www.europarl.europa.eu/sides/getDoc.do?pubRef=-//EP//TEXT+MOTION+B8-2016-0005+0+DOC+XML+V0//EN&language=es

EWG (2015) EWG's Shopper's Guide to Pesticides in Produce. Available at https://www.ewg.org/foodnews

Friedman, M., Rayburn, J.R. and Bantle, J.A. (1992) Structural relationships and developmental toxicity of solanum alkaloids in the frog embryo teratogenesis assay-Xenopus. *Journal of Agricultural and Food Chemistry* 40, 9, 1617–1624. doi: 10.1021/jf00021a029.

Gaffield, W. and Keeler, R.F. (1996) Induction of teratogenesis in hamsters by Solanidane alkaloids derived from *Solanum tuberosum. Chemical Research in Toxicology* 9, 2, 426–433. doi: 10.1021/tx950091r.

Gold, L.S., Slone, T.H., Manley, N.B. and Ames, B.N. (2002) Misconceptions About the Causes of Cancer 1415–1460. *In: Human and Ecological Risk Assessment: Theory and Practice* Paustenbach, D (ed.) New York, John Wiley & Sons.

Gold, L.S., Ames, B.N., Slone, T.H. (2008) Animal Cancer Tests and Human Cancer Risk: A Broad Perspective. http://potency.berkeley.edu/MOE.html

Grabendorfer, S. (2014) Biofumigation – an alternative method to control late blight in organic potato production? 9–12 In: Rahmann, G. and Aksoy U. (Eds.) *Proceedings of the 4th ISOFAR Scientific Conference. Building Organic Bridges*, Istanbul, Turkey.

Guedes, R.N. and Cutler, G.C. (2014) Insecticide-induced hormesis and arthropod pest management. *Pest Management Science* 70, 5, 690–697. doi: 10.1002/ps.3669.

Hart, K. and Pimentel, D. (2002), Public health and costs of pesticides 677–679 In: Pimentel, D. (ed.), *Encyclopaedia of Pest Management* CRC Press.

Henshaw, D.L. and Suk, W.A. (2015) Diet, transplacental carcinogenesis, and risk to children. *British Medical Journal*, 351:h4636 doi: 10.1136/bmj.h4636.

Hlywka, J.J., Stephenson, G.R., Sears, M.K. and Rickey, Y.Y. (1994) Effects of insect damage on glycoalkaloid content in potatoes (*Solanum tuberosum*). *Journal of Agricultural and Food Chemistry*. 42, 11, 2545–2550. doi: 10.1021/jf00047a032.

Holland, K.W. and O'Keefe, S.F. (2010) Recent applications of peanut phytoalexins. *Recent Patents on Food Nutrition and Agriculture* 2, 3, 221–232.

Huber, M., Rembiałkowska, E., Srednicka, D., Bügel, S. and van de Vijver, L.P.L

(2015) Organic food and impact on human health: Assessing the status quo and prospects of research. NJAS – Wageningen *Journal of Life Sciences* 5, 3–4, 103–109. doi: 10.1016/j.njas.2011.01.004.

IARC (2016) Agents classified by IARC monographs Volumes 1-115. Available at http://monographs.iarc.fr/ENG/Classification

IARC (2016a) Press Release 244 15 Monographs evaluate drinking coffee, maté, and very hot beverages. Available at https://www.iarc.fr/en/media-centre/pr/2016/pdfs/pr244_E.pdf

ICH (2016) European Medicines Agency. ICH guideline S1 – Regulatory notice on changes to core guideline on rodent carcinogenicity testing of pharmaceuticals EMA/CHMP/ICH/536328/2013.

Knekt, P., Kumpulainen, J. Jarvinen, R., Rissanen, H., Heliövaara, M., Reunanen, A., Hakulinen, T. and Aromaa, A. (2002) Flavonoid intake and risk of chronic diseases. *American Journal Clinical Nutrition* 76, 3, 560–568.

Kühne, S. (2014) Minimization strategies for copper pesticides in organic potato cultivation 335–338. In: Rahmann, G. and Aksoy U. (Eds). *Proceedings of the 4th ISOFAR Scientific Conference Building Organic Bridges*, Istanbul, Turkey.

Lomborg, B. (2001) *The sceptical environmentalist: measuring the real state of the world*. Cambridge University Press.

Long, M.E. (2007) Predicting carcinogenicity in humans: The need to supplement animal-based toxicology. *AATEX 14, Special Issue*, 553–559. In Proceedings of 6th World Congress on Alternatives and Animal Use in the Life Sciences, Tokyo, Japan. http://altweb.jhsph.edu/wc6/paper553.pdf

Luckey, T.D. (2006) Radiation Hormesis: The Good, the Bad, and the Ugly. *Dose Response* 4, 3, 169–190. doi: 10.2203/dose-response.06-102.

Mallqui, K.S., Vieira, J.L., Guedes, R.N. and Gontijo, L.M. (2014) Azadirachtin-induced hormesis mediating shift in fecundity-longevity trade-off in the Mexican bean weevil (*Chrysomelidae: Bruchinae*). *Journal of Economic Entomology* 107, 2, 860–866.

Matthews, G.A. (2016) *Pesticides: health, safety and the environment* (2nd edn.) Wiley Blackwell.

Mattsson, J.L. (2006) Spray more for safer food. *New Zealand Geographic* 77, 12–16.

Mattsson, J. (2008) Why Mother Nature must be a part of pesticide regulation. New Zealand Food Safety Authority. Food Focus. Available at www.foodsafety.govt.nz/elibrary/food-focus-2008-august/page-17.htm

Mattsson, J.L. (2008) Opinion: Improved Food Safety Requires Integration of Pest, Plant and Pesticide Interactions. *Journal of Consumer Protection and Food Safety* 3, 3, 259–264. doi: 10.1007/s00003-008-0351-7.

Medina-Remón, A., Zamora-Ros, R., Rotchés-Ribalta, M., Andres-Lacueva, C., Martínez-González, M.A., Covas, M.I., Corella, D., Salas-Salvadó, J., Gómez-Gracia, E., Ruiz-Gutiérrez, V., García de la Corte, F.J., Fiol, M., Pena, M.A., Saez, G.T., Ros, E,, Serra-Majem, L., Pinto, X., Warnberg, J., Estruch, R. and Lamuela-Raventos, R.M. (2011) Total polyphenol excretion and blood pressure in subjects at high cardiovascular risk. *Nutrition Metabolism and Cardiovascular Diseases* 21, 5, 323–331. doi: 10.1016/j.numecd.2009.10.019.

Nebert, D. (2015) Organic foods and

feel-good health mythology. Available at www.oregonlive.com/opinion/index.ssf/2015/12/organic_foods_and_feel-good_he.html

Ng, T.B., Ye, X.J., Wong, J.H., Fang, E.F., Chan, Y.S., Pan, W., Ye, X.Y., Sze, S.C., Zhang, K.Y., Liu, F. and Wang, H.X. (2011) Glyceollin, a soybean phytoalexin with medicinal properties. *Applied Microbiology and Biotechnology* 90, 1, 59–68. doi: 10.1007/s00253-011-3169-7.

NIEHS (2016) Organophosphates, Superfund Research Program. Available at http://tools.niehs.nih.gov/srp/research/research4_s3_s5.cfm

Oerke, E.C. (2006) Crop losses to pests. *The Journal of Agricultural Science* 144, 1, 31–43. doi: doi.org/10.1017/S0021859605005708.

OMRI (2015) OMRI Canada Products List. Available at https://www.omri.org/canada-lis

PAN Germany (2012 Pesticides and health hazards – facts and figures. Available at www.pan-germany.org/download/Vergift_EN-201112-web.pdf

PAN UK (2013) Pesticides are a major killer www.pan-uk.org

Parkin, D.M., Boyd, L. and Walker, L.C. (2011) The fraction of cancer attributable to lifestyle and environmental factors in the UK in 2010. *British Journal of Cancer* 105, *Supplement* 2:S77–81. doi: 10.1038/bjc.2011.489.

Paull, J. (2013) The Rachel Carson Letters and the Making of Silent Spring. *Sage Open*, 3, 1–12. http://sgo.sagepub.com/content/3/3/2158244013494861.full.pdf

PDP (2013) Pesticide Data Program. Annual Summary 2013. Available at www.ams.usda.gov/sites/default/files/media/2013%20PDP%20Anuual%20Summary.pdf

PDP (2014) Pesticide Data Program. Annual Summary 2014 Available at https://www.ams.usda.gov/sites/default/files/media/PDP%202014%20Annual%20Summary%20Q%26As.pdf

Pedras, M.S., Yaya, E.E. and Glawischnig, E. (2011) The phytoalexins from cultivated and wild crucifers: chemistry and biology. *Natural Products Reports.* 28, 8, 1381–1405. doi: 10.1039/c1np00020a.

PETA (2006) Chemicals and cancer. What the regulators won't tell you about carcinogenicity testing. Available at https://www.peta.de/mediadb/EUreport300.pdf

Porterfield, A. (2016) Winemakers drop organic status to support environment. Genetic Literacy Project. Available at https://www.geneticliteracyproject.org/2016/03/13/winemakers-drop-organic-status-to-support-environment

PRIF (2012) The expert committee on Pesticide Residues in Food (PRiF) Annual Report 2012. Available at www.panna.org/sites/default/files/23101.01PRiFAR2012WEB.pdf

PRIF (2015) The expert committee on Pesticide Residues in Food (PRiF) Annual Report 2014. http://webarchive.nationalarchives.gov.uk/20151023155227/www.pesticides.gov.uk/Resources/CRD/PRiF/Documents/Results%20and%20Reports/2014/PRIF%20Annual%20Report%202014%20FINAL.pdf

Pubchem (2016) National Center for Biotechnology Information Pubchem the Open Chemistry database. Available at https://pubchem.ncbi.nlm.nih.gov/compound

Randall, W.A., Price, C.W. and Welch, H. (1947) Demonstration of hormesis (increase in fatality rate) by penicillin. *American Journal of Public Health Nations Health* 37, 4, 421–425.

Renwick, J.H., Claringbold, W.D., Earthy, M.E., Few, J.D. and McLean, A.C. (1984) Neural-tube defects produced in Syrian hamsters by potato glycoalkaloids. *Teratology* 30, 3, 371–381. doi: 10.1002/tera.1420300309.

Ritter, L. (1997) Report of a panel on the relationship between public exposure to pesticides and cancer. *Cancer* 80, 10, 2019–2033.

Rodricks, J.V. (2003) Hormesis and toxicological risk assessment. *Toxicological Sciences* 71, 2, 134–136. doi: 10.1093/toxsci/71.2.134.

Rosen, J. (1990) Much Ado About Alar. *Issues in Science & Technology*, 85–90. Available at http://courses.washington.edu/alisonta/pbaf590/pdf/Rosen_Alar.pdf

Sakamoto, K. (2004) Radiobiological Basis for Cancer Therapy by Total or Half-Body Irradiation. *Nonlinearity in Biology. Toxicology. Medicine* 2, 4, 293–316. doi: 10.1080/15401420490900254.

Sanborn, M., Cole, D., Kerr, K., Vakil, C., Sanin, L.H. and Bassil, K. (2004) Pesticides Literature Review. Ontario College of Family Physicians. Available at http://citeseerx.ist.psu.edu/viewdoc/download;jsessionid=763F5F1F2C0E8CA97EA0410F87E5BBAF?doi=10.1.1.689.4719&rep=rep1&type=pdf

Sanborn, M., Cole, D., Kerr, K., Vakil, C., Sanin, L.H. and Bassil, K. (2012) Systematic Review of Pesticide Health Effects. Ontario College of Family Physicians. http://ocfp.on.ca/docs/pesticides-paper/2012-systematic-review-of-pesticide.pdf

Savary, S., Ficke A., Aubertot, J-N. and Hollier, C. (2012) Crop losses due to diseases and their implications for global food production losses and food security.

Food Security 4, 4, 519–537. doi: 10.1007/s12571-012-0200-5.

Southam, C.M. and Ehrlich, J. (1943) Effects of Extract of western red-cedar heartwood on certain wood-decaying fungi in culture. *Phytopathology* 33, 517–524.

Starks, S.E., Gerr, F., Kamel, F., Lynch, C.F., Alavanja, M.C., Sandler, D.P. and Hoppin, J.A. (2012) High pesticide exposure events and central nervous system function among pesticide applicators in the Agricultural Health Study. International Archives of Occupational and Environmental Health 85, 5, 505–515. doi: 10.1007/s00420-011-0694-8.

Trewavas. A.J. (2005) Organic food the hidden agenda. Association of Applied Biologists. Letters Issue 56.

Trewavas, A. (2008) The cult of the amateur in agriculture threatens food security. *Trends in Biotechnology* 26, 9, 475–478. doi.org/10.1016/j.tibtech.2008.06.002.

UNEP (2012) Urgent Action Needed to Reduce Growing Health and Environmental Hazards from Chemicals: UN Report. Available at www.unep.org/newscentre/default.aspx?DocumentID=2694&ArticleID=9266#sthash.A1g1wnU7.dpuf

UNEP (2013) Global chemical outlook 2013 – Towards Sound Management of Chemicals. Available at www.unep.org/hazardoussubstances/Portals/9/Mainstreaming/GCO/The%20Global%20Chemical%20Outlook_Full%20report_15Feb2013.pdf

USDA (2016) USDA Releases 2014 Annual Summary for Pesticide Data Program. Available at https://www.ams.usda.gov/press-release/usda-releases-2014-annual-summary-pesticide-data-program-report-confirms-pesticide

Vladimir-Knežević, S., Blazekovic, B.,

Stefan, M.B. and Babac, M. (2012) Plant Polyphenols as Antioxidants Influencing the Human Health. doi: 10.5772/27843.

Wang, A., Costello, S., Cockburn, M., Zhang, X., Bronstein, J. and Ritz, B. (2011) Parkinson's disease risk from ambient exposure to pesticides. *European Journal of Epidemiology* 26, 7, 547–555. doi: 10.1007/s10654-011-9574-5.

WHO (2004) Pesticides and health. Available at www.who.int/mental_health/prevention/suicide/en/PesticidesHealth2.pdf

Winter, C.K. (2012) Pesticide Residues in Imported, Organic, and Suspect Fruits and Vegetables *Journal of Agricultural and Food Chemistry*. 60, 18, 4425–4429. doi: 10.1021/jf205131q.

Winter. C.K. (2015) Chronic dietary exposure to pesticide residues in the United States. *International Journal of Food Contamination* 2, 11. Springer Open Journal. doi: 10.1186/s40550-015-0018-y.

WCRF (1997) World Cancer Research Fund/American Institute for Cancer Research. Food Nutrition, Physical Health and the Prevention of Cancer: a Global Perspective. Washington DC.

Yang, L, Browning, J.D. and Awika, J.M. (2009) Sorghum 3-deoxyanthocyanins possess strong phase II enzyme inducer activity and cancer cell growth inhibition properties. *Journal of Agricultural and Food Chemistry*. 57, 5, 1797–1804. doi: 10.1021/jf8035066.

Yuan, F. and Qian, M.C. (2016) Aroma Potential in Early- and Late-Maturity Pinot noir Grapes Evaluated by Aroma Extract Dilution Analysis. *Journal of Agricultural and Food Chemistry* 64, 2, 443–450. doi: 10.1021/acs.jafc.5b04774.

Nitrogen: the All–Important Element

Human Requirement for Nitrogen and Where it Comes From with a note on Phosphorus and Potassium

"The principal objective of agriculture is to produce nitrogen in a digestible form"
Justus von Liebig

Nitrogen and man

A key question when comparing organic and conventional farming is: "Is it possible to produce sufficient protein for all people without man-made nitrogenous fertilizers?" The potential effects of a global reduction in the demand for animal protein and of the effects of increased consumption of vegetable protein will complicate any answer and will be considered in Chapter 19.

Most agricultural soils contain insufficient naturally occurring plant-available nitrogen to meet crop needs throughout the growing season. The manner in which N is supplied is central to any comparison of organic and conventional farming. The human daily demand for protein is fundamentally satisfied by the provision of N to crops in fertilizers and from biological nitrogen fixation by legumes supplied either directly or indirectly via manure. Nitrogen is absorbed principally as the anion NO_3 but also as the cation NH_4 regardless of the source.

There are four factors that result in nitrogen being the most important and at the same time the most problematic nutrient element for the well-being of humans, animals and plants.

i.   Nitrogen is, after C, H and O – which are all abundantly available as CO_2 and H_2O in the environment, the nutrient element required in greatest quantities by plants, animals and man. It is the nutrient that is most often the major factor limiting crop growth.

ii.   All proteins contain 16% nitrogen; proteins are not stored in animals – they are either used in growth and repair of tissues, as an energy source or are excreted. A 70 kg man contains about 1.8 kg N in about 11 kg protein.

iii.   Reactive forms of nitrogen are readily lost from the soil. The leaching of nitrates under the influence of rain is of particular concern, as are the emissions of ammonia and nitrous oxide from soils.

iv.   It is difficult to build up and maintain sufficient soil nitrogen in organic matter to supply, after mineralization, enough nitrogen for high yields.

Not surprisingly, nitrogen deficiency is most often the main nutrient limiting crop growth across the world.

The central role of fertilizer N in food production was clearly shown in the mid-19th century by the classical continuous wheat experiment at Rothamsted. Wheat yields were not increased by application of only the minerals P, K and Mg (as claimed by Liebig, who believed that the crops obtained N for the atmosphere); wheat needed fertilizer N. Gaseous nitrogen plays no part in human metabolism; it is an unreactive gas constituting about 78% of the atmosphere. The reactive forms of nitrogen are relatively scarce in natural ecosystems. On the one hand they are vital for crop production and human growth and on the other they can have deleterious environmental effects.

It seems a strange quirk of biochemistry that plants make a massive direct use of the relatively small quantities of CO_2 in the atmosphere (measured in ppm) in photosynthesis whereas it is only with the cooperation of bacteria that they are able to make use of the huge amounts of gaseous nitrogen.

Mineral composition of reference man and plants

Growing plants contain variable amounts of water – vegetables typically contain 75–95% water with organic compounds and minerals comprising the rest. The 70 kg reference man (Snyder *et al.*, 1975) contains 50–60% water and 45% organic and mineral matter, comprising 10.6 kg protein, 13.5 kg fat and 3.7 kg minerals. The organic compounds in plants are derived from photosynthesis whereby atmospheric carbon dioxide is reacted with water using the energy of sunlight absorbed by chlorophyll to produce

Table 16.1 Approximate mineral element composition (kg/dry t) of a typical crop and of the reference man

	Crop	Human
N	20	57
Ca	8	32
P	2	24
K	15	4
S	1	4
Na	2.5	3.2
Cl	0.15	3.0
Mg	2	0.6
Si	4–14[+]	<0.1

[+] Cereals

glucose and oxygen. Sugars form the basis of plant metabolism and thence of animal metabolism. In natural ecosytems 60% of the carbon fixed by photosynthesis is used in the shoots for growth and respiration while 40% is allocated to the roots (Jones, 1993). Up to 14% of the fixed carbon may ultimately be used by microorganisms associated with the rhizosphere.

Of the mineral nutrients nitrogen plays a dominant role in both plants and man (Table 16.1); this is because it is a constituent of all amino acids and therefore of all proteins. Nitrogen is a component of nucleic acids and also of the green pigment chlorophyll – yellowing of leaves is the common symptom of nitrogen deficiency.

In contrast to man, plants use silicon together with organic compounds such as lignin as structural materials, which is evident from the comparative mineral composition. The relatively small requirement for Ca and P in plants contrasts with the human requirement for these nutrients, which are mostly found in bones. In humans most N is found in muscle (43%), with significant proportions being present in skin (15%) and blood (16%).

Global minimum requirements for protein and nitrogen

The global minimum human requirement for protein and nitrogen has been roughly estimated (Tables 16.2 and 16.3) based on population sizes, average body weights and protein requirements according to age for a global population of 7.2 billion (CIA World Factbook, 2014). No attempt has been made to account for the varying protein requirements with age and pregnancy, as for example according to EFSA (2012), and no allowance is made for the varying quality of protein.

The absolute minimum protein daily requirements may be less than the values shown. The survival and longevity of people who had lived for several years on starvation diets as result of being displaced by war or incarcerated in prisoner of war camps indicates that if necessary man can exist on diets containing less than 0.8 g protein/kg body weight per day for considerable periods. Experiments with mice have shown that food deprivation can increase the lifespan.

If all the world (7.2 billion) subsisted on the minimum protein requirement there would be an annual global human need for about 21 Tg of N in protein; a global population of 10 billion

Table 16.2 Global populations, weights and minimum protein requirements according to age

Age	Population	Weight	Protein requirement	
	Billions	kg/capita*	g/kg/day	g/capita/day
<14 years	1.89	10–50	1.1	11–55
Adults	4.74	58–81	0.8	46–65
>65 years	0.57	65**	0.8	52

* Family Practice Notebook (2015).
** estimated.

Table 16.3 Approximate weights, and estimated annual minimum protein and N requirements of a global population (7.2 billion) based on data in Table 16.2 (using weights in parenthesis). Tg = Teragram = million t (10^{12} g)

Age	Global Weight Tg	Global protein requirement Tg/year	Global N requirement Tg/year
<14 years	19–94 (56)	22.5	3.6
Adults	275–384 (330)	96.4	15.4
>65 years	37	10.8	1.7
Total	331–515 (423)	129.7	20.7

would require a minimum of 29 Tg N annually.

Pimentel *et al.* (1997) estimated global human weight at 420 Tg in the 1990s. More recently Walpole *et al.* (2012) calculated global body mass of adults in 2005 at 287 Tg, of which 15 Tg was associated with overweight and 3.5 Tg with obesity.

Actual global protein consumption and potential nitrogen requirements

The data used here is from FAOSTAT (2016), which being based on national food surveys and assessments of food consumption may well be very approximate. The quoted average per capita consumption of protein across the world (79 g protein/day) varies very widely from 103 g in developed countries to 74 g in developing countries (Table 16.4).

In developing countries animal protein constitutes 33% of protein consumption, compared with 58% in developed countries (Table 16.4). The dominant position of animal protein in Europe and the Americas is expected to continue, as is increased consumption of animal protein in the rest of the world concomitant with rising living standards.

According to Boland (2013), 38% of

Table 16.4   Average total and animal protein consumption (g/capita/day 2010–2011) (FAOSTAT, 2016)

Country/Region	Total protein	Animal protein
World	79	31
Asia	75	25
Africa	65	14
Europe	106	66
United Kingdom	102	58
North America	107	64
United States of America	110	71
South America	84	45
Latin America + Caribbean	82	41
Oceania	72	38
Developing countries	74	25
Sub-Saharan Africa	59	13
Developed countries	103	60

the globally consumed protein is derived from livestock and fish (Table 16.5).

It would appear that average consumption of protein in all continents exceeds the absolute minimum requirements of 50–60 g/capita/day (Table 16.4), which is somewhat surprising. The low average protein intake in Africa, especially in sub-Saharan Africa, means that many there will be getting inadequate protein. The apparently good situation in Asia is dominated by the relatively high intakes in China (94 g/capita/day) but there are many countries, including India (59 g/capita/day), with low average intakes. Based on dietary energy intakes, about 842,000 people are thought to be undernourished, mostly in Asia with 552,000 (13% population) and Africa with 226,000 (21% population) (FAO, IFAD and WFP, 2013). It is the unequal access to food rather than inadequate global food production that is responsible.

Table 16.5   Source of global protein consumption as percentage of total (Boland, 2013)

Source	%
Cereals	40.4
Meat	17.8
Milk, excluding butter	10.1
Fish, seafood	6.4
Vegetables	5.7
Pulses	4.9
Eggs	3.4
Oilseed crops	3.4
Starch roots	2.8
Other	5.1

It is calculated that the actual global consumption of N (about 33 Tg) considerably exceeds the global minimum annual N requirement (18–20 Tg) (Table 16.6).

It seems inevitable that the demand for livestock products across the world will grow. WHO (2013) projected that the annual global meat production will increase from 218 million t in 1997–1999 to 376 million t by 2030 as a result of population growth and rising incomes, which correlates strongly with increased consumption of animal protein at the expense of staple foods.

If the current global population of 7.2 billion people consumed protein at the same levels as in Europe and American the global consumption of N would increase from 33 to about 45 Tg (Table 16.7) and a global population of 10 billion would require about 60 Tg N.

On average about two to three times more fertilizer N is used than N consumed by humans (to meet the actual intakes of protein) in all regions but Africa where annual consumption of N exceeds that supplied in fertilizer. The apparent short fall in Africa of about 3 Tg/year is currently being made up by biological N fixation and by the mineralization of soil organic matter. The calculations of Lassaletta *et al.* (2014), discussed later, indicate, after accounting for biological N fixation that the organic matter in many African soils is being exhausted in order to supply the required N.

Nitrogen: where it comes from

The N absorbed by crops is generated by biological fixation and by man, and can be supplied in manures, composts, by decomposing legumes and fertilizers. Mineralization of soil organic matter not only provides N but also other essential nutrients. Although historically farmers have been aware that recycling of manure and the growing of legumes benefited crops they did not known why.

Much energy is required to break the triple bond of the unreactive nitrogen

Table 16.6   Minimum human N requirements and actual consumption of N in protein/capita in different regions 2009–2011 (FAOSTAT, 2013)

	Population Billion	Minimum N[+] Requirement Tg/year	Consumption	
			Actual N g/capita/day	Actual N Tg/year
Asia	4.32	10.42	12.0	18.9
Africa	1.06	2.64	10.4	4.0
Europe	0.82	2.30	17.0	5.1
North America	0.55	1.87	17.1	3.4
South America	0.40	1.04	13.3	1.9
Australasia*	0.05	0.11	11.5	0.2
World	7.20	18.38	12.7	33.5

* Includes Oceania.
[+] based on estimated requirements for children (at 1.1 g protein/kg), and all adults (at 0.8 g protein/ kg) and using weights appropriate for each region which accounts for the variation in total N requirement with that in Table 16.3.

Table 16.7 Calculated human N consumption for 7.2 billion on basis of European and American protein consumption and the use of fertilizer N in 2013 (FAOSTAT, 2013)

	N consumption Tg N/year	Fertilizer N Use Tg N/year
Asia	27.3	55.9
Africa	6.6	3.9
Europe	5.1	14.6
North & Central America	3.4	16.7
South America	2.4	6.7
Australasia +Oceania	0.3	1.8
Total	45.1	99.6

molecules, as is evident by that needed in the Haber–Bosch process when combining gaseous nitrogen with hydrogen at 400–650° C and high pressure with iron as a catalyst to make ammonia (see Chapter 2). However, a few microorganisms, notably *Rhizobia* bacteria, have learnt the trick of using nitrogenase, an enzyme that is peculiar to them. They are able to make ammonia that is then assimilated into organic nitrogen compounds, such as ureides, when living in a symbiotic relationship with legumes. For example, in soybeans the ureides, allantoin and allantoic acid are the major forms of N transported from nodulated roots to shoots (Thomas and Schrader, 1981). Decaying roots, nodules and leaf litter after the harvesting of forage legumes can be valuable sources of nitrogen for subsequent crops, as can ploughed leguminous cover crops; root and shoot residues of seed legumes provide little N.

Smil (2001) makes the point that the Haber–Bosch process is considerably more energy efficient in fixing N than symbiotic bacteria, which may surprise.

Biological nitrogen fixation

The ability of legumes to improve soil fertility has been recognized for at least 2,000 years but it was only in the late 19th century that the symbiotic nitrogen-fixing relationship between legumes and bacteria began to be understood (e.g. Atwater, 1884; Hellriegel and Wilfarth, 1888).The N fixing bacteria in the nodules of legumes receive carbohydrates from the plant to meet their energy needs and in return make a conribution to the N requirement of the host plant.

The contribution of non-symbiotic N fixers may be important in natural ecosystems but only relatively small amounts of N are fixed. Some fungi and primitive single-celled organisms can fix nitrogen. Fungal activity in grassland, in semi-arid soils and also in forest systems, where the fungi effectively connect trees, may thus be a potential source of pollutant nitrous oxide in natural ecosystems (Hayatsu *et al.*, 2008).

Some commercial products consisting of organisms with a nitrogen-fixing capacity have been developed as inoculants for wheat, barley, cotton, canola, sugar cane, maize, and vegetables in India, Egypt and Argentina but so far results appear to be inconsistent (Montañez, 2000).

Before the industrial fixation of N by the Haber–Bosch process all crops were reliant on biologically fixed nitrogen and the recycling of N mainly via manure. Yields were low and remained so until the advent of man-made fertilizers (Table 18.1).

Legumes

The quoted amounts of symbiotic N fixation in agriculture vary widely according to plant species and other factors such as soil pH and P status.

When forage legumes are ploughed in or residues decompose in the soil a substantial part of the contained nitrogen in the litter and roots is mineralized and becomes available for uptake by subsequent crops. N-rich exudates may leak from living legume roots. In contrast, much of the nitrogen fixed by seed legumes is removed from the field and the amounts left for a following crop will be less than indicated in Table 16.8. However, the fixed N, especially that by soybeans, can benefit arable crops via manure from cattle consuming soybean-rich animal feed; such arable crops will often not be in countries where the soybeans are grown.

The amounts of N fixed by lucerne reported by different authorities vary widely from 150 to 500 kg/ha/year (Table 16.8). Haygarth *et al.* (2013) mention the ability of clovers to fix up to 600 kg N/ha/year under favourable conditions. In their booklet on converting arable farms to organic, especially stockless farms, the Home-Grown Cereals Authority in the UK estimates that if 300 kg/ha N is fixed by a green manure legume crop then 100 kg N will be available at once to the following crop, with further N becoming available later in the rotation (HGCA, 2008); there is no provision for determining or estimating the actual amounts of fixed N.

The fact that the quoted amounts of symbiotically fixed N vary widely indicates that it is difficult to predict the amounts of N that a previous legume crop may contribute to meet the N requirement of a following crop. The use of legumes as a source of N is a somewhat inexact science. Beans and peas may fix 50–100 kg N/ha but contribute little to the following crop. How much of the N fixed by restorative grass/clover leys and legume cover crops is used by a following crop will largely depend on the synchronicity of mineralization with crop absorption.

With the exception of several extenson services in the USA (see nitrogen credits), agronomists generally find it difficult to give more than a rough estimate of the amounts of N fixed by any

Table 16.8 Typical rates of biological fixation of N

	kg N/ha/year
Seed legumes	
Beans (*Phaseolus* spp.) [1]	30–50
Beans [4]	154
Broad beans (*Vicia faba*) [1]	80–120
Broad beans [4]	80
Chickpea (*Cicer arietinum*) [1]	40–60
Chickpea [4]	34
Groundnut (*Arachis hypogea*) [1]	60–100
Groundnut [4]	79
Lentils (*Lens culinaris*) [1]	30–50
Peas (*Pisum sativum*) [1]	30–50
Peas [5]	50–100
Soybean (*Glycine max*) [1]	60–100
Soybean [2]	80–150
Soybean [4]	161
Forage legumes	
Clovers (*Trifolium* spp.) [1]	130–170
Clover [4]	222
Red clover (*Trifolium pratense*) [5]	300–400
White clover (*Trifolium repens*) [5]	100–200
Clover/grasses [3]	30–170
Lucerne (*Medicago sativa*) [1]	150–250
Lucerne [1]	200–250
Lucerne [4]	164
Lucerne [5]	to 500
Rice, paddy (*Oryza sativa*) [3,4]	33
Rice, paddy [3]	50–100
Savannas [3]	<10
Sugar cane, free–living [4]	25
Trefoils [5]	100–200
Vetch (*Vicia sativa*) [5]	to 200

[1] Smil (2001), [2] Boddey *et al.* (1997), [3] Cassman and Harwood, (1995). [4] Global average (Lassaletta *et al.*, 2014), [5] HGCA (2008).

particular forage legume let alone give an estimate for a given field on which a fertilizer plan can be based (Smil, 2001; Herridge *et al.*, 2008). This lack of precision is in marked contrast with the way inorganic nitrogen fertilizers can be applied in the desired amounts at the most efficacious time; split applications of fertilizer N to crops will maximize utilization and minimize nitrate leaching. Although in the UK farmers could use the programme Manner–NPK (Sparkes, 2014) to decide how much nitrogen (as fertilizer and manure) to use based on analyses of soil nitrogen at different depths and the desired yield it would appear that decisions are more likely to be made on previous experience, hoped-for yields and market conditions. Organic farmers will also be able to plan on past experience but they cannot use split applications of manure, which has to be incorporated into the soil before preparing the seedbed; organic farmers are afforded some flexibility when applying slurries.

Smil (2001) considered that the best combinations of recycling, rotations including leguminous crops and green manures should be able to supply about 120–150 kg N/ha, which would be sufficient to produce about 200 kg protein/ ha and to feed ten people.

Non-legumes

Legumes are the best known N fixers through their association with *Rhizobia* bacteria. However some non legumes, e.g. Alder (*Alnus spp.*), can enter into a symbiotic relationship with an actinomycete (*Frankia*) that can reduce gaseous nitrogen.

Anabaena azollae, a cyanobacteria living in the cells of *Azolla pinnata* (duckweed fern), also fixes nitrogen. Infected duckweed has been used extensively in paddy rice fields since the 1950s. However, the practice probably goes back to the earliest days of rice cultivation in China and in other Asian countries where *Azolla* is endemic; according to Jia Ssu Hsieh (504), the benefits of *Azolla* were appreciated around 500 AD. *Anabaena/Azolla* is now widely used on paddy rice. Stocks of *Azolla* are maintained in nurseries for adding to the flooded paddies. Most (95%) of the nitrogen fixed becomes available when the *Azolla* decomposes. The Azolla foundation (website) suggests that *Anabaena/Azolla* can typically fix up to 1,100 kg/ha N/year, which it contrasts with 400 kg/ha N fixed by legumes; these values may be optimistic. Vaishampayan *et al.* (2001) quote that up to 600 kg N/ha/year may be fixed *by Anabaena/Azolla*. In contrast, Pabby *et al.* (2003) reported that a single *Azolla* crop can provide 20–40 kg N/ha/ year. They also found substantial yield increases of 112% in rice after using *Anabaena/Azolla* during a previous fallow season: when it was used as an intercrop with the rice it increased yields by 23%. Yield increases of more than 200% followed the combined use of *Anabaena/Azolla* as a monocrop in the previous season and as an intercrop.

Cyanobacteria are photosynthetic and in theory should not require carbon from their host plant. In this respect they have the potential to be superior to *Rhizobia*, which rely on the host plant for their energy. However, it appears that up to 80% of the photosynthate used in N fixation by *Anabaena* is provided by the host (Pabby *et al.*, 2003). *Azolla* is also used as a feed for cattle, fish, pigs and poultry.

Work in Brazil starting in the 1960s principally by Dobereiner and associates, e.g. Boddey (1995), Boddey *et al.* (1995) and Dobereiner *et al.* (1993), showed that several bacteria including *Acetobacter spp.*, *Herbaspirillum seropedicae*, *H. rubrisubalbicans*, *Gluconacetobacter diazotrophicus* and *Burkholderia* spp. can fix nitrogen when growing in association with sugar cane and other grasses such as sorghum and rice. The bacteria live not only on and near roots but also in the roots.

Boddey *et al.* (1995) found that sugar cane in Brazil is capable of obtaining 90 kg N/ha /year from biological nitrogen fixation from bacteria within the roots, which is about 60% of their N requirement. In paddy rice the amounts were 30–60 kg N/ha/year.

Tadra-Sfeir *et* al. (2011) found that *Herbaspirillum seropedicae* can form N-fixing associations with maize, rice, sorghum, sugar cane, banana and pineapple and can increase rice yields equivalent to 40 kg N/ha.

Gluconacetobacter diazotrophicus was originally found able to fix nitrogen in sugar cane roots; it is known to exist in a wide range of crops, including banana, beetroot, carrot, coffee, millet, paddy rice, pineapple, radish, sweet potato and tea (Eskin *et al.*, 2014) but evidence of significant nitrogen fixation is awaited. It mainly colonizes intercellular spaces within the roots and stems of plants and does not form a nodule. It is hoped that nitrogen fixation by *Gluconacetobacter diazotrophicus* may be encouraged on other crops especially cereals.

In contrast to the findings in Brazil, work on field-grown sugar cane in South Africa failed to show that significant amounts of nitrogen are fixed by the free-living bacteria (Hoefsloot *et al.*, 2005).

Cocking *et al.* (2006) successfully inoculated several crops including maize, oilseed rape, rice, tomato, wheat, barley, tomato and white clover with *Gluconacetobacter diazotrophicus* (plus sucrose) taken from sugar cane. No data is available on the amounts of nitrogen that may be fixed. Azotic Technologies licenced the technology known as N-Fix based on *Gluconacetobacter diazotrophicus* in 2012 and are currently working towards commercialization in several countries, including Belgium, Canada, France, Germany, UK and USA (Dent and Cocking, 2015).

Researchers at the University of Missouri are using *Setaria viridis* (a fast-growing grass with a simple genome and C4 photosynthesis) and *Azospirillum brasilense* (a normally free-living bacteria that can fix nitrogen), as a model for identifying the genes that are responsible for the interaction between other C4 plants such as maize, sugar cane and sorghum and nitrogen-fixing bacteria (Meissen, 2015) and (Pankievicz *et al.*, 2015). In laboratory experiments *Setaria viridis* was able to obtain sufficient nitrogen from N fixation by *Azospirillum*

brasilense living on the root surface to promote robust plant growth under nitrogen-limiting conditions.

Symbiotic fixation of N by trees and shrubs

There are several N-fixing trees and shrubs. For example, *Gliricidia sepium*, a medium-sized tree used to shade fruit orchards, cacao, coconuts and sometimes tea, can produce 2–20 t/ha/year of N-rich dry leaf biomass, which on mineralization can provide considerable amounts of N for companion crops and pastures. Patil (1989) found that 1 t dry of *Gliricidia* leaves was equivalent to 27 kg N. Bindumadhava Rao *et al.* (1966) reported that 400 coppiced *Gliricidia* trees grown around the field perimeter could provide sufficient N for 1 ha of paddy rice. In Nigeria Kang and Mulongoy (1987) reported that up to 15 t/ha/year of *Gliricidia* leaves could be produced on good soils providing the equivalent of 40 kg N/ha/year. *Faidherbia albida*, which is indigenous to Africa, is another N-fixing tree that can be grown to provide N to nearby crops by leaf fall. It is used by many farmers in Burkina Faso, Malawi, Mali, Niger, Nigeria, Senegal, Tanzania and Zambia (WAC, 2015) and is grown in association with maize, millet, sorghum; livestock can also benefit from the N-rich leaves and pods.

Pigeon pea (*Cajanus cajan*), a perennial shrub, can be grown as a green manure or as a source of mulch and provide up to 117 kg N/ha when inter-cropped with maize (Adu-Gyamfi *et al.*, 2007).

Free-living biological N fixation

Only the cyanobacteria (e.g. *Anabaena*, *Nostoc* and *Calothrix*) can add significant amounts of nitrogen to cultivated soils possibly by 20–30 kg N/ha/year (Smil, 2001).

There are a few other free-living nitrogen fixing organisms *(e.g. Azotobacteria, Azospirillum* and *Clostridium)* that are capable of nitrogen fixation. They are not associated with any particular crops. The contribution of non-symbiotic fixers is most important in natural eco-systems where relatively small amounts of N (< 3 kg/ha/year) may be fixed; smaller amounts are likely in arable soils (Marschner, 1986). It is possible that the build-up of soil N by free-living bacteria when leaving fields fallow is not of significance due to the limited availability of energy to support the bacteria.

Nitrogen fixation by man

The importance of the factory synthesis of ammonia for the well-being of mankind cannot be overemphasized. As pointed out by Smil (2001), the expansion of the world's population would not have been possible without the synthesis of ammonia. The big challenge for the global organic movement is how to generate sufficient nitrogen to replace the ammonia fixed by man. This is a very tall order. In 2013 the global use of fertilizer N amounted to just under 100 Tg N with 24% used in China, 15% in North America, 15% in Europe and 17% in India.

Globally 740 million ha of forage legumes fixing N at 135 kg N/ha would be required to fix the same amount of N, i.e. about 15% of global agricultural land as defined by FAO (2012).

In Europe the main sources of nitrogen are fertilizers, with manure a close second. However, the situation varies considerably from one country to another. For example, in 1995, mineral fertilizers accounted for at least 50% of total N inputs in Denmark, Germany, Greece, France, Luxembourg, Finland and Sweden whereas in Belgium and the Netherlands livestock manure was responsible for more than 50% of nitrogen inputs (Pau Vall and Vidal, 1999).

Most of the fertilizer N is used by conventional farmers but limited amounts are indirectly used by those organic farmers who import animal feeding stuffs based on crops grown on conventional farms that have used fertilizers. Some organic farmers import manure from conventional farms. Legume fixation of nitrogen can also benefit conventional farmers.

Global estimates of nitrogen sources

There have been several estimates of the global amounts of biologically fixed nitrogen by Smil (2001), Herridge *et al.* (2008), Bouwman *et al.* (2011). Vitousek *et al.* (2013) and Fowler *et al.* (2013), which are in general agreement and which, to some extent, use the same sources. The data presented by Fowler (2013) in Table 16.9 indicate that the amounts of man-made fixed N and biologically fixed N (terrestrial) are roughly the same. In the absence of human activity the only sources of reactive nitrogen would be biological nitrogen fixation (BNF) by a wide range of organisms, and by lightning.

The annual actual human consumption of 33 Tg N (Table 16.6) represents 16% of the anthropogenic generated N; most of the remainder will be lost in the environment with small amounts contributing to soil organic matter.

Bouwman *et al.* (2009) carried out a detailed study of the global nitrogen and phosphorus soil balances for the period 1970–2050. They concluded that the total global nitrogen input to agricultural soils (arable crops and grassland) in 2000 amounted to 249 Tg N, of which fertilizer nitrogen accounted for 83 Tg,

manure 101 Tg, deposition from the atmosphere 35 Tg and BNF for 30 Tg. These values are roughly in line with those in Table 16.9.

Marine organisms are globally responsible for fixing very large amounts of N (140–177 Tg/year) (Voss *et al.*, 2013), which are partially balanced by oceanic denitrification variously estimated at 100–280 Tg N (Fowler *et al.*, 2013, Bouwman *et al.*, 2013).

N fixation in rice paddies (by the nitrogen-fixing cyanobacterium *Anabaena azollae* growing in symbiotic association with the fern *Azolla pinnata*) may account globally for 5 Tg N. Weathering of rocks may release some N but the amounts are likely to be small (Houlton and Morford, 2014). Nitrogen in rocks usually originates as organically bound

Table 16.9 Estimates of annual global amounts of anthropogenic and natural conversion of N_2 to reactive nitrogen in 2005

	Tg N/year
Agricultural BNF (Legumes)	60 ±30%[*]
Factory fixed N[+]	120 ±10%[****]
Combustion (transport and electricity)	30 ±10%[****]
Anthropogenic sources Total	*210* [****]
BNF terrestrial ecosystems	58 ±50% [**]
Lightning	5 ±50% [****]
BNF marine ecosystems	140 ±50% [***]
Natural sources Total	*203* [****]
Grand total	413 [****]

Sources:-
[*] Herridge *et al.* (2008) for agricultural crops and grazed savannas.
[**] Vitousek (2013) pre-industrial terrestrial. This range is much smaller than those of Cleveland *et al.* (1999) based on studies on 23 different natural ecosystems, which suggested pre-industrial BNF in the range 100–290 Tg N/year with tropical savannas, tropical rainforests and arid shrubs accounting for 60%.
[***] Voss *et al.* (2013).
[****] Fowler *et al.* (2013).
[+] about 80% used to manufacture fertilizers.

nitrogen associated with sediments (Morford *et al.*, 2011). Coal usually contains just over 1% N, which on combustion is lost to the atmosphere as NO_x. When coal is heated in the absence of air (as in the manufacture of coke or coal gas) ammonia is released and can be captured as ammonium sulphate.

Estimation of nitrogen fixed by cultivated legumes is reasonably straightforward, being made on the basis of crop areas, their nitrogen contents and estimates of how much plant N is derived from fixation. For soybeans Herridge *et al.* (2008) used values of 80% of crop N being derived from fixation in Brazil and Argentina and 60% in USA (thought to be associated with greater contribution of man-made N). About 60% of nitrogen in pulse crops is generally thought to be derived from symbiotic fixation. In many of the early studies the measurement of fixed nitrogen in roots was difficult and as a result the amounts of residual N values from legumes may have been underestimated. Herridge *et al.* (2008) made corrections for the nitrogen in roots of pasture and fodder legumes on the basis that 50% of the fixed nitrogen was retained below ground, for soybeans 33% and other oilseed legumes 30%.

When considering the contribution that BNF can make to the global N supply it is important firstly to take account of where the fixation takes place, secondly where a nitrogen-rich crop is consumed and thirdly the extent to which residual nitrogen remaining after a crop has been taken (e.g. beans for human or animal consumption) can be utilized by a following crop.

According to Herridge *et al.* (2008), the symbiotic fixation by oilseed legumes and by the pulses dominate biological fixation, globally fixing 18.5 and 2.9 Tg N/year respectively. The pulse-generated fixed N can be of great value in subsistence agriculture where pulses can be vital sources of protein for humans; however, globally only 5% of protein is derived from pulses (Boland, 2013). The situation with the oilseed legumes is more complicated because most of the soybeans, the dominant crop, are grown in three countries with the crop being sold for livestock feed to affluent countries where protein intakes are excessive. Thus the nitrogen fixed by soybeans is of little benefit to crops in the exporting country and may be detrimental to both human health and the environment in the importing countries.

Soybeans represent 50% of the global crop legume area and 68% of global production. Herridge *et al.* (2008) reported that soybeans annually fix 5.7, 4.6 and 3.4 Tg N in the USA, Brazil and Argentina respectively. The average fixation of N by all soybeans in the USA and Argentina appear to be 192 and 177 kg N/ha/year respectively.

Soybeans in the USA, Brazil and Argentina together comprise more than 90% of the global area. They also dominate the global export market with about 50% of Brazilian, 40% of American and 25% of Argentinian soya being exported. Together these three countries make a significant total contribution of 5.4 Tg N to several other countries, notably China and USA, and in Europe.

The amounts fixed by pasture and fodder legumes, by free-living bacteria and by organisms in association with other crops are much more difficult to assess largely due to uncertainties about areas, productivity and rates of fixation.

Recovery and utilization of N by crops

Only a fraction of the N supplied in fertilizer, manure or via legumes will be utilized by crops and find its way to humans. Assessments of the amounts involved can be determined experimentally (e.g. using ^{15}N) and although the experimental results will tend to be specific to the conditions of the experiment they do provide valuable information about how to improve the utilization of N by crops. An alternative approach is to calculate from FAO statistics the proportion of the N inputs that are found in the harvested crops at the national level; in this way the nitrogen use efficiency (NUE) values can be compared between countries and with time.

Uptake studies

Harris *et al.* (1994) compared the fate of labelled ^{15}N derived from red clover with that from $(NH_4)_2SO_4$ in the low-input and conventional cropping systems of the Rodale Farming Systems Trial. In the first year N uptake was measured in maize and in the second year in barley. More fertilizer N than legume N was recovered by the crops (40 vs. 17% of input); more legume N than fertilizer N remained in the soil (47 vs. 17% of input). The amounts of N from both sources that were lost to the environment were similar for the cropping systems (39% of input) over the two-year period. Similar results were reported in Australia by Ladd *et al.* (1983), who measured ^{15}N uptake by wheat following the incorporation of *Medicago littoralis* that had been allowed to decompose for a few months; only about 8% of the N in the medic was recovered in the harvested wheat.

Much has been learnt about the genetic control of nitrogen absorption, translocation and use. The International Service for the Acquisition of Agri-biotech Applications (ISAAA, 2014) lists the several genes (e.g. from tobacco, maize and rice) that have been found to be involved in improving nitrogen use efficiency by crops. Varieties of several crops with improved NUE are being bred in various countries, including maize (USA), wheat (Australia), rice (Colombia), canola (USA), sugar beet (Belgium) and sugar cane (South Africa), according to ISAAA (2014). The key attribute is the ability of the efficient varieties to yield well with much-reduced N applications – often 50% less than required by the less efficient varieties for the same yield.

Nitrogen use efficiency

There are compelling economic and environmental reasons for improving the efficiency of fertilizer N use and of other N inputs across the world. Efficient use of any N input depends on it being applied at the time and place that is optimal for crop uptake, which can be more easily achieved by using fertilizer N than manure or legume N.

Wide variations in the recoveries of fertilizer N in crops were reported by Christensen (2004), namely 11–42% for autumn applications and 42–78% for spring applications. The recovery of N from different manures also varies widely. Recovery from dung and slurries was lower, at 18–31%, whereas 61–88% of N from urine and poultry excreta was recovered.

Global situation

The use of fertilizer N has grown considerably since the Second World War. Conant *et al.* (2013) estimated the global use of fertilizer N applied to crops based on FAOSTAT data at 29, 79 and 82 Tg N/year in 1970, 1995 and 2000 respectively.

Raun and Johnson (1999) reported that 33% of global fertilizer N was recovered in harvested cereals (wheat, rice, maize, barley, sorghum, millet oat and rye) in 1996 (Table 16.10). Nitrogen use efficiencies in 1996 in developed and developing countries cereal were 42% and 29% respectively.

Nitrogen use efficiency (NUE) is

Table 16.10 World fertilizer N use on cereals, the removal of N in grain and the N removed in grain derived from soil and rain Tg N (Raun and Johnson, 1999)

	Tg N
A World fertilizer N use on cereals	49.7
B World cereal grain removal N	33.1
C N from soil and from rain	16.6
Nitrogen use efficiency = (B–C)/A × 100	

measured as the total N removed in crop minus the N coming from the soil plus N deposited in the rainfall as a percentage of fertilizer N applied. Estimates of NUE from green manures were low, e.g. at 20% or less as reported by Ladd *et al.* 1983) and by Harris *et al.* (1994)

Of the total nitrogen manufactured by the Haber–Bosch process, approximately 80% is used in the production of agricultural fertilizers. Unfortunately much of the fertilizer N is lost to the environment (see Chapter 17). It is likely that more nitrogen will be needed to support the growing global population, especially in regions of low crop yields. Although field experiments have shown that the proportion of added N that is recovered in the crop may decline with increasing N input (e.g. Cassman, 1999) the actual situation is one of great variability (Table 16.11 and Appendix). There thus seems to be considerable room for improving NUE, which is clearly a matter of some urgency.

In 2005 it was estimated that of the approximately 100 Tg fertilizer N used in globally, only 17 Tg N was consumed

Table 16.11   Nitrogen use efficiency by crops

Region/Country	Year	%	Source
Global	1963–1967	40	Conant *et al.* (2013)
OECD countries	2007	70	Conant *et al.* (2013)
Non–OECD countries	2007	54	Conant *et al.* (2013)
124 countries	1961	68	Lassaletta *et al.* (2014)
124 countries	2009	47	Lassaletta *et al.* (2014)
China	2007	29	Conant *et al.* (2013)
France	2000	36	Oenema *et al.* (2009)
Netherlands	2000	36	Oenema *et al.* (2009)
Germany	2000	42	Oenema *et al.* (2009)
Italy	2000	44	Oenema *et al.* (2009)

by humans in crop, dairy and meat products (UNEP, WHRC, Paris, 2007). This poor utilization figure was accompanied by a reduction in the global NUE for cereals from about 80% in 1960 to about 30% in 2000.

There are signs in Europe (notably for Belgium, Finland, Greece, Germany, Luxembourg and Turkey) that overall the fraction of nitrogen captured in crops increased from 51% to 59% between 1990 and 2004 (OECD, 2008).

Conant *et al.* (2013) calculated the amounts of N recovered in crops as a percentage of N supplied in fertilizer plus manure and legumes using FAOSTAT data. They investigated changes in the types and amounts of N inputs and how efficiently N inputs were used in the rich OECD (Organization for Economic Co-operation and Development) countries (basically Europe, North America Australia and New Zealand) and, the large emerging economies, namely (the BRICS – Brazil, Russia, India, China, and South Africa), and the developing countries in different parts of the world. In general they found that NUE was greater in countries where more fertilizer N was used. However, they also showed that there had been little change in NUE since the early 1960s and 2007 despite the large increases in yields, growth in N inputs, and substantial shifts in type of N inputs – away from manure and biologically fixed N and toward mineral fertilizer N.

Lassaletta *et al.* (2014) calculated NUE (as N in harvested product as a percentage of N applied from combined fertilizer, manure and symbiotic N fixation) for 124 countries over fifty years from 1961 based on FAO data (FAOSTAT). The 124 countries represented more than 99% of both the global population and cropland area. They found that in 2009 47% of the reactive nitrogen added globally on to cropland is found in the harvested products, compared to 68% in the early 1960s; this means that more than 50% of fertilizer N was lost into the

352 • *Sources of nitrogen for humans*

environment. During this period fertilizer N application increased by a factor
of nine. Their data also suggested that,
in general, those countries using a higher
proportion of N inputs from symbiotic
N fixation rather than from synthetic
fertilizer have a better N use efficiency;
lower yields would be expected in these
countries. Their calculations reveal
considerable country-to-country variation, with some countries improving
their agro-environmental performances,
while in others the increased fertilization has produced little agronomic
benefit and large environmental losses.
NUEs for selected countries are given
in the Appendix.

Smil (2011) calculated that when the
entire global food system down to the
consumer is considered the overall fertilizer nitrogen efficiency is no more
than 15%, much lower that the values
in Table 16.11. In the USA fertilizer
N use efficiency was thought to be no
better at about 12% (Howarth *et al.*,
2002). Earlier, Smil (2001) had reported
increases in NUE (of about 20%) by
wheat in England between 1985 and
1995, when N application was about 190
kg N/ha. Oenema *et al.* (2009) forecasted
improvements in NUE in Europe, which
is reassuring.

China

The pressure to maximize food production in China is intense in view of
the massive population and the limited
arable crop area; the arable land per
capita in China is 0.08 ha compared with

a world average of 0.20 ha in 2013 (WBI,
2013).

In view of the dominant role of China
as the major nitrogen fertilizer user,
NUE in China is of particular concern.
Very low overall NUE of only about 9%
(measured as the percentage of applied
N that reaches the home in food) was
reported by Ma *et al.* (2010) in China in
2005 in studies on the flow of N and P
from crops to animals to processing and
finally to the home. NUEs were estimated at 29% and 11% for crop production and animal production respectively.
The low overall efficiency of 9% is a
result of further inefficiencies during the
processing of food. All forms of new N
were included in the inputs, not only fertilizer N but also BNF and N in irrigation
water.

Only 4.4 Tg N reached households as
food from a total of 48.8 Tg N applied
(mostly 27 Tg fertilizer N, imported
animal feed 14.3 Tg, biological N fixation 4 Tg and atmospheric deposition
2 Tg). The remaining N was lost to the
environment, 23 Tg N to atmosphere
and 20 Tg N to drainage. Ammonia at
57% and denitrified gaseous nitrogen at
33% accounted for most of the atmospheric losses. NO_x and N_2O accounted
for 8% and 2% respectively.

The efficiency of P use showed a
similar trend. Of a total P input of
7.8 Tg, 36.5% was utilized by crops,
5% by livestock with 7% reaching
households.

Utilization of biologically fixed N

Successful use of biologically fixed nitrogen largely depends on providing the required amounts of available nitrogen when the crop needs it.

Nitrogen credits – USA

It was appreciated several years ago in the USA that lucerne was a significant source of N for following maize crops in the Corn Belt. Peterson and Russelle (1991) calculated from published data that fertilizer N applications could be reduced by up to 14% when maize was grown after lucerne. Since then work has been carried out in several states, including Iowa, Minnesota, Nebraska and New York, in order to determine precisely the actual benefits of the N fixed by lucerne to following crops of maize. This has been done by fertilizer experiments designed to determine the yield benefits associated with N from lucerne under different conditions. The amounts, which vary with soil conditions, the excellence and especially the vigour of the lucerne stand, are known as nitrogen credits. They are estimates of the amounts of fertilizer N that can be saved. The relatively low C:N ratios in lucerne crop residues favours their quick decomposition and the release of nitrogen for the next crop.

Extension workers in Wisconsin (2016) advise that 145, 180 and 210 kg N/ha can be supplied by poor, fair and good lucerne stands respectively when growing on loams; on sandy soils the benefits are reduced by 55 kg N/ha. Nitrogen credits for red clover are assessed at 80% of the lucerne credits. Considerable savings of N fertilizer can thus be achieved.

Soybean nitrogen credits for following crops are much smaller than those for lucerne. For example, Zoubek and Nygren (2016) indicate 28–50 kg N/ha according to soil type, a range that is in agreement with the credits advised in Wisconsin and New York. Good crops of peas and beans on loam soils are only likely to provide a credit of 25 kg N/ha and nothing of significance on sands or sandy loam soils (Wisconsin, 2016).

Experimental determination of nitrogen credits for other legumes, especially when grown as cover crops by organic farmers, should be promoted in other countries.

Recycling of organic matter

Recycling of organic matter of all kinds is a very important way of conserving N as well as other nutrients. Until the advent of fertilizers permanent agriculture was largely dependent on the recycling of organic matter. The previously practised shifting cultivation based on slash and burn was very wasteful of nitrogen, much of which was emitted during the burn as NO_x.

Cultivation of virgin soils by improving aeration encourages mineralization, a decline in soil organic matter and the release of N_2O and gaseous nitrogen. Replenishment of soil organic matter became essential. The ploughing in of many crop residues can, however, cause a temporary decline in available soil nitrogen. This is because many residues unfortunately have high C:N ratios (>50), which means that the mineralizing organisms have priority in using available soil N and so immobilizing it.

In developed countries manure is mostly used to recycle nutrients and improve soil structure but in several countries in Asia, notably India, and in Africa it also used as a fuel. The typical composition of manures, which can be very variable, is shown in Table 16.12 (Chambers *et al.*, 2001). More detailed and up-to-date information on the nutrient content of various manures in the UK is now available ahead of the production of a new Nutrient Management Guide (RB209) in 2017 (ADHB, 2016).

The livestock feed may be produced on the farm or it may be imported, as is often the case, for example in Europe when livestock are fed on soybeans and maize imported from North and South America.

The proportions of ingested nitrogen excreted in dung and urine varies from 75 to 85% for dairy cattle, from 65 to 75% for pigs and from 75 to 95% for adult humans (mostly in urine) (Smil, 2001).

Much of the variability in composition of manure is due to the losses, particularly of nitrogen by volatilization and leaching, that can occur before and after application; manure that is stored for three months after deposition in the cattle shed may lose half the N (Chambers *et al.*, 2001); however, storage for at least eight weeks is recommended in order to reduce the transmission of pathogens.

Quite large amounts of manure need to be applied in order to supply significant amounts of N; 17–25 t/ha cattle manure or 13–20 t/ha compost would be required to supply 100–150 kg N/ha. Some 17 tons of cattle manure could be produced by about 1.5 dairy cows in a year (housed for 60% of the time) or by about four beef cattle units. So a mixed farm with say 100 ha of arable crops on which N is applied as manure at 100 kg N/ha would require a dairy herd of 150 dairy cows or 400 beef cattle if the farm is to be self-sufficient (Defra, 2010; ADHB, 2015).

The geographic separation of livestock

Table 16.12 Typical mineral composition of manures and compost (fresh weight basis) (Chambers *et al.*, 2001)

	N %	Readily Available	P % Soluble	K % Soluble	Dry Matter %
Cattle Manure	0.6	10%	0.08	0.6	25
Pig Slurry	0.4	70%	0.04	0.2	4
Poultry litter	3	33%	0.65	1.3	60
Compost	0.75	3%	0.06	0.4	60

and arable farms can be a major factor that limits the availability of manure for crops and thus what can be achieved by recycling. It is only the manure that is produced when the animals are confined that can be used directly for the benefit of arable crops. The nutrients in manure and urine deposited by grazing cattle can only benefit an arable crop after the pasture has been brought under the plough.

Inadequacy of biological N fixation to feed the world

Assessments of where the nitrogen in human diet comes from are required when considering future world food production and whether organic agriculture can play a significant part in feeding the world or as is suggested can be the solution for long-term food sufficiency.

Various estimates have been made of the contribution made by man-made fertilizers to the global diet but it is important to note that these estimates hide major differences between regions.

Smil (1999, 2001) was probably the first to estimate the extent to which mankind was dependent on fertilizers when he calculated for the situation in the mid-1990s that synthetic nitrogenous fertilizers provided just over half of the nitrogen in the world's crops; the rest came from natural and symbiotic legume biological fixation (recycled via manures and crop residues) and atmospheric deposition. His estimates were based on FAO and IFA data on crop production, livestock and fertilizer statistics, their nitrogen contents and the flow of nitrogen through the food chain. He concluded that without the use of nitrogen fertilizers there would not be enough food for the diets of nearly 45% of the world's population, or roughly 3 billion people in the 1990s.

Prevailing diets in Western countries are excessively rich in protein, adequate in China, but inadequate in much of Africa. About 60% of all protein comes from crops, compared with 40% from livestock and fish (Boland, 2013). In China synthetic fertilizers already account for more than 60% of all nitrogen inputs (Ma *et al.*, 2010) as they do in India (Pathak *et al.*, 2010).

Unless there is a big reduction in the global consumption of livestock products more than half of the world's food production will continue to depend on the Haber–Bosch synthesis of ammonia; this share is expected to rise as the global population increases.

Smil calculated globally that to replace the synthetic nitrogen with biologically fixed nitrogen and livestock would require manure production by approximately 7–8 billion additional cattle, roughly a five-fold increase from the current number of 1.3 billion. The United States would have to accept nearly 1 billion additional animals and an added 2 billion acres of leguminous crops to feed the extra animals – an area equal to all the land in the USA except Alaska (Paarlberg, 2010). Clearly these numbers are unattainable.

Stewart *et al.* (2005) later arrived, via a different route, at similar numbers to Smil when they concluded that about

40–60% of food production was attributable to the use of fertilizers in the USA and UK and more than 80% in Brazil and Peru in the second and subsequent years after clearing. The cited increases in crop yield due to fertilizers had been determined in many countries; e.g. for maize, wheat, sorghum and potatoes in long-term experiments in Oklahoma (Magruder plots, from 1892), in Missouri (Sangorn field, from 1889), in Illinois (Morrow plots, from 1876), in Kansas (from 1961) and in the UK (Rothamsted, from 1843). Trials in Brazil and Peru had demonstrated the benefits of fertilizers on maize, rice, soybean and cowpea, although over shorter periods.

Erisman *et al.* (2008) calculated that by 2000, nitrogen fertilizers were responsible for feeding 44% of the world's population and that by 2008 the percentage had risen to 48% for a global population of 6.7 billion. This increasing dependence on fertilizer N will continue unless there is a massive increase in symbiotic and/ or free-living BNF. Such a development would require fundamental changes in forage and grain legume cultivation and consumption. Total legume production could conceivably be increased by two to three times. More forage legumes would be associated with more livestock, which may or may not be desirable. More grain legumes are not particularly attractive propositions except where they are consumed by humans as they only contribute relatively small amounts of fixed N for following crops. This is an unlikely prospect in view of the currently declining consumption of pulses across the world and particularly in India and China (Smil, 2001). The relatively low yields of pulses compared with cereals make them an unattractive proposition for farmers; their appeal to humans is restricted by the toxicity of some pulses and the need for special treatment to remove harmful anti-nutritional components before consuming.

The demand for man-made nitrogenous fertilizers would be much reduced if other crops were able to host nitrogen-fixing bacteria in their roots. Attempts are currently being made, for example at the John Innes Research Centre in the UK, to try to persuade cereal roots to form a symbiotic relation with nitrogen-fixing bacteria. The penalty, hopefully acceptable, would be reduced yields as a result of the bacterial demand for energy from the crop.

Constraints of organic agriculture associated with nitrogen supply

A key difference between conventional and organic systems concerns the control of and the amounts of N available at the time required by the crops. The most relevant factors are the timing and extent of the mineralization of manure and legume residues.

Claims are made that legume N fixation can be sufficient to replace the current use of N fertilizer. For example, Badgley *et al.* (2007) made the case based on published data comparing the yields of organic and conventional crops. Unfortunately much of the data relating to developing countries where the quoted organic yields were higher, by up to four times the conventional, came from studies where the improved yields followed training/extension programs and where fertilizers were not used (discussed in Chapter 13). Extrapolation to claims about legumes and global food production is problematic when based on the organic/conventional yield ratios from such studies.

The virtues of combining manure and fertilizer were examined by Kramer *et al.* (2002) in the sustainable agriculture farming systems study in Davis, California. They found that similar maize yields were achieved by combining the incorporation of a vetch cover crop containing 105 kg N/ha with 330 kg N from composted turkey manure as from 220 kg N as urea. The turkey manure was equivalent to about 47 t of cattle manure, an amount that cannot be supplied sustainably.

The problem of synchronizing the N supply from the mineralization of crop residues was also addressed by Kramer *et al.* (2002) in their study of the uptake of ^{15}N from vetch and fertilizer. They found that fertilizer N was immediately available for the maize early in the growing season whereas uptake of N from the vetch residues was not substantial until between sixty and seventy days after sowing (75–85 days after incorporation). The vetch supplied a sustained and satisfactory level of N later in the growing season. It is suggested that decreased rates of fertilizer N could be accompanied, later in the season, by the release of N from mineralized legume residues and

so reduce the use of fertilizers and possibly N losses also.

Equating fertilizer N with biological fixation of N by forage legumes and soybeans

There are two possible routes whereby biologically fixed nitrogen can substitute for fertilizer N. The first is the growing of forage legumes that can be consumed by livestock or grown as cover crops that are ploughed in. The second is by growing seed legumes, notably soybean, for animal feed followed by the use of manure on arable crops.

Table 16.8 shows the wide variation in fixation rates as determined experimentally, which makes it difficult to assess the amounts of N from a preceding legume crop that may become available for arable crops. However, the nitrogen credits from lucerne as determined in the USA for maize, although varying with conditions, have been found to be very useful and worthy of extension to other legumes and countries.

The areas of forage legumes and soybeans that equate with the amounts of fertilizer N consumed in 2013 have been calculated using annual fixation rates of 135 kg N/ha for forage legumes and 100 kg N/ha for soybeans. Smil (2001) had suggested 120–150 kg N/ha for forage legumes, which may be somewhat pessimistic: on the other hand, the values for soybeans are perhaps optimistic. The use of other values would, however, not materially affect the conclusions.

Forage legumes

The areas of forage legumes that equate with the current use of fertilizer N in several countries and regions are shown in Tables 16.13 and 16.14. It is apparent that the areas required are so massive that they are simply unattainable. In several European countries 30–60% of all agricultural land would be required just to generate via forage legumes the same amount of N as is applied in fertilizers. Agricultural land includes arable crops, temporary permanent pastures as well as permanent crops.

If allowance was made for the greater utilization of N from fertilizers than from legumes the requirement for legumes would be greater than calculated.

Soybeans

In Table 16.15 the areas of soybeans that equate with the current use of fertilizer N are given. In most of the countries listed in this example soybeans are not grown, which implies that the biologically fixed N would have to be imported. In theory more soybeans could be grown in many countries but, as with the forage legumes, the areas required area indicated in Table 16.15 are so massive that they are unattainable. For some countries in this example where relatively little fertilizer N is used, such as Czech Republic, Italy and Sweden, soybean N from the USA could in theory substitute for fertilizer N but this would require an increase in

Table 16.13   Approximate areas of forage legumes (million ha) that equate via N fixation at 135 kg/ha with the annual consumption of fertilizer of N (2013) in various countries (FAOSTAT, 2016)

	Fertilizer N consumed Tg	Forage Legumes Required	
		Area million ha	% agricultural land
Argentina	0.77	6	4
Austria	0.09	0.6	19
Brazil	3.59	27	10
China	24.10	179	35
Czech Rep.	0.33	2.4	57
France	2.00	15	52
Germany	1.67	12	72
India	16.73	124	69
Italy	0.60	4	29
New Zealand	0.30	2	20
Spain	0.96	7	26
Sweden	0.16	1.2	38
Turkey	1.58	12	31
UK	1.06	8	46
Ukraine	1.04	8	19
USA	12.01	89	22

Table 16.14 Approximate areas of forage legumes (million ha) that equate via N fixation at 135 kg/ha with the annual consumption of fertilizer of N (2013) in various regions (FAOSTAT, 2016)

Region	Fertilizer N consumed Tg	Forage Legumes million ha
Africa	3.88	29
Americas	23.44	174
Asia	55.90	414
Europe	14.55	108
Oceania	1.79	13

Table 16.15 Approximate areas of soybeans (million ha) that equate via N fixation at 100 kg/ha with the consumption of fertilizer N in 2013. Current areas in parentheses

	N consumed Tg	Equivalent Soybean million ha
China	24.10	241 (7)
Czech Rep.	0.33	3
France	2.00	20
Germany	1.67	17
India	16.73	167 (11)
Italy	0.60	6
New Zealand	0.30	3
Spain	0.96	10
Sweden	0.16	1
Turkey	1.58	16
UK	1.06	11

the soybean area in the USA by about 10 million ha.

The utilization of N derived from seed legumes such as soybeans is necessarily dependent on consumption by livestock notably cattle. This in turn would require livestock production being concentrated in feedlots in order to permit recycling of N in the manure to arable crops – a situation unlikely to be acceptable to the organic movements and also difficult to attain.

Global production of soybeans is concentrated in three countries with 33% global production in Brazil, 33% in the USA and 19% in Argentina. Globally more than 80% of soybeans are GM, which makes them unattractive to organic farmers.

The stark conclusion is that there is absolutely no way that BNF could ever support the world diets of today. For BNF to sustain mankind there would have to be radical reductions in global population or changes to our diets (to eat pulses rather than livestock products), or scientific breakthroughs on BNF such as persuading cereals to accept N-fixing organisms.

Phosphorus and potassium

Phosphorus and potassium are after nitrogen the nutrients that most often limit crop production but there are no reasons for thinking that organic farming affords a means of obtaining P and K in a way analogous to the biological fixation of N. Organic farmers abhor soluble P and K fertilizers on doctrinal grounds and only accept the use of insoluble ores or rocks. Microorganisms in the soil play an important part in the cycling of soil phosphorus. Organic forms of P usually account for 30–60% total soil

P and are made up of plant, animal and microbial residues. As is the case with N, the mineralization of organic P varies with the C:P ratio; when the C:P ratio exceeds 100 P will be immobilized by the microbes, which may contain 20–30% of the organic P (Haygarth et al., 2013).

The long-term supply of P fertilizers, which will be limited by the accessible ores (mostly of apatite formed in sedimentary deposits), poses a serious problem regarding sustainable food production in probably several hundred years. In contrast, there will be little problem with the supply of K due to its widespread occurrence in many minerals, such as feldspar, biotite, muscovite and some amphiboles found in many rock types; potassium is one of the ten most abundant elements in the Earth's crust.

In the case of P it is suggested, e.g. Haygarth et al. (2013), that there is probably sufficient P ore to make fertilizer for more than 100 years. Others have estimated longer periods. Most estimates are probably based on past use and on known P ore reserves and not on resources, which are two different things. On the other hand, the growing demand for fertilizers of all kinds will hasten the depletion of reserves. The absence of P fertilizers would not immediately restrict food production as there will be a long period before soil reserves run out.

The low-input agricultural systems that supported possibly only 1 billion people in the 19th century depended on a mixed farming system that applied P to arable crops by the recycling of P as manure that would have come from livestock grazing on other parts of the farm; it has been suggested that at least 20 ha of grazing land would have been required to support 1 ha arable crops. Some use would have been made of human excreta. Smit et al. (2009) in the Netherlands presented a very useful report on the global situation regarding the global P cycle, the production and use of P resources and the possibilities for efficient use and reuse of P.

Phosphorus use and utilization by crops

The benefits of P fertilizers on crop production as seen from their use have been most evident for many years in Europe and North America and more recently China, India and Brazil. In contrast, very little P is used in Africa (Table 16.16) and as a consequence crop production has been constrained by the nutrient known for having the ability to unlock soil fertility.

Table 16.16 Total use of phosphatic fertilizers in 2014 and average application rate per ha arable and permanent crop land in 2013 (FAOSTAT, 2016)

	Million t P_2O_5	kg/ha P_2O_5
Africa	1.5	5.6
Europe	3.5	12.0
Oceania	1.4	29.0
Americas	12.2	30.7
Asia	28.1	49.4
World	46.7	29.6

Table 16.17 Cereal yields related to P
application (Smit *et al.*, 2009)

	Total cereals '000 ha	Cereal Yields t/ha	Average P_2O_5 rate kg/ha
Oceania	19	1.2	18
Africa	105	1.4	4
Europe	121	3.3	14
Asia	327	3.5	39
N. America	78	5.9	25

A similar situation was revealed from data linking cereal (several combined) yields with the application rates of fertilizer P, which show very low yields in Africa (Table 16.17); the low application rates in Europe are probably associated with the long-term use and residual effects of P.

The efficiency with which phosphatic fertilizers are used is complicated by the fact that while only small amounts of P from a single application may be utilized there are likely to be considerable residual effects. Wide variation in P uptake in the year of application have been reported from 10–25% (Cooke, 1967), whereas according to Loehr (1974) only about 5–10% of fertilizer P that is added to the soil is taken up by the current crop. Manure is also a source of P, which may have been imported in animal feedingstuffs.

P processes in the soil are mainly controlled by inorganic adsorption and desorption reactions, and the activities of microorganisms. P inputs in natural ecosystems are almost exclusively derived from the weathering of sparingly soluble soil minerals, e.g. apatite in the soil parent material (Vitousek *et al.*, 2010); apatite occurs in granite and many metamorphic rocks. The amounts of P released by such weathering are normally insufficient to support crops. In agricultural systems extra P is needed and was initially supplied in guano, bones and manure and, from the mid-19th century, of superphosphate. In 2014 about 47 million t of phosphate (as P_2O_5) rock was used globally.

Losses of P from soils can occur by soil erosion and removal in excess water of recently applied manure or fertilizer. However, losses of P by leaching of soluble phosphate ions are limited (Haygarth *et al.*, 2013).

Removal of P in crops may be replaced on organic farms by the application of manure, compost or rock phosphate. The suggestion is made, e.g. Nguyen *et al.* (1995), that organic farmers have in fact been mining the soil reserves of slowly available P built up by generations of conventional farmers. A survey in England by Gosling and Shepherd (2005) supported the notion that organic mixed arable systems are mining reserves of phosphorus (and potassium), and that changes to organic management practices to increase inputs of phosphorus and of potassium will be required if long-term declines in soil fertility, and thus, yields are to be avoided.

Oehl *et al.* (2002) showed experimentally that farming organically could maintain an adequate level of available P for twenty-one years and also a satisfactory input–output budget but that organic yields were partly achieved at the expense of soil reserves of residual P from earlier fertilizer applications.

Improving the efficiency of use and reuse of P

Livestock

Animal production systems usually have a P surplus due to the use of cereals in animal feed. There needs to be a balance whereby P brought on to the farm in feed finds its way to arable crops; in the USA and Europe the concentration of livestock in units or regionally and distant to arable crops means that P use is inefficient.

Crops

Most P fertilizers are used on cereals (50%), with China accounting for 30% global use followed by India with 15% and USA 11% (Smit *et al.*, 2009). Insofar as about 30% of cereals across the world are destined for animal feed it is clear that reduction in consumption of animal products would limit the need for P fertilizers; in Europe 60% of cereals are used for animal feed (Smit *et al.*, 2009).

Human excreta

Daily consumption of phosphorus is typically about 3.4 g P_2O_5 per capita, which is well above recommended dietary intakes. Humans excrete 1.1–2.7 kg P_2O_5 /capita/year (globally about 7.5 million t P_2O_5/year) according to Liu

et al. (2008), who estimated that globally about 3.4 million t P_2O_5 per year is actually recycled. This is an amount that will decline as modern sanitation is installed. Such P could be recovered from sewage farms by the Ostara Nutrient Recovery process (see Chapter 12) to produce struvite by application of sewage sludge or of ashes derived from incinerating dried sewage, but the potential recoverable amounts would be small and restricted to countries with sewage farms. The presence of undesirable heavy metal contaminants in sewage poses a major limitation to its use.

Bones

There would appear to be a potential to recycle the P in animal bones as was done in the 19th century and for the production of superphosphate. Could this be extended to human bones in the distant future? A total of about 8 million t P_2O_5 is locked up in the bodies of 7 billion humans.

Mycorrhizae

Work by Gutjahr *et al.* (2009) has shown in dry-land rice, that mycorrhizal colonization can trigger the proliferation of lateral roots and so increase the volume of soil penetrated by roots. This is in addition to the phosphate absorbed by the fungus and translocated to the root cells. The hope is that the mycorrhiza could one day act as a bio-fertilizer that could supplement phosphatic fertilizers (see Chapter 19).

Phosphorus budget and reservoirs

Liu *et al.* (2008) and Smit *et al.* (2009) have estimated global phosphorus budgets that show the global input of P to cropland, in both inorganic and organic forms from various sources, is not sufficient to compensate for the removal in harvests, the losses by erosion and by runoff (Table 16.18). A net loss of phosphorus from the world's cropland is estimated at about 22 million t P_2O_5 each year.

Organic agriculture could play a vital part in conserving P if the loss via soil erosion was halted but it would only postpone the day of reckoning.

Liu *et al.* (2008) itemised the global reservoirs of P, drawing attention to the massive amounts of P in the oceans (62–1900 x10^6 million t P_2O_5), mostly in sediments. Gaining access to such sediments (mostly below 300m) and from the sea is not at present feasible. Eventually, after many millions of years and following tectonic movements, these sediments may be uplifted and exposed on the land surface. They may become low-quality sources of P compared with today's phosphate rock ores.

Living organisms in the sea contain 213,000 million t of P_2O_5 compared with terrestrial plants, 125 million t; terrestrial animals, 90 million t; and humans, 8 million t P_2O_5 (Smit *et al.*, 2009). Could there be a way of harnessing the ability of marine organisms (seaweed) to absorb P from the sea and so return the eroded P to the land?

Table 16.18 Global phosphorus budget for arable crops in 2004 (Liu *et al.*, 2008)

	Million t P_2O_5 per year
Inputs	52.5
Weathering	3.7
Atmospheric deposition	1.0
Synthetic fertilizers	33.7
Organic recycling *	14.1
Removals **	29.1
Losses ***	45.3
Balance	-21.9
Input shares (%)	
Fertilizer application	64.2
Organic recycling	27.1

* Recycling: Crop residues 5.0, Animal wastes 5.7, Human wastes 3.4.
** Removals: Crops 18.8, Crop residues 10.3.
*** Losses: Erosion 44.2, Runoff 1.1.

P and K in animal excreta

A significant proportion of the P and K taken up by plants from the soil will be returned in the form of manure and urine. Cattle are likely to return 75–85% of ingested P and K in excreta. The irregular and concentrated deposition of manure and urine by grazing cattle leads to greater leaching losses of NO_3 than would occur if the manure was applied uniformly.

Sheep droppings may supply 35 kg/ha/year P and grazing cattle may deposit 280 kg/ha/year P. In the case of cattle much of the forage consumed is likely to be grown locally but in the case of high-intensity pig and poultry systems, as in Europe, most of the animal feed will be imported. This can amount to massive amounts of P brought into pig and poultry farms, including organic.

Most of the K in manure is in the form of readily available K cations that behave as those derived from K fertilizers.

Long-term supply of phosphorus and potassium

The supplies of accessible ores that can be used to make phosphorus and potassium fertilizers are finite. There is no immediate likelihood of P reserves running out. The long-term supply will depend on currently unidentified/unquantified resources. Estimates suggesting 100 years or so for phosphorus are based on known reserves; potassium reserves are expected to last much longer than those for P. Accordingly relatively little attention is paid to solving the long-term supply problems. On the other hand, there are suggestions as to how the day when supplies run out can be postponed. For example, Withers *et al.* (2014) proposed four ways of limiting the need for P fertilizers, namely:

i. Investigate how to reduce the crop requirement for P by breeding.
ii. Develop ways to make better use of soil P, the legacy of long-term applications.
iii. Increase the use of recycled and recovered P.
iv. Develop more efficient P fertilizers and application methods to maximize uptake.

Future of P fertilizers

The fact that rock phosphate is a non-renewable P source means that improving the efficiency of P fertilizer use, of conserving soil P and of finding ways to mobilize insoluble P minerals are urgent matters.

Summary

1. Nitrogen plays a major role in crop nutrition. Crops are more often limited by N deficiency than by the shortage of other nutrients.
2. Nitrogen is the nutrient element after C, H and O that is required in greatest quantities by plants, animals and man particularly for the manufacture of proteins, which contain 16% N.
3. A 70 kg man contains about 1.7 kg N in about 11 kg protein.
4. In man proteins are needed regularly for growth and tissue repair. It is generally accepted that adults require about 0.8 g protein/kg body weight every day, although it is possible that less may in fact be adequate.
5. It is calculated that about 21 Tg N is the minimum annual global N requirement for a population of 7.2 billion. This is considerably less than the actual global consumption, which is estimated at about 33 Tg N; Asia accounts for more than 50%.
6. According to FAO, the average consumption of protein across the world (79g protein/capita/day) varies very widely from 103 g in developed countries to 74 g in developing countries. It would appear that average consumption

of protein in all continents exceed the absolute minimum daily requirement of 50–60 g.

7. If, as expected, global protein consumption increases in line with levels in the developed world the annual global N requirements would increase to about 45 Tg N.

8. The current global annual human consumption of N represents only about 15% of annual fertilizer N (110 Tg) plus biologically fixed N (118 Tg).

9. Absorbable N is essentially derived from two sources: biological N fixation and fertilizers. Supply is via manures, composts, fertilizers and from the breakdown of organic matter. Mineralization of soil organic matter not only provides N but also other essential nutrients.

10. The global amounts of man-made N fertilizers used are well documented. The estimates of biologically fixed N vary widely.

11. Significant amounts (globally 5 Tg/year) of N are probably fixed by *Azolla/Anabaena* in rice paddies.

12. The estimated amounts of N fixed by seed legumes vary widely, being dependent on many variables, and much of the fixed N is removed in the harvest.

13. Massive amounts of N can be regularly fixed by forage legumes. The contribution that such crops can make to the nitrogen requirement of a subsequent arable crop is difficult to assess. The organic farmer may not know how much N has been fixed by a previous forage legume crop.

14. Calculation of the nitrogen credits from previous lucerne crops can help conventional farmers in the USA better control their fertilizer N inputs, which is of no consequence to organic farmers who will not use fertilizer N.

15. Wide variations in recovery of fertilizer N in crops have been measured experimentally, for example from 11–42% for autumn applications and 42–78% for spring-time applications. The recovery of N from different manures also varies widely. Recovery from manure and slurries was lower, at 18–31%, whereas 61–88% of N from urine and poultry excreta may be utilized.

16. The calculation of the efficiency with which nitrogen is used (nitrogen use efficiency, NUE) has been made by assessing the output of N in harvested crops as a percentage of all N inputs including biological N fixation and manure. The finding of wide variations between countries indicates there is considerable scope for improving NUE.

17. The somewhat curious finding in countries where little fertilizer N is used that NUE can exceed 100% is worrying as it indicates that the soils are being degraded by the mineralization of organic matter to provide the N, albeit for low-yielding crops.

18. It is estimated that man-made N fertilizers are currently responsible for more than 50% of the nitrogen in the global diet and that without man-made fertilizer N a global population of more than 3 billion would be impossible to sustain.

19. The areas of forage legumes that would be required to replace fertilizer N are massive and simply unachievable.

20. Phosphorus and potassium are, after nitrogen, the nutrients that most often limit crop production. Crop yields are likely to be restricted unless soil fertility is unlocked by P.

21. Unlike N, the supplies of P and K ores are finite. It is inevitable that supplies

will run out – a problem that is more urgent for P than K.

22. The day of reckoning can be postponed by improving the efficiency of using all sources of P, maximising recycling and developing ways of making more use of insoluble minerals.

23. P fertilizers can have long-term residual effects resulting in organic farmers benefiting for many years from the P applied by conventional farmers before conversion to organic farming.

24. Little consideration has been given to global food production when P fertilizers will be in short supply, either at the local individual country level or internationally.

References

ADHB (2015) Agriculture and horticulture development board. Fertilizer handbook RB209. Available www.ahdb.org.uk/documents/rb209-fertiliser-manual-110412.pdf

ADHB (2016) Research Review No. 3110149017. Review of evidence on organic material nutrient supply. Available at www.ahdb.org.uk/documents/RB209_Review-WP2-OrganicMaterials.pdf

Adu-Gyamfi, J.J., Myaka, F.A., Sakala, W.D., Odgaard, R., Vesterager, J. M. and Høgh-Jensen, H. (2007) Biological nitrogen fixation and nitrogen and phosphorus budgets in farmer-managed intercrops of maize-pigeonpea in semi-arid southern and eastern Africa. *Plant and Soil* 295, 1, 127–136. doi: 10.1007/s11104-007-9270-0.

Atwater, W.O. (1884) On the assimilation of amospheric nitrogen by plants. *Rep.*

British Assoc. for the Advancement of Science. 54, 685.

Azolla Foundation. Biofertilizer. Available at http://theazollafoundation.org/azollas-uses/as-a-biofertilizer

Badgley, C., Moghtader J., Qunitero, E., Zakern, E., Chapell, J., Avile´s-Va´zquez, K., Samulon, A. and Perfecto, I. (2007) Organic agriculture and the global food supply. *Renewable Agriculture and Food Systems* 22, 2, 86–108.

Bindumadhava Rao, R.S., Krishnan, R.H., Theetharappan, T.S., Sankaranarayanan, R. and Venkatesan, G. (1966) A note on gliricidia shrubs. *Madras Agricultural Journal* 60, 17–22.

Boddey, R.M. (1995) Biological nitrogen fixation in sugar cane: a key to energetically viable biofuel production. *Critical Reviews in Plant Science* 14, 3, 263–279. doi: 10.1080/07352689509701929.

Boddey, R.M., de Oliveira, O.C., Urquiaga, S., Reis, V.M., de Olivares, F.L., Baldani, V.L.D. and Döbereiner, J. (1995) Biological nitrogen fixation associated with sugar cane and rice: Contributions and prospects for improvement. *Plant and Soil* 174, 1, 195–209. doi: 10.1007/BF00032247.

Boddey, R.M., de Moraes, J.C., Alves, B.J.R., and Urquiaga, S. (1997) The contribution of biological nitrogen fixation for sustainable agriculture in the tropics. *Soil Biology and Biochemistry* 29, 56, 787–799. doi.org/10.1016/S0038-0717(96)00221-0.

Boland, M. (2013) Global protein supply: the world's need for protein. Riddet Institute New Zealand. Available at www.riddet.ac.nz/sites/default/files/content/2013%20Protein%20supply%20Mike%20Boland.pdf

Bouwman, A.F., Beusen, A.H.W. and Billen, G. (2009) Human alteration of the global nitrogen and phosphorous soil balances for the period 1970–2050. *Global Biogeochemical Cycles* 23, 4. doi: 10.1029/2009GB003576.

Bouwman, L., Kees Klein Goldewijk, K.K., Van Der Hoek, K.W., Beusen, A.H.W., Van Vuuren, D.P., Willems, J., Rufino, M.C. and Stehfest, E. (2011) Exploring global changes in nitrogen and phosphorus cycles in agriculture induced by livestock production over the 1900–2050 period. *Proceedings of the National Academy of Science USA* 110, 52, 20882–20887. doi: 10.1073/pnas.1012878108.

Bouwman, A.F., Beusen, A.H.W., Griffioen, J., Van Groenigen, J.W., Hefting, M.M., Smilma, O., Van Puijenbroek, P.J.T.M., Seitzinger, S., Slomp, C.P. and Stehfest, E. (2013) Global trends and uncertainties in terrestrial denitrification and N$_2$O emissions. *Philosophical Transactions of the Royal Society London Series. B*, 368, 20130112. doi: 10.1098/rstb.2013.0112.

Bouwman, A.F., Butterbach-Bahl, K., Dentener, F., Stevenson, D., Amann, M. and Voss, M. (2013) The global nitrogen cycle in the twenty-first century. *Philosophical Transactions of the Royal Society B*, 368, 1621. 20130165. 2, doi:10.1098/rstb.2013.0164.

Cassman, K.G. and Harwood, R.R. (1995) The nature of agricultural systems – food security and environmental balance. *Food Policy* 20, 5, 439454. doi. org/10.1016/0306-9192(95)00037-F.

Cassman, K.G. (1999) Ecological intensification of cereal production systems: Yield potential, soil quality, and precision agriculture, *Proceedings National Academy of Science. USA*, 96, 11, 5952–5959. doi: 10.1073/pnas.96.11.5952.

Chambers, B., Nicholson, N., Smith, K., Pain, B., Cumby, T. and Scotford, I.

(2001). ADAS Booklet 1. Managing Livestock Manures Making better use of livestock manures on arable land. Available at www.rothamsted.ac.uk/sites/default/files/Booklet1ManuresArableLand.pdf

Christensen, B.T. (2004) Tightening the nitrogen cycle 47–67. In: Schjonning, P., Elmholt, S. and Christensen, B.T. (eds.) *Managing Soil Quality: Challenges in Modern Agriculture*, CABI International, Wallingford, UK.

CIA World Factbook (2014) Available at https://www.cia.gov/library/publications/download

Cleveland, C.C. Alan R. Townsend, A.R., Schimel, D.S., Fisher, H., Howarth, R.W., Hedin, L.O., Perakis, S.S., Latty, A.F., Von Fischer, J.C., Elseroad, A. and Wasson, MF. (1999) Global patterns of terrestrial biological nitrogen (N$_2$) fixation in natural ecosystems. *Global Biogeochemical Cycles* 13, 2, 623–645. doi: 10.1029/1999GB900014.

Cocking, E.C., Stone, P.J. and Davey, M.R. (2006) Intracellular colonization of roots of *Arabidopsis* and crop plants by *Gluconacetobacter diazotrophicus, In Vitro Cellular and Developmental Biology* – Plant, 42, 1,74–82. doi: 10.1079/IVP2005716.

Conant, R.T., Berdanier, A.B, and Grace, P.R. (2013) Patterns and trends in nitrogen use and nitrogen recovery efficiency in world agriculture. *Global Biogeochemical Cycles* 27, 2, 558–566. doi: 10.1002/gbc.20053.

Cooke, G.W. (1967) The control of soil fertility. Crosby Lockwood, London

Defra (2010) Fertilizer Manual (RB209) Available at www.ahdb.org.uk/documents/rb209-fertiliser-manual-110412.pdf

Dent, D. and Cocking, E. (2015) Establishment of cereal and non-legume crop symbiotic nitrogen fixation by

Gluconacetobacter diazotrophicus. 23rd North American Conference on Symbiotic Nitrogen Fixation. Ixtapa, Mexico.

Dobereiner, J., Reis, V.M., Paula, M.A. and Olivares, F.L. (1993) Endophytic diazotrophs in sugar cane, tuber plants and cereals 671–676. In: Palacios, R., Mora, J. and Newton, W.E. (eds). *New Horizons in Nitrogen Fixation. Proceedings of the 9th International Congress on Nitrogen Fixation, Cancún, Mexico, 1992. Current Plant Science and Biotechnology in Agriculture* 17 1993. Springer.

EFSA (2012) European Food Safety Authority, Parma, Italy. Scientific Opinion on Dietary Reference Values for protein *EFSA Journal* 10, 2, 2557. doi: 10.2903/j. efsa.2012.2557.

Erisman, J.W., Sutton, M.S., Galloway, J.N., Klimont, Z. and Winiwarter, W. (2008) How a century of ammonia synthesis changed the world. *Nature Geoscience*, 1–4, 636-639. doi: 10.1038/ngeo325.

Eskin, N, Vessey, K. and Tian, L. (2014) Research Progress and Perspectives of Nitrogen Fixing Bacterium, *Gluconacetobacter diazotrophicus, in Monocot Plants. International Journal of Agronomy* 2014. Article ID 208383. doi. org/10.1155/2014/208383.

Family Practice Notebook (2015) Weight Measurement in Children. Available at www.fpnotebook.com/endo/exam/ WghtMsrmntInChldrn.html

FAO (2012) Current world fertilizer trends and outlook to 2016. Available at ftp://ftp. fao.org/ag/agp/docs/cwfto16.pdf

FAOSTAT (2013) Food security. Available at http://faostat3.fao.org/search/ protein%20/E

FAOSTAT (2016) Available at http:// faostat3.fao.org/download/R/RF/E

FAO, IFAD and WFP (2013) The State of Food Insecurity in the World 2013. The multiple dimensions of food security. FAO, Rome.

Fowler, D., Coyle, M., Skiba, U., Sutton, M.A. Cape, J.N. Reis, S., Sheppard, L.J., Jenkins, A., Grizzetti, B., Galloway, J.N., Vitousek, P. Leach, A., Bouwman, A.F., Butterbach-Bahl, K., Dentener, F., Stevenson, D., Amann, M. and Voss, M. (2013) The global nitrogen cycle in the twenty-first century. *Philosophical Transactions of the Royal Society B*, 368, 1621. 20130165. 2, doi: 10.1098/rstb.2013.0164.

Gosling, P. and Shepherd, M. (2005) Long-term changes in soil fertility in organic arable farming systems in England, with particular reference to phosphorus and potassium. *Agriculture, Ecosystems and Environment* 105, 1–2, 425–432. Doi. org/10.1016/j.agee.2004/03.007.

Gutjahr, C., Casieri, L. and Paszkowski, U. (2009) *Glomus intraradices* induces changes in root system architecture of rice independently of common symbiosis signalling. *New Phytologist* 182, 4, 829–837. doi: 10.1111/j.1469-8137.2009.02839.x.

Harris, G.H., Hesterman, O.B., Paul, E.A., Peters, S.E. and Janke, R.R. (1994). Fate of legume and fertiliser ^{15}N in a long term cropping systems experiments. *Agronomy Journal* 86, 5, 910–915. doi: 10.2134/agronj 1994.00021962008600050028x

Haygarth, P.M., Bardgett, R.D. and Condron, L.M. (2013) Nitrogen and phosphorus cycles and their management 132-159. In: Gregory, P.J. and Nortcliff, S.N. (eds.) *Soil conditions and plant growth.* Wiley-Blackwell.

Hayatsu, M., Tago, K. and Saito, M. (2008) Various players in the nitrogen cycle: diversity and functions of the microorganisms involved in nitrification and denitrification. *Soil Science and Plant Nutrition* 54, 1, 33–45. doi: 10.1111/j.1747-0765.2007.00195.x.

Hellriegel, H. and Wilfarth, H. (1888) Examining the nitrogen nutrition of grasses and legumes. Untersuchungen über die Stickstoff-nahrung der Gramineen und Leguminosen. Kayssler & Co., Berlin, Germany. Available at https://archive.org/stream/untersuchungen00hell#page/n5/mode/2up

HGCA (2008) Organic arable farming – conversion options. Available at https://cereals.ahdb.org.uk/media/185287/g43-organic-arable-farming-conversion-options.pdf

Herridge, D.F., Peoples, M.B. and Boddey, R.M. (2008) Global inputs of biological nitrogen fixation in agricultural systems. *Plant and Soil* 311, 1, 1–18. doi: 10.1007/s11104-008-9668-3.

Hoefsloot, G., Termorshuizen A., Watt, D.A. and Cramer, M.D. (2005) Biological Nitrogen Fixation is not a Major Contributor to the Nitrogen Demand of a Commercially Grown South African Sugar cane Cultivar. *Plant and Soil* 277, 1, 85–96. doi: 10.1007/s11104-005-2581-0.

Houlton, B.Z. and. Morford, S.L. (2014) A new synthesis for terrestrial nitrogen inputs. *Soil* 1, 381–497. doi: 10.5194/soil-1-381-2015.

Howarth, R.W., Boyer, E.W., Pabich, W.J. and Galloway, J.N. (2002) Nitrogen use in the United States from 1961–2000 and potential future trends. *Ambio* 31, 2, 88–96.

ISAAA (2014) Nitrogen Use Efficient Biotech Crops (No. 46) Available at www.isaaa.org/resources/publications/pocketk/foldable/Pocket%20K46%20(English).pdf

Jia Ssu Hsieh (504) The Art of Feeding the People. In: Peters, G.A. and Calvert H.E. (1983).The *Azolla-Anabaena azollae*

symbiosis. In: *Algal symbiosis.* In: Goff. L.J. (ed.) Cambridge University Press.

Jones, D.L. (1993) Rhizosphere carbon flow. D.Phil thesis Oxford University UK.

Kang, B.T. and Mulongoy, K. (1987) *Gliricidia sepium* as a source of green manure in an alley cropping system. Special Publication 87, 1, 44–49. In: Withington, D., Glover, N. and Brewbaker, J.L. (eds), *Gliricidia sepium. Management and Improvement*, Proceedings of a workshop at CATIE, Turialba, Costa Rica. NFTA.

Kramer, A.W., Doane, T.A., Horwath, W.R., and van Kessel, C. (2002) Combining fertiliser and organic inputs to synchronize N supply in alternative cropping systems in California. *Agriculture, Ecosystems and Environment* 91, 233–243.

Ladd, J.N., Amato, M., Jackson, R.B. and Butler, J.H.A. (1983) Utilization by wheat crops of nitrogen from legume residues decomposing in soils in the field. *Soil Biology and Biochemistry* 15, 3, 231-238. doi.org/10.1016/0038-0717(83)90064-0.

Lassaletta, L, Billen, G., Grizzetti B., Anglade, J. and Garnier, J. (2014) 50 year trends in nitrogen use efficiency of world cropping systems: the relationship between yield and nitrogen input to cropland. *Environmental Research Letters* 9, 10. doi: 10.1088/1748-9326/9/10/105011.

Liu, Y., Villalba, G., Ayres, R.U. and Schroder, H. (2008) Global phosphorus flows and environmental impacts from a consumption perspective. *Journal of Industrial Ecology* 12, 2, 229–247. doi: 10.1111/j.1530-9290.2008.00025.x.

Loehr, R.C. (1974) Characteristics and comparative magnitude of non point sources *Journal Water Pollution Control Federation.* 46, 8, 1849–1872.

Ma, L, Ma, W.Q., Velthof, G.I., Wang, F.H., Qin, W., Zhang, S. and Oenema, O. (2010). Modeling nutrient flows in the food chain of China. *Journal of Environmental Quality*, 39, 4, 1279–1289. doi: 10.2134/jeq2009.0403.

Marschner, H. (1986) *Mineral Nutrition of Higher Plants*. Academic Press, London.

Meissen, R. (2015) Move over Arabidopsis, there's a new model plant in town. University of Missouri Bond Life Sciences Center. http://bondlsc.missouri.edu/2015/05/28/move-over-arabidopsis-theres-a-new-model-plant-in-town

Montañez, A. (2000) Overview and Case studies on Biological Nitrogen Fixation: Perspectives and Limitations. FAO. Available at www.fao.org/fileadmin/templates/agphome/scpi/SCPI_Compendium/Overview_and_Case_studies_on_Biological_Nitrogen_Fixation.pdf

Morford, S.L., Houlton, B.Z., and Dahlgren, R.A. (2011) Increased forest ecosystem carbon and nitrogen storage from nitrogen rich bedrock, *Nature*, 477, 78–81. doi: 10.1038/nature10415.

Nguyen, M.L., Haynes, R.J. and Goh, K.M. (1995) Nutrient budgets and status in three pairs of conventional and alternative mixed cropping farms in Canterbury, New Zealand. *Agriculture. Ecosystems and Environment* 52, 2–3,149–162. doi: 10.1016/0167-8809(94)00544-O.

OECD (2008). Environmental performance of agriculture in OECD countries since 1990. Organization for Economic Co-operation and Development, Paris. Available at www.oecd.org/tad/sustainable-agriculture/44254899.pdf

Oehl, F., Oberson, A., Tagmann, H.U., Besson, J.M., Dubois, D., Mäder, P., Roth, R.R. and Frossard, E. (2002, Phosphorus budget and phosphorus availability in soils under organic and conventional farming. *Nutrient Cycling in Agroecosystems* 62, 1, 25–35. doi: 10.1023/A:1015195023724.

Oenema, O., Witzke, H.P. Klimont, Z., Lesschen, J.P. and Velthof, G.L. (2009) Integrated assessment of promising measures to decrease nitrogen losses from agriculture in EU-27. *Agriculture, Ecosystems and Environment* 133, 3–4, 280–288. doi.org/10.1016/j.agee.2009.04.025.

Paarlberg, R. (2010) *Food Politics: What Everyone Needs to Know*, Oxford University Press, UK.

Pabby, A., Prasanna, R. and Singh, P.K. (2003) *Azolla–Anabaena* Symbiosis – From traditional agriculture to biotechnology. *Indian Journal of Biotechnology* 2, 26–37. Available at http://nopr.niscair.res.in/bitstream/123456789/11286/1/IJBT%202(1)%2026-37.pdf

Pau Vall, M. and Vidal, C. (1999) Nitrogen in agriculture. Agriculture and the Environment. European Commission 167 to 180. In: Agriculture, environment, rural development: facts and figures – A challenge for Agriculture, Office for Official Publications of the European Communities.

Pankievicz, V.C.S., do Amaral, F.P., Santos, K.F.D.N., Agtuca, B., Xu, Y., Schueller, M.J., Arisi, A.C.M., Steffens, Maria. B.R., de Souza, E.M., Pedrosa, F. O., Stacey, G. and Ferrieri, R.A. (2015) Robust biological nitrogen fixation in a model grass–bacterial association. *The Plant Journal* 81, 6, 907–919. doi: 10.1111/tpj.12777.

Pathak, H., Mohanty, S., Jain, N. and Bhatia, A. (2010) Nitrogen, phosphorus, and potassium in Indian agriculture. *Nutrient Cycling in Agroecosystems* 86, 3, 287–299. doi: 10.1007/s10705-009-9292-5.

Patil, B.P. (1989) Cut down fertilizer

nitrogen need of rice by *Gliricidia* green manure. *Indian Farming* 39, 34–35.

Peterson, T.A. and Russelle, M.P. (1991) Alfalfa and the nitrogen cycle in the Corn Belt. *Journal of Soil and Water Conservation* 46, 3, 229–235. Available at www.jswconline.org/content/46/3/229. abstract

Pimentel, D., Wilson, C., McCullum, C., Huang, R., Dwen, P., Flack, J., Tran, Q., Saltman, T. and Cliff, B. (1997) Economic and Environmental Benefits of Biodiversity. *BioScience* 47, 11, 747–757.

Raun, W.R. and Johnson, G.V. (1999) Improving nitrogen use efficiency for cereal production. *Agronomy Journal* 91, 357–363. Available at www.plantstress.com/articles/min_deficiency_m/nue.pdf

Smil, V. (1999) Nitrogen in crop production: an account of global flows. *Global Biogeochemical Cycles,* 13, 2, 647–662. doi: 10.1029/1999GB900015.

Smil, V. (2001) Enriching the Earth: Fritz Haber, Carl Bosch and the Transformation of World Food Production. MIT Press, Cambridge, Massachusetts.

Smil, V. (2011) Nitrogen cycle and world food production. *World Agriculture* 2, 9–15.

Smit, A.L., Bindraban, P.S., Schroder, J.J. Conjiin, J.G. and van der Meer, H.G. (2009) Phosphorus in Agriculture:Global Resources, Trends and Developments, Report 282. Plant Research.International, Wageningen. Available at http://edepot.wur.nl/12571

Snyder, W.S., Cook, M.J., Nasset, E.S., Karhausen, L.R., Howells, G.P. and, Tipton. I, H. (1975) Report of the Task Group on Reference Man. ICRP Publication 23. Pergamon Press, Oxford.

Sparkes, J. (2014) Making nutrient management easier. MANNER-NPK.

Available at www.adas.uk/News/making-nutrient-management-easier

Stewart, W.M., Dibb, D.W., Johnston, A.E. & Smyth, T.J. (2005) The contribution of commercial fertilizers to food production. *Agronomy Journal* 97, 1–6. doi: 10.2134/agronj2005.0001.

Tadra-Sfeir, M.Z., Souza, E.M., Faoro, H., Müller-Santos, M., Baura, V.A., Tuleski, T.R., Rigo, L.U., Yates, M.G., Wassem, R., Pedrosa, F.O. and Monteiro, R.A. (2011) Naringenin Regulates Expression of Genes Involved in Cell Wall Synthesis in *Herbaspirillum seropedicae. Applied and Environmental Microbiology* 77, 2180–2183. doi: 10.1128/AEM.02071-10.

Thomas, R.J. and Schrader, L.E. (1981) The assimilation of ureides in shoot tissues of soybeans 1. Changes in allantoinase activity and ureide contents of leaves and fruits. *Plant Physiology* 67, 5, 973–976.

UNEP, WHRC, Paris (2007) Reactive Nitrogen in the Environment: Too Much or Too Little of a Good Thing. UN Environmental programme, Paris, 2007

Vaishampayan A., Sinha, R.P., Hader, D.P., Dey, T., Gupta, A.K., Bhan, U. and Rao, A.L. (2001) Cyanobacterial bio-fertilizers in rice agriculture. *Botanical Review* 67, 453–516. doi: 10.1007/BF02857893.

Vitousek., P.M., Porder, S., Houton, B.Z. and Chadwick, O.A. (2010) Terrestrial phosphorus limitations, mechanisms, implications and nitrogen-phosphorus interactions. *Ecological Applications* 20, 1, 5–15. doi: 10.1890/08-0127.1.

Vitousek, P.M., Menge, D.N.L., Reed, S.C. and Cleveland, C.C. (2013) Biological nitrogen fixation: Rates, patterns, and ecological controls in terrestrial ecosystems. *Philosophical Transactions of the Royal*

Society London series B, 368. 1621.
doi: 10.1098/rstb.2013.0119.

Voss, M., Bange, H.W., Dippner,
J.W., Middelburg, J.J., Montoya,
J.P. and Ward, B. (2013) The marine
nitrogen cycle: Recent discoveries,
uncertainties and the potential relevance
of climate change. *Philosophical
Transactions of the Royal Society London
series B*, 368, 1621. doi: 10.1098/
rstb.2013.0121.

WAC (2015) World Agroforestry Center.
Faidherbia albida, Keystone of evergreen
agriculture in Africa. Available at www.
worldagroforestry.org/sites/default/
files/F.a_keystone_of_Ev_Ag.pdf

Walpole, S.C., Prieto-Merino, D.,
Edwards. P., Cleland, J., Stevens, G.
and Roberts, I. (2012) The weight of
nations: an estimation of adult human
biomass. *BMC Public Health* 12, 439. doi:
10.1186/1471-2458-12-439.

WBI (2013) World Bank Institute. World
development indicators. Available at
http://data.worldbank.org/indicator/
AG.LND.ARBL.HA.PC

WHO (2013) Global and regional food
consumption patterns and trends Available
at www.who.int/nutrition/topics/3_
foodconsumption/en/index4.html

Wisconsin (2016) Legume and manure
credits. Available at www.soils.wisc.edu/
extension/materials/CCA_Legume_
Manure_Credits.pdf

Withers, P.J., Sylvester-Bradley, R., Jones,
D.L., Healey, J.R. and Talboys, P.J. (2014)
Feed the crop not the soil: rethinking
phosphorus management in the food chain.
Environmental Science and Technology 17,
48, 12, 6523–6530. doi: 10.1021/es501670j.

Zoubek, G. and Nygren, A. (2016) University
of Nebraska Giving Proper Nitrogen Credit
for Legumes in Corn and Milo Rotations.
Available at http://cropwatch.unl.edu/
giving-proper-nitrogen-credit-legumes-
corn-and-milo-rotations

CHAPTER 17

Behaviour of Nitrogen in the Soil and Environment and the Contribution of Nitrous Oxide and Methane to Greenhouse Gases

"Emissions of nitrous oxide and methane are not laughing matters"

A question of fundamental importance when comparing organic and conventional farming is: "Does N that originates from organic matter behave differently in the soil and environment from that derived from fertilizers?" As far as plant roots are concerned there are no reasons for thinking that the nature or properties of NO_3 anions varies with their origin. Observed differences between organic and conventional farming regarding N effects on yield are likely to be related to the amounts and availability of N and not to its origin. Likewise it would not be expected that soil microroganisms would be able to differentiate between reactive N according to source.

Supply of absorbable nitrogen

Nitrogen is absorbed by most crops as the soluble and readily leached NO_3 anion and to a lesser extent as the NH_4 cation, which can be bound in the expandable inner layers of clay minerals where it takes the place of cations such as Ca and Mg. Such mineral-bound NH_4, which is not readily mobilized, chemically or biologically, may account for a few kg to more than 1 t N/ha (Stevenson, 1986).

Most of the N in soils is held in organic matter, e.g. a UK survey showed that inorganic N in topsoils amounted to only 76 kg N/ha compared with 7,000 kg N/ha in organic matter (Shepherd *et al.*, 1996).

The key biological processes that control the provision of N from soil organic matter are mineralization, immobilization, nitrification and denitrification.

Mineralization

Mineralization is the process whereby soil microorganisms break down organic matter and convert it into inorganic forms that can be absorbed by plants. Most of the N (96–98%) in dead organic matter occurs as complex insoluble organic compounds such as proteins, nucleic acids and chitin, which are too large to pass through microbial membranes. Soil microorganisms produce enzymes that break down the organic matter into smaller water soluble compounds that they can absorb. This material, called dissolved organic N, normally constitutes a large proportion of the total

soluble N pool in natural ecosystems. Even in fertilized grassland soils dissolved organic N concentrations can be equal to or greater than inorganic N concentrations (Bardgett *et al.*, 2003).

Most of the dissolved organic N (much as amino acids, Bardgett *et al.*, 2003) can be utilized by free-living soil microorganisms that make use of both the C and N. When the microorganisms have insufficient C from other sources they use the C in dissolved organic N and excrete NH_4 as a waste product, i.e. organic matter is mineralized. When dissolved organic N is insufficient to meet the N requirements of the microorganisms they will then absorb additional inorganic N (NO_3 or NH_4) from the soil solution, i.e. N will be immobilized and the microbial biomass will effectively reduce the amount of N available for crops.

There is a critical substrate C:N ratio of about 30:1, which is needed to meet the microbial needs for N. When the C:N ratio exceeds 30:1, which can be the case with many plant residues, the microorganisms immobilize inorganic N to the detriment of the crop (Kaye and Hart, 1997). The incorporation of straw into soils with a high C:N ratio can immobilize much NO_3 in the autumn whereas oilseed rape residues can contain sufficient N to facilitate rapid mineralization.

Any dissolved organic N compounds not utilized by microbes may be leached or possibly absorbed by plants as amino acids or peptides, although this is thought to be of litle significance with respect to crop N nutrition in most soils due to the rapid microbial turnover of dissolved organic N (Bardgett *et al.*, 2003).

Nitrification

NH_4 can be absorbed by plants but the major part is converted in two steps to NO_3 by the process known as nitrification. NH_4 is oxidised firstly to nitrite NO_2 (by *Nitrosomonas*) and then to NO_3 (by *Nitrobacter*). Nitrification rates tend to be highest in heavily fertilized agricultural soils. Although nitrification is optimal at 30–35° C it also operates in cold but not frozen soils (Brooks *et al.*, 1996) and can thus lead to the production of leachable NO_3 in the winter. Nitrification can be delayed by certain chemicals that specifically inhibit *Nitrosomonas*. The objective is to try to maintain the applied ammonium nitrogen in its applied form and so minimize the leaching of NO_3 and the subsequent generation of N_2O. The cost of stabilized fertilizers containing nitrification inhibitors are significantly higher than those of conventional fertilizers, which has restricted their use to high value horticultural crops. Likewise the development and use of slow release N fertilizers, for example by coating or by encapsulation, has been limited by high costs of production.

Denitrification

Unlike nitrification, which is dominated by two groups of bacteria, many soil bacteria are able to reduce NO_3 to gaseous NO, N_2O and N_2. The process is enhanced at low oxygen levels, such as in wet soils. Zero tillage can also lead to increased N_2O emissions, as demonstrated by Rochette (2008) on poorly aerated soils. It is easy to forget that nitrogen compounds both terrestrial and aquatic are denitrified and gaseous nitrogen emitted. Oceanic denitrification is estimated at 100–280 Tg/year (Fowler *et al.*, 2013). A study by Galloway *et al.* (2004) indicated that approximately 40% of fertilizer nitrogen lost to the environment in different forms is denitrified back to unreactive atmospheric dinitrogen.

The processes of mineralization, nitrification and denitrification continue throughout the year, regardless of whether a crop is present, and result in losses of NO_3 to drainage and gases to the atmosphere. Such losses will occur between the ploughing in of manure or cover crops and when the establishing crop can make use of the mineralized and nitrified N. In contrast, application of fertilizer N can be better synchronized with the time of crop uptake and at the same time can provide the large amounts of N needed by annual crops, usually over a relatively short period.

Reactive nitrogen

The fundamental human and crop needs for nitrogen mean that globally large amounts of reactive nitrogen (Nr) are required. The molecular form (N_2) is usable by only a small number of organisms. Reactive nitrogen includes all the biologically and chemically active nitrogen compounds in the atmosphere and biosphere. It includes forms of nitrogen, such as ammonia (NH_3) and ammonium (NH_4), nitric oxide (NO), nitrogen dioxide (NO_2), nitric acid (HNO_3), nitrous oxide (N_2O), and nitrate (NO_3), and organic compounds such as urea, amines, amino acids, proteins and nucleic acids. Generally only a small proportion of Nr that crops absorb, mostly as nitrate, reaches human food, which leaves massive amounts of Nr to escape into the environment; Nr can have adverse effects on human health and contribute to greenhouse gas emissions. Nr compounds generally move easily between the air, soil and water.

The problems of the detrimental environmental effects of Nr are exacerbated by human activities other than agriculture, namely power generation and transport that also produce Nr.

Leach *et al.* (2012) calculated for thirty-eight countries (based on assessments of resource consumption) the human N footprint, which is the amount per capita of Nr released into the environment annually as a result of the consumption of food and the use of energy. Values varied from around 20 kg Nr/capita/year (Finland, Israel, Denmark, New Zealand, Belgium, Australia, the Netherlands, Germany, France and Italy) to around 30 kg Nr (UK, Belgium, Austria, Canada, Spain and Norway) and to about 40 kg N (USA, Brazil, China and South Africa). Food consumption was the major component, accounting for about 70% of the Nr in the USA and 90% in the Netherlands. The losses of Nr into the environment are associated with the annual consumption of about 5 kg N per capita (USA 5.2 kg and the Netherlands 4.6 kg N).

Premature deaths, agriculture and pollution

Lelieveld *et al.* (2015) assessed for 2010 the global burden of diseases related to pollution based on epidemiological studies connecting premature mortality to a wide range of causes, including the long-term health impacts of seven emission source categories, one of which was agriculture. Other sources were natural, industry, land traffic, residential energy, power generation and biomass burning (Table 17.1). The effects of ozone and of fine particulate matter with a diameter smaller than 2.5 micron ($PM_{2.5}$) were of prime interest. They used a global atmospheric chemistry model that related air quality with health statistics country by country. They calculated that outdoor air pollution, mostly by $PM_{2.5}$, leads globally to 3.3 (1.6–4.8) million premature deaths each year, mostly in Asia, and mostly due to cerebrovascular and ischaemic heart diseases.

The sources of the most harmful pollutants vary between regions. In India and China it is the emissions from residential heating and cooking that have the largest impact, whereas in the USA and most developed countries it is the agricultural emissions and the emissions from traffic and power generation that make the largest contribution to premature deaths caused by $PM_{2.5}$ (Table 17.1). It is likely that the toxicity of the particles varies with composition; for example it is thought that carbonaceous particles are far more harmful than those of ammonium nitrate and ammonium sulphate formed in the atmosphere following the release of NH_3 from farming and NO_x and SO_2 from other activities into the atmosphere. Such water soluble compounds will be readily washed out of the air. It is unfortunate that in India there is a soaring demand by city dwellers for cow dung patties for domestic fuel, a demand stoked by a desire to reconnect with their former rural life (*The Times*, 29 December, 2015); some sales are via Amazon India.

Although globally agriculture accounts for fewer premature deaths (20% of total) related mostly to particulate matter than combined energy use in the home, power generation and traffic (50%), there is much room for improvement (Table 17.1). The annual number of premature deaths associated with agriculture is unacceptable, e.g. China about 400,000, India 40,000, USA and Germany more than 15,000, France and UK about 7,000.

In marked contrast to the numbers in Table 17.1, the WHO estimated that globally in 2012 there were about 7 million premature deaths related to all air pollution (WHO, 2014a); 3.7 million were associated with outdoor pollution and 4.3 million with indoor pollution (the people who are exposed to both indoor and outdoor air pollution account for the 1 million overlap). Indoor smoke is clearly a serious health risk for some 3 billion people who cook and heat their homes with biomass fuels and coal.

It would appear globally that premature deaths related to agricultural activities are much less than those caused by indoor pollution.

The WHO (2014a) estimated that some 80% of outdoor air pollution-related premature deaths were due to ischaemic heart disease and strokes combined, while 14% were due to chronic

Table 17.1 Total premature deaths and % attributed to different air pollutant sources in 2010. (Lelieveld *et al.*, 2015)

Country	Total million	Agriculture	Residential energy	Traffic + Power generation	All other[*]
China	1.357	29	32	21	18
India	0.644	6	50	19	25
USA	0.055	29	6	52	13
Germany	0.034	45	8	33	14
Japan	0.025	38	12	27	23
Italy	0.021	39	7	32	22
France	0.018	41	14	30	15
UK	0.015	48	5	36	11
World	3.297	20	30	19	31

[*] Natural, Industry and biomass burning.

obstructive pulmonary disease or acute lower respiratory infections and 6% to lung cancer. These deaths are mainly the result of exposure to small particulate matter of 10 microns or less in diameter (PM_{10}), which can cause cardiovascular and respiratory diseases and cancers especially of the lung. Indoor air pollution was responsible for fewer cardiovascular but more respiratory problems.

Particulate matter affects more people than any other pollutant (WHO, 2014). The major components of particulate matter are sulphates, nitrates, ammonia, sodium chloride, black carbon, mineral dust and organic matter. It consists of a complex mixture of solid and liquid particles of organic and inorganic substances suspended in the air. Nanoparticles of magnetite identical with those in air have been found in high concentrations in brains of people living in highly polluted areas; they are being investigated as possible links to neurodegenerative diseases, including Alzheimer's. The most health-damaging particles are those that can penetrate and lodge deep inside the lungs. Small particulate pollutants can have health impacts even at very low concentrations – no threshold has been identified below which no damage to health is observed. The WHO 2005 guideline limits were aimed to achieve the lowest concentrations possible, namely annual mean values < 20 μg/m^3 for PM_{10}, and <25 μg/m^3 for $PM_{2.5}$.

The assessment by the WHO (2014) that in 2014, 92% of the world population was living in places where the WHO air quality guidelines levels were not met indicates that there is much scope for improvement. However, confirmatory data linking the number of premature deaths to pollution would seem to be required.

Release of reactive nitrogen into the environment

The reactive nitrogen compounds of particular concern are NO_3, N_2O, NH_3 and NO_x, of which NO_3 is vital for crop production. Methane (CH_4) will also be considered with Nr because agriculture is a major source and because it is a very important greenhouse gas (with twenty-five times the warming potential of CO_2). As nitrogen oxides (NO_x and NO_2) are mostly derived from fossil fuel combustion (for transportation and electricity production) they are of minor concern with regard to agriculture.

Considerable amounts of Nr, which constitute a large proportion of the total N in circulation/inputs, are involved. For example, Bouwman *et al.* (2013) in their studies on the annual global nitrogen and phosphorus soil balances concluded that of the total global input, to agricultural soils, of 248 Tg N from all sources (of which fertilizer N was 83 Tg), 110 Tg is recovered in harvested crops and grassland, leaving a balance of 138 Tg N that is lost to the environment (ammonia volatilization 34 Tg N, denitrification 47 Tg N and leaching 57 Tg N). On this basis 56% of the total N input is lost to the environment at some point in the food chain.

There is growing concern about the adverse environmental effects of Nr derived from agriculture, industry and human activities, which has acquired the title the "nitrogen problem or crisis". In Europe concern about the nitrogen problem was sufficient to generate a valuable multi-authored, 664 page report entitled The European Nitrogen Assessment (Sutton *et al.*, 2011). The report reviewed the current scientific understanding of reactive nitrogen compounds, their impacts and interactions in order to provide a basis for local and global policies. Water quality, air quality, greenhouse gas balance, ecosystems, biodiversity and soil quality were the major topics reviewed.

When crops depended on biological nitrogen fixation and the recycling of animal and human wastes Nr were scarce in the environment but the situation changed dramatically following the adoption of the Haber–Bosch process to fix atmospheric N, which can be considered as perhaps the greatest single experiment in global geo-engineering that humans have ever made (Sutton *et al.*, 2011).

The rate at which man puts Nr into the environment is continuing to increase in line with agricultural production. Sutton *et al.* (2013), in a recent UN report, call for a major global rethink on how fertilizers are used across the world, so that more food and energy can be produced while pollution is reduced. Greater precision in the application and use of fertilizers and manures is called for. Not only will a growing world population require more food but increasing affluence will be accompanied by demands for more livestock products. More N fertilizers will be required to satisfy this demand, leading to more emissions of Nr. Although the adverse health effects of obesity and the overconsumption of animal protein are widely acknowledged it is wishful thinking that action would ever be taken at a global level to reduce reliance on livestock in order to improve health and help conserve the environment.

Before the industrial era, the creation of reactive nitrogen species from the non-reactive atmospheric N occurred primarily through two natural processes: lightning and biological nitrogen fixation by free-living or symbiotic microorganisms growing in association with legumes. It is likely that balance was achieved by terrestrial denitrification, which is currently estimated at 109 million t (Ciais *et al.*, 2013). By 2010 global man-made fixation of N had approximately equalled that of the total biological fixation.

Once formed, reactive nitrogen can be transferred and moved in the environment in a cascade (Galloway *et al.*, 2003) leading to potentially negative effects on human health, the climate and the environment. The potential for deleterious effect exists until the reactive nitrogen is converted back to gaseous N.

The sources, global amounts and behaviour in the environment of reactive N have recently been reviewed by the Intergovernmental Panel on Climate Change (Ciais *et al.*, 2013). Very wide ranges for all the amounts on reactive N are presented which indicate considerable uncertainty about the validity of the numbers. Emissions of NH_3 and NO_x considerably exceed those of the greenhouse gas N_2O (Table 17.2).

Agriculture is the major source of NH_3, whereas burning of fossil fuels is the main source of NO_x. Anthropogenic generation of N_2O accounts for about one third the global total of N_2O emissions.

Recent estimates of annual emissions as Nr reviewed by Fowler *et al.* (2013) are only roughly in line with those in Table 17.2, e.g. N_2O from soils 13 Tg N, NH_3 from land 60 Tg N and NO_x from combustion 40 Tg N. Between 30 and 40% of fertilizer N totalling 100–120 Tg/year is lost in emissions to the atmosphere and by

Table 17.3 Amounts (Tg N/year) of NH_3, NO_x and N_2O deposited from the atmosphere in 2000s (Lamarque *et al.*, 2010, in Ciais *et al.*, 2013)

Site of deposition	NH_3	NO_x	N_2O
Continents	36.1	27.1	0.4
Oceans	17.0	19.8	0.2

nitrate leaching to rivers and estuaries; it is likely to amount to 40–70 Tg N/year according to Fowler *et al.* (2013). Most (about 70%) of the emitted NH_3 and 60% of NO_x is deposited over land (Table 17.3) and could benefit crop growth.

Nitrous Oxide (N_2O)

N_2O, which is otherwise relatively inert in the atmosphere, is a powerful greenhouse gas. The concentration of N_2O in the atmosphere is currently 20% higher than in pre-industrial times (Ciais *et al.*, 2013). N_2O has a long atmospheric life of more than 100 years (Fleming *et al.*, 2011) and is ultimately broken down by photolysis and reaction with ozone.

Agricultural N_2O emissions are mainly produced during the denitrification of NO_3 present in the soil derived from the dissolution of synthetic fertilizer or from the mineralization of manure and organic materials. The more NO_3 there is available, regardless of its source, for bacteria to work on then the greater the potential for the emission of N_2O: it is the amount of NO_3 that is available that determines how much N_2O is generated and not the origin of the NO_3.

Table 17.2 Sources and global amounts (Tg N/year) of emissions of NH_3, NO_x and N_2O from farming, combustion and natural ecosystems in 2005 (Ciais et al., 2013)

	NH_3	NO_x	N_2O
Agriculture + Sewage	30.4	3.7	4.3
Fossil Fuel Burning	0.5	28.3	0.7
Biomass + Biofuel Burning	9.2	5.5	0.7
Man-made Total	40.1	37.5	5.7
Natural Ecosystems	2.4	7.3	6.6
From Sea + Estuaries	8.2	–	3.8
Lightning	–	4.0	0.6 *
Natural Total	10.6	11.3	11.0
Total	50.7	48.8	16.7

* chemical reaction in atmosphere

Improving the efficiency of use of fertilizer N, especially by synchronizing the time of application with crop uptake, has the potential to limit the amounts of NO_3 available for denitrification and so reduce N_2O emissions. Improving the efficiency of use of N from manure, composts and ploughed-in cover crops is difficult because of the need to incorporate such materials into the soil. Slurry applications of N can be better synchronized with crop demand.

Manure deposition and handling accounts for 51% N_2O of agricultural emissions compared with 28% from fertilizer use (Table 17.6) (FAOSTAT, 2016). There is substantial uncertainty about the average emission of N_2O from fields, ranging from 1% to 4.5% of fertilizer N applied (Sutton *et al.* 2011). Waste management and burning are of minor importance (Table 17.2).

According to Smith *et al.* (2008), agriculture is a major source of N_2O, accounting for 84% of global anthropogenic emissions, a little more than indicated in Table 17.2 (Ciais *et al.*, 2013). Biomass burning also generates N_2O. About half the N_2O is produced during nitrification and denitrification of NO_3 in terrestrial and aquatic environments including rivers estuaries and the open sea (Freing *et al.*, 2012). Agriculture accounted for 80% of the N_2O emissions in the UK, according to Defra (ADHB, 2015) in 2009.

Ammonia (NH₃)

Globally agricultural activities account for around 76% of the ammonia

emissions to the air, according to Ciais *et al.* (2013); in the UK the proportion was 90% (ADHB, 2015) in the 2000s.

NH_3 emissions associated with farming come largely from the breakdown of urea in fertilizers and of urine in manures and slurries. Ammonia is volatilized from manure as soon as it is produced, during storage and when it is spread on the land. Between 60–80% of the N fed to cattle is excreted mainly as urine, and much of it is rapidly converted to ammonia (Webb *et al.*, 2006). The total amount of N excreted by livestock in Europe peaked at about 11 Tg N in the late 1980s, which was similar to the 12 Tg N used as fertilizer (Oenema *et al.*, 2007).

More than half the manure collected from housed cattle in Europe is managed in the form of slurry or liquid, while the remainder is managed in a solid form (Jarvis, 2011). There are wide regional variations in manure management systems in Europe (Menzi, 2002); slurry-based systems are dominant in the Netherlands and Denmark (>90%), while separate collection of slurries and solids dominate in UK, France and Central/Eastern Europe (<50% slurry). Most of the slurries are stored in tanks, some with covers.

Modelling studies (Jarvis, 2011) indicate that in 2000 almost 30% of the N excreted by housed animals in Europe was lost during storage; approximately 19% via NH_3 emissions, 7% via nitrification and denitrification and 4% via leaching and runoff.

Of the synthetic fertilizers, urea has the greatest potential for emitting NH_3 to the atmosphere, especially from alkaline soils. Bouwman *et al.* (2002) estimated that globally the loss of NH_3 averaged 14% (10–19%) of fertilizer N and 23% (19–29%) of animal manure N (based on 33 Tg N applied globally as manure and 78 Tg N as fertilizer).

Nitrate leaching

Regardless of the source, nitrate leaching is undesirable, not only because it is wasteful and may be harmful in drinking water, but also because it causes eutrophication of waterways, particularly coastal seas, promotes toxic algal blooms, kills fish, disturbs ecosystems and contributes to the decline in coastal fishing. Whereas the positively charged NH_4 cations can be adsorbed on clay minerals, soils have little capacity to adsorb the negatively charged NO_3 anions, which are free to move in the soil water and thus be leached.

Bouwman *et al.* (2013) estimated that in 2000, 57 Tg N/year were lost globally by leaching and runoff following the total input of 248 Tg N from all sources, which was in line with an earlier figure of 55 Tg N/year (van Drecht *et al.*, 2003).

In China Ma *et al.* (2010) estimated that 41% of fertilizer N applied (48.8 Tg N) was lost by leaching. These massive, wasteful amounts will hopefully be reduced by better synchronization of N application and crop requirement.

The use of nitrogen fertilizers in both Europe and the USA plateaued at about 11 Tg N/year around 1994. In Europe, after a peak of about 15 Tg N in the mid-1980s, use has declined after the Nitrates Directive of 1991 (Oenema *et al.*, 2011). The directive restricts the use of fertilizer and manure in situations where there is a high risk of leaching and runoff, and it sets limits to the use of N in manure per hectare of agricultural land. In the USA N use is also influenced by acts relating to water quality (van Grinsven *et al.*, 2015).

Nitrate leaching can also reduce soil fertility and pH as a result of the leaching of the accompanying charge-balancing cations, notably of Ca, K and Mg with the NO_3 anions.

Leaching by winter rain when crop uptake of nitrates is minimal is of significance. Minimizing winter leaching by synchronizing the supply of nitrates with the time of maximum uptake is easier to achieve when using fertilizer N than with manure or compost. Wheat, for example, needs to take up a lot of N very quickly in spring. Spring applications of soluble fertilizer readily satisfy that requirement. Mineralization of organic matter does occur in spring but is not usually rapid enough to meet crop demand. Mineralization continues throughout the year and beyond the growing period, permitting NO_3 loss. Ideally manure should be applied in late winter or spring in order to minimize leaching losses but this is not possible for autumn-sown crops and is difficult to achieve, except for slurries, before the sowing of spring crops.

The possibility that unused fertilizer N was a major source of nitrate that could be leached after harvest was investigated by Macdonald *et al.* (1989). [15]N labelled fertilizer was applied at rates between 47 and 234 kg N/ ha in spring in eleven experiments on winter wheat. On average, 17% of the N from spring-applied labelled fertilizer remained in the 0–23 cm soil layer at harvest (range 7–36%). Only 1–3% of the unused fertilizer N, amounting to no more than 5 kg N/ha, was in inorganic forms (NH_4 and NO_3). Between 79 and 98% of the inorganic N in soils at harvest was unlabelled, being derived from the mineralization of organic matter rather than from unused fertilizer. It was concluded that almost all the nitrate at risk to leaching over the following winter had come from mineralization of organic matter and not from unused fertilizer applied even after large applications of N.

Comparisons of leaching of NO_3 on five organic farms and three conventional farms in northern France showed wide ranges associated with crop rotation, soil and rainfall variability in 2012–2013; leaching losses were 10 to 36 kg N/ha/year for organic farming and 26 to 41 kg N/ha/ year for conventional farming (Benoit *et al.*, 2014). As data on the amounts of N applied and on crop productivity was not provided it is not possible to assess whether the observed variations were of significance and how they related to the farming system.

Field and lysimeter studies on leaching over six years in Sweden (Bergström *et al.*, 2008) showed, contrary to popular belief, that organic crop production uses agricultural soils less efficiently as a result of lower yields, without any benefit of reduced leaching or improved water quality. In two lysimeter experiments, leaching of N derived from either poultry manure or a red clover (*Trifolium pratense* L.) green manure were compared with that from fertilizer N, all labelled with [15]N. Over three years 32% of N applied as poultry manure leached, whereas only about 3% leached of that supplied as ammonium nitrate. Over the six-year crop rotation on a sandy soil, average annual N leaching was 39 kg N/ha in the organic rotation (clover/grass ley, cereals, pea/barley and potato) and 25 kg N/ha in the conventional rotation using fertilizer N (cereals, rape and sugar beet). Leaching was minimal when a ryegrass cover crop was used in the conventional system, which also had the largest crop yields.

Tampere *et al.* (2015) in Estonia studied leaching of N and K from three different grass swards consisting of timothy (*Phleum pratense*), perennial ryegrass (*Lolium perenne*) and smooth meadow-grass (*Poa pratensis*), mixed with either white clover (*Trifolium repens*) or lucerne (*Medicago sativa*) over three years. Application of N, at rates up to 180 kg N/ha, as NPK fertilizer was compared with the same amounts as cattle slurry. It was found that the better grass growth when given NPK also resulted in less leaching of N than when slurry was applied.

A good example of the value of precision application of N was provided by work comparing organic and conventional production of vegetables in greenhouses in Israel (Dahan *et al.*, 2013).

Leaching losses of NO_3 were much greater when composted manure, incorporated into the soil prior to planting, was the only source of nutrients than when liquid fertilizer was applied through drip irrigation – a common practice in conventional intensive agriculture in Israel. Accurate fertilization methods that supply the nutrients through the irrigation system, according to plant demand, during the growing season can also dramatically reduce the potential for groundwater contamination as well as improving the efficiency of nutrient utilization.

Zero tillage and conservation agriculture systems can considerably reduce nitrate leaching (Macdonald *et al.*, 1989); leaving the soil undisturbed also reduces mineralization and the subsequent production of nitrates. Cover crops can absorb NO_3 after harvest and so reduce leaching losses but their use may be restricted by the need to grow cash crops.

The concerns about the potential involvement of nitrates with the blue baby syndrome (methaemoglobinaemia) and with nitrosamine metabolism in the gut, together with the possible beneficial effects of ingested nitrates in food and drinking water to reduce blood pressure, are discussed in Chapter 10. The regulations governing maximum nitrate levels in different countries and the designation for example in the UK of nitrate vulnerable zones should promote improved practices that will reduce the wasteful loss of nitrates. Large areas of land (currently about 60% in England and Wales) have been identified as being at risk of exceeding the permitted concentration of 50 mg NO_3/l drinking water. In these zones the amounts and timing of applications of manure and fertilizers are regulated.

Recovery of nitrogen from fertilizers and organic materials

The potential wasteful losses of residual fertilizer nitrate and of nitrate from mineralization of organic matter in autumn and winter are important issues because of possible environmental hazards (greenhouse gas production and eutrophication).

Powlson *et al.* (1989) found, in the long-term spring barley experiment at Rothamsted that over winter the manure-treated soils contained much more inorganic N than did fertilizer-treated soils due to the continued mineralization of existing soil organic matter. Very little N was mineralized from manure ploughed in in the autumn until the following March. Goulding *et al.* (2000) also showed that, on winter wheat (Broadbalk) more inorganic N was likely to be lost by leaching from soils with a long history of manure addition.

In cereals given sufficient N for economically optimum grain yields it was found that the amount of inorganic N remaining in the soil at harvest was often only a little larger than that present when no fertilizer N was applied (Glendining and Powlson, 1995; Glendining *et al.*, 1997). Uptake studies on spring barley with ^{15}N labelled fertilizer applied at

48, 96 and 144 kg N/ha showed that irrespective of the rate of N application about 51% of the labelled N was present in above-ground crop and weeds at harvest with about 30% in the top 70 cm. It appeared that the uptake of N was blocked once the capacity of the crop to absorb N was exceeded (Glendining *et al.*, 1997). This experiment confirmed that there is no evidence that a spring-sown cereal is more likely to leave unused fertilizer N in the soil than an autumn-sown one.

Macdonald *et al.* (1989) found in wheat on a sandy loam soil that at harvest only about 3 kg N/ha of the fertilizer applied (140 kg N/ha) was present as inorganic N in the top 50 cm of soil compared to about 13 kg N/ha derived from the mineralization of soil organic matter.

Wide variations in recovery of fertilizer N in crops were quoted by Jarvis (2011), namely 11–42% for autumn applications and 42–78% for spring-time applications a clear indication of the importance of synchronizing application with crop uptake. These values are to be compared with only 18–31% for manure (solid) applied in spring and 61–88% for urine and poultry excreta. Similar recovery values for fertilizer N of 60% on most soils, 70% on sandy and silty soils and 55% on shallow soils over chalk are used in the UK when assessing N requirements of winter wheat (HGCA, 2009). Jarvis also reported recoveries of 8–13% N from mature crop residues incorporated in the autumn and of 70% from grass incorporated in spring. Limited (2–6%) recovery of N from manure in the second year after application can be expected.

Crop residues from growing legumes in the rotation provide an important source of N in many farming systems as well as in organic farming. Crews and Peoples (2005) reported recovery values of 10–30% for N incorporated as legume N residues, which is considerably less than that from fertilizer N. The release of labile organic N into the soil from roots and from the nodules of legumes during growth (rhizodeposition) may also be important sources of N, and could represent from 25–43% of total N recovered in a following crop (Russell and Fillery, 1996). Mayer *et al.* (2003) showed that N rhizodeposition by grain legumes (beans, peas and lupins) can make a significant contribution.

The classical experiments, Broadbalk, Hoosfield and Park Grass at Rothamsted have provided useful information on N cycling using ^{15}N. On average about 50% of the applied fertilizer N was recovered by the crop, 25% remained, as organic N, in the soil and 25% was not accounted for. Most of the nitrate present in the soil profile in the autumn, and therefore at risk of being leached, was derived from soil organic matter, not from unused fertilizer N (Rothamsted, 2012).

Greenhouse gas emissions associated with agriculture

Methane and Nitrous oxide

There is evidence from analysis of gas bubbles in Greenland ice cores of small increases in CH_4 in the atmosphere in Roman times and also during the Han dynasty in China from about 100 B.C. (Sapart *et al.*, 2012). It is thought that the keeping of cows, sheep and goats by the Romans and the expansion of rice cultivation in China were probably the responsible factors. Although the amounts of CH_4 were not sufficient to have a greenhouse warming effect, it is surprising that man was altering the composition of the atmosphere even though the global population was probably only about 300 million. A further small rise in atmospheric CH_4 occurred in Medieval times coinciding with a warm period from 800 to 1200 A.D. CH_4 concentrations rose from about 600 ppb around 2,000 years ago to 700 ppb by 1800 but since then they have risen to about 1,800 ppb (Sapart *et al.*, 2012).

Agriculture is a major source of CH_4 accounting for about 52% of global anthropogenic CH_4 according to Smith *et al.* (2008); later data (Ciais *et al.*, 2013). indicated a value of 30–40% CH_4 from agriculture. CH_4 is produced when organic materials are decomposed in oxygen-deprived conditions, notably during digestion by ruminant livestock, stored manures and rice grown under flooded conditions (Mosier *et al.*, 1998). Hydroelectric dams constitute another source of CH_4 from the decomposition of organic matter. It is thought that dams emit 1–2% of globally emitted CH_4 and that the contribution has been underestimated (Deemer, 2016). The quantity of CH_4 emitted will vary according to the amount of organic matter.

Globally rice is a major contributor to CH_4 emissions (Table 17.6). The warm, waterlogged soils of rice paddies provide ideal conditions for bacterial methanogenesis, and though some of the CH_4 produced can be oxidized by bacteria in the shallow overlying water, the vast majority is released into the atmosphere. Methanogenesis is promoted by the exudation of sugars from rice roots. The potential for reducing such leakage and thereby minimizing CH_4 generation has been indicated by the breeding work of Su *et al.* (2015); they inserted a single gene from barley into rice which resulted in more photosynthates being allocated to the shoots; not only did this increase rice yields but the activity of the CH_4 generating bacteria that congregated around the roots was reduced. As they used genetic modification techniques there is likely to be opposition to the new rice varieties being used in breeding programmes. If all rice in the world used this technology, it would be the equivalent of closing down 150 coal-fired power stations or removing 120 million cars from the road. CH_4 is also emitted from a number of natural sources, mainly from wetlands where bacteria generate CH_4 when decomposing organic materials anaerobically. Smaller sources include termites, oceans, sediments, volcanoes, and wildfires (Ciais *et al.*, 2013).

Considerable amounts of CH_4 and N_2O are generated during composting and manure handling (Hao *et al.*, 2001). If reducing emissions of greenhouse gases is perceived as a major concern it may be preferable to produce CH_4 from composts and manure under controlled conditions in anaerobic digesters and to use the gas to generate electricity. The residual fibrous and liquid digestate could be used to supply limited amounts of N, P and K that will vary in composition according to the raw material. Liquid digestate from Agrivert in the UK can contain about 3 kg N, 0.7 kg P_2O_5 and 2.7 kg K_2O per t. (Agrivert, 2016).

A basic problem for both conventional and organic agriculture is how to remain productive and so meet the food requirements of a growing global population while at the same time minimizing emissions of the greenhouse gases CH_4 and N_2O. It is the ability of CH_4 and N_2O to act as greenhouse gases (GHG) that is important not the actual amounts emitted, so they are assessed in terms of their CO_2 equivalent (CO_2eq) based on CH_4 being twenty-five times and N_2O being 298 times more powerful than CO_2 as a greenhouse gas; emissions of 1 t CH_4 and 1 t N_2O respectively are equivalent to emissions of 25 and 298 t of CO_2.

The agricultural production of the individual GHG firstly need to be considered in relation to the total quantities of anthropogenic GHG and secondly in order to assess the contribution of different agricultural activities.

Edenhofer *et al.* (2014) report (for IPCC) that in 2010 49,100 ($\pm$ 4,500) Tg CO_2eq were emitted globally, with CO_2 from fossil fuel burning for power generation and transportation dominating (Table 17.4). The World Bank reports similar numbers for 2012, namely CO_2 35,800 Tg, CH_4 8,000 Tg CO_2eq and

Table 17.4   Total Global Emissions 2010 according to gas (Edenhofer *et al.*, 2014)

Gas	Tg CO_2eq	%
CO_2 Non-agricultural	32,000	65
CO_2 deforestation, land use	5,400	11
CH_4	7,800	16
N_2O	2,900	6
F gases	1,000	2
Total	49,100	100

N_2O 3,100 Tg CO_2eq (WB, 2016), giving a total of 46,900 Tg CO_2eq.

According to Edenhofer *et al.* (2014), all agricultural activities account for 24% of global GHG (Table 17.5) namely 11,800 Tg CO_2eq with the crops and livestock sector and deforestation and the land–use change sector providing about the same amounts.

The emissions of N_2O and CH_4, expressed as CO_2eq associated with different agricultural activities, are regularly

Table 17.5   Global Emissions 2010 according to sector (Edenhofer *et al.*, 2014)

Sector	Tg CO_2eq	%
Crops and livestock	5,300	11
Deforestation, land use change	6,500	13
Electricity and heat production	12,200	25
Industry	10,300	21
Transportation	6,900	14
Buildings	3,000	6
Other energy	4,900	10
Total	49,100	100

Table 17.6 Agricultural emissions according to agricultural activity Tg CO_2eq in 2014 (FAOSTAT, 2016)

	Total	CH_4	N_2O
Enteric fermentation	2,085	2,085	0
Manure management	352	205	147
Rice cultivation	524	524	0
Synthetic fertilizers	660	0	660
Manure applied to soils	192	0	192
Manure left on pasture	846	0	846
Crop residues	212	0	212
Cultivation of organic soils	133	0	133
Burning crop residues	30	22	8
Burning savanna	213	91	122
Total Agricultural 2014	5,247	2,927	2,320
Total Agricultural 2010	5,077	2,864	2,213

reported by FAO (FAOSTAT, 2016) (Table 17.6). For details of the methodology see http://faostat3.fao.org/mes/methodology_list/E. Many types of livestock apart from dairy and beef cattle are included, namely buffaloes, camels, goats, horses, llamas, mules, pigs and sheep.

The variations found between different reviewers of the precise contribution of agriculture to the atmospheric GHG load are likely to be related to the components included and the age of the data. Here the IPCC data for 2010 and the FAO data for 2014 are used; the individual crop and livestock activities are set against a grand agricultural total of 5,247 Tg CO_2eq of CH_4 and N_2O.

Emissions in Asia contribute 44% of the global agricultural greenhouse gas emissions, of which China and India together account for 60% (Table 17.7). In China N_2O emissions exceed those of CH_4, whereas in India the reverse is true, which is probably related to the differences in use of fertilizers and the numbers of cattle. Nearly 70% of the emissions in the Americas are contributed by Brazil, USA and Argentina in that order. In Brazil CH_4 emissions exceed those of N_2O, whereas in the USA emissions (as CO_2eq) of CH_4 and N_2O are roughly similar.

Total GHG emissions from UK agriculture in 2009 amounted to 50 Tg CO_2eq comprising 28 Tg of N_2O, 18 Tg CH_4 and 4 Tg CO_2 (ADHB, 2015a). There had been a small reduction (about 17% over the previous twenty years) in agricultural GHG from a total exceeding 60 Tg CO_2eq with CH_4 playing the major part. CO_2, with 8% of the total, plays a relatively small part.

Table 17.7 Agricultural emissions of combined N_2O and CH_4 as proportion of total global* in different continents in 2014. (FAOSTAT, 2016)

	Tg CO_2eq	%
Asia	2,314	44
Americas	1,323	25
Europe	589	11
Africa	833	16
Oceania	188	4
Total	5,247	100

* excluding emissions from deforestation, land clearing and soil degradation.

CH_4

Agricultural CH_4 emissions account for between 50 and 60% of total agricultural GHG emissions (5,247 Tg CO_2eq) with ruminants being the main source (Table 17.6). In contrast, N_2O from the use of synthetic fertilizers accounts for 13% total agricultural GHG emissions and about 30% of total agricultural emissions of N_2O. Thus reducing livestock numbers has a greater potential for limiting greenhouse gas emissions than optimizing the way fertilizers are used but in view of the likelihood of increased global demand for animal protein such reductions are unlikely. Anthropogenic emissions account for 50 to 60% of total emissions of CH_4 (Ciais *et al.*, 2013).

Production of CH_4 in the gut of ruminants accounts for 71% of agricultural CH_4 emissions followed by release from paddy rice soils (18%) (Table 17.6). Emissions from manure handling and burning of crop residues are relatively small. Bridgham *et al.* (2013) pointed out that rice paddies are a significant source of man-made CH_4.

Organic farms that rely on livestock and the recycling of nutrients in manure are responsible for more CH_4 emissions than livestock-free arable farms using fertilizers that have no direct impact on CH_4 generation. Many ruminants contribute to CH_4 emissions with dairy and beef cattle being responsible for more than 70%, buffaloes for 10% and sheep 7% (FAOSTAT, 2013).

The FAO Organic Agriculture Programme set up in 1999 seems to ignore the contribution of livestock to CH_4 and N_2O emissions (FAO, 2014). The FAO organic programme is confident that soil emissions of N_2O and CH_4 from arable and pasture lands can be avoided by improving organic management practices and that organic agriculture can sequester large amounts of CO_2 from the atmosphere in the soil. It is possible that organically raised livestock may generate less CH_4 in their gut than conventionally raised livestock because antibiotics are not used; Hammer *et al.* (2016) found that cowpats produced from antibiotic-treated cattle gave off twice as much CH_4 as cowpats from untreated cattle. It was also found that microbes in the gut of associated dung beetles were affected, which suggests that there may be unintended antibiotic effects beyond the treated animal.

N_2O

In the period between 2000 and 2009 agriculture accounted for about 25% of global N_2O emissions, natural sources about 65% with the remainder coming from fossil fuel and biomass burning. (Ciais *et al.*, 2013) (Table 17.2).

Asia is the major contributor to global N_2O generation (Table 17.8), which parallels its position with regard to CO_2 emissions.

Livestock, the dominant factor

The massive amounts of CH_4 produced by cattle and the emissions, particularly of N_2O, associated with manure

Table 17.8 Emissions of N_2O from the application of man–made N fertilizer according to continent (FAOSTAT, 2016)

	Tg CO_2eq	%
Asia	388	57
Americas	154	22
Europe	101	15
Africa	28	4
Oceania	11	2
Total	682	100

mean that livestock farming is responsible for far more emissions of greenhouse gases than arable farms devoid of cattle (Table 17.9). As organic farms are likely to be more reliant on livestock (for recycling nutrients) than conventional mixed farms they will, as a group, emit more greenhouse gases than conventional farms. It can be seen from Table 17.9 that 3,475 Tg CO_2eq (66% total) emissions of CH_4 and N_2O are associated with cattle and their manure compared with 660 Tg CO_2eq (13% total) associated with fertilizers. Despite these numbers from FAO, it is believed that

Table 17.9 Combined emissions of CH_4 and N_2O associated with cattle, manure and fertilizers as percentage of total agricultural emissions of these gases

	Tg CO_2eq	%
Cattle	2,085	40
Manure use and handling	1,390	26
Synthetic fertilizers	660	13
All other+	1,112	21
Total	5,247	100

+ See Table 17.6

organic farming is preferable for mitigating climate change because the non–use of N fertilizers leads to a reduction of N_2O (e.g. IFOAM, 2015).

Livestock and their manure are conveniently forgotten, e.g. by FAO (2010), when the benefits of the greater sequestering of carbon in the soils farmed organically is highlighted.

Comparisons on the basis of N_2O emissions alone show that manure handling and use associated with livestock accounted for 57% (1,185 Tg CO_2eq) of total N_2O emissions, whereas about 28% (660 Tg CO_2eq) was linked to synthetic fertilizer use (Table 17.6).

If organic farming was able to generate the same amount of NO_3 at the right time as can be achieved by conventional farming using fertilizer N then it would be expected that similar amounts of N_2O would be produced during denitrification, but this is clearly not the case.

A study involving almost 40,000 participants over twelve months in the UK (Scarborough *et al.*, 2014) showed that eating more meat is associated with high greenhouse gas emissions (Table 17.10).

Table 17.10 Meat consumption and associated emissions of GHG (CO_2, CH_4 and N_2O) as kg CO_2eq per capita per day (Scarborough *et al.*, 2014)

	g meat/day	kg CO_2eq
High meat eater	>100	7.2
Medium meat eater	50–99	5.6
Low meat eater	<50	4.7
Fish eaters	–	3.9
Vegetarians	–	2.9

It led to the conclusion that reducing meat consumption would have environmental as well as personal health implications, findings that are in line with the analysis based on CH_4 and N_2O emissions.

Williams *et al.* (2006), in their life cycle analyses on wheat, potatoes, oilseed rape and tomatoes in the UK, assessed that the risks of eutrophication and global warming potential emanating from organic farms are greater than the risks from conventional farms per unit of production. Organic crop production was generally less energy consuming, except for poultry meat, eggs and tomatoes, than conventional farming but this benefit was offset by the lower productivity on organic farms. Despite such assessments, it is often assumed that organic farming has a smaller impact on the environment than conventional farming.

Whether livestock are a major threat to climate change is questioned, e.g. by Idel (2011) and Schäfer (2013). Livestock can also be seen, at least in developing countries, as making a positive contribution regarding climate change as a result of their use as draught animals, so reducing the use of fossil fuel.

Environmental impacts of organic and conventional farming

The environmental impact of farming has been the subject of concern, especially since the book *Silent Spring* (Carson, 1962) was published. It has become a mantra of the organic movement that organic farming is much kinder to the environment than conventional but despite many comparative studies little supportive data has been presented.

It is clear that organic farming does not have the capacity to generate fewer GHG than conventional farming (Tables 17.6 and 17.9) and so has the potential for a greater influence on the greenhouse effect and thereby climate change.

Mondelaers *et al.* (2009) conducted a meta-analysis of studies on the environmental impacts of organic and conventional farming. This was extended by Tuomisto *et al.* (2012), who brought together seventy-one European studies. They considered many aspects, including soil organic matter, NO_3 leaching, N_2O and NH_3 emissions, phosphorus losses, the land area and energy used. Clear statistically significant differences were not found due to the wide data variability, which is often an inherent problem in meta-studies. However, the results indicated that organic farming generally leads to higher soil organic matter contents, which as discussed in Chapter 11 have the potential for several benefits. Some conventional farming systems, as a result of higher crop residues, will also have the potential to achieve similar or even higher soil organic matter levels – especially when some manure is also applied.

The meta-analyses showed that conventional farming had lower impacts than organic farming on NO_3 leaching, NH_3 and N_2O emissions when comparisons were made on the basis of production. It is only when comparisons were made on the basis of area of land used that organic farming appeared to have fewer environmental impacts, a finding related to the lower nutrient applications and the associated lower yields (75% on average) of organic crops. Nitrate leaching was 49% greater per unit of product from organic farming compared to conventional farming. Comparisons on the basis of land used would only be acceptable if there was no pressure on land required for food production. Median N_2O emissions found in the meta-analysis were 8% higher and NH_3 emissions were 11% higher per unit of product on organic farms but were lower when assessed on the basis of land area (Tuomisto *et al.*, 2012).

It is the quantity of NO_3 and NH_3 and not their source that is relevant when considering their possible detrimental effects on the environment. Provided the same amounts of NO_3 and NH_4 are made available for crop uptake, whether arising from fertilizers, manure or biological fixation, it would be expected that similar amounts of NO_3 would be leached and N_2O emitted. A positive relationship between total nitrogen input and N_2O emissions was reported by Jones *et al.* (2007) and by Petersen *et al.* (2006). Several workers, notably in Scandinavia, have explained that the observed lower NO_3 leaching levels from organic farming (on a unit area basis) were due to the smaller amounts of N applied (Hansen *et al.*, 2000; Korsaeth, 2008; Torstensson *et al.*, 2006). Higher leaching levels of NO_3 found on some organic farms is likely to be due to the poor synchrony between the time of manure application and the time of crop N uptake (Aronsson *et al.*, 2007). There is also a risk of poor synchrony after incorporation of grass leys, which can exacerbate NO_3 losses (Syväsalo *et al.*, 2006).

Organic farming would be expected to show more emissions of NH_3 from the handling of manure than that associated with conventional farming reliant on fertilizer N (Tuomisto *et al.*, 2012). Organic farming systems showed on average 21% lower energy consumption (per product unit) than conventional but the variation was very wide; from 63% less energy to 40% more energy used in organic systems (Tuomisto *et al.*, 2012). The higher energy inputs in conventional farming were mainly due to the energy needed for production and transport of fertilizers.

In a debate on Swedish Radio on 23 November 2014, scientists at the Swedish University of Agricultural Sciences said organic production of the same amount of food as is currently produced conventionally in Sweden would increase significantly the generation of greenhouse gases and that environmental goals would not be achievable (Radio Sweden P6, 2014).

The deleterious effects of reactive nitrogen compounds on the environment can be minimized by reducing their creation and by increasing the efficiency with which N is used by crops.

Biodiversity

It would be expected that vigorously growing crops by successfully competing for light and nutrients will reduce weed numbers; insofar as organically grown crops are usually not as vigorous as conventional they will harbour more weeds and so be said to be increasing biodiversity. The differences will be widened by the non-use of herbicides on organic crops. Insects are more likely to prosper on organic farms, thereby promoting bird and other animal life. However, the situation is perhaps not so straightforward, as indicated by Gabriel *et al.* (2015); they studied the relationship between winter cereal (wheat, barley, oats) yields and the abundance and density of farmland plants, bumblebees, butterflies, solitary bees and epigeal arthropods in England. Grain production was 54% lower in organic than conventional fields. Biodiversity was found to be negatively related to crop yield and organic farming per se did not have an effect other than via reducing yields. With the exception of farmland plants (not specified) there were generally few or no diversity gains through organic farming when compared with conventional farming at similar yields. It was concluded that any modest gains in biodiversity cannot be justified in the light of the substantial reductions in yields. They suggested that conservation efforts should be concentrated on soils of low-productivity or on non-agricultural land where they are likely to be more cost-effective. In contrast, Bengtsson *et al.* (2005) had earlier suggested that positive effects of organic farming on species richness can be expected in intensively managed agricultural situations but not where small-scale enterprises were common.

The loss of rural biodiversity during the last decades has been associated with the change in agricultural landscapes. In Europe, formerly heterogeneous landscapes with a mix of small arable fields, semi-natural grasslands, wetlands and hedgerows have been replaced in many places by homogeneous areas of intensively cultivated fields. This has resulted in the decline of many animal and plant populations and losses of species (Krebs *et al.*, 1999).

The assessment of the effects of different farming systems on biodiversity is compounded by factors such as the proximity of natural vegetation which can lead to considerable variation. For example, a meta-analysis that compared biodiversity in organic and conventional farms by Bengtsson *et al.* (2005), found organic farms generally had 30% more species and 50% more organisms than conventional farms; the differences were most notable for plants, birds and predatory insects but curiously not for earthworms. The hypothesis that the greater diversity and abundance of natural enemies could contribute to pest control on organic farms was supported. However, there was a wide variation between different studies with 16% of the studies showing a negative effect of organic farming on species richness.

Acidification potential

The major acidifying pollutants from agriculture are ammonia (NH_3) and sulphur dioxide (SO_2). Acidification of soils can increase the loss of nutrient cations and impair biological activity. Globally agriculture, namely that associated with livestock production and the use of urea, accounts for more than 90% of ammonia emissions (e.g. 94% in 2009 [FAOSTAT, 2016]). Ammonia was predicted to be the largest acidifier in Europe by 2010 (Webb *et al.*, 2006); this followed the considerable reductions in SO_2 in the atmosphere. It was proposed that the most effective reductions in ammonia emissions could be achieved by replacing urea with ammonium nitrate, immediate incorporation of manures and slurries, storing all farmyard and poultry manures before spreading on the land and applying slurries to grassland by a trailing shoe machine (deposits the slurry into a slit in the soil surface).

As was the case with other impact categories, the generally lower NH_3 emissions per unit of land area in organic systems (livestock and crops) were due to lower nitrogen inputs, whereas the acidification potential per unit of production was higher due to lower crop and animal yields (Tuomisto *et al.*, 2012).

Carbon sequestration

The sequestration of carbon in soil organic matter (SOM) is often advocated as a means of counteracting the deleterious effects of CO_2 on the climate but the situation is far from straightforward.

There is no doubt that SOM is a valuable commodity. In the USA, as elsewhere, SOM is seen as an important indicator of soil quality because of its role in biological, chemical, and physical processes. Examination of the long-term fertilizer experiments in the USA showed the benefits of manure, of adequate fertilization and of crop rotations on SOM (USDA, 2001). Maintaining SOM levels is, however, difficult; conventional cropping systems generally result in a steady decline of SOM despite the use of manure and the rotating of crops. The USDA advises that SOM levels should be increased by growing cover crops, by applying manure, by improving crop productivity, by reducing the loss of organic matter by controlling erosion and reducing tillage. Farming organically is not singled out as being a particularly valuable system. The incentive of American farmers to improve their soils may be limited by the fact that many farms are rented.

The build-up of SOM in the soil depends upon the supply of organic matter in the form of manure, compost, crop litter and decaying roots, exceeding the rate of microbial decomposition; SOM can be enhanced by increasing such inputs and/or by reducing decomposition rates principally by minimizing soil cultivation. Opening up the soil by mechanical cultivation, especially of warm damp soils, enhances decomposition of organic matter and leads to the release of CO_2.

It might be expected that applying manure and composts and having grass/clover leys in the rotation would always build up SOM levels. There is, however, a limit to the increases in SOM even when massive amounts of manure are applied. For a given climatic regime and management system, soil organic carbon in most permanent agricultural systems tends toward an equilibrium level possibly after decades or even centuries, as described by Paustian *et al.* (2000), Johnston *et al.* (2009), Chung *et al.* (2010) and Powlson *et al.* (2012).

SOM levels in the long-term wheat and barley experiments at Rothamsted plateaued after about eighty years of the annual application of 35 t/ha manure. Soil organic carbon levels were increased from about 30 t C/ha to 70 t C/ha (in the top 23 cm soil) after eighty years. Increases that were rapid in the first twenty years (1 t C/ha/year) declined to around 0.2 t C/ha/year after sixty years. A total of 35 tons of farmyard manure could be obtained from 2.3 dairy cows housed for 60% of the year. When an arable silt clay loam soil was converted to permanent pasture it took about 100 years for SOM levels to reach a new higher equilibrium level (Johnston *et al.*, 2009).

The equilibrium level of SOM appears to be specific to the farming system, soil type, and climate. Generally in similar climates, for a given cropping system, the equilibrium level of SOM in a clay soil will be larger than that in a sandy soil, and for any one soil type the equilibrium level will be larger under permanent grassland than under continuous arable cropping (Johnston *et al.*, 2009).

The existence of an equilibrium level of SOM for any soil/climate/farming system suggests that microbial decomposition of organic matter is keeping pace with organic inputs after a certain SOM level has been reached.

Despite the limitation imposed by the finite nature of building up of soil carbon, Paustian *et al.* (2000) concluded that agricultural soils have the capacity eventually to act as significant sinks for CO_2, in view of the long timescale involved; appropriate changes in management practices will be needed. These practices include increasing the area of grassland at the expense of annual cropping, reducing mechanical cultivation and maintaining no-till systems. Organic farmers may find it difficult to minimize soil cultivation and particularly operate no-till systems in view of their non-use of herbicides. Whether the net effect will be positive or negative on the greenhouse gas situation is problematic, especially if increased grassland is accompanied with increased cattle numbers and the inevitable generation of more CH_4.

Rodale in the USA, hypothesises, based on recent data from farming systems in the USA, Egypt and Iran and on pasture trials around the globe, that more than 100% of current annual CO_2 emissions could be captured by switching to organic management practices, known by it as "regenerative organic agriculture" (Rodale, 2014). Its own studies, namely the Farming Systems Trial and the similarly named Tropical Farming Systems Trial in Costa Rica, are also quoted in support.

Comparisons of SOM levels on organically managed and conventional farms often show contradictory results, as pointed out by Andrén *et al.* (2008). In studies on the effects of different farming systems they developed a detailed model covering thirty years of the changes in SOM under cereal crops (winter wheat and spring barley) in Sweden. They concluded that agricultural practices that increase photosynthesis and crop yields will also increase the amount of C stored as organic matter in soil. This means that producing cereals organically reduces soil C stocks relative to conventional cereals simply because the poorer crops leave fewer crop residues to be turned into SOM. Furthermore, reduced crop growth leads to less transpiration and higher soil moisture contents, which will promote the rate of organic matter decomposition. Moreover, the lower cereal yields in organic production need to be balanced by additional cereal production elsewhere, requiring additional production areas. The scenario in Sweden if all cereals were grown organically would require all current fallow and some current forests being converted to agriculture; at the same time there would be less SOM equivalent to the annual emissions of 1.6 Tg CO_2.

Gattinger *et al.* (2012) concluded from a meta-analysis (seventy-four studies) that organic farming increases soil carbon levels relative to conventional farming and can thus mitigate climate change through carbon sequestration in soil. However, these conclusions were refuted by Leifeld *et al.* (2013), who pointed out that the Gattinger study compared systems with high stocking densities and considerable carbon inputs to systems with low stocking densities (receiving about one quarter the carbon), which indicates that the findings were more associated with livestock numbers than with organic farming. Gattinger did not consider CH_4.

Conventional arable farms are often geographically distant from livestock farms, resulting in a concentration of manure production in some areas and an absence in others. The conversion of conventional farms to lower yielding organic farms with livestock would necessitate more land (if available) being brought into cultivation, which in turn would lead to increased CO_2 as well as CH_4 emissions.

The potential for mitigating the effects of CO_2 by sequestering carbon in the soil was perceived as being important in relation to climate change at the Lima–Paris Agenda for Action (LPAA, 2015) – COP21. It was proposed that all countries should voluntarily aim to increase soil organic matter levels annually by 0.4% in order to prevent further increases in atmospheric CO_2 and by doing so improve soil fertility and yields; this is called the four per 1,000 initiative.

If SOM levels are to be built up more organic matter is needed, which in turn requires sufficient nitrogen being available to promote plant growth and so mop up atmospheric CO_2. It was suggested by the IPCC that considerable and significant amounts of carbon could be sequestered in soil organic matter (and also in trees) and thus counterbalance man-made emissions of carbon dioxide (Ciais *et al.*, 2013). This scenario is championed by the organic movements, e.g. Rodale and the Organic Trade Association in USA and by the Soil Association in the UK. However, Hungate *et al.* (2003) point out in relation to earlier IPCC estimates that more reactive nitrogen (from biological fixation and deposition) would be required to promote the IPCC carbon storage scenarios for terrestrial ecosystems in 2100 than will be available. Moreover, the supply of other nutrients, notably P and K, could also limit growth.

Summary

1. Most N is absorbed by crops as the soluble and easily leached NO_3 anions, which are not adsorbed on clay particles. Most N in soils is held in organic matter.
2. The mineralization of organic matter by soil microorganisms results in the release of NH_4, which will mostly be oxidised in a process called nitrification to NO_3 by two groups of bacteria.
3. Many bacteria can reduce NO_3 to gaseous N_2O and to nitrogen in the process of denitrification. There is growing concern about the adverse effects on the environment of reactive nitrogen compounds such as NO_3 and of N_2O and NH_3 generated by agriculture and released into the atmosphere.
4. Reactive forms of N are implicated in causing premature deaths as a result of the formation of tiny particles of $(NH_4)_2SO_4$ and NH_4NO_3 in the atmosphere.
5. Annually there are several million premature deaths across the world, of which 20% are linked to agriculture. Particulate matter less than 2.5 microns diameter that can reach the inner surfaces of the lungs are believed to be the most harmful. Globally the biggest contributor to premature deaths is indoor cooking and heating, which accounts for 30% deaths.
6. NO_3 may leach into groundwater and thence to drinking water. Leaching of NO_3 is not only wasteful but it causes eutrophication and is seen as a hazard in drinking water.
7. Leaching of NO_3 can be minimized by synchronizing the supply of nitrates with the time of maximum crop uptake. This is easier to achieve when using N fertilizers than manure as manure needs to be incorporated in the soil before sowing, which means there is a longer time for leaching especially under the influence of winter rain.
8. Most of the NO_3 leached after harvest of crops given fertilizer is likely to originate from mineralized organic matter and not from unused fertilizer.
9. Ideally NO_3 should be supplied by drip irrigation in amounts according to crop demand.
10. Considerable amounts of NH_3 are emitted by agricultural activities with over 50% being deposited on land. NH_3 emissions come from the breakdown of urea in fertilizers, urine, manure and slurries.
11. Although more N_2O may be generated in natural ecosystems than by agriculture the amounts of N_2O associated with agriculture are of significance.
12. Agricultural emissions of N_2O arise during denitrification of NO_3. If yields of organic crops are limited by the availability of NO_3 during the growing season

N_2O emissions will be limited. On the other hand, mineralization when crop growth is minimal will provide NO_3 that can lead to N_2O generation.

13. N_2O, together with CH_4, adds to the burden of greenhouse gases in the atmosphere.

14. Agriculture is a major source of CH_4, a potent greenhouse gas, contributing 30–40% of global CH_4 emissions. Fermentation in the rumen of livestock is responsible globally for 70% agricultural emissions of CH_4.

15. The agricultural generation of both CH_4 and N_2O is important. Together they are thought to be responsible for about 11% of total global greenhouse gas emissions.

16. In terms of equivalent CO_2, CH_4 and N_2O make similar contributions to the global agricultural greenhouse gas burden.

17. The massive amounts of CH_4 produced by cattle and the emissions of N_2O associated with manure mean that livestock farming is responsible for the emissions of far more greenhouse gases than arable farms devoid of cattle.

18. As organic farms are likely to be more reliant on livestock for recycling nutrients than conventional mixed farms as a group they will emit more greenhouse gases.

19. Any increase in demand for livestock products will, in the absence of finding ways to limit CH_4 generation by ruminants, exacerbate the greenhouse gas problem.

20. Recent meta-analyses have provided clear evidence that while the low level of nitrogen inputs in organic farming can result in less nitrate leaching and lower energy use per unit area of land, the lower productivity of organic farming means that on the basis of crop production organic farming is potentially more harmful than conventional farming in many respects.

21. Much attention is given to the sequestering of carbon as soil organic matter. However, there are limits to the amounts of soil organic matter that can be built up in soils. For any given soil type/climate/agricultural system an equilibrium level soil organic will eventually be reached. Appreciable build-up of soil organic matter may, especially in temperate climates, take many years of continuous application of massive amounts of manure or after converting to permanent pastures.

References

ADHB (2015) Agriculture and horticulture development board. Fertilizer handbook RB209. Available at www.ahdb.org.uk/documents/rb209-fertiliser-manual-110412.pdf

ADHB (2015a) Greenhouse gas action plan of the agriculture Industry in England: progress report. Available at www.ahdb.org.uk/projects/documents/GHGAPprogressreportApril2012.pdf

Agrivert (2016) Biofertiliser NPK values. Available at www.agrivert.co.uk/wp-content/uploads/Biofertiliser-NPK-values.pdf

Andrén, O., Kätterer, T. and Kirchmann, H. (2008). How will conversion to organic cereal production affect carbon stocks in Swedish agricultural soils? 161–172. In: Kirchmann, H. and Bergström, L. (eds) *Organic Crop Production – Ambitions and Limitations*. Springer, Dordrecht, The

Netherlands. Available at http://pub. epsilon.slu.se/3512/1/Organic_Crop_ Production_Chapter8_2008.pdf

Aronsson, H., Torstensson, G., Bergstrom, L. (2007) Leaching and crop uptake of N, P and K from organic and conventional cropping systems on a clay soil. *Soil Use and Management* 23, 1, 71–81. doi: 10.1111/j.1475-2743.2006.00067.x.

Bardgett, R.D., Streeter, T.C. and Bol, R. (2003) Soil microbes compete effectively with plants for organic nitrogen inputs to temperate grasslands. *Ecology* 84, 5, 1277–1287.

Bengtsson, J., Ahnström, J. and Weibull, A-C. (2005) The effects of organic agriculture on biodiversity and abundance: a meta-analysis. *Journal of Applied Ecology* 42, 261–269.

Benoit, M., Garnier, J., Billen, G., Mercier, B. and Azougu, A. (2014) Nitrogen leaching from organic agriculture and conventional crop rotations 283–286. In: Rahmann, G. and Aksoy, U. (Eds.) *Proceedings of the 4th ISOFAR Scientific Conference. "Building Organic Bridges"*, at the Organic World Congress, Istanbul, Turkey. Available at http://orgprints. org/23831/1/23831_MM.pdf

Bergström, L., Kirchmann, H., Aronsson, H., Torstensson, G. and Mattsson, L. (2008). Use Efficiency and Leaching of Nutrients in Organic and Conventional Cropping Systems in Sweden 143–159. In: Kirchmann, H. and Bergström, L. (eds) *Organic Crop Production – Ambitions and Limitations*. Springer. doi: 10.1007/978-1-4020-9316-6_7.

Bouwman, A. F., Boumans, L.J.M. and Batjes, N.H. (2002) Estimation of global NH$_3$ volatilization loss from synthetic fertilizers and animal manure applied to arable lands and grasslands. *Global Biogeochemical Cycles* 16, 2, 8.1–14 doi: 10.1029/2000GB001389.

Bouwman, L., Goldewijk, K.K., Van Der Hoek, K.W., Beusen, A.H.W., Van Vuuren, D.P., Willems,. Rufino, M.C. and Stehfest, E. (2013) Exploring global changes in nitrogen and phosphorus cycles in agriculture induced by livestock production over the 1900–2050 period. *Proceedings of the National Academy of Sciences* 110, 52, 20882–20887. doi: 10.1073/pnas.1012878108.

Bridgham, S.D., Hinsby, C.Q., Jason, K.K. and Zhuang, Q. (2013) Methane emissions from wetlands: biogeochemical, microbial, and modelling perspectives from local to global scales. *Global Change Biology* 19, 5, 1325–1346. doi: 10.1111/ gcb.12131.

Brooks, P.D., Williams, M.W. and Schmidt, S.K. (1996) Microbial activity under alpine snowpwacks, Niwot Ridge Colorado. *Biogeochemistry*. 32, 93–113.

Carson, R. (1962) Silent spring. Houghton Mifflin, Cambridge, Massachusetts.

Chung, H., Ngo, K.J., Plante, A.F., Six, J. (2010). Evidence for carbon saturation in a highly structured and organic-matter-rich soil. *Soil Science Society of America Journal* 74, 130–138. doi: 10.2136/sssaj2009.0097.

Ciais, P., Sabine, C., Bala, G., Bopp, L., Brovkin, V., Canadell, J., Chhabra, A., DeFries, R., Galloway, J., Heimann, M., Jones, C., Le Quéré, C., Myneni, R.B., Piao, S. and Thornton, P. (2013) Carbon and Other Biogeochemical Cycles 465–570 In: *Climate Change 2013: The Physical Science Basis. Contribution of Working Group I to the Fifth Assessment Report of the Intergovernmental Panel on Climate Change.* (Stocker, T.F., Qin, D., Plattner, G.-K., Tignor, M., Allen, S.K, Boschung, J., Nauels, A., Xia, Y., Bex. V. and Midgley,

P.M. (eds.). Cambridge University Press, Cambridge, UK.

Crews, T.E. and Peoples, M.B. (2005) Can the synchrony of nitrogen supply and crop demand be improved in legume and fertilizer-based agroecosystems? A review. *Nutrient Cycling in Agroecosystems*, 72, 101–129. doi: 10.1007/s10705-004-6480-1.

Dahan, O., Babad, A., Lazarovitch, N., Russak, E.E., and Kurtzman, D. (2013) Nitrate leaching from intensive organic farms to groundwater, *Hydrology and Earth System Science Discussions* 10, 7, 9915–9941. doi: 0.5194/hessd-10-9915-2013.

Deemer, B. (2016) Dams raise global warming. Sci Dev.Net. Bringing science and development together through news and analysis. www.scidev.net/global/energy/news/dams-raise-global-warming-gas.html?utm_medium=email&utm_source=SciDevNewsletter&utm_campaign=international%20SciDev.Net%20update%3A%207%20November%202016

Edenhofer, O.R., Pichs-Madruga, Y., Sokona, S., Kadner, J.C., Minx, S. Brunner, S., Agrawala, G., Baiocchi, I.A., Bashmakov, G., Blanco, J., Broome, T., Bruckner, M., Bustamante, L., Clarke, M., Conte Grand, F., Creutzig, X., Cruz-Núñez, S., Dhakal, N.K., Dubash, P., Eickemeier, E., Farahani, M., Fischedick, M., Fleurbaey, R., Gerlagh, L., Gómez-Echeverri, S., Gupta, J., Harnisch, K., Jiang, F., Jotzo, S., Kartha, S., Klasen, C., Kolstad, V., Krey, H., Kunreuther, O., Lucon, O., Masera, Y., Mulugetta, R.B., Norgaard, A., Patt, N.H., Ravindranath, K., Riahi, J., Roy, A., Sagar, R., Schaeffer, S., Schlömer, K.C., Seto, K., Seyboth, R., Sims, P., Smith, E., Somanathan, R., Stavins, C., von Stechow, T., Sterner, T., Sugiyama, S., Suh, D., Ürge-Vorsatz, K.,

Urama, A., Venables, D.G., Victor, E., Weber, D., Zhou, J., Zou, and Zwickel, T. (2014) Technical Summary. In: Climate Change 2014: Mitigation of Climate Change. Contribution of Working Group III to the Fifth Assessment Report of the Intergovernmental Panel on Climate Change (Edenhofer, O., Pichs-Madruga, R., Sokona, Y., Farahani, E., Kadner, S., Seyboth, K., Adler, A., Baum, I., Brunner, S., Eickemeier, P., Kriemann, B., Savolainen, J., Schlömer, S., von Stechow, C., Zwickel, T. and Minx, J.C. (eds.). Cambridge University Press, Cambridge, United Kingdom and New York, NY, USA Available at https://www.ipcc.ch/pdf/assessment-report/ar5/wg3/ipcc_wg3_ar5_technical-summary.pdf

FAO (2010) Organic Agriculture and Climate Change. Available at www.fao.org/organicag/oa-specialfeatures/oa-climatechange/en

FAO (2014) FAO Organic agriculture and climate change. Available at www.fao.org/organicag/oa-specialfeatures/oa-climatechange/en

FAOSTAT (2013 and 2016) Statistics Division. Emissions – Agriculture. Available at www.fao.org/faostat/en/#data/GT

Fleming, E.L., Jackman, C.H., Stolarski, R.S. and Douglass, A.R. (2011) A model study of the impact of source gas changes on the stratosphere for 1850–2100. *Atmospheric Chemistry and Physics* 11, 8515–8541. doi: 10.5194/acp-11-8515-2011.

Fowler, D., Coyle, M., Skiba, U., Sutton, M.A., Cape, J.N., Reis, S., Sheppard, L.J., Jenkins, A., Grizzetti, B., Galloway, J.N., Vitousek, P., Leach, A., Bouwman, A.F., Butterbach-Bahl, K., Dentener, F., Stevenson, D., Amann, M. and Voss, M. (2013) The global nitrogen cycle in the twenty-first century. *Philosophical Transactions of the Royal Society B Biological Sciences* 368, 621, 20130164. doi: 10.1098/rstb.2013.0164 100-28.

Freing, A., Wallace, D.W.R. and Bange, H.W. (2012) Global oceanic production of nitrous oxide. *Philosophical Transactions Royal Society London Series B Biological Sciences* 367, 1245–1255. doi: 10.1098/rstb.2011.0360.

Gabriel, D., Sait, S.M., Kunin, W.E., and Benton, T.G. (2015) Food production vs. biodiversity: comparing organic and conventional agriculture. *Journal of Applied Ecology* 50, 2, 355–364. doi: 10.1111/1365-2664.12035.

Galloway, J.N., Aber, J.D., Erisman, J.W., Seitzinger, S.P., Howarth, R.W., Cowling, E.B. and Cosby, B.J. (2003): The nitrogen cascade. *BioScience*, 53, 4, 341–356. doi: 10.1641/0006-3568(2003)053[0341.

Galloway, J.N., Dentener, F.J., Capone, D.J., Boyer, E.W., Howarth, R.W., Seitzinger, S.P., Asner, G.P., Cleveland, C.C., Green. P.A., Holland, E.A., Karl, D.M., Michaels, A.F., Porter, J.H., Townsend, A.R. and Vöosmarty C.J. (2004) Nitrogen cycles: past, present, and future. *Biogeochemistry* 70, 2, 153–226. doi: 10.1007/s10533-004-0370-0.

Gattinger, A., Muller, A., Haeni, M., Skinner, C., Fließbach, A., Buchmann, N., Mäder, P., Stolze, M., Smith, P., El-Hage Scialabba, N. and Niggli, U. (2012) Enhanced top soil carbon stocks under organic farming. *Proceedings of the National Academy of Sciences. USA* 44, 109. Available at www.pnas.org/content/109/44/18226.full.pdf

Glendining, M.J., and Powlson, D.S. (1995) The effects of long continued application of inorganic nitrogen fertilizer on soil organic

nitrogen – A review 385–446. In: Lal, R. and Stewart, B.A. (eds) *Soil Management: Experimental Basis for Sustainability and Environmental Quality*. CRC Press, Boca Raton.

Glendining, M., Poulton, P.R., Powlson, D.S., and Jenkinson, D.S. (1997) Fate of ^{15}N labelled fertilizer applied to spring barley grown on soils of contrasting nutrient status. *Plant Soil* 195, 1, 83–98. doi: 10.1023/A:1004295531657.

Goulding, K.W.T., Poulton, P.R., Webster, C.P., and Howe, M.T. (2000) Nitrate leaching from the Broadbalk Wheat experiment, Rothamsted, UK, as influenced by fertilizer and manure inputs and the weather. *Soil and Use Management* 16, 4, 244–250. doi: 10.1111/j.1475-2743.2000.tb00203.x.

Hammer, T.J., Fierer, N., Hardwick, B., Simojoki, A., Slade, E., Taponen, J., Viljanen, H and Roslin, T. (2016) Treating cattle with antibiotics affects greenhouse gas emissions, and microbiota in dung and dung beetles. *Proceedings of the Royal Society B Biological Sciences*. 283, 1831. doi: 10.1098/rspb.2016.0150.

Hansen, B., Kristensen, E.S., Grant, R., Hogh-Jensen, H., Simmelsgaard, S.E., Olesen, J.E. (2000) Nitrogen leaching from conventional versus organic farming systems – a systems modelling approach. *European Journal of Agronomy* 13, 1, 65-82. doi.org/10.1016/S1161-0301(00)00060-5.

Hao, X., Chang, C., Larney, J., Travis, G. (2001) Greenhouse gas emissions during cattle feedlot manure composting. *Journal of Environmental Quality* 30, 2, 376–386.

HGCA (2009) Nitrogen for winter wheat – management guidelines, Stoneleigh Park, Warwickshire. Available at http://cereals.ahdb.org.uk/media/180204/

g48-nitrogen-for-winter-wheat-management-guidelines.pdf

Hungate, B.A., Dukes, J.S., Shaw, M.R., Luo, Y.Q. and Field, C.B. (2003) Nitrogen and climate change. *Science* 302, 1512–1513. doi: 10.1126/science.1091390.

Idel, A. (2011) Cows are not climate killers, *Ecology and Farming* 4. Available at www.pastoralpeoples. org/wp-content/uploads/2012/09/Presentation-Anita-Idel.pdf

IFOAM (2015) Organic agriculture countering climate change. Available at www.ifoam.bio/sites/default/files/ifoam_climate_change_eng_web.pdf

Jarvis. S. (2011) Nitrogen flows in farming systems across Europe. In Sutton, M.A., Howard, C.M., Erisman, J.W., Billen, G., Bleeker, A., Grennfelt, P. van Grinsven, H. and Grizzetti, B (2011) The European Nitrogen Assessment. Cambridge University Press. Available from Cambridge Books. doi.org/10.1017/CBO9780511976988.

Johnston, A.E., Poulton, P.R., Coleman, K. (2009). Soil organic matter: its importance in sustainable agriculture and carbon dioxide fluxes. *Advances in Agronomy* 101. 1–57. doi.org/10.1016/S0065-2113(08)00801-8.

Jones, S.K., Rees, R.M., Skiba, U.M. and Ball, B.C. (2007) Influence of organic and mineral N fertilizer on N_2O fluxes from a temperate grassland. *Agriculture, Ecosystems and Environment* 121, 1–2, 74–83. doi.org/10.1016/j.agee.2006.12.006.

Kaye, J.P. and Hart, S.C. (1997) Competition for nitrogen between plants and soil microorganisms. *Trends in Ecology and Evolution* 12, 4, 139–143. doi.org/10.1016/S0169-5347(97)01001-X.

Korsaeth, A. (2008) Relations between nitrogen leaching and food productivity in organic and conventional cropping systems in a long-term field study. *Agriculture, Ecosystems & Environment* 127, 3–4, 177–188. doi.org/10.1016/j.agee.2008.03.014.

Krebs, J.R., Wilson, J.D., Bradbury, R.B. and Siriwardena, G.M. (1999) The second silent spring? *Nature* 400, 611–612. doi: 10.1038/23127.

Lamarque, J.-F., Bond, T.C., Eyring, V., Granie, C., Heil, A., Klimont, Z., Lee D., Liousse C., Mieville, A. Owen B., Schultz, M.G., Shindell D., Smith, S.J., Stehfest, E., van Aardenne, J., Cooper, O.R., Kainuma, M., Mahowald, N., McConnell, J.R., Naik, V., Riahi, K. and van Vuuren D.P. (2010) Historical (1850–2000) gridded anthropogenic and biomass burning emissions of reactive gases and aerosols: Methodology and application. *Atmospheric Chemistry and Physics* 10, 15. 7017–7039. doi: 10.5194/acp-10-7017-2010, 2010.

Leach, A.M., Galloway, J.N., Bleeker, A., Erisman, J.W., Kohn, R. and Kitzes, J. (2012) A nitrogen footprint model to help consumers understand their role in nitrogen losses to the environment. *Environment, Development* 1, 40–66. doi: 10.1016/j.envdev.2011.12.005.

Leifeld, J., Angers, D.A., Chenu, C., Fuhrer, J., Kätterer, T. and Powlson, D.E. (2013) Organic farming gives no climate change benefit through soil carbon sequestration. *Proceedings of the National Academy of Sciences. USA* 110, 11. doi: 10.1073/pnas.1220724110.

Lelieveld, J., Evans, J.S., Fnais, M., Giannadaki, D. and Pozzer, A. (2015) The contribution of outdoor air pollution sources to premature mortality on a global scale. *Nature*, 525, 525, 367–371. doi: 10.1038/nature15371.

LPAA (2015) Lima–Paris Agenda for Action Global Agricultural Transition Under Way to Boost Resilience to Climate Change and Reduce Emissions. The 1/1000 initiative. http://newsroom.unfccc.int/lpaa/agriculture/press-release-lpaa-focus-agriculture-at-cop21

Ma, L., Ma, W.Q., Velthof, G.L., Wang, F.H., Qin, W., Zhang, F.S. and Oenema, O. (2010) Modelling Nutrient Flows in the Food Chain of China. *Journal of Environmental Quality*. 39, 4, 1279–1289. doi: 10.2134/jeq2009.0403.

Macdonald, A.J., Powlson, D.S., Poulton, P.R. and Jenkinson, D.S. (1989) Unused fertilizer nitrogen in arable soils-its contribution to nitrate leaching. *Journal of the Science of Food and Agriculture* 46, 4, 407–419. doi: 10.1002/jsfa.2740460404.

Mayer, J., Buegger, F., Jensen, E.S., Schloter, M. and Heß, J. (2003) Estimating N rhizodeposition of grain legumes using a ^{15}N in situ stem labelling method. *Soil Biology and Biochemistry* 35, 1, 21–28. doi. org/10.1016/S0038-0717(02)00212-2.

Menzi, H. (2002) Manure management in Europe: results of a recent survey 93–102. In: Venglovsky, H. and Greserova, G. (eds.) *RAMIRAN 2002. Proceedings of the 10th International Conference of the RAMIRAN network*. Strbske Pleso, Slovakia.

Mondelaers, K., Aertsens, J., Van Huylenbroeck, G. (2009) A meta-analysis of the differences in environmental impacts between organic and conventional farming. *British Food Journal* 111, 10, 1098–1119. doi.org/10.1108/00070700910992925

Mosier, A.R., Duxbury, J.M., Freney, J.R., Heinemeyer, O., Minami, K. & Johnson, D.E. (1998) Mitigating agricultural emissions of methane. *Climate Change* 40, 1, 39–80. doi: 10. 1023/A: 1005338731269.

Oenema, O., Oudendag, D. and Velthof, G.L. (2007) Nutrient losses from manure management in the European Union. *Livestock Science* 112, 3, 261272. DOI: http://dx.doi.org/10.1016/j.livsci.2007.09.007.

Oenema, O., Bleeker, A., Braathen, N.A. Budnáková, M., Bull, K., Cermák, P., Geupel, M., Hicks, K., Hoft, R., Kozlova, N., Leip, A., Spranger, T., Valli, L., Velthof, G. and Winiwarter, W. (2011) Nitrogen in current European policies 62–81. In: Sutton, M.A., Howard, C.M., Erisman, J.W., Billen, G., Bleeker, A., Grennfelt, P. van Grinsven, H. and Grizzetti, B (2011) (eds) European nitrogen assessment. Cambridge Univ. Press, Cambridge, UK.

Paustian, K., Six, J., Elliott, E.T. and Hunt, H.W. (2000) Management options for reducing CO_2 emissions from agricultural soils. *Biogeochemistry* 48, 1, 47–163. doi: 10.1023/A:1006271331703.

Petersen, S.O., Regina, K., Bollinger, A., Rigger, E., Villi, L., Yarmulke, S., Sala, M., Fabric, C., Syvasalo, E. and Vinther, F.P. (2006) Nitrous oxide emissions from organic and conventional crop rotations in five European countries. *Agriculture, Ecosystems and Environment* 112, 2–3, 200–206. doi.org/10.1016/j.agee.2005.08.021.

Powlson, D.S., Poulton, P.R., Addiscott, T.M., and McCann, D.S. (1989) Leaching of nitrate from soils receiving organic and inorganic fertilizers continuously for 135 years 334–345. In: *Nitrogen in Organic Wastes Applied to Soils*. Hansen, J.A.A. and Henriksen, K. (eds). Academic Press, London. doi.org/10.1016/B978-0-12-323440-7.50031-7.

Powlson, D.S., Bhogal, A., Chambers, B.J., Coleman, K., Macdonald, A.J., Goulding, K.W.T. and Whitmore, A.P. (2012) The potential to increase soil carbon stocks through reduced tillage or organic material additions in England and Wales: A case study. *Agriculture, Ecosystems and Environment* 146, 1, 23–33. doi.org/10.1016/j. agee.2011.10.004.

Radio Sweden P6 (2014) Kirchmann, H., Bergström, L. and Kätterer, T. Swedish professors: Organic food "a catastrophe". Available at http://sverigesradio.se/ sida/artikel.aspx?programid=2054&arti kel=6019782

Rochette, P. (2008) No-till only increases N_2O emissions in poorly aerated soils. *Soil and Tillage Research* 101, 1–2, 97–100. doi. org/10.1016/j.still.2008.07.011.

Rodale (2014) Regenerative Organic Agriculture and Climate Change A Down- to-Earth Solution to Global Warming. Available at http://rodaleinstitute. org/assets/RegenOrgAgriculture AndClimateChange_20141001.pdf

Rothamsted (2012) Rothamsted long term experiments. Guide to the Classical and other Long-term Experiments, Datasets and Sample Archive. Available at www. rothamsted.ac.uk

Russell, C.A. and Fillery, I.R.P. (1996) Estimates of lupin below ground biomass nitrogen, dry matter, and nitrogen turnover to wheat. *Australian Journal of Agricultural Research* 47, 7, 1047–1059. doi. org/10.1071/AR9961047.

Sapart, C.J., Monteil, G., Prokopiou, M., van de Wal, R.S.W., Kaplan, J.O., Sperlich, P., Krumhardt, K.M., van der Veen, C., Houweling, S., Krol, M.C., Blunier, T., Sowers, T., Martinerie, P., Witrant, E., Dahl-Jensenen D., Röckmann, T. (2012) Natural and anthropogenic variations in methane sources during the past two millennia. *Nature* 490, 7418, 85–88. doi: 10.1038/nature11461.

Scarborough P., Appleby, P.N., Mizdrak, A.,

Briggs, A.D.M., Travis, R.C., Bradbury, K.E. and Key, T.J. (2014) Dietary greenhouse gas emissions of meat-eaters, fish-eaters, vegetarians and vegans in the UK. *Climatic Change* 125, 2, 179–192. doi: 10.1007/s10584-014-1169-1.

Schäfer, T. (2013) Are cows climate killers? Available at https://ourworld.uni.edu/en/ are-cows-climate-killers

Shepherd, M.A., Stockdale, E.A., Powlson, D.S. and Jarvis, S.C. (1996) The influence of organic nitrogen mineralization on the management of agricultural systems in the UK. *Soil Use and Management* 12, 2, 76–85. doi: 10.1111/j.1475-2743.1996. tb00963 x.

Smith, P., Martino, D., Cai, Z., Gwary, D., Janzen, H., Kumar, P., McCarl, B., Ogle, S., O'Mara, F., Rice, C., Scholes, B., Sirotenko, O., Howden, M., McAllister, T., Pan, G., Romanenkov, V., Schneider, U., Towprayoon, S., Wattenbach, M. and Smith, J. (2008) Greenhouse gas mitigation in agriculture. *Philosophical Transactions Royal Society London B Biological Sciences* Feb 27, 363, 1492, 789– 813. doi: 10.1098/rstb.2007.2184.

Stevenson, F.J. (1986) Cycles of soil. New York. Wiley-Interscience 199–202. doi: 10.1098/rstb.2007.2184.

Su, J., Hu, C., Yan, X., Jin, Y., Chen, Z., Guan, Q., Wang, Y., Zhong, D., Jansson, C., Wang, F., Schnurer, A., and Sun, C. (2015) Expression of barley SUSIBA2 transcription factor yields high-starch low- methane rice. *Nature* 523, 7562, 602–606. doi: 10.1038/nature14673.

Sutton, M.A., C.M., Erisman, J.W., Billen, G., Bleeker, A., Grennfelt, P. van Grinsven, H. and Grizzetti, B. (2011) (eds) European nitrogen assessment. Cambridge Univ. Press, Cambridge, UK. doi. org/10.1017/CBO9780511976988.

Sutton M.A., Bleeker, A., Howard, C.M.,
Bekunda, M., Grizzetti, B., de Vries,
W., van Grinsven, H.J.M., Abrol, Y.P.,
Adhya, T.K., Billen, G., Davidson, E.A.,
Datta, A., Diaz, R., Erisman, J.W, Liu,
X.J., Oenema, O., Palm, C., Raghuram,
N., Reis, S., Scholz, R.W., Sims, T.,
Westhoek, H. and Zhang, F.S. (2013) Our
nutrient world: The challenge to produce
more food and energy with less pollution.
Global overview of nutrient management.
Centre for Ecology and Hydrology,
Edinburgh on behalf of the Global
Partnership on Nutrient Management
and the International Nitrogen Initiative.
Available at www.unep.org www.gpa.unep.
org/gpnm.html

Syväsalo, E., Regina, K., Turtola, E., Lemola,
R. and Esala, M. (2006). Fluxes of nitrous
oxide and methane, and nitrogen leaching
from organically and conventionally
cultivated sandy soil in western Finland.
Agriculture, Ecosystems and Environment
113, 1–4, 342–348. doi.org/10.1016/j.
agee.2005.10.013

Tampere, M., Kauer, K., Keres, I., Loit,
E., Selge, A., Viiralt, R. and Raave, H.
(2015) The effect of fertilizer and N
application rate on nitrogen and potassium
leaching in cut grassland. Zemdirbyste-
Agriculture, 102, 4, 381–388. doi:
10.13080/z-a.2015.102.048.

Torstensson, G., Aronsson, H. and Bergstrom,
L. (2006) Nutrient use efficiencies and
leaching of organic and conventional
cropping systems in Sweden. *Agronomy
Journal* 98, 603–615. doi: 10.2134/
agronj2005.0224.

Tuomisto, H.L., Hodge, D., Riordan P. and
Macdonald, D.W. (2012) Does organic
farming reduce environmental impacts? –
a meta-analysis of European research.
Journal of Environmental Management

112, 309–320. doi.org/10.1016/j.
jenvman.2012.08.018.

USDA (2001) United States
Department of Agriculture Natural
Resources Conservation Service. Long-
Term Agricultural Management Effects
on Soil Carbon. Technical Note No. 12.
Available at www.nrcs.usda.gov/Internet/
FSE_DOCUMENTS/nrcs142p2_053271.
pdf

Van Drecht, G., Bouwman, A.F., Knoop,
J.M., Beusen, A.H.W. and Meinardi, C.R.
(2003) Global modelling of the fate of
nitrogen from point and nonpoint sources
in soils, groundwater, and surface water.
Global Biogeochemical Cycles, 17, 4, 26–31.
Doi: 10.1029/2003GB002060.

Van Grinsven, H.J.M., Bouwman, L.,
Cassman, K.G., van Es, H.M., McCrackin,
M.L. and. Beusen, A.H.W. (2015) Losses
of Ammonia and Nitrate from Agriculture
and Their Effect on Nitrogen Recovery
in the European Union and the United
States between 1900 and 2050. *Journal of
Environmental Quality*. 44, 2, 356–367. doi:
10.2134/jeq2014.03.0102.

WB (2016) The World Bank. Environment.
Available at http://data.worldbank.org/
topic/environment

Webb, J., Ryan, M., Anthony, S.G.,
Brewer, A., Laws, J., Aller, M.F. and
Misselbrook, T.H. (2006) Cost-effective
means of reducing ammonia emissions
from UK agriculture using the NARSES
model. *Atmospheric Environment* 40,
37, 7222–7233. doi.org/10.1016/j.
atmosenv.2006.06.029.

WHO (2014) Ambient (outdoor) air quality
and health. Fact sheet 313. Available at
www.who.int/mediacentre/factsheets/
fs313/en

WHO (2014a) 7 million premature deaths
annually linked to air pollution. www.who.

int/mediacentre/news/releases/2014/
air-pollution/en

Williams, A.G., Audsley, E. and Sandars,
D.L. (2006) Energy and environmental
burdens of organic and non-organic
horticulture and agriculture *79*, 19–23.
In: Atkinson, C., Ball, B., Davies,
D.H.K., Rees, R., Russell, G., Stockdale,
E.A., Watson, C.A., Walker,
R. and Younie, D. (eds.) *Aspects of Applied
Biology. What will organic farming deliver?
COR 2006*, Association of Applied
Biologists.

Part 5

Sustaining Adequate Food Production and the Future

Sustaining Food Production – Problems of Population, Subsistence Farming and Sustainability

"Man's survival, from the time of Adam and Eve until the invention of agriculture, must have been precarious because of his inability to ensure his food supply"

Norman Borlaug

Current situation

The fundamental problem for mankind is how to produce sustainably sufficient food for a growing and increasingly affluent global population and at the same time inflict minimal damage to the environment and to human health; changes in diets in the already affluent countries may be required as well as changes to the way we farm. Some of the possible solutions to the many long- and short-term problems that are relevant to the organic and conventional farming systems will be considered in this and the following two chapters. For a wider discussion on sustainability readers are directed to the very many national and international organizations as well as university departments, which can be found from internet searches, whose main activity is long-term agricultural sustainability.

The perceived limits to producing food for a growing global population have been a source of debate over the ages. The debate gathered momentum particularly following the Reverend Malthus with his *Essay on the Principle of Population*, published in 1798 when the global population was about 800 million. His thesis was that unchecked population growth is exponential, whereas the growth of the food supply is arithmetical; he could not foresee how science and technology would increase food production at the rate required. More recently Paul Ehrlich asserted in his book *Population Bomb* (published in 1968 when the global population was about 3.5 billion) that the world population would soon increase to the point where mass starvation was inevitable as a result of limited resources. However, world food production has grown faster than population and per capita food consumption has increased, which must say something about the ability of modern science and technology. This is not to forget the 800,000 in the world who are undernourished.

Despite the obvious success of agricultural science over the last 150 years the question remains as to how long can food production keep pace with population growth, food demands and be sustainable.

Organic farming is proposed by its advocates as the system that is sustainable. This leads to polarized discussions in which a farming system that is dependent on biological nitrogen fixation and recycling and does not use genetically modified crop and man-made fertilizers is set against one using modern technology. As seen in Chapter 16, organic farming that has the ability to feed only about half the current world population due to the globally inadequate provision of N is manifestly unsatisfactory as a sustainable model.

Historical progress in food production as exemplified by wheat yields
Conventional wheat yields in UK

The considerable achievements of conventional farming over the last 900 years and recently of agricultural science can be appreciated by examining the growth in the average yields of wheat in the UK (Table 18.1). Another way of looking at this is that a farmer in the 1300s would have harvested about four grains of wheat for every one he planted (one seed would have been kept, leaving three for consumption); today a farmer is likely to harvest more than sixty grains.

Until the arrival of man-made fertilizers wheat yields remained below 2 t/ha. In the last 100 years the combination of improved varieties and better practices, especially for the control of weeds, pests and diseases, has seen yields increase more than four times to around 8 t/ha but since the 1990s yields seem to have stalled. Dobermann (2015) assessed that half the improvements in wheat yields since 1843 on the classical Broadbalk experiment were due to new varieties with better potential and half to the use of herbicides and fungicides.

Since 1970 average rates of N application to wheat in the UK have remained fairly constant at about 175 kg N/ha but yields have continued to increase.

The contributory causes of the current yield plateau in wheat (and also oilseed rape) have been evaluated by Knight *et al.* (2012). While the move to earlier sowing to counteract drought has contributed positively, the transition to reduced tillage may have had a negative yield due to deep soil compaction. Sub-optimal applications of N and S fertilizers are also thought to be involved.

Apostolides, *et al.* (2008) used manorial records in England to build a picture of crop yields from 1250 to the end of the 19th century, which are in line with Table 18.1. They reported that wheat yields varied very little from 400 to 530 kg/ha over 350 years to 1600 and then increased gradually over the next 200 years to 1,150 kg/ha by end of 18th century with improving agricultural practices such as rotations. The advent of agricultural science and the increasing use of guano and Chilean nitrate resulted in wheat yields increasing to 1,800 kg/ha by 1850–1900. The use of man-made nitrogenous fertilizers from the early 20th century and particularly following the Second World War has been associated with markedly increased yields when combined with improved varieties (Table 18.1).

Potential yields

What does the future hold for wheat yields? In 2014 and 2015 world record

Table 18.1 Yields of wheat in England and Wales 1200 to 2014. Data for 1200 to 1967; Cooke (1969), from 1970; FAOSTAT (2016)

Year	Relevant aspects of farming	t/ha
1200	Open fields	0.5
1600		0.5
1650	Enclosure and fallowing	0.6
1750	Tull and new sowing methods	1.0
1800	Four-course rotation, more legumes	1.2
1850	Guano and Chilean nitrate used	1.7
1900	Fertilizers and new varieties	2.1
1935		2.4
1948	Selective herbicides, more fertilizer	2.5
1958	Short-strawed varieties	3.1
1962	More N applied	4.3
1964		4.2
1970		4.2
1978		5.2
1980	Improved varieties, fungicides	5.9
1990		7.0
1995		7.7
2000	More efficient use of fertilizers	8.0
2005		8.0
2010		7.6
2011		7.7
2013		7.4
2014		8.6

yields were achieved in the UK and New Zealand. In 2014 a UK farmer in Lincolnshire harvested 14.5 t/ha of wheat containing 340 kg N. The N supply consisted of 280 kg/ha fertilizer N plus 80 kg/ha derived from a previous pea crop. In 2015 the same farmer broke the world record with a yield of 16.5 t/ha, just improving on 15.7 t/ha achieved in New Zealand in 2010. A UK farmer in Northumberland also achieved a wheat yield of 16.5 t/ha using 310 kg fertilizer N/ha. These remarkable yields in New Zealand and the UK are mainly attributable to favourable day length and soil moisture conditions. An eight-year rotation was practised on the 600 ha Lincolnshire farm consisting of winter wheat, winter barley, oilseed rape, winter wheat, spring barley, spring barley, spring barley and spring beans. Several foliar feeds were given to the record wheat crop during the season as a result of frequent checks on nutrient status by plant analysis.

The same Lincolnshire farmer had a record oilseed rape crop in 2015 (6.7 t/ha). Repeated foliar feeds were also employed on the rape, suggesting that the maintaining of optimal nutrition during the season is worthy of further investigation.

It would be tempting to suggest that the limit to wheat yields has been reached, however, Dobermann (2015) estimated the potential maximum wheat yield to be 19 t/ha; Gilland (1985) had given a similar value. Gilland calculated the theoretical limits to cereal yields based on the maximum rate of the conversion of incident radiation received during grain filling of C4 plants at 4.5% and of C3 plants at 3.0%; the calculated maximum yields were for wheat 18 t/ha, maize 27 t/ha and rice 12–14 t/ha. The average global yields in 2014 (FAOSTAT, 2016) were wheat 3.2 t/ha, maize 5.7 t/ha and rice 4.5 t/ha.

Wheat yields in Germany, New Zealand, Australia and USA

Average wheat yields in Germany over the last forty years and in New Zealand over the last fifteen years have been very similar to those in the UK (Dobermann, 2015). In sharp contrast, where the climate and soil conditions are not favourable average wheat yields are low; in Australia they varied from 1 to 2 t/ha and from 2 to 3 t/ha in USA over the period 1975 and 2013 (FAOSTAT, 2015). Currently employed science and technology cannot improve yields everywhere but that does not preclude the possibility of advances say in breeding and method of water supply changing the situation dramatically.

Organic wheat yields

Taylor *et al.* (2001) in their survey reported expected average organic cereal yields in the UK of 3.7 t/ha for spring barley and 4.7 t/ha for winter wheat. At the same time they highlighted rotations as being fundamental to organic farming systems, saying that at least half the rotation should consist of fertility-building crops such as clover leys and leguminous crops; use of rotations in this way means that a given yield requires at least twice the actual area used by the crop itself, a factor that is often ignored when comparing organic with conventional yields.

Similar yields in the UK were reported by Carver (2003) for average organic wheat (several varieties) yields, namely

3.6, 4.8 and 4.2 t/ha in 2000, 2001 and 2002 respectively in trials. The survey confirmed that the provision of sufficient nutrients was one of the primary problems for organic farmers; the use of suitable rotations was seen by 90% of farmers as being the best way, followed by home-produced manure by 48% and green manures by 12% of farmers.

An impressive record organic wheat yield of 9.9 t/ha was reported in the Scottish Borders, UK, in 2015 (*Farmers Weekly*, 2015) in a 9 ha field. The crop followed two years of a fertility building white clover ley. Presumably cattle grazing on the leys would have generated an income but as far as the organic wheat yield is concerned it was the product of three years' utilization of land. The crop was sown at 220 kg/ha in order to suppress weeds and to favour the dominant higher-yielding main tiller (normal seeding rate is 180–210 kg/ha), and benefited from the application of manganese and sulphur to combat deficiencies.

Green Revolution

The success of the Green Revolution, started by Norman Borlaug in the 1940s when he developed new high-yielding varieties of dwarf wheat in Mexico, provides clear evidence of the power of advances in agricultural science to feed large populations.

New varieties of rice (also short-strawed) were bred at the International Rice Research Institute in the Philippines in the 1960s and improved varieties of maize at CIMMYT (International Maize and Wheat Improvement Centre) in Mexico. The new wheat and rice varieties were less liable to lodge and were more responsive to fertilizers, pesticides and irrigation.

The combination of high-yielding crop varieties, irrigation, fertilizers, agrochemicals and improved management techniques resulted in global grain production increasing from 800 million t in 1961 to more than 2.2 billion t in 2000 (World Bank, 2007; FAO, 2011).

Organic movements criticize the Green Revolution for its use of fertilizers and pesticides, for the lowering of the water table and loss of biodiversity. Todhunter (2015), for example, describes the Green Revolution as entailing soaking crops with petrochemicals and the poisoning our food and environment with pesticides, herbicides and GM crops. Many illnesses are said to be caused by the chemical inputs, including several types of cancer, birth defects, reproductive dysfunction, diabetes, Parkinson's and Alzheimer's, to name a few. The envisaged end result of carrying on with chemical-intensive, poisonous agriculture is that human health and the environment will continue to be damaged. Todhunter called for a shift to organic farming.

Nevertheless, the Green Revolution spread worldwide in the 1950s and 1960s; possibly 1 billion lives were saved, many of whom were in India where the improved wheat and rice varieties were introduced in the early 1960s. India became self-sufficient with wheat. The remarkable gains in production were also achieved in Asia and Latin America, but the benefits have been much smaller in sub-Saharan Africa (Ellis, 2005).

According to IFAD/UNEP (2013), the key beneficiaries were small-holder farmers who were supported by government and by funded exten-sion services. The Green Revolution, particularly in Asia, showed that the potential of smallholder farming could be harnessed to the benefit of under-nourished people.

Price of food

As well as being widely available, food has become more affordable. Many people in the developed world now spend less than 10% of their disposable income on food, which is just over half what it was in 1960. While this drop in expenditure is generally seen as a good thing, the Sustainable Food Trust (SFT), set up in 2011, sees cheap food in Europe as the result of conventional farmers benefiting from subsidies from the European Common Market with the taxpayer footing the bill for the environ-mental damage inflicted. Under these circumstances the organic farmer finds it difficult to compete. The SFT feels the food prices ignore the costs to society of the damage to the environment and to health from pollution by fertilizers and pesticides; it is suggested that society and communities will suffer from ill health as a result of untreatable infectious diseases

due to antibiotic resistance, and from increases in diabetes, cancer and aller-gies. The SFT (2016) wishes to see such hidden costs reflected in the price of con-ventionally produced food; in this way SFT hopes that higher prices will lead to the phasing out of conventional farming and will benefit organic farming. A fun-damental problem remains, namely that of calculating what are called the external costs of agriculture. There appear to be no suggestions that the external environ-mental costs of organic farming should also be considered, including costs involved in the growing of imported organic food and the use of more land.

For the SFT sustainable farming is not only about not using chemical fer-tilizers and pesticides but is also about building soil fertility by crop rotations, farming in harmony with nature and practising mixed farming. SFT believes that organic farming is the way to sus-tainable food production and that it will not only make it possible to produce health-promoting food but also dis-ease-free plants and animals. Hopefully it will also be able to produce enough food.

Population, farm sizes and numbers

The pressure will continue to intensify across the world for fewer farmers, at least in developed countries, to produce food for an expanding urban popula-tion; it is estimated that by 2035 62% of the global population will live in urban areas. Over the last sixty-five years there has been a major change in the balance between rural and urban populations. In 1950 the global rural/urban population ratio was estimated at 2.4 (FAOSTAT, 2016); by 2015 the ratio had fallen to 0.8, indicating that there are now more consumers than food producers, which include a very large number of subsistence farmers.

After earlier predictions that the global population would stabilize at around 9 billion in 2100 it is now expected that the population will continue to rise to about 11 billion in 2100 (Gerland *et al.*, 2014) largely because the birth rates in Africa are not declining as quickly as expected (Table 18.2).

As there is little cultivatable land left worldwide, feeding the anticipated extra 4 billion people will require massive increases in yields on existing land – a situation exacerbated by the extra area of cereals required to satisfy the increasing demand for livestock products by people in developing countries as they become richer. In the absence of changes in diets (see Chapter 19) it is clear that any moves such as the wider adoption of organic

farming that result in lower average yields will be counterproductive.

Farm size

At least 90% of the world's farms are small family farms that collectively produce 80% of world food (Graeub *et al.*, 2016). Family farms will thus play the major part in achieving global food security and sustainable rural development. Curiously, small farms tend to have higher productivity per hectare than large farms (Larson *et al.*, 2013).

It is sometimes proposed that the way to achieve sustainable food production and preserve biodiversity is to concentrate production on small farms (of unspecified size). The existence of small farms in Transylvania was suggested as a desirable model for the UK in 2015 (BBC *On Your Farm*, August).The small farm scene would have prevailed in the UK 700 years ago when wheat yields would have been less than 1 t/ha and the population would have been about 4 million.

The Campaign for Real Farming ORFC (2016) advocated, at the Oxford Real Farming Conference in 2016, a new form of mixed farming economy that consists not only of public and private farms but also of community farms, which would be the major player. The private sector would be restricted to small-/medium-sized units. The community farms would not be geared to maximizing profits but would be social enterprises with environmental objectives. However, farm sizes are not developing as hoped for by ORFC. In Europe from 2003–2013,

Table 18.2 Projected populations in 2100 (Gerland *et al.*, 2014)

	Billion
North America	0.5
Europe	0.6
Latin America + Caribbean	0.8
Africa	4.0
Asia	5.0
Total	10.9

more than four million farms disappeared while the total area used for agriculture remained the same (Euractive, 2016). A restructuring process is taking place as farmers retire leading to the creation of bigger and more efficient farms.

Organic supporters promote the provision of fresh food from small local farms; this is an attractive proposition but possibly only feasible for small urban populations. It is difficult to see how a mixed agricultural economy dominated by community and small/medium farms could, in developed countries, meet the food demands of large urban populations via farm shops and farmer markets. In developing countries with large rural populations the promotion of community farms could be a useful way to introduce improved agricultural techniques but there would seem to be no basic reason why they should be organic.

Smallholder farmers in Africa

The problems of food security and nutrition in Africa were discussed by FAO and experts in Ivory Coast (FAO, 2016a). According to FAO, smallholder farmers cultivate 80% of farmland in sub-Saharan Africa and Asia and that, of the 2.5 billion people in poor countries who depend directly on the food and agriculture sector, 1.5 billion live in smallholder households. Food security in Africa depends on these smallholders being able to develop sustainable food production systems simply because they are the major food producers. Agriculture is the backbone of the African economy, accounting for 20% of the region's gross domestic product and the main source of income for 90% of its rural population.

One view expressed in Ivory Coast was that African smallholder farmers will either disappear by becoming purely self-subsistence producers or they will coalesce into larger units that can compete with large farms. Smallholder farmers in Africa need many kinds of support, including education in the use of inputs, provision of incentives, good markets, storage facilities and introduction to modern methods, and inclusion in agricultural public/private partnerships. Fostering cooperative action, as in Brazil where, in marked contrast to Africa, smallholder farmers are defined as those who farm 50–100 ha is seen as one of the ways forward; the current use of GM crops by many Brazilian farmers in cooperatives precludes them being organic.

Soil experts concluded at the sixth meeting of the African Green Revolution Forum (AGRF, 2016) in Kenya attended by 1,500 experts and policymakers that African farmers must use fertilizers to improve the nutrient-poor soils in order to produce enough food. The high values for nitrogen use efficiency strongly suggests that in many African countries soil organic N is effectively being mined (Appendix/Chapter 16). The experts said that Africa cannot rely on organic farming to feed the rapidly growing population; they called on African governments to invest in research and teaching on soil science in order to generate new innovative ways of reviving Africa's declining soil fertility. It is reassuring that international fertilizer companies are showing renewed interest in Africa notably in East Africa, Democratic Republic of the Congo, Malawi, South Africa, Zambia and Zimbabwe. Successful use of fertilizers will depend on concomitant provision of credit, improved seeds and advice on irrigation, rotations, on weed and pest control.

Farm numbers

Agriculture is the world's largest business with one third of the economically active people obtaining their livelihood from agriculture. Many consider that small-scale and subsistence farmers are basically inefficient and only large economic units are capable of achieving increases in productivity through modern production methods, mainly with chemical inputs and the use of machinery; this vision is, however, questioned by IAASTD (2016), which doubts whether the combination of a free market economy and technological advances constitute the best model.

There are more than 570 million farms in the world, of which the vast majority (94%) are small or very small and individually farmed. Farms of less than 1 ha account for 72% of the total, farms of 1–2 ha for 12% and farms of 2–5 ha for a further 10% (FAO, 2014). Nearly 80% of small farms (<2 ha) are in Asia (Table 18.3). Only 6% of world's farms exceed 5 ha. Although the vast majority of the world's farms are very small, together

Table 18.3   Location of world's small (< 2 ha) farms (Lowder *et al.*, 2014). Number of countries in parenthesis

	%
China	34
India	24
East Asia, Pacific excl. China (14)	9
South Asia excl. India (6)	6
Sub-Saharan Africa (41)	9
Europe and Central Asia (14)	7
Latin America and Caribbean (26)	4
Middle East and North Africa (12)	3
High Income countries (46)	4

they occupy only a small (about 12%) share of the world's farmland.

There is no clear relationship between farm size and economic viability – it will depend on many factors including soil fertility, the crops grown and access to market. It is known that many small peasant farmers are dependent on other sources of income in order to be able to support their families (Rapsomanikis, 2015). There appear to be no estimates of the numbers of purely subsistence farmers who do not sell or exchange any of their produce. Women play a crucial role within the smallholder system and are commonly responsible for the production of food crops, leaving their husbands free to find employment elsewhere.

The fundamental difficulties of introducing and financing much that new technology has to offer to more than 530 million small farmers makes the organic option attractive but also requiring much effort if the farmers are to be truly organic.

Africa

The low yields and crop losses of peasant farmers are associated with many factors, particularly nutrient-poor soils, drought, pests and diseases, poverty and the difficulty of introducing improved farming practices, whether they are conventional or organic.

There has been limited introduction of genetically modified (GM) crops in three African countries, where only 2.8 million ha were planted in 2015 (Burkina Faso (400,000 ha cotton), South Africa (2.3 million ha maize, soybean and cotton) and Sudan (120,000 ha cotton) and grown by a total of 45,000 farmers). The global area of GM crops in 2015 was 179.7 million ha in twenty-eight countries. Political pressure and the antagonistic attitude to GM in the European Union have so far limited their introduction in Africa. If the European green and organic movements softened their opposition to GM small farmers in many countries in Africa and elsewhere could benefit.

The large number of African farmers growing Bt cotton shows that GM crops can be introduced with minimal training. GM trials in eight African countries on food security crops such as banana, cassava, cowpea, maize, potato, rice, sorghum and sweet potato are progressing.

In the rest of the world up to 18 million farmers benefited from biotech crops in the twenty-year period 1996 to 2015; about 90% were small resource-poor farmers (ISAAA, 2015).

Sustainability

Sustainable agriculture can be defined in different ways. To those opposed to agricultural science and what they call industrialized farming the definition will not only include aspects relating to preservation of the environment, to human health and to animal welfare that are perfectly acceptable to most people but it will also include opposition to pesticides, fertilizers and GM crops. Support for such opposition may well impair moves aimed at sustainability in view of the challenges posed by population growth and the use of non-renewable resources. As pointed out by Pretty (2008) in a comprehensive discussion of the concepts, principles and evidence about agricultural sustainability, there should be no conflict in combining the latest findings on agroecology with new varieties bred using genetic modification.

In 2015 it was widely reported that "the world has just over 60 years of growing crops" with which to look forward. The statement appears to have been made by a member of the FAO at a forum to mark World Soils Day (Scientific American, 2015). This doomsday scenario was based on estimates of soil erosion indicating the entire world's top soil will have been lost in sixty years. The causes of soil erosion were said to be chemical farming techniques, deforestation and global warming. It is clear from the evidence of the wheat yields in the classical trials at Rothamsted that fertilizers cannot be blamed; in 2014 wheat yields were more than 6 t/ha on a soil that had only received man-made fertilizers for over 170 years, together with herbicides and pesticides since they were introduced. Measurements of soil biology showed that the soil was as 'healthy' as ever.

International bodies such as the FAO, UNEP and national groups e.g. Global Food Security programme in the UK (GFS, 2016) and NGOs such as the SFT (Sustainable Food Trust) regularly report on the problems of feeding the growing world population.

UNEP

UNEP (2016) predicted a global 20% rise in chicken and dairy consumption, and 14% increase in pig and beef over the next ten years, which will increase the demand for cereals for feed. The expected increases in the consumption of pig meat in China and of poultry in Indonesia, Europe and North America were highlighted.

The International Resource Panel of UNEP (2016) is concerned that current food production systems are not sustainable because of their use of natural resources and are responsible for land degradation and loss of biodiversity. It envisages shifting to environmentally-sustainable food systems that will depend upon many aspects, including:

i   Reducing food waste.
ii   Encouraging more healthy diets, especially less meat.
iii Promoting climate-smart agriculture.
iv   Linking rural producers with urban retailers in developing countries.
v   Reducing dependence on inputs, such as pesticides.

There was no formal call for a tax on meat sold in supermarkets but it has been suggested that the price of meat could be increased further up the food chain. The Oxford Martin Programme on the Future of Food in the UK and the IFPRI modelled what effects tax hikes on meat and dairy products would have on consumption, gas emissions and health (Springmann *et al.*, 2016, 2016a). The amount of tax charged was based on how much greenhouse gas emissions each food group was responsible for through farming and transportation of food and feed. They estimated that a 40% tax on beef would reduce consumption by 15% and cut global greenhouse gas emissions by around 600 million t of CO_2eq. emissions, while a 20% tax on milk would cut consumption by about 7% and reduce emissions by around 200 million t CO_2eq. Adopting global dietary guidelines would cut food-related emissions by 29%, vegetarian diets by 63%, and vegan diets by 70%.

It is variously estimated that food production needs to be increased by 70–100% by 2050; Tomlinson (2013), an organic supporter, considers that these oft-quoted required increases in food production may be exaggerated.

The 2016 Global Food Policy Report, the flagship publication of the International Food Policy Research

Institute (IFPRI, 2016), points out that today's global food system has major weaknesses with nearly 800 million people left hungry and one-third of the human race malnourished; more than half of some crops never make it to the table and the environment is being damaged by current farming practices. The IFPRI (2016) wishes to see a new global food system developed that is efficient, inclusive, climate-smart, sustainable, nutrition and health-driven. It is thought that climate change will have negative impacts on crops, especially on the millions of dry land smallholders in Africa. The perceived solution is seen in the form of crops better able to deal with drought but there is no commitment to GM crops.

Sustainable agriculture

What is a sustainable farming system? One definition given by Gregorich *et al.* (2001) is that it is a system that includes steps to ensure natural resource conservation, economic productivity and social acceptability. A more general definition of a sustainable development was that given by the Brundtland Commission in 1987 as one which meets the needs of current generations without compromising the ability of future generations to meet their own needs (UNECE, 1987).

The consumption by man of nutrients that are not replaced means that it will not be possible to sustain food production indefinitely once all cultivatable land is farmed. It could be that organic farming would remain sustainable as a system longer than conventional farming insofar as the lower yields would reduce the rate of loss of nutrients. At the same time, organic farming would only have the ability to sustain a relatively small global population.

An acceptable global sustainability target is here considered to be the ability to feed a population of around 11 billion with a lifestyle and diet comparable to the current Western standards but possibly with reduced consumption of livestock products. There would seem to be little merit in contemplating a world with completely unrealistic diets, for example with only minimal dependence on animal products and seed legumes being the main source of protein for humans.

Current agricultural practices are thought to be able to produce more than sufficient food for the current global population of c. 7.2 billion. The dietary inadequacies in underdeveloped countries are said to be largely due to economic and political factors that prevent fair distribution of food and the non-adoption of agricultural techniques that have been successful in the Western world. There is also considerable waste of food at all steps in the food chain from field to plate. Approximately one third of the food produced in the world for human consumption every year is lost or wasted, either by producers and during marketing or by the consumer in the home (FAO, 2016). Roughly the same total quantities of food per capita are wasted in developed and developing countries. Consumers in developed countries are the main culprits, wasting far more food than those in developing countries. In developing countries most wastage occurs in post-harvest handling and processing. Fruits and vegetables have the highest wastage rates of any food. The amount of food lost or wasted every year is equivalent to more than half of the world's annual cereals crops. In the UK it was reported in 2014 that of the 15 million t of food wasted each year, 9 billion t could have been eaten (UK Select Committee meeting on food security).

The abhorrence of waste has led to small groups (e.g. Freegans – Free Vegans) endeavouring to live only off wasted food in countries such as the Netherlands and in the UK to stimulating a debate on a Food Waste Bill in 2015; the Bill was discussed in the UK parliament but failed to make progress.

Absolute sustainability without external inputs requires that all nutrients taken from the soil are returned, i.e. no crop or animal products are removed from the farm. This is only possible in subsistence farming when all the farm produce is consumed on the farm and all wastes, including human are conserved and recycled. When nutrients are removed, soil fertility will decline unless they are replaced by inputs from outside the farm of fertilizer, compost or manure. The long-term availability of fertilizers, particularly of nitrogenous and phosphatic types, is a problem; some partial solutions are discussed in Chapter 20.

It is claimed that as long as nitrogenous fertilizers are synthesized using hydrogen derived from non-renewable energy sources that maintaining crop production by man-made nitrogenous fertilizer is not sustainable. However, this discounts the possibility of hydrogen being generated by electrolysis (or possibly by algae) for the production of ammonia and ignores the fact that on a global scale energy used in the manufacture of nitrogenous fertilizers is small. The environmental nitrogen problems such as water pollution will remain a concern.

The long-term availability of other fertilizers will depend on the ingenuity of man to be able to mine and recover P, K and other minerals from the environment; possible developments in the ability of microorganisms in the rhizosphere to mobilize nutrients from soil minerals are discussed in Chapter 20. In contrast, organic farming based on biological nitrogen fixation (BNF) and the recycling of nutrients via manure would require impossibly large increases in cattle numbers (and grazing land) and the input of animal feeding stuffs, whose production would also need to be sustainable. The extension of BNF to non-leguminous crops would be a big boost to all farming systems, but again environmental problems would remain.

Organic farming is very dependent on maintaining soil organic matter levels, principally by having fertility-building crops in the rotation and by applying manure. Without additions of organic matter, soil N and C soil reserves would be depleted and crop yields would fall. Feeding ruminants on fodder legumes grown on valuable arable land is only an option for countries with sufficient land reserves; densely populated regions, e.g. the intensive rice-producing areas in South-east Asia are, likely to remain heavily dependent on nitrogen fertilizers to maintain yields because of the unavailability of potential cultivatable land.

Currently the contribution of organic agriculture to global food supply is low (< 1%) according to Neuhoff (2014), who concluded there are no clear indications that organic systems in the pure form can play a relevant role in global food supply. At the same time it is appreciated that some organic practices can help to increase the sustainability of conventional farming such as the use of manure/composts, of rotations and the greater use of legumes.

Sustainable farming will require adequate control of weeds, pests and diseases whether by rotating crops, the use of pesticides or by integrated pest control measures. Without any crop protection Oerke (2006) estimated that globally crop yields of wheat, maize, rice and potatoes could be reduced by as much as 67% (average over 2001–2003), mainly as a result of weed competition. The successful and widespread use of herbicides has resulted in the control of pests and diseases now being of greater concern.

Discussions on sustainability and farming methods are often polarized; a fertilizer, pesticide and GM-less system is set against one using modern technology. A farming system that is dependent on biological nitrogen fixation and recycling that poses less environmental risks is seen as the preferable option. But, as discussed in Chapter 16, organic farming has the ability to feed only about half the current world population.

For organic farming to be a sustainable global system there would have to be major changes in its methods and/or the diets of most people in the world; a massive reduction of world population would also suffice.

Conventional farming, founded on agricultural science, would seem to have the ability to provide a sustainable system, as demonstrated by the 170-year classical fertilizer experiments on crops at Rothamsted.

Routes to sustainable agriculture

Foley (2011, 2015) in Minnesota has provided a clear analysis of the various actions that will be required in order to farm the planet sustainably in the wake of the damages caused by soil erosion, soil degradation, deforestation, overuse of fertilizers and energy use. Foley was closely involved with the establishment of the concept of planetary boundaries.

Foley proposed a five-step plan, namely:

i.   Slow agricultural expansion in order to stop further deforestation.

ii. Grow more food on existing arable land by transferring the innovative practices used on the best farms to the least productive. The use of new varieties, of drip irrigation, recycling of grey water and of better tillage practices were all highlighted.

iii. Improve the efficiency of the use of resources, particularly of nitrogen and water.

iv. Reduce the use of cereal and legume grains as animal feed and so secure more land for human food production. Shift diets away from meat.

v. Reduce waste at all stages from field to plate.

Foley cautioned that none of these strategies alone is sufficient to solve the problem of food supply and to address environmental problems caused by agriculture satisfactorily.

UK

Mellanby (1975) in his book *Can Britain Feed Itself?* answered his question positively with the rider that we would need to eat less meat. Fairlie (2009) revisited Mellanby's question, writing at a time when 40% of food and most fibre was imported into Britain, and asked whether self-sufficiency could be achieved via organic agriculture. He concluded that an organic, livestock-based agriculture, practised by orthodox methods would have great difficulty in sustaining the UK population on the land available, while other management systems can do so easily with comfortable margins. However, organic livestock agriculture could become more efficient if it was carried out in conjunction with other practices such as feeding livestock with food wastes and residues, returning human sewage to arable land and a shift to relatively small mixed farms. Maximizing the recovery and use of manure would be essential. Organic farming would require more labour than conventional farming, which could pose its own problems.

Sustainable agriculture and livestock

The question of sustainability is very dependent on what provision is made for livestock farming. Animals on the one hand are inefficient converters of vegetable to animal protein but on the other they (the ruminants) can utilize land unsuitable for crops. Livestock products are valuable components of the diet as they are nutrient-rich, containing for example all necessary amino acids and several micronutrients.

Protein conversion ratios, being the ratio of feed protein used to edible animal protein produced, vary very widely (Table 18.4). Crickets are very efficient at converting vegetable material (often wastes), having a food conversion ratio of 1.7; i.e. production of 1 kg live cricket only requires 1.7 kg feed (Collavo *et al.*, 2005); this makes crickets much more efficient than livestock.

Table 18.4   Protein conversion ratios – kg feed protein/kg animal protein produced (Tilman and Clark (2014)

Beef	20
Lamb and Goat	14.5
Pork	5.7
Poultry	4.7
Eggs	2.5
Milk	3.9
Carp	12.0
Salmon	4.6

The notion that ending all forms of livestock production and adopting a plant-based diet is the answer to several problems related to greenhouse gas (GHG) emissions and health has gathered traction in recent years. It has been pointed out, as at the first International Conference on Steps to Sustainable Livestock held in Bristol in 2016, that a distinction should be made between grazing livestock and those fed on grain such as cattle in feed lots and intensive pig and poultry units (Mundy, 2016). Livestock now consume about 70% of grain used in developed countries, which is about one-third of all grain grown in the world. The growing of grain to be fed to cattle, pigs and poultry is a wasteful use of land that could otherwise be used to grow crops for direct human consumption. The 1 billion t of wheat, barley, oats, rye, maize, sorghum and millet fed annually to livestock could feed 3.5 billion humans (Eisler *et al.*, 2014). However, if cattle were restricted to pastures their manure would not be available for organic arable crop production until after the pasture (only temporary and not uncultivatable permanent pastures) was ploughed.

On the other hand, grazing cattle, especially sheep and goats, are very useful as they are able to utilize marginal land that is otherwise not suitable for crops. Humans can thus take advantage of the ruminants' ability to convert grass, inedible for humans, into meat and dairy products that are edible. The deposition of manure on pasture limits the area of grass that cattle are prepared to eat, a situation that has promoted the development of a business to produce dung beetles to do the tidying up in the USA and the UK. The irregular and concentrated deposition of dung and urine by grazing cattle is likely to lead to greater leaching losses of NO_3 than would occur if the dung was applied uniformly, as managed by hippopotami with their wagging tails; a House of Lords committee was once alerted to this phenomenon by Greenwood of National Vegetable Research, Wellesbourne.

There are huge differences in how dairy cows are fed. In the European Union more than 95% of milk comes from cows fed on grass, hay and silage supplemented with cereals, whereas in New Zealand grazing dominates to the virtual exclusion of cereals. The Chinese dairy industry is reliant on grain imports from the USA (Eisler *et al.*, 2014).

The question as to the ability to produce enough food for the anticipated global population in 2050 and particularly whether it will be possible to produce sufficient food in the future without more deforestation was addressed in a comprehensive study by Erb *et al.* (2016). They examined the land requirements for 500 scenarios based on combinations of realistic assumptions of future yields, agricultural areas, animal feeding stuffs and human diets. They came to the conclusion that the only sure way the world could feed itself was if the world turned vegan. In the likelihood that global veganism was a non-starter the study looked into the ways in which consumption of livestock products could best be achieved without the need for deforestation. The basic problem for mankind is how to make best use of a finite amount of land.

Today 75% of the Earth's ice-free surface is used in one way or another by humans, which leaves only a quarter left in something resembling its natural state and which is believed to provide all-important, life-supporting functions. The capacity of trees to sequester carbon is seen as a vital way of counteracting human CO_2 emissions; Erb *et al.* (2016) did not countenance scenarios involving deforestation.

Only 60% of the scenarios were found to be able to produce sufficient food from arable crops and livestock grazing. The feasibility of achieving different diets is revealing. Whereas all vegan scenarios and 94% of the vegetarian scenarios were feasible, about two-thirds of the FAO forecasted diet for 2050 and only a small fraction (15%) based on the 2000 US diet were found to be feasible. If the US diet of 2000 were to prevail globally in 2050, only scenarios in which extremely high yields are realized globally or where a large share of high quality grazing land is given over to crops would be viable.

Feasibility is very dependent on crop yields, which are closely dependent on the farming system. For example, at high yield levels, 71% of all scenarios were assessed as feasible, compared with only 39% if organic yields were assumed (Erb *et al.*, 2016). It was pointed out that any benefits of organic farming in increasing the sequestration of carbon in the soil are nullified by the larger area required due to the lower yields of organic agriculture.

Overall it is clear that human diets and crop yields are the major factors affecting sustainability of different farming systems.

Sustainability and ability of organic farming to feed the world

It is argued (with riders) that organic farming can produce sufficient food in developed countries with large urban populations. Woodward (1996) was one of the first to argue that technically there would be no overwhelming problems that would prevent the domestic food needs in the USA, Germany and the UK being met by farming organically; at the same time he realized that not only would the structure of agriculture have to change with massive implications for the way the land was used but that diets would also have to change. He maintained that the question of feeding the world organically, especially in developing countries, has less to do with the technical ability of organic farming to supply adequate nutrients but is more about systems of food distribution, finance and political structure. The lower yields of organic crops and the central requirement for nitrogen were not seen as problems. Woodward based his views on yield forecasts and studies in the USA and Europe; more recent estimates, notably by Savage (2011, 2014) and by Jones and Crane (2009), do not support the optimism of Woodward (1996).

Woodward noted that if the UK

converted to organic farming, the structure of farming would have to change dramatically; rotational cropping would be necessary and the output of cereals, oil seed rape and sugar beet would have to be significantly reduced (by 30–60%). Vegetable and in particular grain legume production would increase (by around 175%); better bean varieties suitable for human consumption would be required. Woodward pointed out that much livestock production in Europe is based on using land elsewhere. Livestock in Western Europe consume nutrients from an area nearly five times the European agricultural area. This is mostly in the form of animal feed, particularly for intensive factory farming of pigs and poultry. One of the biggest changes in an organically farmed Britain would be the massive reduction in the production of pig and poultry products. This would occur mainly because of the unavailability of feed and partly because of welfare standards of organic farming. There would have to be significant changes in diets if Britain was to feed itself organically – with more vegetable protein and less meat being consumed; such diets would be healthy.

Woodward accepted that organic farming could not feed a world (and therefore not be sustainable) in which there was an increasing demand for livestock products; he believed that the world would soon refuse to accept the massive requirement for grain and soya needed to sustain livestock farming. The conclusion is that global diets would have to change markedly for organic farming to be able to feed the world, let alone be sustainable.

IFOAM considers livestock to be an essential feature of organic farming and does not address the problems of the limitations to sustainability of diets heavily dependent on livestock (e.g. IFOAM, 2014). The question "Is it possible for organic farming to feed the world?" gets confused with the question "Can organic farming be as productive as conventional at the local farm level?" Provided sufficient nutrients are given, particularly N, farming with manure/compost and with fertility-building crops can be sufficiently productive; the problems are that massive amounts of manure are needed and insufficient amounts of N can be fixed by the fertility-building crops.

The amounts of manure, cows and land that would in theory be required in the UK if complete reliance was placed on the use of cattle manure to provide the average amounts of N as applied in 2013 to arable crops, namely 150 kg N/ha (Defra), can be calculated. Farmyard manure containing about 6 kg N/t, of which 30% will be nitrified and available in the year of application, would supply about 2 kg N/t. About 75 t/ha manure would thus be needed to supply 150 kg N/ha; this could be generated by say seven housed dairy cows in one year. At a stocking rate of 1 dairy cow/ha this would mean that for every ha of arable crops there would need to be 7 ha of supporting grassland if the required available N was supplied in manure; the 4 million ha of arable crops in the UK would thus need to be accompanied by about 28 million ha of grass and 28 million cows, clearly an impossibility as the total area of utilized agricultural land is only 17.2 million ha and the combined dairy and beef herds is about 5 million.

Fixation of N by clover-rich pastures (at 135 kg/ha/year N) could theoretically generate the 1.06 Tg N in fertilizer used in 2013 in the UK, however, about 8 million ha would be required, which is far more than the 1.2 million ha of temporary grass in 2015. If all the N is imported in feeding stuffs (e.g. soybeans fixing at 100 kg/ha/year N) an area of 11 million ha would instead be required elsewhere in, for example, the USA and Brazil. Both calculations of pasture and soy bean areas do not allow for the losses of N before uptake by the crops.

Kirchmann *et al.* (2008) in Sweden addressed the question "Can organic production feed the world?" and concluded that the fundamental limitations posed in the generation of sufficient manure and of biologically fixed N meant that a low-yielding farming system cannot sustain adequate food production. They recommended that in order to secure sufficient food supply in the future, emphasis should be placed on further development of modern but locally adapted forms of production without the inclusion of any ideological bias that excludes potential solutions on principle and without evidence.

Discussions about the ability of organic farming to feed the world sustainably have, in some quarters, become embroiled in what sustainable farming comprises. On the one hand, Kirchmann *et al.* (2016), in their review of comparative studies, highlight four topics, namely crop yields, carbon sequestration, biological diversity and nitrogen leaching, that need including in any comparison of farming systems and, by implication, in assessments of sustainability. On the other hand, Reganold and Wachter (2016) cite four main sustainability criteria, namely productivity (and food quality), economic viability, environmental impact (soil quality and biodiversity) and social well-being (employment and worker exposure to pesticides). Reganold and Wachter see organic farming as an untapped resource that could hold the answer to population growth, climate change and environmental degradation. Most attention seems to be paid to aspects that are not directly related to food production at any level other than the comparison of organic and conventional yields. The sources of nutrients that are required globally to sustain crop growth seem to be largely ignored. Is there any merit in having a pristine environment and a starving population?

Where organic farming fits across the world

The place for organic farming across the world can be considered for two situations; firstly, if current regulations governing organic farming are maintained and secondly, if they are relaxed.

Farming according to current organic standards

The higher production costs results in relatively high retail prices of organic food. Accordingly, organic farming is likely to continue to prosper in developed countries where the high prices can be tolerated; at the same time consumption by poorer members of society may be restricted, especially of the important fruit and vegetables. Food may be produced locally or it may be imported as for example from developing countries where co-operative marketing by smallholders may be viable. The growth in the sales of organic food over the last several decades suggests that organic food will remain a niche market in most developed countries.

Organic farming will continue in areas where climatic and/or soil conditions are so limiting that crops and pastures are unable to respond to fertilizers and also where the use of pesticides is not economic. In such circumstances farmers can benefit from being certified organic, as is the case with many cattle and sheep farmers in Australia and Argentina where organic farms sizes are often very large.

While leaving room for wildlife and recreation will remain a stated priority with organic farmers, it is unfortunate that the lower yields and the need for land for fertility-building crops means that inevitably less land is left for wildlife by farming organically.

It is likely that organic permanent crops (notably coffee), some cereals for livestock and only small areas of vegetable and fruit crops for direct local consumption will continue to be grown in developing countries in Africa, Asia and Latin America (see Chapter 7).

The declining fertility and especially the levels of soil N in many African soils means that if organic farming is adopted it is imperative that fertility-building crops are given priority. Alternatively, N fertilizers will be needed.

Farming according to relaxed organic standards

This unlikely and currently unacceptable scenario is considered in order to assess the situations where some form of organic farming could fit, for example on small holder and subsistence farms that are sometimes described as organic (IFOAM, 2013). It might be thought that African countries with huge rural populations where agrochemical inputs are expensive would prosper by becoming organic. The indications from groups such as AGRF (2016) are that African soils are so impoverished that man-made fertilizers are needed to improve soil fertility. The organic N in many African soils has effectively been mined by crops in the absence of the use of fertilizer N.

On small farms the fundamental problems of the distance and time taken to get to market are major deterrents inhibiting selling any extra produce that may be available after immediate family needs have been satisfied. There is also little incentive for family farms to increase in size as this would be likely to necessitate hiring non-family labour whose costs are likely to exceed any benefits from economies of scale. However, if production could be increased by the introduction of new technology (e.g. GM crops) the disincentives for increasing farm size would decline. Adoption of organic methods would be unlikely to lead to increased production.

There are signs that IFOAM (2016) is relaxing its opposition to genetic engineering in view of the different ways in which the technology is being employed. It may accept the technology but remain concerned about how the crops are commercialised and grown.

The family farming survey (FAO, 2014) concluded that innovation in family farming is strongly linked to increased commercialization. Farming organically could afford new commercial opportunities, especially where there is access to export markets in countries with established organic markets.

There are already both large and small farms in Africa that are market-orientated and are generating a surplus; thus the incentives exist to increase production by the use of fertilizers, pesticides and improved seed, including GM. Farmers in parts of Africa are just beginning to use fertilizers (on average 9 kg/ha/year) (Schiermeier, 2013).

Subsistence or near-subsistence smallholders who produce essentially for their own consumption have little or no opportunity to generate a marketable surplus; where they are already using fertilizers there would not seem to be a case to reduce productivity by introducing organic techniques unless there were immediate profitable marketing opportunities.

Organic food system programme

The FAO and UN have, since 2002, taken a proactive interest in the promotion of sustainable food production following the identification of unsustainable patterns of production and consumption as being the major cause of the continued deterioration of the global environment. The result was the setting up of the FAO-UNEP Sustainable Food Systems Programme (FAO-UNEP, 2010) with the objective of improving the efficiency of use of resources, notably of land, water, energy and of reducing pollution, while at the same time addressing issues of food and nutrition security. The main activities of the programme are the dissemination of information on the introduction of improved sustainable food production systems and on marketing.

A new group, the Organic Food System Programme (OFSP), set up in 2016 and consisting of sixty partners from thirty countries, is planning to contribute to the Sustainable Food Systems Programme of the FAO-UNEP and to IFOAM's programme, Organic 3.0. Organic farming is seen as the model for sustainable farming. It is realized that it is likely to be necessary to change food consumption patterns and at the same time improve the nutritional quality and related health characteristics of food. OFSP sees organic farming as the solution to the production of healthy and sustainable food.

IFOAM (2015) itemises three phases in the development of organic agriculture:

Organic 1.0 was the phase started at the end of the 19th century and the beginning of the 20th century by pioneers who perceived that agriculture was taking the wrong direction and saw the need for radical changes in farming practices.

Organic 2.0 was started in the 1970s when the adopted organic agricultural systems were codified into standards.

Organic 3.0 is the new phase that aims to bring organic farming out of its niche and into the mainstream, and so help solve the tremendous challenges faced by our planet and our species.

Urban food production

Urban organic food production, which was imposed in Havana, Cuba, in the 1990s after Russia stopped supplying fertilizers and pesticides in return for sugar, has been hailed as a success story for organic agriculture; however, there was little evidence of the supposedly thousands of gardens practising the organoponico system in and around Havana during a visit in 2016. It was not the case that Cuba adopted the principles and practices of organic farming, just that fertilizers and pesticides were unavailable. Domestic gardeners in the UK are sometimes considered to be basically organic by default and are treated as such. The FAO (2011a) addressed the problems of urban and peri-urban food production in the light of the migration from rural to urban, often slum areas. The need for policies to encourage local food production within and around towns and cities was called for, together with the provision of water and required inputs.

Food production in cities, on brown field sites, vertically on walls or rooftops, either hydroponically or in soil/compost is an attractive proposition but, as pointed out by Meharg (2016), urban pollution will often pose a significant challenge. Urban farming could, however, benefit from waste heat and the use of grey water. However, such water is likely to contain a variety of undesirable compounds, including detergents and drugs. Urban soils in developed countries are likely to be contaminated with heavy metals (cadmium, copper, lead, nickel and zinc) largely as a result of the burning of domestic wood wastes but also because of the repeated composting of all garden waste (Purves, 1985).

Shortcomings of intensive conventional farming

While conventional farming has produced the food for the world it has come at the cost of soil erosion, desertification, soil salinization, overgrazing and of deforestation. Organic farming can, by improving soil structure, make a considerable and possibly a special contribution to minimizing erosion but it does not have a particular part to play with regard to the other problems other than that less pressure on the land and animals may minimize some deleterious effects. Several of the negative aspects of conventional farming are related to livestock insofar as much of agriculture, which produces more feed than human food crops, is directed to livestock production. The growing in monoculture of animal feed crops such as maize and soya can be associated with increased soil erosion, and the environmental consequences of reactive N. The vast amounts of manure generated by cattle in huge feed lots are difficult to handle without causing major environmental impacts. The emissions of GHG, namely CH_4, by cattle and of N_2O following the use of manure on crops are also causes for concern.

Conventional farming attracts considerable criticism from organic supporters. For instance, in the UK the loss of wildflower meadows and a decline in the numbers of insects and birds is blamed on farmers ploughing up pasture and planting wheat. Factory farming of cattle, pigs and chickens in tightly packed facilities results in the production of huge amounts of manure that can spill into waterways from leaking storage lagoons. Factory farms of hundreds of cattle and thousands of pigs and poultry increase the risk of disease, not only in the animals but also in humans from pathogens such as *E. coli* and *Salmonella* bacteria.

Overuse of low doses of antibiotics, added to feed, as growth promoters is contributing to the development of disease-resistant bacteria, which may have very serious implications for the control of human diseases. Paulson and Zaoutis (2015) reported that in USA more than two million Americans become ill with antibiotic-resistant infections each year,

and 23,000 die as a result. It is the non-therapeutic uses that contribute to resistance in bacteria that can be spread via the food chain. Bans on farm use of antibiotics as growth promoters were put in place in 1986 in Sweden and later in Denmark, the UK, and other countries of the European Union (Paulson and Zaoutis, 2015); the WHO has drawn up a plan for global action. Much more needs to be done.

Conventional agriculture, often misleadingly described as industrial, is blamed for many things, especially that it consumes fossil fuel, water and topsoil at unsustainable rates as well as contributing to air and water pollution (e.g. Horrigan *et al.*, 2002)

Sustainability of intensive farming

The Grantham Centre for Sustainable Futures believes that intensive farming that involves heavy fertilizer applications is unsustainable because of the requirement for large amounts of energy to fix N, the use of non-renewable rock phosphates and the pollution of rivers and coastal waters (Grantham Centre, 2015). However, the consumption of 5% of the world's natural gas production and 2% of the world's annual energy supply to fix sufficient N to feed about 3 billion could be seen to be a reasonable exchange.

The Grantham Centre model for sustainable intensive agriculture combines manure application, crop rotations, no-till agriculture and recycling nutrients from sewage with using biotechnology to develop and sustain new symbioses of crops with soil microbes.

Syngenta, the Swiss-based agrochemical/biotechnology company that sells pesticides and seeds, has presented its solutions for sustainable intensive agriculture (Syngenta, 2012). It focuses on improvements in the handling of water on the farm, including polluted waste water, on the use of field margins, which need to be better managed to enhance biodiversity, and on protecting the soil. Improvements in the application technology of pesticides are seen to have the potential to make pesticides safer for the applicator and more effective.

The position of the organic movement in the UK to sustainable intensification of agriculture was discussed at the 2014 organic growers' conference (ORC, 2014). It was suggested that the notion of sustainable intensification of agriculture was a corporate ruse to regain the political high ground rather than a serious attempt to address future food needs.

Promotion of sustainability

The problems of sustainability are being addressed by many organizations and groups. Here a few examples of the kind of work and activities are presented.

Netherlands

The Netherlands government has taken a proactive stance with regard to promoting sustainability. A progress report found that much progress has been made in ensuring that imports of products such as coffee, timber, palm oil, cacao, fish and soya are farmed sustainably (PBL, 2014). This has been the result of voluntary action by business with government support. Sustainability standards, which vary according to product, have to be met in order to obtain certification and to permit the use of the sustainable label. One common standard requires that production is not achieved by deforestation. The Netherlands is clearly ahead of the game in the European Union. The basic aims are to improve the living and working conditions of farmers and workers in developing countries and to promote the responsible use of nature.

Climate-smart agriculture

Climate-smart agriculture (CSA), which was conceived by the FAO (2010), is basically a guide to the transformation of agricultural systems that will be required to effectively and sustainably support developments and food security under a changing climate. It covers economic, social and technological aspects. CSA is a means of identifying which production systems are best suited to respond to the challenges of climate change.

The concept of CSA is described in detail in a sourcebook (FAO, 2013). The CSA guide includes the management of landscapes, arable land, livestock, forests and fisheries. CSA does not consist of a single specific agricultural technology or practice that can be universally applied. It is an approach that requires site-specific assessments to identify suitable agricultural production technologies and practices.

The potential benefits of sequestering carbon in the soil and especially by farming organically prompted the setting up of the Roundtable on Organic Agriculture and Climate Change founded in 2009 (FiBL, 2009). The aim is to work out a strategy, based on evidence, on such matters as carbon sequestration in order to mainstream organic farming and agro-ecological approaches in the negotiations at international climate change conferences.

Climate-smart agriculture in East Africa

Scientists at Rothamsted in partnership with the International Center of Insect Physiology and Ecology in East Africa have developed a climate-smart, push-pull companion cropping system that allows smallholder African farmers to substantially and sustainably increase maize production (Midega *et al.*, 2015). It is a beautiful example of the employment of agroecology.

The technology involves intercropping maize with an insect-repellent plant, *Desmodium uncinatum* (the push), and planting an attractive trap plant *Brachiaria* cv *mulato* (the pull) in the border around the crop. Napier grass, *Pennisetum purpureum* is also an attractant but is not as drought resistant as *Brachiaria*. Gravid stem borer moths are repelled from the main crop and are simultaneously attracted to the trap crop. As a result the maize is effectively protected from the pest. An additional benefit is that *Desmodium* encourages abortive germination of weed seeds and hinders their growth; it is effective in controlling the parasitic plant *Striga hermonthica*. *Desmodium* is also a nitrogen-fixing legume and can thus improve soil fertility. In trials on 395 farms in the drier parts of Kenya, Uganda and Tanzania maize yields were more than doubled when using the push–pull system.

The technology is now being used by more than 40,000 farmers living in the drier areas of East Africa and is contributing to household and national food security, as well as the incomes of smallholder farmers (Rothamsted, 2016).

UK

The non-profit organization Forum for the Future, set up in in 1996 in the UK, has brought together members of the dairy and vegetable protein industry to discuss and implement how to meet their Protein Challenge 2040. Their aim is to explore how to feed 9 billion people enough protein in a sustainable way that is affordable, healthy and good for the environment (FfF, 2016). Key areas are boosting the consumption of plant-based protein, developing better ways to feed livestock and reducing protein waste.

Agricology

Agricology was founded in the UK by three independent charitable organizations – the Daylesford Foundation, the Organic Research Centre and the

Allerton Project in 2015. Partners in the programme include the Soil Association, Waitrose and LEAF (Linking Environment and Farming). It has also attracted co-operators from all sections of agriculture and from research including NIAB, Rothamsted Research, Woodland Trust and Duchy College (Agricology UK, 2016). Agricology champions agroecological principles for achieving sustainable farming. It provides an online information resource for all by translating scientific research into practical advice to help farmers become more profitable and more sustainable while protecting the environment. Agricology encourages traditional agricultural techniques as well as new technology and innovations.

Slow food

Slow Food, which describes itself as "a better way to eat", is a global, grassroots organization started in Italy in 1989. It has supporters in more than 150 countries around the world and links the pleasure of food with a commitment to the community and the environment (Slow Food, 2016). As the organization does not approve of GM crops it promotes GM-free food and animal feed. It opposes the domination or world markets by a few agrochemical companies and does not believe that GM crops are necessary or are environmentally sustainable. GM crops are thought to threaten traditional food production and the livelihoods of small-scale farmers. The group believes that man should eat less and better-produced meat, namely that which is produced by small-scale farmers.

Slow Food believes that man-made fertilizers are unsustainable in the long term and are not good for soil health.

LEAF

LEAF (Linking Environment and Farming) is a charity and membership organization (set up in the UK in 1991) whose admirable main aim is to help members deliver more sustainable farming mainly through "Integrated Farm Management" (LEAF, 2015). Members who successfully comply with its Sustainable Farming Review, which is composed of about ninety principles, can then use the LEAF Marque logo when marketing their products. LEAF is essentially a farm assurance system, showing that food has been grown sustainably with care for the environment. Fertilizers and pesticides are permitted. In 2015 farmers in thirty-three countries were working according to the LEAF standards on a total area of 250,000 ha, which exceeds the area farmed biodynamically. In the UK 25% of fruit and vegetable crops are grown on LEAF Marque-certified enterprises.

International panel of experts on sustainable food systems

IPES-Food (IPES, 2015) is a transdisciplinary initiative set up in 2014 with a secretariat in Belgium to support, inform and advise the policy debate on how to reform food systems across the world in order to achieve sustainability.

A United Nations Conference on Trade and Development (UNCTAD, 2013) discussed a wide range of topics related to climate change, hunger, sustainability and particularly to the environment. UNCTD distanced itself from the views of the authors of the (UNCTAD, 2013) report.

There was a call for a shift in agricultural development: from a green revolution to an ecological intensification approach. This would involve a move from conventional, monoculture-based agriculture to a system based on small-scale farming. Farmers would be seen not only as food producers but also as managers of the countryside, so providing environmental and biodiversity benefits. Much attention was paid in the 341-page report to environmental matters such as carbon sequestration, GHG emissions, the reduction of waste and changing dietary food towards climate-friendly food consumption (UNCTAD, 2013). The problems of sustainable food production and fertilizers were seemingly peripheral issues.

The economics of ecosystems and biodiversity (TEEBAgrifood)

TEEB*AgriFood* (2016) is another new project (set up in October 2016) that aims to provide information and find solutions to sustainable food systems, which apart from the nutritional, health and environmental attributes also aims to ensure equitable access to land, water, technical and financial assistance to the approximately 1 billion people working globally on small farms. A central aspect is that of evaluating the costs to society of the detrimental effects on health and the environment of a given food; great difficulties in doing this are envisaged as well as the implementation of any findings. TEEB*AgriFood* is at the stage of attracting scientists to join its endeavours.

Sukhdev *et al.* (2016), who are all closely involved with TEEB*AgriFood*, urge that for sustainable, equitable nutrition we must count the true global costs and benefits of food production. Sustainable food production must also encompass not damaging the environment and providing fair access to all inputs.

They place emphasis on the 1 billion workers on small farms (more than 500 million) for whom there would be no work available if they were displaced by highly productive industrial farms. According to the International Labour Organization (ILO, 2015), the world is already short of about 200 million jobs; major industries such as steel and car manufacture employ only 6 million and 9 million people worldwide respectively, making it unlikely that an industrial solution would be available.

The inevitable conclusion would seem to be that in developed countries where there is sufficient economic activity that conventional productive farming is the answer that at the same time will permit more land being left for use by wildlife. In developing countries where there is no alternative employment, food production will best be left in the hands of the small-holders who, it must be remembered, are the ones largely responsible for the undernourished 800 million around the world. Organic farming will be acceptable for the smallholder provided he can grow sufficient food and at the same time improve soil fertility by growing fertility-building leguminous crops. However, it is unlikely the latter objective can be achieved without the use of fertilizers, especially in Africa.

Planetary boundaries for mankind

Mankind received a severe warning from Rockström *et al.* (2009) about the consequences of man's potentially catastrophic disruption of the environment unless global solutions are found. A bleak future was foreseen in the absence of collaborative efforts.

Planetary boundaries (PBs) comprise the central concept in an earth system framework proposed by a group of environmental scientists led by Rockström from the Stockholm Resilience Centre and Steffen from the Australian National University (Steffen *et al.*, 2015). The setting of PBs is an attempt to define the safe operating space in which humanity can not only survive but globally can provide adequate healthy nutrition for all regardless of race or creed (Rockström, TEDtalks).

The fact that during most of the Holocene epoch (from 11,700 years ago) man had no or only minimal effect on the atmosphere and the environment poses the questions as to what were the conditions like then in the pre-industrial era and could these conditions be used to set the PBs?

PBs are meant to be used by the international community, governments, international organizations and the private sector as a precondition for sustainable development. They are based on the fact that over the last 200 years human actions have gradually become the main driver of global environmental change. It is asserted that once human activity has passed certain thresholds or tipping points, defined as the planetary boundaries, there is a risk of irreversible and catastrophic environmental changes.

Nine earth system processes have been identified that have boundaries that should not be crossed. The proposed planetary boundaries (Table 18.5) are inevitably somewhat arbitrary but by ascribing numbers it is possible to direct attention to the major dangers and also to identify changes with time and hopefully of improvements. For instance, some of the boundaries have already been crossed, namely those dealing with extinction rates, CO_2, the use of man-generated reactive N and deforestation.

It is often incorrectly said that the oceans are becoming more acidic. The pH of sea water exceeds pH7 (normal range 7.5–8.4), indicating that the oceans are alkaline: following the absorption of CO_2 they are becoming less alkaline. It is misleading to imply that the oceans are acidic and are becoming more acidic.

Table 18.5 Planetary boundaries and values before industrial revolution, in 2009 and 2015 (Rockström *et al.*, 2009 and Steffen *et al.*, 2015)

Earth System Process	Measure	Proposed Boundary	Before Ind. Rev.	Status 2009	Status 2015
Climate change	CO_2 ppm	350	280	387	396
Biodiversity loss	E/MSY[1]	10	0.1–1	>100	100–1000
Nitrogen fixation	Tg N fixed/year[2]	62[3]	0[4]	121	150
Phosphorus cycle	Tg P to oceans	11	1	9	22
Ozone thickness	Dobson unit[5]	276	290	283	200
Ocean acidification	Saturation aragonite[6]	2.75	3.44	2.90	2.31
Global water use	km[3]/year	4,000	415	2,600	2,600
Change in land use	% converted to crop	75[7]	–	12	62

[1] E/MSY = extinctions per million species years.
[2] Global industrial and agricultural biological fixation of N.
[3] Rockstrom *et al.* (2009) initially proposed a boundary value of 35 Tg N/year which was defined as the amount of N removed from the atmosphere for human use. The boundary is seen as a global 'valve' limiting introduction of new reactive N to the Earth System, Range of uncertainty 62–82 Tg N/year.
[4] No allowance for legume N fixation.
[5] Ozone in stratosphere measured by Dobson units.
[6] Carbonate ion concentration, average global surface ocean saturation state with respect to aragonite.
[7] Increased from 15% in 2009.

PBs for aerosol loading of the atmosphere and for chemical pollution have not been agreed.

Nitrogen

The PB for N is of concern. Initially in 2009 the boundary was set at 35 Tg N/year but in 2015 it was increased to 62 Tg N. It is not possible to reconcile such values with 33 Tg N being the actual annual global dietary consumption, 120 Tg N being the biological fixation (half in farming and half in natural ecosystems) and the 120 Tg N fixed by man (see Chapter 16).

In order to achieve a PB of 62 Tg it would be necessary to reduce legume fixation from 60 Tg to 30 Tg N and at the same time man-made fixation from 120 Tg to 32 Tg N. The global population supported by such fixation of N would be about one quarter the current level.

While the ways of minimizing inefficient and maximizing efficient use of N, including dietary changes, are discussed it is beyond the scope of this book to attempt to quantify the possible effects of different scenarios of N:diet and so attempt to estimate a realistic PB for N that would permit mankind to eat.

Summary

1. Despite the voices of doom that food production would be unable to keep pace with population growth it is apparent that much progress has been made in food production over the last 100 years.

2. Up to 1900 wheat yields in the UK increased gradually to about 2 t/ha. Since then the use of fertilizers and new varieties with greater potential have increased yields to about 8 t/ha on average. Average organic wheat yields in the UK are assessed at just less than 5 t/ha without any allowance for the area of fertility-building crops.

3. There are an unacceptable number of people in the world who are undernourished but this is partly due to political, social and economic factors unrelated to food production.

4. Record wheat yields of around 16 t/ha have been achieved by farmers in UK and New Zealand. A record organic wheat yield of just less than 10 t/ha following two years of fertility building legumes has been achieved in the UK.

5. The introduction, from the 1960s, of new wheat, rice and maize varieties combined with improved agricultural practices in the Green Revolution permitted huge increases in global population and the eradication of famines.

6. There will continue to be a need for fewer farmers to produce food for the increasing urban populations.

7. In developed countries where farm sizes are large it seems inevitable that the only way to produce food for the foreseeable future is for farmers to continue to use the advances provided by agricultural science.

8. In developing countries where most (72%) farms are extremely small (<1 ha) the basic problem is that of incentivising subsistence farmers to become suppliers of surplus produce to the market. There would seem to be no reason why organic farming should be the answer other than by default due to fertilizers and pesticides being, in the absence of support, too costly.

9. It is proposed by environmentalists in the UK that small and community farms should take the place of the large industrial farms.

10. Most (80%) of global food production is in the hands of 90% of world's farmers, a clear indication that they need to be the focus of international bodies with regard to ensuring food security.

11. There are about 535 million farms, mostly in China and India, that are smaller than 5 ha out of a total of 570 million farms worldwide.

12. Smallholders are the major food producers in Africa. It is likely that consolidation of properties will result in larger units, which will be better placed to benefit from support, improved marketing as well as technology. It has become accepted that African farmers have no choice but to use fertilizers to improve the many impoverished nutrient-poor soils.

13. History suggests that food production can be sustained for a global population of 11 billion by continuing to farm conventionally and making use of new scientific advances.

14. Sustainable agriculture means different things to different people. To many in developed countries who perceive no food shortage the sustaining of the environment and the non-use of chemicals are the major concerns. On the global scale sustainable agriculture is a system that primarily supports humans and is dependent on the use/availability of nutrients, notably N and P.

15. Sustainability is ultimately dependent on crop yields and human diets.

16. The complete conversion to an organic farming system dependent on livestock

producing manure for recycling and/or on biological N fixation by legumes would require impossibly large numbers of cattle and pastures and/or of areas of nitrogen-fixing cover crops.

17. The main challenge for organic farming in striving for sustainability is how to increase yields without requiring more land for animals or for fertility-building crops.

18. Organic agriculture will continue to satisfy niche markets in developed countries.

19. In Asia the increasing population pressure combined with limited land will necessitate higher crop yields than those achievable by organic farming.

20. Conventional farmers face their own challenge to increase food production while improving soil quality, using fewer pesticides and at the same time protecting the environment.

21. It is not necessary to contemplate a ruminant-free agriculture as part of the solution to sustainability as ruminants have the capacity to make use of land that is unsuitable for crops.

22. An agricultural system that is sustainable forever is difficult to envisage as it would require all nutrients removed in crops to be returned to the soil and so obviate the need for non-renewable minerals; as long as civilization can generate power, nitrogen fixation in the factory will provide N fertilizers.

23. Sustainability into perpetuity could be more achievable by subsistence farming insofar as non-renewable resources would last longer.

24. Several NGOs and the FAO have proposed guides to more sustainable agriculture and one, LEAF, has developed a marketing logo with the objective of linking the consumer to sustainable production systems.

25. Most ruminants are inefficient converters of vegetable protein but their use of low-grade land is in their favour. As pigs and poultry are highly dependent on cereals and soya they are effectively in competition with humans for about 1 billion t of cereals that could feed 3.5 billion humans.

26. Some organic supporters accept that organic agriculture would be unable to feed the current global population (on current diets) and that it would not be sustainable unless there were major changes in diet involving the consumption of markedly fewer livestock products.

27. Organic farming has an established position in several developed countries with an emphasis on livestock products and limited vegetable and fruit products. In developing countries in Africa, Asia and Latin America there are only limited areas of organic vegetables and fruit, indicative of tiny local markets.

28. The conviction that mankind is in danger of creating irreversible and catastrophic damage to the environment has led to estimations of boundaries within which man must live to ensure a secure future; these are known as Planetary Boundaries.

29. Of the ten earth systems identified, PBs have been set, somewhat arbitrarily, for eight, namely N fixation by man and legumes, CO_2, extinctions, P loss to oceans, ozone thickness, ocean pH, water use and deforestation.

30. Initially the PB for N was set at what has here been estimated as the actual global human consumption of N, making the PB untenable.

31. The upward adjustment of the PB for N still leaves it adrift from the amounts of N fixed by legumes and by man. To reach the desired PB for N would require reductions in legume fixation to about half the current value and of man-made fixation to about one quarter. If achieved such reductions in N would be accompanied by a reduction of the global population to about 2 billion, or a quarter the current level.

References

AGRF (2016) African Green Revolution Forum 2016. The Nairobi communique. Available at https://agrf.org/articles/the-nairobi-communique

Agricology UK (2016) Practical, sustainable farming regardless of labels. Available at www.agricology.co.uk

Apostolides, A., Broadberry, S., Campbell, B., Overton, M. and van Leeuwen. (2008) English agricultural output and labour productivity, 1250–1850: some preliminary estimates. From the project Reconstructing the National Income of Britain and Holland. Available at www.basvanleeuwen.net/bestanden/agriclongrun1250to1850.pdf

Carver, M.F.F. (2003) Production of organic wheat: trials on varieties, seed rate, weed control and the use of permitted products HGCA Project report 304. Available at https://cereals.ahdb.org.uk/media/367004/pr304-final-project-report.pdf

Collavo, A., Glew, R.H., Huang, Y.S., Chuang, L.T., Bosse, R. and Paoletti, M.G. (2005) House cricket small-scale farming 519–544. In: Paoletti, M.G. (ed.) *Ecological Implications of Mini-livestock: Potential of Insects, Rodents, Frogs and Snails*. CRC Press.

Cooke, G.W. (1969) The carrying capacity of the land in the year 2000. In: Taylor, L.R. (ed.) *The optimum population for Britain. Proceedings, Symposium, Royal Geographical Society, London.* Academic Press 1970.

Dobermann, A. (2015) Yields – why are they static – is it soil management? RRA Summer Meeting, 01 July 2015, RRes. Available at www.rothamsted.ac.uk/sites/default/files/users/baverst/Achim.pdf

Eisler, M.C., Lee, M.R.F. and colleagues. (2014) Agriculture: Steps to sustainable livestock. *Nature*, 507, 32–34. doi: 10.1038/507032a. Available at www.nature.com/news/agriculture-steps-to-sustainable-livestock-1.14796

Ellis, F. (2005) Small farms, livelihood diversification, and rural–urban transitions: Strategic issues in sub-Saharan Africa 135–150. In: IFPRI (2005) The future of small farms. Proceedings of a research workshop organized by the International Food Policy Research Institute. Wye, UK. Washington DC. Available at http://iri.columbia.edu/~jhansen/Sonja/10.1.1.139.3719.pdf

Erb, K.-H., Lauk, C., Kastner, T., Mayer, A., Theurl, M.C. and Haberl, H. (2016) Exploring the biophysical option space for feeding the world without deforestation. *Nature. Communications* 7, 11382. doi: 10.1038/ncomms11382.

Euractive (2016) A new and better food supply chain for a new and better EU. Available at www.euractiv.com/section/agriculture-food/opinion/a-new-and-better-food-supply-chain-for-a-new-and-better-eu

Fairlie, S. (2009) Can Britain feed itself? *The Land* 4, 2007–2008.

FAO (2010) Climate-Smart Agriculture

Hague Conference on agriculture, food security and climate change. Available at www.fao.org/fileadmin/user_upload/newsroom/docs/the-hague-conference-fao-paper.pdf

FAO (2011) *The State of the World's Land and Water Resources for Food and Agriculture* SOLAW – Managing systems at risk. Food and Agriculture Organization of the United Nations, Rome.

FAO (2011a) The place of urban and peri-urban agriculture (UPA) in national food security programmes. Integrated food security support service (TCSF) policy and programme development support division technical cooperation department. Available at www.fao.org/docrep/014/i2177e/i2177e00.pdf

FAO (2013) Climate-Smart Agriculture Sourcebook. Available at www.fao.org/docrep/018/i3325e/i3325e.pdf

FAO (2014) Family farming, State of Food and agriculture. Available at www.fao.org/3/a-i4040e/i4040e02.pdf

FAO (2016) Save Food: Global Initiative on Food Loss and Waste Reduction. Available at www.fao.org/save-food/resources/keyfindings/en

FAO (2016a) FAO Regional Conference for Africa (ARC). Available at www.fao.org/about/meetings/regional-conferences/arc29/documents/en

FAO-UNEP (2010) The FAO-UNEP Sustainable Food Systems Programme. Available at www.fao.org/fileadmin/templates/ags/docs/SFCP/SustainableFoodSystemsProgramme.pdf

FAOSTAT (2013) and (2015) Statistics Division. Available at http://faostat3.fao.org/home/E

FAOSTAT (2016) FAO Statistics Division. Available at http://faostat3.fao.org/download/Q/QC/E

Farmers Weekly (2015) Organic winter wheat yields 9.9t/ha in Scottish Borders. Available at www.fwi.co.uk/arable/organic-winter-wheat-yields-99tha-in-scottish-borders.htm

FfF (2016) The future of protein. Available at https://www.forumforthefuture.org/sites/default/files/The_Protein_Challenge_2040_Summary_Report.pdf

FiBL (2009) The Round Table on Organic Agriculture and Climate Change. Available at www.fibl.org/en/service-en/news-archive/news/article/round-table-on-organic-agriculture-and-climate-change-established.html

Foley, J.A. (2011) Can we feed the World and Sustain the planet? A five-step global plan could double food production by 2050 while greatly reducing environmental damage. *Scientific American* 305, 62–65. Available at www.geog.psu.edu/sites/default/files/Scientific%20American%20Article.pdf

Foley, J.A. (2015) 5 Steps to Feed the World and Sustain the Planet. *Scientific American Special Edition 24, Issue 2s.* Available at https://www.scientificamerican.com/article/5-steps-to-feed-the-world-and-sustain-the-planet

Gerland, P., Raftery, A.E., Sevčíková, H., Li, N., Gu, D., Spoorenberg, T., Alkema, L., Fosdick, B.K., Chunn, J., Lalic, N., Bay, G., Buettner, T., Heilig, G.K. and Wilmoth, J. (2014) World population stabilization unlikely this century. *Science* 10, 346 (6206), 234–237. doi: 10.1126/science.1257469.

GFS (2016) Global Food Security Champion, Professor Tim Benton. Available at www.foodsecurity.ac.uk/programme/governance/champion.html

Gilland, B. (1985) Cereal yields in theory and practice. *Outlook on agriculture* 14, 2, 56–60. doi: 10.1177/003072708501400201.

Gregorich, E.G. Turchenek, L.W. Carter, M.R. and Angers, D.A. (2001) *Soil and Environmental Science Dictionary*. CRC Press, Boca Raton, Florida.

Graeub, B., Chappell, J., Wittman, H., Ledermann, S., Batello, C. and Gemmill-Herren, B. (2016). The state of family farmers in the world: global contributions and local insights for food security. *World Development 87*, 1–15. doi: 10.1016/j.worlddev.2015.05.012.

Grantham Centre (2015) Grantham Centre for Sustainable Futures Briefing note: A sustainable model for intensive agriculture. Available at http://grantham.sheffield.ac.uk/wp-content/uploads/2015/12/A4-sustainable-model-intensive-agriculture-spread.pdf

Horrigan, L., Lawrence, R.S. and Walker, P. (2002) What's wrong with industrial agriculture? How sustainable agriculture can address the environmental and human health harms of industrial agriculture. *Environmental Health Perspectives* 110, 5. Available at https://www.organicconsumers.org/old_articles/organic/IndustrialAg502.php

IAASTD (2016) International Assessment of Agricultural Knowledge, Science and Technology for Development. Reports on Industrial Agriculture and Small-scale Farming. Available at www.globalagriculture.org/report-topics/industrial-agriculture-and-small-scale-farming.html

IFAD/UNEP (2013) *Smallholders, food security, and the environment*. The International Fund for Agricultural Development and United Nations Environmental Programmes. Available at www.unep.org/pdf/SmallholderReport_WEB.pdf

IFOAM (2013) Our Earth, Our Mission. Annual report. Available at www.ifoam.bio/sites/default/files/annual_report_2013_web.pdf

IFOAM (2014) Norms for Organic Production and Processing. Version 2014. Available at www.ifoam.bio/sites/default/files/ifoam_norms_version_july_2014.pdf

IFOAM (2015) Organic 3.0 – The Next Phase of Organic Development for truly sustainable farming and consumption. Available at www.ifoam.bio/sites/default/files/organic3.0_v.2_web_0.pdf

IFOAM (2016) Public Consultation on the position of IFOAM – Organics International on Genetic Engineering and Genetically Modified Organisms. Available at www.ifoam.bio/en/news/2016/02/26/public-consultation-position-ifoam-organics-international-genetic-engineering-and

IFPRI (2016) Global Food Policy Report 2016 How We Feed the World is Unsustainable. Press release. Available at https://www.ifpri.org/news-release/gfpr-2016-press-release

ILO (2015) International Labour Organization *World Employment Social Outlook: Trends 2015*. International Labour Office. Available at www.ilo.org/global/research/global-reports/weso/2015/lang--en/index.htm

IPES (2015) International panel of experts on sustainable food systems. Available at www.ipes-food.org

ISAAA (2015) International Service or the Acquisition of Agri-biotech Applications, Brief 51-2015. Available at www.isaaa.org/resources/publications/briefs/51/executivesummary/default.asp

Jones, P. and Crane, R. (2009) CAS Report 18 Centre for Agricultural Strategy, University of Reading England and Wales under organic agriculture: how much food could be produced?

Kirchmann, H., Bergström, L., Kätterer, T., Andrén, O. and Andersson, R. (2008). *Can organic crop production feed the world?* 39–72. *In:* Kirchmann, H. and Bergström, L. (eds *Organic crop production – ambitions and limitations*.). Springer, Dordrecht, The Netherlands.

Kirchmann, H., Kätterer, T., Bergström, L., Börjesson, G. and Bolinder, M.A. (2016) Flaws and criteria for design and evaluation of comparative organic and conventional cropping systems. *Field Crops Res.* 186, 99–106.

Knight, S., Kightley, S., Bingham, I., Hoad, S., Lang, B., Philpott, H, Stobart, R., Thomas, J., Barnes, A. and Ball, B. (2012) Project Report No. 502 Desk study to evaluate contributory causes of the current 'yield plateau' in wheat and oilseed rape. Available at https://cereals.ahdb.org.uk/media/198673/pr502.pdf

Larson, D., Otsuka, K., Matsumoto, T. & Kilic, T. (2013) Should African rural development strategies depend on smallholder farms? An exploration of the inverse productivity hypothesis. Policy Research Paper No. 6190. World Bank, Washington, DC. Available at http://elibrary.worldbank.org/doi/abs/10.1596/1813-9450-6190

LEAF (2015) Linking Environment And Farming. Available at www.leafuk.org/leaf/farmers/LEAFmarquecertification.eb

Lowder, S.K., Skoet, J. and Singh, S. (2014) What do we really know about the number and distribution of farms and family farms in the world? Background paper for The State of Food and Agriculture 2014. Available at www.fao.org/docrep/019/i3729e/i3729e.pdf

Meharg, A.A. (2016) Perspective: City farming needs monitoring. *Nature* 531, S60, 7594. doi: 10.1038/531S60a.

Mellanby, K. (1975) *Can Britain Feed Itself?* Merlin Press, London.

Midega, C.A.O., Bruce, T.J.A., Pickett, J.A., Pittchar, J.O., Murage, A. and. Khan, Z.R. (2015) Climate-adapted companion cropping increases agricultural productivity in East Africa. *Field Crops Research* 180, 118–125. Available at http://dx.doi.org/10.1016/j.fcr.2015.05.022

Mundy, P. (2016) First International Conference on Steps to Sustainable Livestock. Sustainability is complex: There is no single diet solution, Sustainable Food Trust. Available at http://sustainablefoodtrust.org/articles/sustainability-is-complex-there-is-no-single-diet-solution

Neuhoff, D. (2014) An agronomic approach to yield comparisons between conventional and organic cropping systems 253–255. In: Rahmann, G. and Aksoy U. (Eds.) *Proceedings of the 4th ISOFAR Scientific Conference, Building Organic Bridges.* Istanbul, Turkey.

Oerke, E. (2006) Crop losses to pests. *Journal of Agricultural Science.* 144, 1, 31–43. doi: http://dx.doi.org/10.1017/S0021859605005708.

ORC (2014) Organic Research Centre. Organic Producers' Conference 2014. Available at www.organicgrowersalliance.co.uk/ConferenceNov14_13_301014.pdf

ORFC (2016) Oxford Real Farming Conference. Available at www.oxfordrealfarmingconference.org/

Paulson, J.A. and Zaoutis, T.E. (2015) Nontherapeutic Use of Antimicrobial Agents in Animal Agriculture: Implications for Pediatrics. *Pediatrics* 136, 6, 1670–1677. doi: 10.1542/peds.2015-3630.

PBL (2014) Netherlands Environmental Assessment Agency. *Sustainability of international Dutch supply chains. Progress, effects and perspectives* PBL publication 1289. Available at www.pbl.nl/en

Pretty, J. (2008) Agricultural sustainability: concepts, principles and evidence. *Philosophical Transactions of the Royal Society London B Biological Science.* 12; 363, 1491, 447–465. doi: 10.1098/rstb.2007.2163.

Purves, D. (1985) *Trace element contamination of the environment.* Elsevier.

Rapsomanikis, G. (2015) The economic lives of smallholder farmers, An analysis based on household data from nine countries. FAO, Rome.

Reganold, J.P. and Wachter, J.M. (2016) Organic agriculture in the twenty-first century. *Nature Plants,* 2, 2, 15221 DOI: 10.1038/NPLANTS.2015.221.

Rockström, J., Steffen, W., Noone, K., Persson, A., Chapin, F.S. 3rd., Lambin, E.F., Lenton, T.M., Scheffer, M., Folke, C., Schellnhuber, H.J., Nykvist, B., de Wit, C. A., Hughes, T., van der Leeuw, S., Rodhe, H., Sörlin, S., Snyder, P.K., Costanza, R., Svedin, U., Falkenmark, M., Karlberg, L., Corell, R.W., Fabry, V.J., Hansen, J., Walker, B., Liverman, D., Richardson, K., Crutzen, P. and Foley, J.A. (2009) A safe operating space for humanity. *Nature.* 461, 7263, 472–5. doi: 10.1038/461472a. Available at http://pubs.giss.nasa.gov/docs/2009/2009_Rockstrom_ro02010z.pdf

Rothamsted (2016) A new climate-smart companion cropping system allows African farmers to substantially their yields. Available at www.rothamsted.ac.uk/news-views/new–climate–smart-companion-cropping-system-allows-african-farmers-substantially-increase

Savage, S.D. (2011) A comparison of 2008 organic row crop yields to historical yield trends. Available at www.scribd.com/doc/48257015/Organic-Historical-Comparison-2-5-11

Savage, S.D. (2014) The Organic Yield Gap, An independent analysis comparing the 2014 USDA Organic Survey data with USDA – NASS statistics for total crop production. Available at https://www.scribd.com/doc/283996769/The-Yield-Gap-For-Organic-Farming

Schiermeier, Q. (2013) Farmers dig into soil quality. *Nature.* 502, 7473, 607. doi: 10.1038/502607a.

Scientific American (2015) Only 60 Years of Farming Left If Soil Degradation Continues. Available at www.scientificamerican.com/article/only-60-years-of-farming-left-if-soil-degradation-

SFT (2016) Sustainable Food Trust. Patrick Holden wants to shed light on the true cost of American food. Available at http://civileats.com/2016/03/29/true-cost-patrick-holden/#sthash.sPMFFa7M.dpuf

Slow Food (2016) Slow Food in the UK. Available at https://www.slowfood.org.uk/about/about/what-we-do

Springmann, M., Godfray, H.C., Rayner, M. and Scarborough, P. (2016) Analysis and valuation of the health and climate change co-benefits of dietary change *Proceedings of National Academy of Science USA* 113, 15, 4146–4151. Available at www.pnas.org/content/113/15/4146.full

Springmann, M., Mason-D'Croz, D., Robinson, S., Wiebe, K., Godfray, C., Rayner, M. and Scarborough P.D. (2016a) Mitigation potential and global health impacts from emissions pricing of food commodities. *Nature Climate Change.* doi: 10.1038/nclimate3155.

Steffen, W., Richardson, K., Rockström, J., Cornell, S.E., Fetzer, I., Bennett, E.M., Biggs, R., Carpenter, S.R., De Vries, W., De Wit, C.A., Folke, C., Gerten, D.,

Heinke, J., Mace, G.M., Persson, L.M., Ramanathan, V., Reyers, B. and Sorlin, S. (2015). Planetary boundaries: Guiding human development on a changing planet. *Science*. 347, 6223. doi: 10.1126/science.1259855. Available at www-ramanathan.ucsd.edu/files/pr210.pdf

Sukhdev, P., May, P. and Müller, A. (2016) Fix food metrics. *Nature*, 540, 7631. Available at www.nature.com/news/fix-food-metrics-1.21050 doi: 10.1038/540033a

Syngenta (2012) Bringing plant potential to life. Annual review 2012. Available at www4.syngenta.com/~/media/Files/S/Syngenta/documents/ar-2012/syngenta-annual-review-2012-english.pdf

Taylor, B.R., Watson, C.A., Stockdale, E.A., Mckinlay, R.G., Younie, D. and Cranstoun, D.A.S. (2001) Cereals and oilseeds project report 45 (HGCA) Current Practices and Future Prospects for Organic Cereal Production: Survey and Literature Review. https://cereals.ahdb.org.uk/media/288268/rr45 final project report.pdf

TEEB*AgriFood* (2016) Call for Evidence. TEEB for Agriculture and Food "Scientific and Economic Foundations". Available at www.teebweb.org/agriculture-and-food/call-for-evidence-oct2016

Tilman, D. and Clark, M. (2014) Global diets link environmental sustainability and human health. *Nature*. 515, 518–522. doi: 10.1038/nature13959.

Todhunter, C. (2015) The world must step off the chemical farming treadmill. Available at www.theecologist.

org/blogs_and_comments/commentators/2986062/the_world_must_step_off_the_chemical_farming_treadmill.html

Tomlinson, I. (2013) Doubling food production to feed the 9 billion: A critical perspective on a key discourse of food security in the UK. *Journal of Rural Studies* 29, 81–90.

UNCTD (2013) Trade and environment review Number 2013. Make agriculture truly sustainable now for food security in a changing climate. Available at http://unctad.org/en/PublicationsLibrary/ditcted2012d3_en.pdf

UNECE (1987) World Commission on Environment and Development. Sustainable development – concept and action. Available at www.unece.org/oes/nutshell/2004-2005/focus_sustainable_development.html

UNEP (2016) Food Systems and Natural Resources. A Report of the Working Group on Food Systems of the International Resource Panel. Westhoek, H., Ingram, J., Van Berkum, S., Özay, L. and Hajer, M.

Woodward, L. (1996) Can organic farming feed the world? Elm Farm Research Centre For Organic Principles and Best Practice. Available at www.efrc.com/manage/authincludes/article_uploads/art001.pdf

World Bank (2007) World Development Report 2008: Agriculture for development. Washington, DC. Available at https://siteresources.worldbank.org/INTWDR2008/Resources/WDR_00_book.pdf

CHAPTER 19

Sustaining Food Production–Supply of Animal Protein, the Nitrogen problem, Lessons of Organic Farming and Importance of Technology and Innovation

"We cannot solve our problems with the same thinking we used when we created them"

Albert Einstein

Supply of animal protein

Implications of reduced consumption of livestock products

It is suggested that organic farming would be able to feed the world if the consumption of livestock products was markedly reduced in developed countries and curtailed in developing countries. Health and environmental benefits are cited as motivating factors.

Humans with their short and simple digestive tracts have evolved as meat eaters. As some of the countries with longest life expectation (Japan, Sweden and Norway) are also meat eaters there would not seem to be a case for worldwide vegetarianism (Smil, 2014). However, it is becoming generally accepted in developed countries that the consumption of livestock products should be reduced in order to improve health and reduce greenhouse gas emissions (GHG).

IARC/WHO (2015) issued warnings that consumption of as little as 350 g/week of processed meats (sausages, bacon) can increase the risk of cancer by 18% and that red meat is probably carcinogenic. The independent policy institute

Chatham House London (Wellesley *et al.*, 2015) considered that meat should be subject to a carbon tax as a means of reducing GHG emissions. However, meat production is expected to rise by 76% by 2050 to meet the demands of the growing global population and the adoption of Western style diets (Alexandratos and Bruinsma, 2012).

The PBL (Netherlands Environmental Assessment Agency) studied the implications of dietary changes within Europe (Westhoek *et al.*, (2014). It concluded that if all inhabitants of the European Union (EU) reduced their consumption of meat, dairy and eggs by 50% there would be a 40% reduction in reactive nitrogen emissions, a 25–40% reduction in GHG emissions from agriculture and 23% (per capita) reduction in the cropland needed for food production. Such dietary changes would also lower health risks. The EU would become a net exporter of cereals and the import of soymeal would be reduced by 75%. The nitrogen use efficiency of the food system would increase from the current 18% to 41–47%.

Other studies at PBL (van Grinsven *et al.*, 2014) examined the ways to reduce reliance, in Europe, on the use of man-made fertilizers on which livestock production is dependent. They proposed relocating current production systems for feed and livestock production from north-western to central and eastern Europe. This would allow a reduction of N use on cereals in north-west Europe by 30% (by 50 kg N/ha). Another option is to change towards growing legumes for animal feed in order to decrease dependence on N fertilizer and nitrogen-rich feed imports. It is recognized that the greatest challenge for Europe is how to reduce the demand for animal feeding stuffs, and thus for land and N. The findings of the PBL are likely to apply to many countries where there is overconsumption of livestock products.

There are signs that countries are beginning to pay attention to the health and environmental implications of reducing meat consumption. For example, recently revised dietary guidelines in Sweden, the Netherlands and the UK have included recommendations for reduced meat consumption on health grounds. In Sweden it is recommended that no more than a total 500 g of meat from cows, pigs, lambs, reindeer and game is consumed in one week (Swedish National Food Agency, 2016). The Netherlands Nutrition Centre recommends that not only should Dutch citizens adopt a more plant-based diet but also that they reduce weekly meat consumption to less than 500 g (Voedingscentrum, 2016). In the UK it is recommended that weekly consumption of red or processed meat should not exceed 490 g (Public Health England, 2016); additionally, in the UK, it is suggested that at least two portions of 140g of fish a week, including one portion of oily fish, are consumed.

In the USA there have been no recent changes in advice about the consumption of protein; the current recommendation for someone on a healthy diet (2,000 calories) is that the weekly consumption of meats, eggs and poultry should not exceed 740g (US DH and HS, 2015). Actual intake levels by most American males exceed the recommended level by 50%.

There are also moves in China to limit weekly meat consumption to 280–525 g (Xiaodong, 2016). The stated aim is to improve the health and well-being of an increasingly sedentary population that has a growing appetite for high-protein, high-calorie and high-fat foods; the Chinese government is apparently not thinking about the benefits of reducing GHG emissions, which would accompany a successful programme.

In a survey of meat consumption in Europe only 3% of people said that they never consumed meat but 80% said they would be willing to eat less meat and 75% would be willing to replace beef or pork with poultry or fish (72%) in the interests of protecting the environment EC (2013); 26,573 respondents were involved who knew what environmentally friendly products were. There are thus promising signs that moves to reduce the consumption of livestock products could be effective.

Huge benefits to health and from reducing GHG emissions were forecast by Springmann *et al.* (2015) if the fraction of animal-sourced foods in our diets was limited. They concluded that the impacts of dietary changes toward less meat and more plant-based diets would vary greatly between regions. If the diets were more in line with standard dietary guidelines global mortality could be reduced by 6–10% and food-related GHG emissions by 29–70% compared with a reference scenario postulated for 2050; this would be at a time when many in the developing world are expected to be eating much more meat. It was accepted that it would be difficult if not impossible to bring about the necessary changes in farming practices and in dietary patterns. The possibility of putting a tax on meat has been suggested by several groups and organizations to sit alongside the taxes on smoking and alcohol aimed at deterring consumption (see Chapter 18).

Sustainable meat production

Smil (2014) considered whether adequate amounts of meat could be produced globally without the need for further deforestation and without feeding cereals and soya to cattle. Concentrated cereal-based feeds would be better reserved for the more efficient converters, pigs and poultry. He calculated that an annual production of about 190 Tg of meat was possible by combining 40 Tg of ruminant meat from grazing cattle with 40 Tg from feeding forages and crop residues with 70 Tg chicken meat and 40 Tg pork from feeding with highly nutritious crop processing residues. Annual global consumption of 190 Tg meat would amount to a supply of about 50 Tg protein and 8 Tg N. In comparison, annual global minimum requirement for N was calculated at 18 Tg N and actual consumption at 33 Tg N (Table 16.6).

The restriction of cattle, sheep and goats to land only suitable for grazing could reduce the need for large areas of feed crops but, as Smil points out, globally much grassland has already been so degraded that 25% of the currently grazed area needs to be rested. The conversion of more forests to grassland is not acceptable for global environmental and loss of diversity reasons.

It would be possible to reduce meat intake by the consumption of more dairy products, more eggs and especially of fish (by aquaculture); the latter substitution will partly depend on the successful replacement of the wild fishmeal currently fed to cultured fish with plant-derived protein meals rich in omega-3 fatty acids. By using animal feeding stuffs more efficiently and by substitution Smil considered that up to 25% of today's meat supply in affluent Western countries with high levels of meat consumption could be displaced (15% by dairy products, 5% each by eggs and seafood). In Asian countries with moderate meat intakes and traditional preference for freshwater fish, as much as 40% could be displaced (20% by dairy products and 10% each by eggs and seafood).

The global annual production of 190 Tg meat envisaged by Smil and an absolute maximum of 300 Tg are considerably lower than the long-term forecasts of FAO, namely of 450 Tg in 2050, which would require the undesirable expansion of current cereal and soya areas (Alexandratos and Bruinsma, 2012). Other, even higher, forecasts of the demand for meat in 2050 based on growth in GDP have been made,

for example Elam (2006) forecasted a demand of more than 620 Tg meat.

The alternative approaches, namely voluntary meatless diets or the consumption of ersatz meat made from plant protein, do not seem currently to be acceptable; the marketing of mock meats (imitation meat cuts and burgers) and of tofu in the USA has not been successful despite them being available for many years. The selling (early 2017) of ersatz meat products alongside meat by a major UK supermarket was seen as a threat to livestock farmers and was accordingly challenged.

Grazing

The many merits of mixed farms to improve soil fertility, increase yields, reduce pests, to strengthen local communities and provide employment are extolled by Harvey (2016) in his manifesto for grazing livestock. He envisages that by converting arable land back to pasture, erosion will be minimized, the soil will be able to store more carbon and the long-term health of the planet improved. Harvey is critical of industrial farming and foresees benefits to human health as well as the environment from a return to grass-based farming. He associates the incidence of coronary heart disease and type 2 diabetes directly with industrial agriculture. Harvey links intensive arable cropping with the decline of half of the plant species, one third of insect species and four-fifths of bird species in the UK. He refutes the idea that industrial crop production methods are required to feed a growing world population by quoting IAASTD (2009), which said that industrial farming is not fit for purpose.

The 606-page IAASTD (2009) report focused attention on how agricultural knowledge can be used to reduce hunger and poverty, improve rural livelihoods and human health, and facilitate equitable environmentally, socially and economically sustainable developments in developing countries. The problems experienced in adopting new methods by smallholders, who globally are responsible for growing most food, were appreciated. A later study (IAASTD, 2016) doubted whether a free market economy combined with technological advances is the best model for the huge numbers of small-scale farmers, particularly in Africa. It is necessary to distinguish between developed and developing countries where conditions are such that inputs are uneconomic and the provision of advice difficult because of the number and location of the farms.

Röös *et al.* (2016) in Sweden examined the implications of restricting livestock production to grazing and to the consumption of by-products that humans can't or don't want to eat and to wastes; the arable land formerly used to produce animal feed can then be utilized for crops directly consumed by humans. In this way livestock production will be constrained by the availability of grassland. The several hypothetical low-meat Swedish diets (only 11% protein from animal sources) reduced the impact on the environment and diversity but were found to be unsatisfactory in that more N and P inputs were required than allowed for under the concept of planetary boundaries (Steffen *et al.*, 2015). According to Röös *et al.* (2016), N inputs would have to be reduced by half and P inputs by two-thirds for all diets to stay within the planetary boundaries; more efficient nutrient recycling from society back to agriculture would also be needed. Precision agriculture, better crop rotations and efficient use of catch crops and new crop varieties with improved nutrient use efficiency are seen as promising tools that can help maintain high crop yields with reduced fertilizer inputs.

In the UK a government-funded project entitled the Defra Wheat Genetic Improvement Network project (WGIN) initially involving the John Innes Centre, Rothamsted Research and the University of Nottingham was started in 2008. Several aspects of wheat genetics are being studied but much effort is focused on the analysis of the genetic variation in nitrogen use efficiency.

No case can be made for the continued high levels of meat consumption in developed wealthy countries; there are too many detrimental effects on health and the environment. Likewise, industrial feedlot production units yield too many negative outcomes, notably the requirement for large quantities of imported feed and the amounts of GHG emissions. Meat production from grazing animals should ideally be integrated into local farming systems.

Insects for feed and food

Edible insects are a promising and intriguing alternative to livestock as sources of protein, either for direct human consumption or for indirect use as feedstock, because of their efficiency in converting protein in organic wastes into edible protein. The high efficiency of insects is basically due to the fact that as they are cold-blooded they do not waste energy in heating their bodies. The so-called 'yuck' factor that will inhibit direct human consumption in many countries suggests that insects will first be used for animal feed and will partially replace soya beans and maize; more of these crops will then be available for direct human consumption. If cattle were only fed on grass and insect-derived feed the numbers of cattle could be kept within sustainable limits dictated by the amount of grazing land available. However, the successful use of insects as animal feed would require massive and widespread developments globally to have any impact.

Out of 2,000 edible insect species recorded worldwide, only around twenty are farmed, of which the 'golden five' – crickets, mealworms, silkworms, grasshoppers and black soldier flies – dominate. Farmed insects are small and transportable. They require less energy, water and land than conventional livestock. According to an FAO paper (van Huis *et al.*, 2013), mass farming of insects is likely to use 50–90% less land per kg protein, 40–80% less feed per kg of edible weight and emit far less GHG than livestock.

Much remains to be learnt about the hazards and risks of using insects for feed and food. However, a report by the European Food Safety Authority (EFSA, 2015) considers that farming insects on a large scale poses no more hazards than livestock farming. The substrates to be used as feed by the insects will need to be studied closely. Ideally waste organic materials will be used such as unwanted human food, products from slaughter houses, waste food, animal manure and sewage sludge. EU regulations governing substrates for insects are already in place, for example human faeces are not allowed even though they are in other parts of the world (EFSA, 2015).

Insects are already consumed by humans in many countries in Africa (n=34), in Asia, notably India, China, Laos and in South/Central America notably Brazil, Colombia, Ecuador, Mexico, Peru and Venezuela (FCRN, 2016). Around 2 billion people on the planet are thought to consider insects as acceptable food. At present most, about 90%, of the consumed insects are collected in the wild and thus are only available seasonally. While insect farming is growing in developing nations it is rare in developed countries. Thailand has more than 20,000 registered cricket farms and about 5,000 palm weevil larvae farms (FCRN, 2016). China has a large number of farms raising insects for food and feed. In Europe and North America there is little insect farming, consisting mainly of crickets for pets, mealworms (for pet reptiles) and some for humans.

In the UK an organic cricket flour (from Canada) is available on the market (Gryllo Co, 2016) that is said to contain all the essential amino acids, more iron than spinach and is rich in both calcium and micronutrients; it is promoted as producing much less methane than cattle. A wide range of insect products are marketed in Europe, USA, Australia and Asia (from Thailand) including crickets (many flavours), beetles, caterpillars, mealworms, bees, wasps and ants, grasshoppers and locusts (Next-Food, 2016).

Insect farming has a considerable potential to increase the supply of protein but it will have to compete with the wild insect market.

CABI and partners has set up the Proteinsect project, which aims to facilitate the use of insects as an alternative source of protein for animals (CABI, 2016). The project will develop and optimize maggot production systems for animal feed in Europe, China and West Africa, determine safety and quality criteria, and evaluate the performance of protein extracts. A consumer survey by Proteinsect found that more than 70% of respondents would eat fish, chicken or pork from animals fed insect protein. Consumers of free-range chickens have been doing so for a long time; naturally occurring insects may constitute 30% of the free-range chicken's diet.

The nitrogen problem and greenhouse gas emissions

Despite the fact that man–made nitrogenous fertilizers have been and are responsible for feeding about half the world population (Chapter 16) they continue to get a bad press related to pollution and environmental damage (Chapter 17). For example, the US fertilizer industry was described as evil by Ecowatch (2016), which blamed it for polluting the environment, disrupting the climate and damaging public health. It is hoped that by focusing on such issues people will be provided with a reason to boycott chemically derived food production and choose organic and grass-fed animal products instead. It is said that two-thirds of US drinking water is contaminated with high levels of nitrates and that some waters are dangerous (causing cancer) and even deadly for children (methaemoglobinaemia). Anhydrous ammonia, a popular fertilizer in the USA, is described as dangerous to farm workers and neighbouring communities, and also because it poses an explosion and fire risk.

One solution to reducing the impact of undesirable reactive N compounds is to improve the efficiency of the use of nitrogen (Chapter 16); this can be achieved by timing the application of the N to synchronize with the period of maximum uptake and by varying the amount of N applied according to soil fertility across the field using GPS technology. Synchronization and precision application are more easily achieved when using fertilizer N than with N supplied in manure, compost or green manures. Precision application is feasible using drip irrigation but such use is limited at present to high value horticultural crops such as tomatoes, cucumbers and peppers. The incorporation of slurry into soil rather than on the surface and the rapid incorporation of fresh manure into the soil are ways to reduce losses NH_3.

Studies on root morphology and development of string beans *Phaseolus vulgaris* and of maize has led to the breeding of varieties with an enhanced capacity to absorb N and P (Gilbert, 2016). Beans with long root hairs that are associated with better phosphorus absorption were crossed with beans with strong roots near the base of the stem, which resulted in progeny with higher yields (Miguel *et al.*, 2015). With maize better N utilization was achieved by breeding from plants with few long lateral roots that penetrated deeply (Zhan and Lynch, 2015). Traditional breeding techniques have succeeded where genetic modification (GM) technology has so far failed.

If, as discussed in Chapter 20, it should prove possible to breed staple non-leguminous crops that are able to fix N symbiotically at rates sufficient to maintain current yields, pollution problems would still remain due to the leaching of nitrates following the mineralization of N-rich crop remains before uptake by the next crop and as a result of denitrification. Unabsorbed reactive N is a potential problem regardless of its origin.

Agricultural activities related to livestock farming and the use of manure are responsible for more GHG emissions than fertilizer use (Chapter 17).

Mitigation of greenhouse gas emissions

In 2008 Smith *et al.* (2008) estimated that agriculture could only reduce total global emissions of CO_2 eq by no more than 20%. They highlighted several agricultural practices that are responsible for GHG emissions, notably when and how nitrogen is applied and the type of animal feed. Changing the diet of ruminants can influence CH_4 generation, for example feeding maize silage instead of grass silage can reduce CH_4 emission. Other attempts aimed at reducing CH_4 emissions through feed modification include supplementing feed with garlic (in Wales) and with oregano (in Denmark). The possibility of reducing CH_4 generation by wetland rice by periodic drying of the paddies was proposed, although this might lead to increased emission of N_2O.

Greenhouse gas action plan (GHGAP) in UK

An example of the way the agricultural industry can take collective action to reduce GHG emissions is that taking place in the UK where in 2011 a Greenhouse Gas Action Plan (GHGAP, 2011) was launched by a coalition of industry partners, including the independent Agriculture and Horticulture Development Board, ADAS, the

National Farmers' Union and the Organic Research Centre, Elm Farm. A few thousand advisers have been trained. The aim is to reduce GHG emissions by 3 Tg CO_2eq per year from 2018–2022, which amounts to about 6% of total agricultural GHG emission each year in the UK. GHGAP (2016) comprehensively covers all the possible measures that can be taken and at the same time assesses the progress being made. Emissions from UK agriculture in 2010 were more or less stable at about 50 Tg CO_2eq, being about 8% of the UK total. Agriculture is a major source of N_2O (75%) and of CH_4 (38% of the total) although emissions have decreased by 20% and 19% respectively since 1990 (GHGAP, 2016).

Progress has been reported in several mitigating areas including crop nutrient management, soil and land management, manure handling, livestock nutrition, health and fertility and energy efficiency and use of renewables.

It is estimated that more than 60% of farmers have improved the efficiency and accuracy of N fertilizer application and more than 60% of contractors are following advice about precision fertilizer application. These improvements have been achieved by (i) optimizing the quantity of nitrogen applied as fertilizer and manure, taking into account N from previous crops and ensuring that all other crop inputs (including other nutrients, lime and crop protection products) are sufficient and (ii) by applying nitrogen from fertilizers and organic manures at times that match crop uptake and by

avoiding applying nitrogen when the soil is waterlogged, frozen or when the crop/grass is not growing.

More farmers are taking an interest in cover crops and in grassland nutrition. There is also a greater awareness of the importance of looking after soil structure especially by appropriate cultivations, minimizing wheelings, by drainage and by adding organic matter. Farmers are advised to store and spread manure in such a way that will reduce volatilization of NH_3 and leaching of NO_3. Slurry spreading techniques that minimize emission are being promoted.

Diets and feeding regimes are planned to match more closely animal requirements. More farms (62%) are growing temporary pastures, adopting high sugar grasses in order to improve efficiency of rumen microorganisms and reducing dietary protein levels for dairy cattle. As a result of proactive health planning and monitoring of performance of all livestock (e.g. through weighing) there has been an increase in the number of dairy herds covered by mastitis control plans; 74% of livestock farmers have a farm health plan.

One third of farmers have invested in renewables. At the same time it is appreciated that agriculture can also make a big contribution to mitigating climate change by storing carbon in soil organic matter and in vegetation.

USA

In the USA GM crops have contributed to reducing CO_2 emissions as a result of farmers using less energy and pesticides and by enhancing soil carbon sequestration (ISAAA, 2014). These benefits have been achieved by growing herbicide-resistant crops that facilitate zero or no-till practices, which significantly reduce the loss of soil carbon, and by insect-resistant crops requiring fewer pesticide sprays.

Europe

In Europe Oenema *et al.* (2009) modelled the effects of measures first implemented in the 1990s to reduce the losses of reactive N to the environment. The three most promising measures identified were (i) balancing N fertilization with crop need, (ii) low-protein animal feeding, and (iii) abatement of ammonia emissions by covering manure, redesigning livestock sheds and by the quick incorporation of manure and slurries. All the measures are connected to current EU environmental policy, e.g. the EU Nitrates Directive. Their modelling suggested that with no further intervention or additional changes in practices N use efficiency (NUE) in crop production could be increased from 44% in 2000 to 48% in 2020, while total N losses could be reduced by 10%. Full strict implementation of the promising measures could increase NUE further to 51–55%, and decrease NH_3 emissions by up to 23%, cut N_2O emissions by up to 10% and reduce NO_3 leaching losses by up to 35%. However, it was also revealed that implementation of all measures markedly reduced farm incomes.

FAO

The FAO (2016) in its report on the state of food and agriculture said that the pledge to eradicate hunger and poverty must go hand-in-hand with rapid changes to farming systems to cope with a warmer world and that agriculture has a big role to play in curbing greenhouse gas emissions. Owing to the predominance globally of the large number of farmers in developing countries it was stressed that solutions had to be tailored to their situations.

The adoption of climate-smart practices, such as the use of nitrogen-efficient and heat-tolerant crop varieties, zero-tillage and integrated soil fertility management would boost both productivity and farmers' incomes. Proposals by the FAO for the mitigation of GHG emissions included water-conserving alternatives to the flooding of rice paddies in order to cut methane emissions by up to 45%, and the adoption of more efficient practices for feeding livestock aimed at reducing methane emissions by up to 41%. Sequestering of carbon in soils and forests was advocated.

The FAO report highlighted the policies and strategies that would be required in order to incentivise farmers to be able to adopt the necessary changes in agricultural practices. Subsidies to promote greater use of inputs, the provision of advice and assistance with marketing would all be required.

Lessons of organic farming

Organic farming has many merits associated with crop rotations, with biological N fixation, with weed and pest control and with the adoption of agro-ecological practices but are there any ground-breaking procedures exclusive to organic farming that should be adopted by conventional agriculture? Manure and legumes were valued in antiquity and rotations appreciated for a few hundred years. In the UK there are signs that the organic movement wishes to engage with conventional farming to share ideas. In many ways the organic movements across the world have played a valuable part in drawing attention to such aspects as the value of rotations, the misuse of pesticides and the importance of improving soil conditions to conserve soil and minimize erosion.

The Soil Association (2012) in the UK, in its document "The Road to 2020, towards healthy, humane and sustainable food, farming and land use" issued an updated strategy with the basic objective of reaching out to a wider audience. It wishes, commendably, to engage with others not directly involved with organic farming to find ways of getting the balance right between its standards and other farming and land use systems.

The association would like to pioneer new ways of tackling climate change, of enhancing biodiversity and of improving animal welfare, all admirable aims. The Soil Association wants to become better known for what it stands for rather than for what it is against, such as man-made fertilizers and pesticides; consumer perceptions, and thus the market for organic food, has been influenced by this opposition (Chapter 4). The Soil Association is convinced that the most effective way of feeding the world is via small-scale, agro-ecological systems. Small-scale farmers numerically dominate the global agricultural scene and by so doing play by far the biggest part in feeding the world but there is no reason why they should not benefit from fertilizers, pesticides and the advances of agricultural science. Moreover, where the soils are impoverished, as in many African soils, it is essential to improve soil nutrient status. This can be achieved most easily by fertilizer inputs but fertility-building legumes are also ideally required.

Further evidence of the extension of the objectives of the organic movement was provided by their concern about sustainability at an organic producers' conference in the UK in November 2014. Apart from aiming to cooperate with conventional farmers in the pursuit of sustainability they are helping to develop methods of assessing sustainability, starting from the assumption that organic farming is more sustainable than conventional farming. The Organic Research Centre (2014) is working on using farm data such as soil nitrogen, nutrient inputs and outputs, energy use and biological diversity as measures of sustainability in a project called the Public Goods Tool (PG Tool). The areas of public good covered by the PG Tool are very wide, including soil management, bio-diversity, landscape and heritage, water management, manure management and nutrients, energy and carbon, food security, social capital and animal health and welfare.

Organic farming is extolled as being kind to the environment, to livestock and for promoting diversity. At the local level diversity can be enriched and environmental damage restricted but globally organic farming is deficient in needing more land than conventional farming and so reducing the area of land available for wildlife and natural vegetation that have to be safeguarded. Efforts to minimize the ravages of the Brazilian rain forest and of the jungles of South-East Asia face major challenges posed essentially by global population growth. In developed countries that have the luxury of not needing to produce all their own food, hedges have been extended and the land area set aside for wildlife increased.

The application of manure and compost to maintain soil biological activity as well as organic matter is very important to organic farmers but manure is similarly valued by conventional farmers as a source of nutrients and for improving soil structure.

In contrast, organic farming has been on its own when it comes to pest, disease and weed control because of the non-use of man-made pesticides. Where there is political pressure to restrict the use of pesticides the organic lessons on how best to use crop rotations, cultivation and integrated pest control are very valuable.

Costs of use of pesticides

Between 2 and 3 billion kg of pesticides are applied globally each year. The economic and practical benefits of pesticides are appreciated by farmers. However, the hidden costs of pesticide use are seldom considered. In the USA the national costs of using pesticides were investigated by Pimentel (2005). He estimated that annually costs reached $10 billion in environmental and societal damages. The major costs were associated with pesticide impacts on public health ($1.1 billion), pesticide resistance ($1.5 billion), crop and crop product losses ($1.4 billion), bird losses ($2.2 billion) and ground water contamination ($2.0 billion). In strictly economic terms these costs should be set against the annual $40 billion saved in the USA by the protection afforded by the pesticides, costing $10 billion (Pimentel and Greiner, 1997). There is clearly a strong economic case to minimize the use of pesticides.

On the other hand, in France Guillemaud and Bourguet (2016) concluded based on sixty-one papers published between 1980 and 2014 that the real cost of pesticides may far outweigh their benefits. They looked at four categories of costs: regulatory, human health, environmental costs and defensive expenditures (by organic growers). Regulatory and health costs were the major components and the researchers advocated that a more accurate evaluation of the cost of pesticide effects on health is required. Partially as a consequence of this work the French Agency for Food,

Environmental and Occupational Health and Safety (ANSES) has been charged with finding alternatives to pesticides. In 2015 ANSES became responsible for issuing the marketing authorizations for all biocidal products on the French market.

The Pesticide Action Network have been surveying pesticide poisoning in farming communities, agricultural workers in developing countries for many years PAN (2010); surveys in countries in Africa, Asia and Latin America revealed widespread ill health and many associated symptoms.

Weed control

Organic farmers may tolerate small amounts of weeds, which are seen as hosts for beneficial insects and food for birds; detrimental effects on crop yield are accepted. The organic farmer has several options for controlling weeds including cultivation, false or stale seedbeds, delaying sowing dates, increasing seeding rates, under-sowing crops and practising crop rotations to break the life-cycle of weeds. Mechanical cultivation with harrows, mowers, weed-surfers and by hand is commonly practised. It has been calculated that in the USA that 55 million labourers would be required to replace herbicides and mechanical weed control (Gianessi, 2011). Hand weeding of crops such as lettuce, sugar beets and celery in California became a contentious issue after the banning in 1975 of hand hoeing with a short hoe (known as el cortito with a 30–40 cm handle) and of hand pulling of weeds (banned in 2004)

because of back injuries. Organic farmers obtained exemption from these laws because they said that they would suffer significant yield and profit losses without hand weeding with the short hoe.

The use of deep cultivation to control weeds is a problem for organic farmers because of the damage to soil structure, the loss of soil moisture and increased GHG emissions.

There are attempts to develop natural phytotoxic substances as weed killers that organic farmers would be allowed to use. For example, pine oil and D-limonene were tested on soybean in Brazil but with only limited success, giving about 50% weed control; yields were intermediate (2.7–3.6 t/ha) between 4.07 t/ha from weed-free crops and 2.6 t/ha for untreated crops (Giepen *et al.*, 2014).

Rotations

The merits of the age old practice of rotating crops to control weeds and diseases was evident on the long-term classical experiments at Rothamsted when the labour shortages during and after the First World War limited hand hoeing. The extra yield of 2 t/ha wheat grown after a two-year break (oats followed by maize) was attributed to reductions in the effects of soil-borne pests and disease, particularly take-all (*Gaeumannomyces graminis var. tritici*) (Rothamsted, 2012).

Integrated pest management

Starting in the 1960s a different approach to pest control called integrated pest management (IPM) was proposed. IPM aims to keep pests at economically insignificant levels by using crop production methods that discourage pests. Beneficial predators or parasites that attack pests are encouraged and pesticide applications are timed to coincide with the most susceptible period of the pest's life cycle. Examples of beneficial predators include ladybirds and their larvae (aphids and mealy bugs eaten), lacewing larvae (one can eat 600 aphids before reaching maturity), hoverfly larvae (aphids eaten) and parasitic wasps (lay egg in aphids).

Eradication of pests is not necessarily a goal or even desirable in some cases because this may also result in the loss of the beneficial predators or parasites that depend on the pest in order to survive. Successful IPM programmes are dependent on comprehensive information on the life cycles of pests and their interaction with the environment; it allows for the judicious use of pesticides that would not be permitted for organic farmers. IPM is not seen as a substitute for using pesticides but as a way of improving their efficacy and hopefully of reducing their use. Pesticides are, however, often the only way to deal with emergency pest outbreaks.

Agroecology

Agroecology is the study of the ecological processes involved in agriculture; the term may refer to a science, or a movement. It is not associated with any one particular farming system but is perhaps more closely linked to organic than to conventional farming. According to the website Agroecology (2016), agroecology is the application of ecology to the design and management of sustainable agro-ecosystems.

The basic features of agroecology are (Agroecology, 2016):

i. Use of renewable energy, biological N fixation, natural materials and recycling of farm nutrients.
ii. Reduce or eliminate the use of potential toxic substances.
iii. Use practices that minimize pollution.
iv. Conserve energy, save seed, use local and traditional varieties.
v. Manage pests, diseases and weeds instead of controlling them.
vi. Use intercrops and cover crops.
vii. Encourage beneficial insects and soil microbes.
viii. Reduce mechanical disturbance of the soil and use mulches.
ix. Encourage diversification by undisturbed zones and by rotational grazing.

There is much to approve of in these features of agroecology. However, the reliance on biological N fixation and the apparent restriction to traditional varieties would impair the prospects of agroecological practices being able to feed more than half the current global population. In this respect agroecology is similar to organic farming. The proponents of agroecology believe that conventional/industrial agriculture is responsible for the loss of biodiversity, climate-changing emissions and the entrapment of small farmers by a declining number of global agrochemical companies that sell seed and pesticides.

Genetically modified crops are singled out for disapproval largely on the basis that they enable the monopolistic vendors to dominate world farming.

Agroecology was given the blessing of the UN in 2011 (de Schutter, 2010) when the work reported in UNEP-UNCTAD (2008) was used in support; the same report was used to demonstrate the superiority of organic farming. As discussed in Chapter 4, it was the training programmes that introduced the improved farming methods including mixed farming, agroforestry and improved varieties that were responsible for the better yields after training. The improvements were not demonstrated by experiments comparing farming systems but were the differences in yield before and after training.

International agrochemical companies are blamed for limiting the choice of seed varieties for farmers all over the world. In this context the promotion of genetically modified varieties are singled out for approbation. Concern is expressed that long-term disasters to crop diversity are on the horizon. Many countries have created their own seed banks of their own crops and varieties. And if that was not enough the Norwegian government has built, in a converted coal mine on Svalbard, an international seed bank that contains more than 800,000 varieties of crops from most countries. Low temperature (-18°C) storage should ensure

the current seed will remain viable for 20,000 years, a time span longer than that from the beginning of farming!

The choices about how to achieve sustainability were examined by McKenzie and Williams (2015). They were very clear in their conclusion that it is necessary to shift the emphasis from productivity while reducing environmental impacts, to the establishment of ecological sustainability as the entry point for all agricultural development. Practical solutions for intensification of agriculture and greater food production would have to comply with the ecological imperatives. Such a proposal appears ideal but is it acceptable to countenance starvation while the environment survives? McKenzie and Williams are optimistic about crop productivity and believe that there is still potential arable land that can be brought into cultivation as the population grows.

Importance of technology and innovation

Most of the food consumed in developed countries is produced on conventional farms using the advances provided by agricultural science. It is the industrial nature of such farming that is unacceptable to many of those looking for natural ways of producing food. Farmers in developing countries have also benefited from agricultural science, primarily from new varieties as in the Green Revolution and more recently from GM crops that in 2015 were grown by about 18 million farmers in twenty-eight countries on 180 million ha (James, 2015). In comparison, in 2014 mainstream organic farming was practised by about 2.3 million farmers on about 44 million ha, of which 21 million ha was comprised of extensive rangeland and pampas in Australia, Argentina and Uruguay.

It was agreed at a meeting of agriculture ministers from the world's twenty major economies at their 2016 meeting in China that science, technology and innovation will continue to play an important and leading role in sustainable agricultural growth across the world (G20, 2016). Innovation and collaboration were also key topics discussed at the annual meeting of the Forum for the Future of Agriculture FFA (2016), which dealt with how to meet the sustainable development goals of the UN. The FFA was created in 2006 by the European Landowners' Organization and Syngenta in an endeavour to focus attention on future problems of food security.

These positions are in marked contrast to the view expressed at the Oxford Real Farming Conference in 2016 that farming should return to labour intensive, mixed farming systems (ORFC, 2016). The European Parliament endorsed a Green Party report (June 2016) that did not accept the findings of the G20 ministers, saying that intensification of agriculture in Africa and the use of GM crops would be a mistake; in Europe it was said that intensive agriculture had destroyed family farming and reduced biodiversity.

Research in farming practices and food technology are expected to lead to innovations similar to the examples presented below.

Improvement of food quality
Bio-fortification

Improvements to the nutritional value of crops in relation to human health will continue to be targets of crop breeding. A recent example is the super broccoli with enhanced levels of the cholesterol-lowering glucoraphanin that was introduced on to the market in 2011 after twenty-seven years of conventional breeding; this followed the finding of the responsible genes in wild varieties of broccoli in Sicily (Armah *et al.*, 2015). Breeding wheat containing more iron to help combat iron deficiency, a common and widespread nutritional disorder, is being carried out at the John Innes Centre (BBSRC, 2016) using mutagenesis. Work on the genetically simpler rice (with only one pair of chromosomes compared to wheat with three pairs) in China and Japan has already, by using GM technology, managed to increase iron content by about six times; with traditional breeding strategies iron levels were only increased about three times (Wirth *et al.*, 2009).

Apple

Okanagan Specialty Fruits in Canada has produced, by genetic engineering, and has registered for use in the USA and Canada three apple varieties (Arctic Granny, Golden and Fuji) that do not brown when cut open (OFS, 2016).

Maize – mycotoxins

The potential benefit of Bt maize's reduction of mycotoxin damage was demonstrated several years ago by Wu (2006) but has been virtually ignored despite the serious health problems that can be caused by mycotoxins such as fumonisin and aflatoxin found in corn. Mycotoxins can be produced by fungi, which colonize and prosper in crop tissues that have been damaged by pests; Bt maize wards off the corn borer, thereby reducing feeding damage, fungal growth and mycotoxin production.

Tomato

Purple tomatoes have been genetically engineered at the John Innes Centre that are rich in anthocyanins, compounds that give purple fruit such as blueberries and blackberries their colour (BBSRC, 2016). The purple tomatoes have potential health benefits such as anti–inflammatory properties, and have been shown to slow the progression of soft-tissue carcinoma in cancer-prone mice. They also have double the usual shelf life, which can help reduce food waste. There is interest in developing a purple tomato juice product for the American market but the opposition to GM technology will inhibit developments in Europe despite the product having potential health-promoting properties.

Golden rice

Vitamin A deficiency is a very serious problem in several parts of the world, causing death and blindness. Rice, a staple for millions in South-east Asia, does not contain ß carotene, the precursor of vitamin A. Using GM technology genes initially from a narcissus were transferred to rice, which enabled it to produce ß carotene. A more efficient gene from maize was later transferred to create golden rice, which can then be crossed with local elite varieties. In contrast, sweet potato (widely consumed in Africa but not South-east Asia) does contain ß carotene and so can be improved by conventional breeding without the need for gene transfer.

New oil crop

The development of new oil crops, particularly stem, leaf and root crops rather than seed crops, as advocated by Reader (2015) is an example of work in progress. He focused attention on the wax-berries *Myrica cerifera* and *Myrica pensylvannica* (shrubs), which are fatty-acid secreting temperate species that could potentially be bred to produce substantial quantities of oil. The fact that their roots bear nodules containing *Actinomycetes* bacteria that can fix nitrogen is an added attraction. Unfortunately the presence of desirable but flammable aromatic compounds in the leaves, stem and branches make *Myrica* a fire hazard.

Livestock

Food improvements are not restricted to crops. Workers at the University of Reading have shown that feeding dairy cows with more oily seeds such as rapeseed and linseed can reduce the saturated fats level and increase the monounsaturated fat content of milk (BBSRC, 2016); at the same time they found a marked reduction in the amount of methane produced per litre of milk (Beauchemin *et al.*, 2009).

The breeding for resistance to *Campylobacter, Salmonella* and *E. coli* infections in livestock, particularly in poultry, with the aim of reducing the incidence of foodborne infections in humans is being studied at the Roslin Institute, University of Edinburgh (BBSRC, 2016). It is hoped, because of European objections, that genetic engineering need not be involved and that the identification of the genes associated with resistance will inform marker-assisted selection in breeding programmes.

Salmon

Conventional breeding techniques of salmon following the discovery of the genes responsible for resistance to the devastating infectious pancreatic necrosis disease have been successful (Roslin, 2015).

Atlantic salmon, in Canada, have been genetically engineered by transferring growth hormone genes from Pacific Chinook salmon, which has enabled them to grow year-round instead of only during spring and summer (AquaBounty, 2016). The growth rates have been increased so that the fish, AquAdvantage salmon, grow

to market size in sixteen to eighteen months rather than the three years taken by fish that have been improved by traditional selective breeding. The salmon are only raised in onshore facilities where escape to the sea is impossible; this is primarily in order to achieve certification in the USA in 2015 (thirteen years after application) and in Canada in 2016. Trials in Argentina, Brazil and Panama are under way.

Omega-3 fish oils are known to have human health benefits. Fish do not produce these oils but obtain them from eating algae and plankton. Farmed fish such as salmon are fed food pellets containing fish oils derived from wild fish stocks and so are dependent on and put pressure on wild fish populations. Although many plants also possess omega-3 fatty acids they are not the same long-chain types derived from marine sources. This situation has led to work (over at least fifteen years) at Rothamsted to develop, by genetic modification, *Camelina sativa* plants (false flax) that have the ability to produce the key fatty acids that are found in marine omega-3 fish oils (BBSRC, 2016). Provided field trials continue to be successful and the many regulatory, safety and approval steps satisfied this advance could be a major boost both for farmed fish and for the preservation of wild fish stocks.

Canola has been genetically modified using genes from algae to produce the omega-3 fatty acids eicosapentaenoic acid (EPA) and docosahexaenoic acid (DHA) in commercially relevant amounts; this is the first time this has been achieved in a widely grown crop (Walsh *et al.*, 2016). In an ideal world canola oil would be a very good source of omega-3 fatty acids when directly consumed by humans but it is likely that regulatory and political barriers will inhibit this. Instead, the GM canola is more likely destined for feeding to fish.

In 2012 42% of all fish directly consumed by humans worldwide (136 million t) was produced by fish farming (inland aquaculture 26% and marine aquaculture 16%) (FAO, 2014); captured fish accounted for 58% of the total. The proportion of farmed fish is predicted to rise.

It may even be possible in the future to enrich the diets of other animals such as pigs and poultry and improve the quality of their products by increasing the provision of omega-3s to the human diet.

Innovation in weed, disease and pest control

Weeds

There are signs that it may be possible to control weeds by the application of microbes. A bacterial product developed in Washington State is being assessed by the US regulatory authorities. The bacteria acts by limiting the root growth of grass weeds but does not harm cereals. A major drawback is that the bacteria takes two to three years to establish in the soil and can take up to five years to eliminate the weeds.

A new method of physical weed control called 'Propelled Agricultural Grit Management' (PAGMan), which uses grit made from waste products (e.g. dried corn cob bits) or organically approved fertilizers to abrade weed seedlings in row crops, is being developed in the USA (Forcella, 2012). The high cost, possibly several times that of herbicides, may limit the process, which involves blasting the grit at weeds on either side of a row crop.

Diseases

Multinational agrochemical companies are looking into the potential of using bacteria and fungi to fend off pathogens. The basic idea is to coat seed with the appropriate organism that will, by living on the surfaces of the plant, provide a defence.

Workers in Denmark have for the first time demonstrated that one species of a beneficial bacterium, *Pseudomonas fluorescens*, can stimulate *Arabidopsis* to generate cytokinins that have the ability to afford protection against a disease caused by another species of *Pseudomonas syringae* (Großkinsky *et al.*, 2016); the defence mechanisms against the pathogen are activated. The application potential within integrated plant protection strategies is significant for both organic and conventional farming.

Algae are being called into the battle against fungal diseases on grapes (*Vitis vinifera*) by a European-funded project (ProEcoWine). The objective is to replace the copper fungicides commonly used by organic farmers in preference to synthetic fungicides (Bilbao *et al.*, 2014). Strains of microalgae with antifungal properties have been cultivated in Hungary and processed into a plant protection product enriched with micronutrients. The algae are grown in a photo-bioreactor. Strains with the most efficient control of at least 90% over downy mildew and *Botrytis* were identified and selected. Field trials have been successfully carried out in Germany.

Pests

The possibility of making use of the pesticidal properties of locally available plants was investigated in Tanzania (Mkindi *et al.*, 2015). Extracts made from four common weeds, *Tithonia diversifolia, Tephrosia vogelii, Vernonia amygdalina* and *Lippia javanica*, controlled important pest species on common beans (*Phaseolus vulgaris*), comparable to a synthetic pyrethroid. The plant pesticide treatments were more cost-effective than the synthetic pesticide.

Increasing disease resistance and tolerance to adverse environments
Banana

Cavendish bananas, which replaced the variety Gros Michel when it succumbed to Panama disease caused by the soilborne fungus *Fusarium oxysporum* in the 1950s, has in the Far East been found to host a new strain of the same fungus. It is likely to reach Latin America and the Caribbean and could cause a major disruption to the global market. In Africa, where bananas are an important food, the situation may not be so serious because the many cultivars grown there may afford some protection. Progress in creating bananas fully resistant to the fungus by classical breeding has so far been limited. The wild Asian banana seems to be resistant, and researchers in Australia are experimenting with transferring its resistance genes into the Cavendish variety.

A genetically modified banana with six times the normal level of vitamin A has been developed in Queensland, Australia, with the intention of introducing it in Uganda in 2020 (Rebgetz, 2014). Bananas in Uganda also suffer from a devastating and widespread bacterial wilt for which there is no treatment. Progress has been made, using genetic engineering, in developing cooking banana varieties that are resistant to bacterial wilt (NARO, 2016).

Cassava

Cassava, a very important food crop in Africa, suffers from several pests and diseases (USAID, 2015). Scientists in Uganda and Kenya are working, in combination with researchers in Missouri, USA on genetically engineered cassava resistant to brown streak virus, a disease that is spreading across sub-Saharan Africa. Traditional breeding methods have been slow to help cassava crops develop resistance to brown streak, partly because cassava is commonly propagated by planting sections of the stalk and not by seed.

Wheat

The prospects of breeding for resistance to more than one pathogen in wheat have been given a boost by the work of Steuernagel *et al.* (2016) at the John Innes Centre. They have developed a new technology called "MutRenSeq" that accurately pinpoints the location of disease resistance genes in large plant genomes and which has reduced the time it takes to clone these genes in wheat from five to ten years down to just two. The possibility now exists to develop new varieties of wheat with strong resistance to more than one pathogen, such as stem rust and yellow rust.

Other traits under development using GM techniques with wheat include *Fusarium* resistance, drought resistance, and protein quality (Wilson and Dahl, 2016).

Salinity tolerant crops

Salinity is a major stress affecting crop plants worldwide. In Australia the problem is only going to get worse, with 51% of Western Australian farms already affected in some way by saline soils. Hundreds of lines of GM wheat and GM barley modified using genes obtained from wheat, barley, maize, *Arabidopsis*, moss and yeasts have been studied in Australia (ISAAA, 2014). Several genes influencing salt tolerance have been found. Some of the candidate genes may prove feasible in developing salt tolerance in sugarcane, rice, barley, wheat, tomato and soybean.

Drought tolerant crops

GM crops carrying different drought tolerant genes are being developed in rice, wheat, maize, sugarcane, tobacco, *Arabidopsis*, groundnut, tomato, potato and papaya (ISAAA, 2014). In Africa drought-tolerant/water-efficient maize varieties are being developed through marker-assisted breeding. A drought-tolerant GM maize, MON 87460, has been approved in the USA. The genetic control of heat shock proteins that are associated with recovery of plants after heat stress is being examined with a view to providing protection against protein damage; such work is another example of the way in which GM technology can facilitate advances that would be very difficult with traditional mutagenesis.

Recycling human waste

The attention paid to the recycling of nutrients in manure has not been paralleled with regard to the disposal of human excreta. The problems of dealing with human excreta, especially in developing countries, will increase as the population rises. At present according to the World Bank, about 2.5 billion people globally lack access to adequate sanitation and more than 700,000 children die every year as a direct result of the consumption of food and water contaminated with human waste. This is an unacceptable and constant health hazard. Moreover, there is little opportunity to recycle the nutrients involved.

Workers at Colorado University are one of sixteen teams around the world, largely funded by the Bill and Melinda Gates Foundation, charged to find solutions by reinventing the toilet (The Conversation, 2015). They have developed a solar-powered system (the Sol-Char toilet) to render human excreta biologically harmless by heating it to 300°C using parabolic mirrors with the heat transferred using fibre-optics. Unfortunately the very high cost ($12,000 per unit) will limit its use. The end product when applied to the soil increases the cation exchange capacity and improves soil structure. The potassium and phosphorus content will be valuable but there will be minimal if any nitrogen recovered.

Innovation: hazards, exposure, risks and vulnerability

Risks are attendant to any innovation whether in the field of nuclear energy generation, the use of electricity or in developing new drugs, pesticides or in farming practices that may be vital for improving health and economic growth. Hazards are generated that may operate locally, such as the misuse of chemicals, or internationally, as in the emissions of greenhouse gases. Many innovations can both benefit and cause harm; the risks of the hazards arising from innovation need to be managed and not necessarily avoided.

It is important to be specific when considering the risks associated with innovations. For example, it is not sensible or helpful to talk about the risks of a technology such as nanotechnology or genetic engineering; it is the risk of a particular hazard (such as the exposure to a nanoparticle or to edited genes) that needs to be addressed. Assessment is required of the exposure, in terms of amounts and the time of exposure, to the hazard such as a pesticide (natural or man-made) in food in relation to the experimentally determined hazards on test animals (see HERP values in Appendix). In this way a judgement can be made on the likely risks to the health and well-being of people in general and also of people who because of say their age or occupation are particularly vulnerable to the hazard.

It is unfortunate that too few, whether in the media or decision-making circles in governments, seem able to distinguish between hazard, exposure and risk, probably because of the confused use of the terms. It does favour opponents of, for example, pesticides not to consider them individually but to group them together and misguidedly call a hazard a risk.

People living in advanced economies seem to have become more risk averse compared to previous generations. The removal of a hazard can lead to undesirable consequences elsewhere. For instance, the effective worldwide ban on DDT in the 1970s satisfied the opponents of pesticides as forward-thinking but it led to millions of deaths by malaria in underdeveloped countries; no global body was able to consider such a scenario.

The United Kingdom has a strong, possibly world-leading, position in the assessment and management of the risks facing the population, having a publicly available National Risk Register (UK 2014, 2015). The register is mainly concerned with civil emergencies, natural hazard and terrorism. Chemicals are only mentioned in the context of weapons!

Industrialization of agriculture

A strong dislike of industrial farming was one of the main drivers of the organic movement in the UK in the 1930s. This antipathy is still alive in the UK and was evident as the central theme of the Oxford Real Farming Conference (ORFC, 2016), which was set up in 2010 with the aim of promoting a wholly new farming economy that is better able to meet current needs. The ORFC considers that current farming practices are designed to maximize financial returns rather than to grow food and that farming systems needs to be changed.

It is envisaged by the ORFC that there will be a renaissance in farming designed to deter the development of big farms by reducing subsidies and with the savings promote small mixed, preferably organic, low input and labour-intensive farms. Farming based on agroecology is advocated where the farms are part of the biosphere and are sustainable. Hi-tech modern conventional farms, which are abhorred, are said to be in crisis and effectively run for the benefit of a few large companies. The seed industry is used as the classic example. In the USA in the 1930s more than 150 companies sold maize seed, whereas the market is now served by a handful of large companies.

It is difficult to see how a mixed agricultural economy dominated by community and small/medium farms could meet the dietary demands of a burgeoning global population, especially in developed countries with large urban populations. In developing countries where rural populations dominate, the concept of community farms could be an attractive proposition but there would seem to be no reason why they would benefit from being organic.

Across the world, many scientists are looking at ways to improve our food and its production. The developments are likely to be mainly concerned with improving efficiency of production, minimizing damage to the environment, improving the quality of food and reducing the need for labour. Conventional farming will have to solve the problems of sustainable food production when using fewer chemical inputs. Organic agriculture will have to learn how to increase biological nitrogen fixation and to compensate for reductions in livestock numbers. It is likely that much of the research being carried out across the world will be perceived as promoting the further industrialization of farming and will be anathema to green and organic movements.

Envisaged developments in the UK (BBSRC, 2016, 2016a) include:

i. Plant breeding will use gene editing, gene insertion and the search for useful genes in wild varieties of our crops. Varieties better adapted to the varying climatic conditions will be required as well as crops with improved nutritional attributes.

ii. Precision farming with robots that will be able to detect weeds and pests photographically and then apply the herbicide or pesticides as appropriate will become popular. It is anticipated that one robot weeder will take the place of seven to fourteen men.

iii. Fertilizer application will be in accordance with the varying fertility status across the field using geopositional satellite technology.

iv. Areas of weed, pathogen or insect infestation will be identified by drones in order to achieve more effective and lower rates of pesticide application. Drones can facilitate inspection of sheep, especially at lambing time, and so identify ewes in trouble, as currently practised on sheep farms in New Zealand.

v. Hydroponics will be used more, possibly including vertical farming in urban areas.

A vision of the future

Pirie, an economist and founder of the Adam Smith Institute, painted an optimistic picture of what Britain would look like in 2050 in his monograph *Britain and the World* (Pirie, 2016), which is a refreshing antithesis to doom-ridden forecasts. He looked at trends in scientific research in order to make predictions about how new technology will change people's lives and solve current energy, environmental and health problems.

Pirie envisaged that energy will be derived from gas, solar and nuclear power but not wind. Solar is likely to play a minor part in the winter in countries in northerly latitudes such as Britain. Fast-growing trees will capture much CO_2. Agriculture will have experienced a new Green Revolution by benefiting from GM crops that will not only have the ability to withstand pests and diseases but will also be able to grow well in poor soils and in extreme climates. Pirie sees future crops being self-fertilizing, which may apply to N but not to the other nutrients normally derived from the soil.

Production of meat substitutes in factories will replace much of livestock farming, so freeing up much land for vegetable production, wildlife and leisure. Pirie considered it is better to deal with environmental problems by investing in new technology rather than by trying to change people's eating habits: this should apply to the problems linked to livestock farming but it is difficult to imagine that laboratory-produced meats or edible insects will be a sufficient or acceptable solution.

Savage (2011) was also optimistic that by following a few practices it would be possible in the USA to produce sufficient food in the future without increased damage to the environment. The practices advocated were minimum tillage, cover cropping, controlled wheel traffic, precision fertilization and integrated pest management.

Summary

1. There is increasing awareness in developed countries that the consumption of livestock products should be reduced in order to improve health and to reduce GHG emissions. On the other hand, in developing countries meat consumption is expected to continue to rise.

2. It will be difficult for the recommended meat intake levels of no more than 500 g/week for adults, as is proposed in some European countries, to be achieved. Limiting the consumption of processed meat is singled out because of links with cancer.

3. Much land would be freed for the production of food for direct human consumption if there were reductions in the consumption of livestock products.

4. A major reduction in the global requirement for N for the production of animal feed means that organic farming would be better equipped to feed more people in a world that was less reliant on animal products.

5. The restriction of cattle, sheep and goats to land only suitable for grazing could also reduce the need for large areas of feed crops.

6. Edible insects are a promising alternative to livestock as sources of animal protein, either for direct human consumption or for indirect use in animal feeding stuffs, because of their efficiency in converting protein in organic wastes into edible protein.

7. Several insects are being farmed globally but collection from the wild provides most of those currently consumed by humans. Of the twenty farmed species, five are dominant, namely crickets, mealworms, silkworms, grasshoppers and black soldier flies.

8. The feeding of nitrogen-rich insect products to livestock in animal feed would reduce the demand for crop-derived protein.

9. Man-made nitrogenous fertilizers continue to get a bad press, mainly related to pollution and environmental damage, despite their essential role in food production; there is a need to improve their use and to minimize pollution.

10. Improving the efficiency of use of nitrogenous fertilizers can be achieved by timing the application of the N to synchronize with the period of maximum uptake and by precision application designed to vary the amount of N according to soil fertility; synchronization and precision farming are more easily achieved when using fertilizer N than with N in manure, compost or green manures.

11. Agriculture is a major source of N_2O and of CH_4 emissions. In the UK efforts to mitigate emissions are being successful by improving efficiency of N use, by better handling of manure and slurries, by modifying livestock feeding and by increasing the use of temporary pastures.

12. Organic farming has many merits associated with the use of crop rotations, weed and pest control and has played a valuable part in drawing attention to agro-ecological practices that can benefit all types of farming.

13. At the local level organic farming can encourage a greater diversity of wildlife than conventional farming but because the yields are lower, organic farming requires more land, so reducing the land available for wildlife and increasing deforestation.

14. Integrated pest management combines discouraging pests by encouraging beneficial predators or parasites that attack pests with hopefully limited use of pesticides.
15. The ecological processes involved in agriculture – agroecology – are attracting attention. Most of the principles of agroecology would be approved of by conventional farmers but the reliance on biological fixation of N and on local crop varieties would limit food production and inhibit the breeding of improved varieties.
16. There is agreement amongst scientists and politicians that science, technology and innovation will continue to play an important and leading role in achieving sustainable food production across the world.
17. Improvements in crop quality with regard to the nutritional value of food for human health will continue to be targets of crop breeding in which GM technology is playing a major part.
18. Breeding for disease resistance is making progress. In Africa the introduction of new GM varieties of banana and cassava that are resistant to various diseases is imminent.
19. The dislike of industrial farming remains one of the drivers of the organic movements. Farming based on agroecology and small and community farms is preferred; however, it is difficult to see how such a system could meet the dietary demands of an increasing global urban population.
20. In the UK it is envisaged that plant breeding will benefit from gene transfer and gene editing, and that the application of fertilizers and pesticides will be made more precise by using robots and geo-positioning technology.

References

Agroecology (2016) Available at www.agroecology.org/Principles_List.html

Alexandratos, N. and Bruinsma, J. (2012) World agriculture towards 2030/2050: The 2012 revision. ESA Working paper No. 12-03. Rome, FAO. Available at www.fao.org/docrep/016/ap106e/ap106e.pdf

AquaBounty (2016) AquaBounty Technologies. Low impact fish farming. Available at https://aquabounty.com/sustainable

Armah, C.N., Derdemezis, C., Traka, M.H., Dainty, J.R., Doleman, J.F., Saha, S., Leung, W., Potter, J.F., Lovegrove, J.A. and Mithen, R.F. (2015) Diet rich in high glucoraphanin broccoli reduces plasma LDL cholesterol: Evidence from randomised controlled trials. *Molecular Nutrition and Food Research* 59, 5, 918–926. doi: 10.1002/mnfr.201400863.

BBSRC (2016) Foods for the future. Available at www.bbsrc.ac.uk/news/food-security/2016/160406-f-ways-research-is-changing-food-of-the-future

BBSRC (2016a) The farms of the future. Available at www.bbsrc.ac.uk/news/food-security/2016/160223-f-farms-of-the-future

Beauchemin, K.A., McGinn, S.M. Benchaar, C. and Holtshausen, L. (2009) Crushed sunflower, flax, or canola seeds in lactating dairy cow diets: Effects on methane production, rumen fermentation, and milk production. *Journal of Dairy Science.* 92, 5, 2118–2127. doi: 10.3168/jds.2008-1903.

Bilbao, J., Tseng, E., Furst, M., Erb-Brinkmann, M., Schmid-Staiger, U. Ördög, V. and Egner, S. (2014) ProEcoWine: Development of a novel plant protection product to replace copper in organic viticulture 539–5442. In: Rahmann, G. and Aksoy, U (Eds.) *Proceedings of the 4th ISOFAR Scientific Conference. Building Organic Bridges,* Organic World Congress 2014, Istanbul, Turkey.

CABI (2016) *Insects as a source of protein.* Available at www.cabi.org/projects/project/33056?utm_source=CABI&utm_medium=email&utm_campaign=6716721_CABI%20News%20Update%20January%202016

Ecowatch (2016) 8 Disturbing Facts About Monsanto's Evil Twin – The Chemical Fertilizer Industry. Available at http://ecowatch.com/2016/04/12/facts-fertilizer-industry

EC (2013) Flash Eurobarometer 367. Attitudes of Europeans Towards Building The Single Market for Green Products. Available at http://ec.europa.eu/public_opinion/flash/fl_367_en.pdf

EFSA (2015) Risk profile related to production and consumption of insects as food and feed. *EFSA Journal* 13, 10, 4257. doi: 10.2903/j.efsa.2015.4257.

Elam, T.E. (2006) Projections of global meat production through 2050. Available at www.farmecon.com/Documents/Projections%20of%20Global%20Meat%20Production%20Through%202050.pdf

FAO (2014) The State of World Fisheries and Aquaculture Opportunities and challenges. Available at www.fao.org/3/a-i3720e.pdf

FAO (2016) The State of Food and Agriculture. Climate change, agriculture and food security. Available at www.fao.org/publications/sofa/sofa2016/en

FCRN (2016) Overview of insect farming; Where, What Species and Areas for Future Research. Food Climate Research Network. Available at http://www.fcrn.org.uk/fcrn-blogs/wendylumcgill/part-2-edible-insects-food-and-feed-series-overview-insect-farming-where

FFA (2016) Media statement 9th Forum for the Future of Agriculture. Not business as usual – UN SDGs and Forum demand action and collaboration. Available at www:forumforagriculture.com/wp-content/uploads2016/03/FFA2016_Media_statement.pdf

Forcella, F. (2012) Air-propelled abrasive grit for post emergence in-row weed control in field corn. *Weed Technol.* 26: 161–164.

G20 (2016) China G20 Agriculture ministers communique. Available at http://foodsecurityportal.org/china-g20-agriculture-ministers-communique

GHGAP (2011) Agricultural Industry Greenhouse Gas Action Plan Framework. Available at www.ahdb.org.uk/projects/documents/GHGAPDeliveryPlan04April2011.pdf

GHGAP (2016) Agricultural Industry Greenhouse Gas Action Plan (2016) Promoting GHG activity and productivity Progress report and Phase III strategy and activities. Available at www.cfeonline.org.uk/ghgap-progress-report-2015-v820

Gianessi, L. (2011) Weeds, herbicides and human rights. Available at https://www.youtube.com/watch?v=8ySixmE4SH4

Giepen, M., Neto, F.S. and Köpke, U. (2014) Controlling weeds with natural phytotoxic substances (NPS) in direct seeded soybean 469–472. In: Rahmann, G. and Aksoy U. (Eds.) *Proceedings of the 4th ISOFAR Scientific Conference. Building Organic Bridges,* Istanbul, Turkey.

Gilbert. N. (2016) The race to create super-crops. Old-fashioned breeding techniques are bearing more fruit than genetic engineering in developing hyper-efficient plants. *Nature* 533, 7603, 308–310. doi: 10.1038/533308a.

van Grinsven, H.J., Spiertz, H.J., Westhoek, J.H.J., Bouwman, A.F. and Erisman, W. (2014) Nitrogen use and food production in European regions from a global perspective. *Journal of Agricultural Science* 152, S1, 9–19.

Großkinsky, D.K., Tafner, R., Moreno, M.V., Stenglein, S.A., García de Salamone, I.E., Nelson, L.M., Novák, O., Strnad, M., van der Graaff, E. and Roitsch, T. (2016) Cytokinin production by *Pseudomonas fluorescens* G20-18 determines biocontrol activity against *Pseudomonas syringae* in Arabidopsis, *Scientific Reports* (2016). doi: 10.1038/srep23310.

Gryllo Co (2016) The-future of food. Available at http://gryllo.co/the-future-of-food

Guillemaud, T. and Bourguet, D. (2016) The hidden and external costs of pesticide use. *Sustainable Agriculture Reviews* 19, 35–120. Available at https://www.researchgate.net/publication/295813785_

Harvey, G. (2016) *Grass-Fed Nation*. Icon Books Ltd, London.

van Huis, A., van Itterbeeck, J., Klunder, H., Mertens, E., Halloran, A., Muir, G. and Vantomme, P. (2013) Edible insects: future prospects for food and feed security. FAO Forestry Paper 17. Available at www.fao.org/docrep/018/i3253e/i3253e.pdf

IAASTD (2009) International Assessment of Agricultural Knowledge, Science and Technology for Development. Agriculture at a crossroads. McIntyre, B.D., Herren, H.R., Wakhungu, J. and Watson. R.T (Eds.) Available at www.unep.org/dewa/agassessment/reports/IAASTD/EN/Agriculture%20at%20a%20Crossroads_Synthesis%20Report%20(English).pdf

IAASTD (2016) International Assessment of Agricultural Knowledge, Science and Technology for Development. Reports on Industrial Agriculture and Small-scale Farming. Available at www.globalagriculture.org/report-topics/industrial-agriculture-and-small-scale-farming.html

IARC/WHO (2015) International Agency for Research on Cancer. Monographs evaluate consumption of red meat and processed meat. Press release 240. www.iarc.fr/en/media-centre/pr/2015/pdfs/pr240_E.pdf

ISAAA (2014) International service for the acquisition of agri-biotech applications. Pocket K No. 43: Biotechnology and Climate Change. Available at https://www.isaaa.org/resources/publications/pocketk/document/Doc-Pocket%20K43.pdf

James, C. (2015) 20th Anniversary (1996 to 2015) of the Global Commercialization of Biotech Crops and Biotech Crop Highlights in 2015. ISAAA Brief No. 5-2015. ISAAA: Ithaca, New York. Available at http://isaaa.org/resources/publications/briefs/51/executivesummary/default.asp

McKenzie, F. C., Williams, J. (2015) Sustainable food production: constraints, challenges by 2050. *Journal of Food Security* 7, 221–233. doi: 10.1007/s12571-015-0441-1.

Miguel, M.A., Postma, J.A. and Lynch, J.P. (2015) Phene Synergism between Root Hair Length and Basal Root Growth Angle for Phosphorus Acquisition. *Plant Physiol.* 167, 4, 1430–1439.

Mkindi, A.G., Mtei, K.M., Njau, K.N. and Ndakidemi, P.A. (2015) The Potential of Using Indigenous Pesticidal Plants for Insect Pest Control to Small Scale Farmers in Africa *American Journal of Plant Sciences*, 2015, 6, 3164–3174. Available at http://file.scirp.org/pdf/AJPS_2015121414314346.pdf

NARO (2016) National Agricultural Research Organization starts trials on banana bacterial wilt resistant varieties. Available at www.monitor.co.ug/Magazines/Farming/Naro-starts-trials-on-banana-bacterial-wilt-resistant-varieties/689860-3129076-klok1u/index.html

Next-Food (2016) Insect. Available at www.next-food.net/wholesale-rates/insect

Oenema, O., Witzke, H.P. Klimont, Z., Lesschen, J.P. and Velthof, G.L. (2009). Integrated assessment of promising measures to decrease nitrogen losses from agriculture in EU-27. *Agriculture, Ecosystems and Environment* 133, 3–4, 280–288.

OFS (2016) Arctic®Fuji: the newest member of the Arctic® family. Available at www.arcticapples.com/arctic-fuji-newest-member-arctic-family

ORFC (2016) Oxford Real Farming Conference. http://orfc.org.uk/archives/orfc-2016

PAN (2010) Pesticide Action Network. Communities in Peril: Global report on health impacts of pesticide use in agriculture. Available at www.panna.org/sites/default/files/PAN-Global-Report_0.pdf

Pimentel, D. and Greiner, A. (1997) Environmental and socio-economic costs of pesticide use 51–78. In: Pimentel, D. (ed.), *Techniques for Reducing Pesticide Use: Environmental and Economic Benefits*, John Wiley & Sons. Chichester, UK,

Pimentel, D. (2005) Environmental and economic costs of the application of pesticides primarily in the United States. *Environment, Development and*

Sustainability 7, 229–252. doi: 10.1007/s10668-005-7314-2.

Pirie, M. (2016) Britain and the world in 2050. Adam Smith Institute.

Public Health England (2016) *The Eatwell Guide*. Available at https://www.gov.uk/government/uploads/system/uploads/attachment_data/file/528200/Eatwell_guide_booklet.pdf

Reader, M. (2015) Plant science should focus on stem, leaf and root crops. East Anglian Daily Times. 8 August.

Rebgetz, L. (2014). Genetically modified Queensland bananas to join fight against catastrophic results of vitamin A deficiency in Africa. Available at www.abc.net.au/news/2014-06-14/genetic-banana-benders-fight-vitamin-a-deficiency/5523088

Röös, E., Patel, M., Spångberg, J., Carlsson, G. and Rydhmer, L. (2016) Limiting livestock production to pasture and by-products in a search for sustainable diets. *Food Policy* 58, 1–13. Available at http://dx.doi.org/10.1016/j.foodpol.2015.10.008

Roslin (2015) Salmon breeding to benefit from gene study of disease resistance. Available at www.roslin.ed.ac.uk/news/2015/05/22/salmon-breeding-to-benefit-from-gene-study-of-disease-resistance

Rothamsted (2012) Rothamsted long-term experiments. Guide to the Classical and other Long-term Experiments, Datasets and Sample Archive. (Updated reprint of 2006 edition) Available at www.rothamsted.ac.uk

Savage, S.D. (2011) What I Hope Will Be The Future Of Sustainable Farming. Available at http://appliedmythology.blogspot.co.uk/2011/03/what-i-hope-will-be-future-of.html

de Schutter, O. (2010) Agroecology outperforms large-scale industrial farming for global food security says UN food expert. Available at www.srfood.org/images/stories/pdf/press_releases/20100622_press_release_agroecology_en.pdf

Smil, V. (2014) Eating meat: Constants and changes. *Global Food Security* 3, 2, 67–71. doi.org/10.1016/j.gfs.2014.06.001.

Smith, P., Martino, D., Cai, Z., Gwary, D., Janzen, H., Kumar, P., McCarl, B., Ogle, S., O'Mara, F., Rice, C., Scholes, B., Sirotenko, O., Howden, M., McAllister, T., Pan, G., Romanenkov, V., Schneider, U., Towprayoon, S., Wattenbach, M. and Smith, J. (2008) Greenhouse gas mitigation in agriculture. *Philosophical Transactions of the Royal Society London B Biological Science*. 363, 1492, 789–813. doi: 10.1098/rstb.2007.2184.

Soil Association (2012) www.soilassociation.org

Springmann, M., Godfray, H.C.J., Rayner, M. and Scarborough, P. (2015) Analysis and valuation of the health and climate change co-benefits of dietary change. *Proceedings of the National Academy of Sciences of the United States of America* 113, 15, 4146–4151. doi: 10.1073/pnas.1523119113.

Steffen, W., Richardson, K., Rockström, J., Cornell, S.E., Fetzer, I., Bennett, E.M., Biggs, R., Carpenter, S.R., De Vries, W., De Wit, C.A., Folke, C., Gerten, D., Heinke, J., Mace, G.M., Persson, L.M., Ramanathan, V., Reyers, B. and Sorlin, S. (2015) Planetary boundaries: Guiding human development on a changing planet. *Science*. 347, 6223. doi: 10.1126/science.1259855. Available at http://science.sciencemag.org/content/347/6223/1259855

Steuernagel, B., Sambasivam, K Periyannan, S.K., Hernández-Pinzón, I., Witek, K.,

Rouse, M.N., Yu, G., Hatta, A., Ayliffe, M., Bariana, H., Jones, J.D.G., Lagudah, E.S. and Wulff, B.B.H. (2016) Rapid cloning of disease-resistance genes in plants using mutagenesis and sequence capture. *Nature Biotechnology*, 34, 652–655. doi: 10.1038/nbt.3543.

Swedish National Food Agency (2016) Swedish dietary guidelines – risk and benefit management report. Available at www.livsmedelsverket.se/globalassets/rapporter/2015/rapp-hanteringsrapport-engelska-omslag--inlaga--bilagor-eng-version.pdf

The Conversation (2105) Toilet talk: meeting one of the world's grand challenges with innovation. Available at https://theconversation.com/toilet-talk-meeting-one-of-the-worlds-grand-challenges-with-innovation-50646

UK (2014) Annual Report of the Government Chief Scientific Adviser 2014. Innovation: Managing Risk, Not Avoiding It. The Government Office for Science, London. Available at https://www.gov.uk/government/uploads/system/uploads/attachment_data/file/381906/14-1190b-innovation-managing-risk-evidence.pdf

UK (2015) National Risk Register of Civil Emergencies 2015 edition. Cabinet office. Available at https://www.gov.uk/government/uploads/system/uploads/attachment_data/file/419549/20150331_2015-NRR-WA_Final.pdf

UNEP-UNCTAD (2008) Organic Agriculture and Food Security in Africa, United Nations, New York and Geneva. Available at www.unep.ch/etb/publications/insideCBTF_OA_2008.pdf

USAID (2015) Agency for International Development. Genetically modified crops under research and development in Africa.

Available at http://nepad-abne.net/wp-content/uploads/2015/07/Biotech-in-Africa.pdf

US DH and HS (2015) Dept. of Health and Human Services 2015–2020 Dietary Guidelines for Americans. 8th Edition. Available at http://health.gov/dietaryguidelines/2015/guidelines

Voedingscentrum (2016) Netherlands Nutrition Center dietary guidelines. Fish, legumes, eggs nuts and milk. Available at www.voedingscentrum.nl/nl/gezond-eten-met-de-schijf-van-vijf/hoeveel-en-wat-kan-ik-per-dag-eten-/vis-peulvruchten-vlees-ei-noten-en-zuivel.aspx

Walsh T.A., Bevan. S. A., Gachotte, D.J., Larsen, C.M., Moskal, W.A., Merlo, P.A., Sidorenko, L.V., Hampton, R.E., Stoltz, V., Pareddy, D., Anthony, G., Bhaskar, P.B., Marri, P.R., Clark, L.M., Chen, W., Adu-Peasah, P.S., Wensing, S.T., Zirkle, R. and Metz, J.G. (2016) Canola engineered with a microalgal polyketide synthase-like system produces oil enriched in docosahexaenoic acid. *Nature Biotechnology* 34, 8, 881–887. doi: 10.1038/nbt.3585.

Wellesley, L., Happer, C. and Froggatt, A. (2015) Changing Climate, Changing Diets: Pathways to Lower Meat Consumption. Chatham House. Available at https://www.chathamhouse.org/publication/changing-climate-changing-diets

Westhoek, H., Lesschen, J.P., Rood, T., Wagner, S., De Marco, A., Murphy-Bokern, D., Leip, M., van Grinsven, H., Sutton, M. A. and Oenema, O. (2014) Food choices, health and environment: Effects of cutting Europe's meat and dairy intake. Lower meat and dairy consumption benefits climate and health. *Science Direct* 26, 196–205. http://dx.doi.org/10.1016/j.gloenvcha.2014.02.004.

Wilson, W.W. and Dahl, B. (2016) Potential Economic Impacts of Low Level Presence (LLP) in the Global Wheat Market 241–256. In: Kalaitzandonakes, N., Phillips, P.W.B., Wesseler, J. and Smyth, S.J. (eds.) *The Coexistence of Genetically Modified, Organic and Conventional Foods Volume 49 of the series Natural Resource Management and Policy.*

Wirth, J., Poletti, S., Aeschlimann, B., Yakandawala, N., Drosse, B., Osorio, S., Tohge, T., Fernie, A.R., Günther, D., Gruissem, W. and Sautter, C. (2009) Rice endosperm iron biofortification by targeted and synergistic action of nicotianamine synthase and ferritin. *Plant Biotechnology Journal* 7, 7, 631–644. doi: 10.1111/j.1467-7652.2009.00430.x.

Wu, F. (2006) Mycotoxin reduction in Bt corn: potential economic, health, and regulatory impacts. *Transgenic Research* 15, 3, 277–289. doi: 10.1007/s11248-005-5237-1.

Xiaodong, W. (2016) China Daily. 14 May. Ministry tweaks eating guidelines. Available at www.chinadaily.com.cn/china/2016-05/14/content_25269469.htm

Zhan, A. and Lynch, J.P. (2015) Reduced frequency of lateral root branching improves N capture from low-N soils in maize. *Journal of Experimental Botany* 66, 7, 2055–2065. doi: 10.1093/jxb/erv007.

CHAPTER 20

Sustaining Food Production – Basic Constraints and Potential of Plant Breeding and Mycorrhiza

"Prediction is very tough, especially about the future"
Niels Bohr

In this final chapter consideration is given to some of the basic constraints to crop growth and to ways that long-term sustainable food production might be promoted. Many accompanying parallel advances in agronomy, physiology, entomology, pathology and ecology will be required.

Nutrient supply

The central and dominant role of nitrogen in crop production has been emphasized here and little attention has been given to the other essential nutrients.

Without man-made nitrogenous fertilizers global populations would, in the absence of developments in biological nitrogen fixation discussed in this chapter, be limited to about 3 billion people; this is not a scenario that is contemplated here as it would imply that mankind has been unable to generate sufficient electrical power after fossil fuels run out. Provided man can still fix nitrogen, using electric power to generate hydrogen, the spotlight would fall on other essential nutrients and particularly on phosphorus and potassium. The availability of phosphorus ores is likely to run out long before potassium-containing ores and of potassium–rich rocks that could be used. Regardless of how long the phosphatic ores last, it may well be that the lower yields when farming relies on recycling of phosphorus will limit global population to the approximately 1 billion before phosphatic fertilizers were introduced.

Efforts to improve the efficiency of use of phosphatic fertilizers will postpone the day of reckoning but will not solve the basic problem. Sustainable farming in the long term (i.e. several hundred years) is likely to be primarily dependent on the availability of P inputs/fertilizers, but if a way is found to recover P from marine sediments, sustainable food production could be imagined into perpetuity.

Basic constraints to plant growth

Two fundamental constraints to plant growth are firstly the limits imposed by photosynthesis and secondly the availability of sufficient nitrogen. Improving the efficiency with which plants use radiant energy in photosynthesis and increasing and extending biological fixation of N would have major impacts on food security and on long term sustainability.

Photosynthesis

The conversion by radiant energy of carbon dioxide and water to glucose and oxygen is arguably the most important chemical reaction in the world, and is one of the oldest. It evolved in cyanobacteria,

which operated in stromatolites about 3.5 billion years ago when the world's atmosphere contained much less oxygen than today. The oxygen produced caused firstly the oxidation of reduced iron compounds in the oceans and then the generation of ozone in the troposphere, which would later protect life on earth against the lethal UV radiation.

Zhu *et al.* (2008) calculated the potential maximum efficiency of photosynthesis from light capture to carbohydrate synthesis; they reported that the maximum conversion efficiency was 4.6% for C3 photosynthesis at 30°C and today's 380 ppm atmospheric CO_2, and 6% for C4 photosynthesis. This advantage of C4 over C3 is expected to disappear when CO_2 levels exceed 700 ppm.

Estimates of average photosynthetic efficiency of the use of radiant energy range from 1–4% for most crops; sugar cane has been recorded at 7% (Hyperphysics, 2016) so there would seem to be scope for improvements.

Research designed to improve photosynthesis are following three paths: firstly the transfer of C4 traits into C3 crops, secondly via the enzyme rubisco and thirdly via the photoprotection mechanism.

C4 and C3

Photosynthesis involves the oxidation of water, the reduction of CO_2 and the release of oxygen. There is a light reaction and a dark reaction when the CO_2 is reduced.

Plants fall into three categories with regard to photosynthesis, namely C3, C4 or CAM (such as the cacti and bromeliads) (Hyperphysics, 2016). The first stable compound produced in the C3 plants is a compound containing three carbon atoms (3 phosphoglyceric acid) – hence the name. In the C4 plants the first stable compound contains four carbon atoms (oxaloacetic acid).

Most, about 85%, of plants are C3 plants including wheat, rice, barley, oats, most grasses, groundnuts, cotton, sugar beet, tobacco, spinach and soya beans. C3 plants are generally not as efficient as C4 plants mainly because in hot dry conditions the closing of stomata is likely to limit the concentration of CO_2 in the chloroplasts and lead to a process called photorespiration, which wastes energy already collected. In C4 plants the CO_2 level is kept high – as in CAM plants. CAM plants are interesting insofar as they absorb CO_2 at night rather than during the day; the CO_2 is held as malic acid until daylight, when it is released internally. C4 plants are all tropical, including maize, sugar cane, sorghum, millet and Bermuda grass (*Cynodon dactylon*). C3 plants can, however, outperform C4 plants if there is adequate water. There are some plants, such as *Panicum spp.* (a grass) and *Flaveria spp.* (a composite), that have a photosynthetic process intermediate between C3 and C4.

Transferring the useful traits from C4 plants into C3 plants is an attractive proposition. However, the study of the genetic control has been limited by the lack of a rapid cycling model system in which to study C4 photosynthesis. The rapid cycling *Setaria viridis*, a C4 plant, is now being used as a model for C4 photosynthesis (Brutnell *et al.*, 2010).

Rubisco

Research on improving photosynthetic efficiency is being carried out at several centres across the world. Lin *et al.* (2014) transferred genes for an efficient rubisco enzyme from a blue-green alga into tobacco to replace the inefficient form with excellent results. The rate of photosynthesis increased, as did growth. As the genes responsible for C4 photosynthesis are unlikely to be present (or even repressed) in C3 plants the transfer of C4 genes would seem to be essential for success.

The efficiency of photosynthesis in a wheat variety has been increased by genetic modification, which boosted yields by 15–20% when grown in the glasshouse at Rothamsted (Le Page, 2016); field trials will start in 2017. The modification followed the identification of the genes responsible for an enzyme (SBPase) involved in photosynthesis; the improvements depend on increasing the levels of this enzyme in chloroplasts. Objections to the trial have been made by the Sustainable Food Trust and the Soil Association on the basis that GM crops are not needed to feed the world and that the public do not want them (SFT, 2016); taking care of the soil is said to be a much better way of increasing production than using genetic engineering.

Photoprotection

Another approach to improving the efficiency of photosynthesis has followed the finding that probably all plants are able to protect their chloroplasts against overexposure to light by dissipating some light as heat (Zhu *et al.*, 2004); it was calculated that this mechanism could reduce photosynthesis by 30% when it operated. In dim light the photoprotection mechanism is inoperative. It is quickly turned on by bright light but is slow to turn off, during which time photosynthesis is not optimal. The genetic control has been studied in *Arabidopsis thaliana*. By transferring the genes responsible for relaxing the protection from *Arabidopsis* into tobacco, recovery was speeded up and normal photosynthesis quickly resumed (Kromdijk *et al.*, 2016): leaf growth was increased by about 15% in field trials, under fluctuating light. Work continues with gene insertion into elite lines of rice and maize and also on achieving the same results using gene editing in order to circumvent the ideological opposition to gene transfer.

Artificial photosynthesis

Work is being carried out in several countries to mimic photosynthesis in the laboratory, which if achieved could have dramatic effects on food production, energy supply as well as the sequestration of CO_2; it would be a game changer. Following the initiative of Faunce (2011) a Joint Center for Artificial Photosynthesis (JCAP, 2016) was set up in the USA in 2010 with the aim of finding effective ways to produce fuels using only sunlight, water and CO_2. Progress has been made (Davey, 2016) in creating the first photosynthetic bio-hybrid system (PBS) using silicon semi-conductors and bacteria to absorb solar energy and create a chemical product from water and CO_2 while releasing oxygen. Initially butanol was produced but with a very low conversion efficiency of solar radiation. By changing the bacteria it has been possible to create methane with a conversion efficiency of 10%, which is greater than natural photosynthesis. However, a high concentration of CO_2 is required by PBS, which would pose its own problems of delivery. The target now is to produce a liquid rather than a gas. Turan *et al.* (2016) in Germany reported that they have produced the first practical design for photo-electrochemical splitting of water to generate hydrogen efficiently in a way analogous to that of thin-film photovoltaic technologies; an overall efficiency of 3.9% in the utilization of solar radiation was reported.

Nitrogen fixation by legumes and non-legumes

The FAO declared 2016 as the "Year of Pulses" with the objectives of promoting their consumption and their cultivation as part of sustainable food production by improving soil fertility. The Africa Soil Health Consortium (CABI, 2016) has projects that encourage smallholders to grow legumes because of their nitrogen-fixing qualities. Pulses can be sold for much higher prices than cereals, making them valuable crops to help lift smallholders out of poverty. Grain legumes grown by smallholders include common beans, chickpea, cowpea, groundnut, pigeon pea and soya bean.

Gene manipulation techniques have indicated ways of improving commercial inoculant strains of rhizobia following studies showing the genetic determinants and the regulation pathways of most of the microbial functions. Genes that control nodulation, nitrogen fixation, host range and energy utilization have been identified, which has led to better inoculant strains (Montañez, 2000).

Nitrogen fixation by non-legumes

Persuading non-leguminous staple crops to fix nitrogen would constitute a major advance and is being studied by various researchers. In the seemingly unlikely event that non-legumes possess the appropriate modifiable genes, the approach will probably have to be via the transfer of genes. If new symbiotic nitrogen-fixing relationships are developed it is inevitable that yields will be restricted due to the host plant having to satisfy the energy requirements of the nitrogen-fixing organism; hopefully this will be an acceptable exchange.

Prospects for developing symbiotic nitrogen fixation by bacteria that will grow in association with cereals were given a boost by work of Pankievicz *et al.* (2015) with a grass (*Setaria viridis*) showing that the grass could obtain sufficient nitrogen to support robust growth from the bacteria *Azospirillum brasilense* growing on the root surfaces. Earlier work (Tien *et al.*, 1979) had associated the improved growth of pearl millet when inoculated with *Azospirillum brasilense* to increased rooting exploration (auxin induced) and not to nitrogen fixation.

Setaria viridis is now being used as a model for research into biological nitrogen fixation by examining whether a similar relationship could benefit other crops, notably wheat, maize, rice and sugar cane. *Setaria viridis* is a small plant that goes through its life cycle very quickly, has a pretty simple genome and can serve as a model for research in much the same way *Arabidopsis thaliana* is used in gene research.

The ability of *Anabaena*, a cyanobacteria living in association with *Azolla* (a duckweed fern), to fix nitrogen could be extended as a green manure and used on crops other than rice. The fact that *Anabaena* contains neurotoxins could be a problem but its extensive use on paddy rice suggests otherwise.

Production of man-made fertilizers, notably nitrogenous, will, in the long term depend on the availability and economics of energy supply and a hydrogen source that at present comes from natural gas; in the future hydrogen could be produced by electrolysis or by algae. It has been known since the 1990s that certain algae if deprived of sulphur will photosynthetically generate hydrogen instead of oxygen but only transiently; this property of algae is being examined in several countries, including Israel, USA, UK and Sweden. Liran *et al.* (2016) in Israel studied the biochemistry of hydrogen generation. Commercial viability would seem to be several years away.

If sufficient and relatively cheap energy should not be available, civilization as in developed countries would effectively be destroyed; we would become a world dependent on the biological fixation of nitrogen and the population would decline.

At the International Rice Congress in 2014 it was confidently forecast that by 2030 farmers would be planting nutritious C4 and nitrogen-fixing rice varieties (IRC, 2014) and that this would mark a new Green Revolution.

Plant breeding and genetic engineering

It is evident that much of the historic improvements in crop yields have come from plant breeding. According to FAO/AEA (2016a), these have amounted to 40% over the last fifty years. There are no reasons to suppose that breeding will not continue to benefit both organic and conventional food production. Organic farming could be given a boost if, for instance, more disease- and pest-resistant crops were bred and conventional farmers could benefit by using fewer pesticides. In the past the methods used by plant breeders were only of interest to the breeders. Now some groups are objecting to the latest methodologies, particularly those involving genetic engineering. These groups have been successful in a few countries but notably in Europe in inhibiting developments not only locally but also in Africa.

It is useful to consider how plant breeding has progressed from traditional crossing to the use of mutagenesis, to marker assisted breeding, to genetic engineering involving the transfer of genes and to gene editing (CRISPR-Cas9) in order to appreciate the limitations of the old methods and the potential of the new technologies.

The ultimate aim is to breed crops adapted to the many varied climatic and soil conditions across the world. Ideally an improved variety such as golden rice would be used locally in crosses with good varieties in order to introduce the beneficial genes.

Before the 1940s plant breeders were limited to selecting potential parent plants on the basis of observed traits. So a wheat plant showing resistance to a particular disease could be crossed, by the manual transfer of pollen, with a known higher yielding but susceptible variety. The breeder could only make use of the genes present in the two parents and they had to be expressed. This is the gene pool available to the breeder. The gene pool can be widened by traditional cross breeding with wild varieties, relatives of those that were originally domesticated. Wheat yield increases of 30% are being forecast from the introduction of genes from a wild grass into modern wheat via intermediary wheat (a cross of the wild grass and ancient wheat) by the National Institute of Agricultural Botany (NIAB, 2013).

Traditional breeding is time-consuming, taking many years (often more than ten) to develop a new variety.

Mutagenesis

In the late 1920s scientists discovered that exposing plants to different types of radiation (e.g. gamma or X-rays) or to certain chemicals damaged the plant's DNA and increased the rate of the formation of hopefully beneficial mutations, possibly by 1,000 times; it was only after the Second World War that mutagenesis was widely used when the techniques of the nuclear age were extended to biology. The irradiation dose rate is now designed to limit the mutational load and not cause extensive disruption to an already elite genotype. Mutations caused by cosmic or terrestrial radiation and by mistakes during DNA replication are the basis for all biological variation throughout evolution; they can be thought of as one of the driving forces of evolution.

China has since 1987 irradiated seeds in space to take advantage of cosmic radiation, with the result that they have developed more than forty mutant varieties with higher yields and better disease resistance in crops such as rice, wheat, aubergines, cucumber and peppers (China, 2012).

A detailed (596-page) account of all aspects of plant mutation breeding and biotechnology is available (FAO/IAEA, 2011), as is a database of mutant varieties (FAO/IAEA, 2016). Mutagenesis can cause hundreds of genes to be deleted or rearranged randomly with unknown consequences. Plants exhibiting new useful traits found after irradiation were selected for further breeding, which may take several years of continued crossing to separate preferred traits from undesirable ones; the ability to examine the genome is now helping to make mutagenesis a more precise and quicker process. In the past the breeder was only able to select by visible traits; there was no way of knowing and therefore of regulating for other random or unexpected changes that may have affected, for example, the chemical composition.

Batista *et al.* (2008) examined the expression of thousands of genes in rice to find out if there were any unintended consequences in mutagenic and in transgenic crops (i.e. using genes from an unrelated species). They compared transgenic and mutagenic rice strains with their closest non-modified strains and found that in all the strains there were unintended changes in the expression of genes that are related to plant stress or defence. Although there were unintended consequences in gene expression in both groups, the genetically modified transgenic strains had fewer changes. Catchpole *et al.* (2005), working with potato, found similar results, namely that gene transfer technology was far less perturbing than mutagenesis.

The widespread use of mutagenesis in plant breeding programmes throughout the world has generated a few thousand novel crop varieties; 3,200 mutant varieties of 200 different plant species had been released and registered in more than ninety countries by 2013 (FAO/AEA, 2016). Crops include varieties of rice, wheat, barley, pears, peas, cotton, peppermint, sunflowers, peanuts, grapefruit, sesame, bananas, cassava and sorghum. Mutant wheat has been used for bread and pasta, as has a mutant barley variety (Golden Promise) for organic beer and organic whiskies by brewers and distillers in Ireland and UK.

Food from crop varieties produced via mutagenesis has been sold for decades without any identifying label or common knowledge about the mutagenic origin of the crops. These varieties may be organic and none has been subjected to testing or regulation. None of the crop varieties derived by mutagenesis has been systematically tested for allergens and toxins as are GM varieties. None has been tested in chronic animal toxicology studies and none has been tested to see if there is any transfer of the new genes to other closely related plant species.

Although it is considered old technology, mutagenesis is becoming popular again thanks to new techniques known as tilling, which allow researchers to identify mutations in specific genes rapidly.

Many food scientists are frustrated at the restrictions imposed on the far more precise method of breeding, namely genetic engineering. The restrictions are largely due to activists who apparently accept the random and unexamined effects of radiation and chemical mutagenesis without the need for regulatory barriers. The arduous approval process for GM crops in the EU has resulted in seed breeding companies returning to mutagenesis to find desirable mutations. Fortunately mutagenesis has not attracted opponents on the basis that the technology was unnatural.

It is difficult to understand why crop plants with random mutations, from mutagenesis, can be used in breeding programmes without regulation whereas GM crops, with fewer genetic changes, have a much more rigorous assessment.

Genetically modified (GM) crops

The developments achieved from modifying the genetic make-up followed advances in knowledge about how genes control the way cells and organisms function at the molecular, biochemical and physiological levels; the genes responsible for desirable traits can now be identified, selected for and constructed. Genetic engineering is not a universal panacea for solving world food problems but is just another, albeit powerful and fairly precise, tool to further crop improvements.

Several commercial GM crops have been generated by the insertion of desirable genes into the DNA of elite host plants. The first GM crops were those tolerant of a particular herbicide glyphosate. In order to understand how this came about it is necessary to understand that the herbicide was an essential part of the procedure and has become an integral part of the methodology involved in the transfer of genes.

The genes to be inserted may be taken from the same species (cisgenic), from unrelated species or organism (transgenic) or they can also be constructed in the laboratory. The transfer is achieved either by using *Agrobacterium* carrying the genes to do the job, or by shooting the genes attached to microscopic gold particles usually into a soup of cells. This is a hit and miss affair. The few cells (as few as 1 in 1,000) that successfully receive the introduced gene need to be separated from the rest in order to avoid the impossibly massive amount of cell culture work to grow whole plants from the cells. Luckily a gene that conferred tolerance to the herbicide glyphosate (Monsanto) and another to the herbicide glufosinate (Bayer) were known. Initially the herbicide tolerance genes were inserted on their own but later it has become routine to add them together with other desired genes. After insertion the application of the appropriate herbicide to the soup of treated cells killed off the cells that did not contain the tolerance gene, leaving only the protected cells to be cultured. The antibiotic sulfonylurea was initially used by BASF in an analogous way but human health concerns ruled out using antibiotics. So began the dominance of Monsanto with its glyphosate tolerant soya, maize, oilseed rape, sugar beet and cotton.

It is the precision of gene transfer that is attractive to plant breeders as it allows identified beneficial genes that were not present or expressed in the gene pool of the elite plants to be added relatively easily. Compared with mutagenesis, gene transfer causes fewer changes to the original DNA but it is theoretically possible that the introduction of one or two genes might alter the performance of existing genes and so could turn out to be deleterious.

The first GM crops that were commercialized were ones (e.g. soya bean and maize) that tolerated a single herbicide – the consequence of the GM methodology. Then insect resistance was transferred into maize, cotton and potatoes from a bacterium (*Bacillus thuringiensis*), which resulted in the production of an insecticidal protein in the crops themselves (now called Bt crops). It is somewhat illogical that organic farmers, who are not allowed to grow Bt crops, are permitted to use insecticidal sprays and dusts derived from *Bacillus thuringiensis*. The gene for making this insecticidal protein could not be introduced by conventional breeding or by gene editing, or generated via mutagenesis.

Crops

There has been a remarkable acceptance of GM crops since 1996, when globally there were just under 2 million ha of GM crops. In 2015 GM crops were grown on 179.7 million ha by about 18 million farmers in twenty-eight countries (James, 2014, 2015). The area is evenly split between developing and developed countries. The countries with the most GM crops were USA, Brazil, Argentina,

Canada and India (Table 20.1). The global area of GM crops in the twenty-eight countries exceeds, by more than six times, that of organic crops in all countries when the Australian GM area is excluded.

The area of GM crops in Europe is limited to about 143,000 ha of maize, mostly grown in Spain and to a small extent in Portugal, Romania and Slovakia. As well as the cultivation of crops the import of GM products is heavily regulated in Europe and even when applications for the use and import of GM products have been successful approvals are not ratified or acted upon.

Four GM crops account for almost the entire global area, namely soya bean (with 83% of total global area), maize (30%), cotton (75%) and canola (25%) (Table 20.2).

Insofar as soya beans are a very important protein source for animal feeding stuffs the dominance of GM soya bean poses supply problems for organic farmers who are not allowed to use GM crops. In the UK, lupins and peas are being considered as alternatives to soya bean.

The traits that have so far been bred into GM crops include the basic herbicide tolerance (mostly to one herbicide), insect protection (many Bt crops), drought tolerance (maize), improved digestibility (lucerne), blight resistance (potatoes), reduced browning on exposure (apples), improved nutrient quality (e.g. ß-carotene-enriched rice, potato

Table 20.1 Areas (million ha) of GM crops (in 2015, James, 2015) and of organic crops (in 2014)

	GM	Organic
USA	70.9	2.18
Brazil	44.2	0.71
Argentina	24.5	3.06
India	11.6	0.72
Canada	11.0	0.90
China	3.7	1.92
Paraguay	3.6	0.05
Pakistan	2.9	0.02
South Africa	2.3	0.07
Uruguay	1.4	1.31
Bolivia	1.1	0.11
Philippines	0.7	0.11
Australia	0.7	17.15
Burkina Faso	0.4	0.01
Myanmar	0.3	0.01
Mexico	0.1	0.50
Spain	0.1	1.71
Colombia	0.1	0.03
Sudan	0.1	0.13
Honduras	<0.1	0.02
Chile	<0.1	0.02
Portugal	<0.1	0.21
Cuba	<0.1	0.01
Czech Republic	<0.1	0.45
Romania	<0.1	0.29
Slovakia	<0.1	0.18
Costa Rica	<0.1	0.01
Bangladesh	<0.1	<0.01
Global total	179.7	43.7

Table 20.2 Areas (million ha) of GM and global total soya bean, maize, cotton and canola (James, 2015)

	GM	Total
Soya bean	92	111
Maize	54	185
Cotton	24	32
Canola	9	36

and bio-fortified bananas), resistance to virus (ringspot on papaya in Hawaii and China) and potatoes with lower levels of acrylamide (a potential carcinogen).

The area of crops with more than one trait (stacked traits) is growing with about 58 million ha planted in 2014; for example, soya beans possessing herbicide tolerance and insect protection were grown in Brazil, Argentina, Paraguay and Uruguay in 2014. Other stacked GM crops, which cover small areas, include brinjal (eggplant), lucerne, papaya, squash, sugar beet, sweet pepper and tomato.

A comprehensive global meta-analysis in 2014 of 147 published biotech crop studies on soya bean, maize and cotton over twenty years (1995 to 2014) confirmed the significant and multiple benefits of biotech crops (ISAAA, 2014; Klümper and Qaim, 2014). On average, growing GM crops has reduced pesticide use by 37%, increased crop yields by 22%, and increased farmer profits by 68%.

Objections to GM crops

The organic movements were in the 1990s, along with Greenpeace, very early opponents of genetically modified crops when very little information was in the public domain. It became fashionable to describe the produce of GM crops as frankenfood and to describe the technology as playing God. Field experiments designed to study GM crops were destroyed by white-coated activists who were demanding more information! It is not clear what precisely motivated the GM opponents at this time other than their basic dislike of the industrialization and commercialization of agriculture, and that they viewed the technology as unnatural. There was probably also a mistrust of agricultural science following mad cow disease (bovine spongiform encephalopathy), and variant Creutzfeldt–Jakob disease (vCJD) in the UK in the 1980s.

The dislike and avoidance of GM crops also provided the organic movements with another peg on which to hang their marketing story. One company in particular, Monsanto, came in for much criticism because it sold the herbicide to which its GM crops were tolerant. It may not have been appreciated by the GM objectors that herbicide tolerance was an inevitable consequence of the GM methodology.

The fact that nearly all the released GM crops are produced by a few companies is a major concern to organic movements. The supposed claim of GM enthusiasts that GM crops will revolutionise farming and solve world hunger was mischievously ridiculed.

However, there appears to be softening of the opposition to genetically modified crops by the organic movement as is apparent in media interviews in which the opposition seems to be based more on how the technology is used rather than with the technology itself, as was originally the case. There are fears for the impact of GM crops on the environment and on the effects of GM-derived material in the food chain. Time will tell if the organic movements across the world remain implacably opposed to GM crops.

In the absence of any evidence that the presence of small quantities of GM material in food poses any risk to human and animal health the opposition based on contamination may be of interest to organic supporters but of no relevance for the vast majority; the loss of the potential benefits arising from the opposition to all aspects of GM technology is self-inflicted by the organic movement.

In their organic standards (Soil Association, 2016a) farmers are told that genetically modified organisms must not be used directly or indirectly (there are forty-one instructions about GMOs in the standards). GMOs are not seen as being in line with the principles of organic agriculture and that once they have been released into the environment they cannot be recalled. Potential risks to the environment and human health are postulated.

Evidence to support the notion of environmental and health risks do not appear to be available. The image of frankenfoods was so successfully planted in the misinformed public's mind in the 1990s that it is claimed people do not want GM technology used.

The position of IFOAM to GM crops is under review following developments in genetic engineering technology as well as the growing presence of GMOs in the world and the perceived potential of new crop introductions.

IFOAM (2002) presented its comprehensive opposition to genetic engineering, saying that it was a danger to the biosphere and posed risks for organic producers. It felt that the practices of genetic engineering were incompatible with the principles of sustainable agriculture and posed unacceptable threats to human health. At the time it called for a total ban on GM crops and organisms. However, IFOAM (2016) is having second conciliatory thoughts; it has issued a public consultation document in which it takes a new look at its opposition to genetic engineering. It would no longer object on principle but would now allow the use of GM crops and the new breeding techniques provided they are used responsibly and are based on clear evidence of benefits. Objections to the IFOAM proposals quickly surfaced (Wickson *et al.*, 2016); it was felt that the status quo should be maintained on the grounds that important socio-economic, political and cultural dimensions were not considered by IFOAM. It seems that ideological objections will take a long time to dissipate.

Greenpeace continues to be resolutely opposed to GM, believing there are health risks to GM crops and food; supporting evidence is difficult to find.

Friends of the Earth (FoE, 2016) has traditionally disliked GM crops and it appears to have two main objections. The first is that the promised increases in yield have not occurred and that in any case most GM crops have little relevance to world hunger as they are concerned with just four crops: maize, soya bean, oilseed rape and cotton. The second objection concerns the vulnerability of farmers to the activities of seed and chemical companies and the power they wield. There is no doubt that seed production in the USA is now in the hands of very few companies. The convergence of ownership of agrochemical and seed firms is another worrying trend, especially in view of the combining of seed and agrochemical sales. It is easy to forget that farmers (in the USA or smallholders in the rest of the world) are not obliged to buy GM or other seeds from a particular company and they will only repeat a purchase if they obtain a benefit.

There is a growing body of evidence from across the world that GM crops, in conjunction with conventional agricultural practices, can offer a safe and effective technology that contributes to a better, cost-effective, sustainable and productive agriculture; this was apparent some years ago (James, 2002). Nevertheless, the self-imposed regulations governing organic farming continue to prevent the growing of GM crops in several countries. Apart from IFOAM (2016), the entrenched position taken many years ago shows little sign of changing.

Conner *et al.* (2003) presented a detailed overview of ecological risks of GM crops. The main perceived concerns were about the impact of GM crops on the environment by gene flow, effects on biodiversity and the impact of the presence of GM material in other crops/products. The movement of genes could result in super weeds resistant to particular herbicides but this is not a new situation since herbicide-resistant plants have followed traditional plant breeding and can arise by entirely natural means, particularly overuse of a single herbicide. The view that GM crops are unnatural has contributed to the perception that widespread use of such plants will lead to secondary or indirect ecological effects with undesirable consequences. The widespread cultivation of GM crops with pest or disease resistance has raised concerns that this will impose intense selection pressure on pest and pathogen populations to adapt to the resistance mechanism. This might result in the development of super pests and super diseases that could be difficult or impossible to control. Another environmental concern about the introduction of GM crops is that they will affect and/or destroy biodiversity and have the potential to contaminate non-GM crops by the transfer of genes.

There have been no cases, after more than twenty years, of any health or environmental problems associated with GM crops but despite this and the wide-scale use there is still considerable opposition orchestrated by the organic movements, Greenpeace and Friends of the Earth. This opposition is very evident in Europe, where the only GM crop currently grown is maize; this is despite the fact that the European Food Safety Authority (EFSA) has assessed GM crops as being as safe as conventional crops.

The attitude in France shows how politics can trump science (Boris, 2012; France, 2016). The ban against cultivating GM maize in France has repeatedly (in 2011, 2013 and 2014) been found to be illegal in the French courts, as well as in the European Court of Justice in 2011. The courts found no evidence of risks in the four studies presented by the French authorities; GM maize is not grown (2016) in France.

Food production continues to live on the legacy of mutagenesis. Although organic breeders may have abandoned using mutagenesis, most of the varieties they have grown and are growing will have been derived either directly or indirectly from mutagenesis.

It remains to be seen whether the organic movements soften their hostility to GM crops and permit organic farmers to benefit from using GM technology to solve their own problems, such as improved resistance to pests and diseases.

The Canadian authorities' approach is to examine the safety of consuming food from new crop varieties rather than the method used to breed the crop. The National Academy of Science in the USA and its committees have consistently emphasized that the properties of a genetically modified organism should be the focus of risk assessments, not the process by which it was produced. Politicians and lobbyists in Europe could take note.

In Australia there are signs that the greens are thinking about changing their policy about GM crops in a move that is linked to their political aims of attracting rural voters (Farmonline, 2016). The leader of the Green party pointed out that genetic modification had been employed in medicine for many years

and that there was no basis for a blanket ban on GM crops.

Six counties in California have passed legislation banning the growing of GM crops, to the delight of organic supporters (Loria, 2016). Others are wondering whether pests and weeds will now be more difficult to control. The fact that there have been no cases of humans or animals being harmed by GM food or feed and that pesticide use has declined when GM crops are grown are not considered. It is said that the opposition to GM is part of an orchestrated marketing campaign for organic food and is designed to keep GM crops controversial.

Genetic engineering – a natural phenomenon

It is revealing that man's use of the soil bacterium *Agrobacterium tumefaciens* to be the carrier of the genes to be inserted had already been used by nature; it is now known that sweet potato (*Ipomoea batatas*) has been modified in this way. *Agrobacterium tumefaciens* DNA has been found in about 300 varieties of sweet potato (Kyndt, *et al.*, 2015). It is suggested that the *Agrobacterium tumefaciens* insertion occurred during the evolution of sweet potato, possibly 8,000 years ago. *Agrobacterium tumefaciens* has thus been nature's own genetic engineer for thousands of years; this occurs when it causes crown gall disease (large swellings) on the crown of plants as a result of the bacterium transferring part of its DNA to the genome of the host plant.

As GM technology using *Agrobacterium* is a natural process it is clearly not correct for those who object to the gene transfer technology to continue to call it unnatural. *Agrobacterium* has been used for gene transfer in many crops, including soybean, cotton, maize, sugar beet, lucerne, wheat, oil seed rape and rice (golden rice).

Marker-assisted breeding

Marker-assisted breeding makes use of genome analysis (and of associated visible markers) to identify genetic sections linked to the desired traits and by doing so reduce the waiting time in traditional breeding for the stable visible or measureable appearance of the traits. Repeated backcrossing is required to remove undesirable traits but the process can be speeded up by analysing the genome long before plants reach maturity. This technique has attracted attention as being natural.

Marker-assisted breeding, which only uses available genes, is not an alternative to gene transfer because GM crops can acquire new traits not otherwise obtainable. However, the political and activist-driven objections to GM mean that marker-assisted breeding will continue to be misguidedly proposed as an alternative to GM.

Gene editing (CRISPR-Cas9)

Gene editing (GE) can be achieved by a few procedures, of which the CRISPR-Cas9 system is the one in the limelight. GE does not generally involve introducing genes from another unrelated organism. GE is the latest genetic modification technique that may help agrochemical companies bypass the regulatory difficulties imposed by authorities. The ability to edit the genes that make up the genome has been made possible firstly by recognizing and understanding the activity of gene sequences (the function of the CRISPR bit) and secondly the capacity to cut and/or affect the genome at precise locations (using the Cas9 bit, which is basically an enzyme). The DNA sequences at targeted locations may be replaced by sequences brought with the CRISPR-Cas9 package. Genes may also be turned on or off; new traits can thus be created with minimal alteration to the genome. The procedure suffers from the title CRISPR (clustered regularly interspaced short palindromic repeats).

Time will tell if the regulatory bodies are able to come to terms with and understand the differences between mutagenesis, genetic engineering by transferring genes and gene editing. Hopefully the opposition to biotechnology will not inhibit research and progress using gene editing. The first signs of an evidence-based assessment of the modern methods of plant breeding became apparent when Niggli, the director of the Switzerland-based Research Institute of Organic Agriculture (FiBL), spoke in favour of CRISPR-Cas9 in an interview (6 April 2016) with the German-language newspaper *Die Tageszeitung* (FiBL, 2016), saying that CRISPR-Cas9 has great potential. FiBL is an independent, non-profit, research institute with the aim of advancing cutting-edge science in the field of organic agriculture. It was founded in Switzerland in 1973 and has offices in Germany and Austria.

One example Niggli gave was that if organic farmers did not use GE technology it might mean conventional farmers having blight-resistant potatoes that would need no pesticides while organic farmers would still have to apply toxic copper salts or other pesticides – in his view an unbearable situation. GM blight-resistant potatoes had already been bred before Niggli spoke.

Niggli mentioned other advantages of the CRISPR-Cas9 technology, namely the low cost compared to the high cost of classical plant breeding and the speed of getting results. He quoted, for the control of downy mildew in several crops, that conventional breeding would require thirty to forty years without any guarantee that, for example, the targeted pathogens would remain unchanged during that period. CRISPR-Cas9 can be applied by small breeders and would thereby remove the influence of the big agrochemical companies – another bone of contention of the organic movement.

It remains to be seen whether this balanced view will spread to other sectors of the organic movement and particularly whether or not the opposition to the modern methods of plant breeding continues as a marketing tool, as a matter of principle and because they are perceived as not being natural.

In Europe the anti-GM lobby is (in 2016) heavily campaigning against any regulation that would allow breeders to bring CRISPR-Cas9-generated plants on the market (FiBL, 2016). In France the French High Council for Biotechnology is considering a proposal that crops derived via CRISPR-Cas9 should be exempt from the European Directive on Genetically Modified Organisms.

In the USA the National Organic Standards Board (NOSB) has advised (November 2016) the prohibition of the use of gene editing (CRISP-Cas9) by organic farmers; ratification by the USDA is awaited. The NOSB, which consists of members of the organic movement, advises the government and is responsible for the list of allowed and prohibited substances.

There are signs in the USA, following the release of a pre-publication recommendation by USDA Animal and Plant Health Inspection Service (APHIS), that because some aspects of gene editing result in plant varieties that are essentially equivalent to varieties developed through more traditional breeding methods they should be treated accordingly (ISAAA, 2017).

Recent developments with CRISPR-Cas9 concern the legal wrangling, in the USA, over patent rights; in Europe patent applicants are awaiting rulings (Ledford, 2016). In the USA Monsanto has a licence to use CRISPR-Cas9 technology and plans to use the technique to make maize and soybeans more resistant to diseases and drought.

Comparison of different breeding techniques

Andersen *et al.* (2015) considered in detail the feasibility of new breeding techniques for organic farmers when they looked into different methods of rewilding a modern crop with the aim of introducing traits from its ancestors; such introduced traits may or may not be useful, it all depends on the genes provided by the wild relative. Cross-breeding of the domesticated crop and the wild relative of interest is followed by repeated backcrossing to the domesticated crop to erase as much of the unwanted genetic material from the wild relative as possible while keeping the genes and traits of interest. Markers can be used to track the trait of interest through the crosses – marker-assisted breeding. The process is likely to take several years and is technically challenging when more than one gene is being selected and when it is difficult to get rid of closely linked undesired genes.

Alternatively, an identified desirable gene from an unrelated organism or from the wild relative can be transferred into the domesticated crop by genetic engineering, a process that does not require further time-consuming backcrossing. As the transfer process can be restricted to a single gene, the method avoids co-transfer of genetically linked genes that could have undesired effects.

Gene editing (CRISPR-Cas9) offers even greater precision than transferring genes, the destination of which cannot be selected. This technique can be used to create specific changes and has the potential to generate plants that differ from the original plant at a single position and otherwise carry no other signs of the modification. Precision breeding is particularly useful when the wild type gene differs from the gene in the domesticated crop at a single location. Andersen *et al.* (2015) concluded that the most efficient methods of rewilding are based on modern GM (gene transfer) technology, which has yet to be accepted by organic movements.

Invasive species

The escape of GM-generated genes and the contamination of existing species is often posed as a basic objection to GM crops. In theory genes might escape from GM crops, as they can from non-GM crops, but in the absence of any substantiated hazards from the GM genes there appears to be little reason to worry. Likewise, genes could escape, for example in Britain, from the few thousand plants that have been imported by accident or by design over many years. There has been no cause for concern about such foreign genes as no deleterious effects related to gene transfer have been found. The potential for gene transfer has been significant considering that there are about 1,500 native British plant species and about 3,000 imports carrying foreign genes (Kinver, 2013). Some of these imports are a problem in their own right because they are highly competitive, damage the environment and reduce biodiversity, e.g. the notorious Himalayan balsam (*Impatiens glandulifera*) and Japanese knotweed (*Fallopia japonica*). They do not seem to have passed on their unwelcome genes.

Digital Darwinism

There is currently much selective suspicion of science, with agricultural science being frequently criticized; this contrasts with advances in medical science that are welcomed with open arms. The contrast is particularly evident with regard to genetic engineering, which for example in Europe is accepted for the preparation of drugs but not in crop breeding. Environmentalists have successfully influenced European politicians to obstruct research. Searching for "organic food" on YouTube reveals about 2 million entries, "organic farming" about half a million and GM crops 20,000.

Probably because of easy internet access, there has been a blurring of the distinction between the qualified and informed professional and the uninformed and unqualified amateur. This was pointed out by Trewavas (2008), who asked whether a butcher would be preferred as a source of advice to that of an experienced heart surgeon, or a bus driver on how to fly a jumbo jet rather than a qualified pilot. Keen (2008) had earlier suggested that there is a danger that we are entering a phase of digital Darwinism in which the views of the loudest and most opinionated survive in a world where everyone has an equal say, with the words of a wise man counting

for no more than those of an uninformed person.

It is often the case regarding agriculture and farming that expert scientific knowledge is seemingly given no more weight than that of well-meaning but possibly unqualified environmentalists. Policies must be based on scientifically established evidence and not on opinions, which only have merit if they are based on qualification and experience (Trewavas, 2008).

Regulations

The non-scientific hands of politicians and bureaucrats, in Europe strongly influenced by activist NGOs, are likely to continue to be impediments to breeding programmes. Opposition to GM in Europe has limited research and development of new varieties using biotechnology in several countries in Africa. However, there are signs, in some African countries, that observed benefits of GM crops will lead to their adoption. Various projects are under way to develop new GM varieties for African farmers, ranging from drought-resistant maize to varieties of cassava, cotton, banana, sorghum, cowpea and sweet potato with resistance to pests and disease. The countries involved include Burkina Faso, Cameroon, Egypt, Ethiopia, Ghana, Kenya, Malawi, Mozambique, Nigeria, South Africa, Tanzania and Uganda. In particular, Nigeria is poised following the Nigerian Academy of Science declaring (November 2016) that genetically modified foods are safe for consumption.

Several trials have been sanctioned on insect-resistant Bt cotton, cowpea and corn, on disease-resistant and vitamin A-enriched cassava and on nitrogen and water-efficient rice.

The opposition to GM crops is likely to continue from the local and/or national organic organizations, resulting in organic farmers having to rely on the slow, traditional plant breeding techniques.

Mycorrhiza

It is curiously fitting to give attention to mycorrhizae here in the last chapter in view of the significant part they played in the evolution of mainstream organic farming.

Fungi have probably lived in associations with the roots of plants since they evolved and may well have played a significant part in the weathering of rock minerals. When fungi grow together with algae they are known as lichens and as mycorrhizae when in association with the roots of higher plants.

Mycorrhizae and organisms in the rhizosphere have many properties that can influence nutrient uptake which make them worthy of consideration and study in the context of long-term sustainability.

That there is considerable interest in mycorrhizal research was evident from the symposium organised by the *New Phytologist* in 2014 entitled "Networks of Power and Influence: ecology and evolution of symbioses between plants and mycorrhizal fungi"; much was reported on mycorrhizae in natural ecosystems but little on their involvement in crop production.

Mycorrhizae figured prominently in the early days of the mainstream organic movement after the publication by Rayner (1927) and her work with Howard. Rayner found mycorrhizae on many samples provided by Howard, which led him to propose that the fungal hyphae were the conduit whereby health-promoting factors were passed from mineralizing compost and soil organic matter to crops. Balfour was attracted to mycorrhizae (Balfour, 1940); she hoped to be able to show that there was a link between humus and health of crops, animals and man and that it was due to the mycorrhizae; no evidence of such a link has yet been found.

There are four main types of mycorrhizal associations with plants, of which two are the most important; namely the endotrophic, particularly the vesicular arbuscular mycorrhizal fungi (VAM), whose hyphae enter the root and form arbuscules in the root cells, and the ectotrophic mycorrhiza (ECM), which largely remain on the surface but may also enter in the space between root cortical cells.

VAM types are the most common; the arbuscules, which are short lived (few days), are the sites of solute exchange. Generally the fungus is dependent on the host plant to provide sugars for the arbuscules or as root exudates. The host plant can benefit from the hyphae, which extend into the soil conducting nutrients to the root. The plant may or may not benefit depending on circumstances, notably soil fertility. VAM fungi belong mainly to four genera: *Acaulaspora*, *Gigaspora*, *Glomus* and *Scleroscystis*, of which *Glomus* is probably the most abundant (Marschner, 1986). ECM occur mainly on the roots of woody plants and only occasionally on herbaceous and perennial grasses (Marschner, 1986).

The two other types of mycorrhiza are those associated with *Orchidaceae* and *Ericaceae*, which are not relevant here.

Most, about 80%, of terrestrial plants are mycorrhizal (Harrier, 2002). The *Brassicaceae* are a notable exception in not forming mycorrhizal associations (Wright, 2009; Marschner, 1986), which will limit the application of any improvements in mycorrhizal activity on nutrition of several vegetables.

Mycorrhizae usually play an important role when soil fertility is low as in low input farming. For example, Mader *et al.* (1999) found that colonization of roots of wheat, vetch/rye and grass/clover by VAM as measured by the length of root colonized, was greater (by 30–60%) in plants grown in soils from low-input organic and biodynamic systems than in those grown in conventionally farmed soils. The extent of colonization was affected most by soluble P levels in the soil. Other studies have shown increased colonization on soils low in P, N, Mn and Zn in the UK, USA and Australia (Russell, 1973).

Benefits of mycorrhizae

Much attention has been focused on the ability of VAM to facilitate the absorption of P and on the very common occurrence of mycorrhizae in natural ecosystems where they are important in the conservation of nutrients. It is doubtful, as pointed out by Ryan and Tibbett (2008), whether mycorrhizae play an important part in crop production. VAM activity is restricted when sufficient P is available; when VAM are active it is possible that any benefits may be outweighed by the provision of energy by the host.

Tinker showed that as a rule mycorrhizal plants can adsorb two to three times more P per unit length of root than non-mycorrhizal plants (Marschner, 1986). VAM plants will also normally have higher Cu and Zn contents than non-mycorrhizal plants. Little appears to be known about the role of VAM in the uptake of Mg and S. Probably the main benefits of VAM are due to the extension of the absorptive area of the root, which are likely to be more evident in soils of low fertility and containing nutrients of low mobility, notably P. It is possible for mycorrhizae to mobilize other nutrients such as K and Ca in natural ecosystems but there is no evidence of any meaningful effects in arable crops.

Mycorrhizae may also enhance the plant's resistance to pathogenic fungi (Newsham *et al.*, 1995) and to chemical stresses of heavy metals (Ni, Pb and Zn) (Ricken and Hoefner, 1996). Subramanian *et al.* (1995) reported that mycorrhizae can help maize withstand drought. Enhanced water absorption has been reported by both ECM- and VAM-colonized plants (Marschner, 1986), which is more likely to be related to improved soil structure rather than to the transport of water down the tiny hyphae, which is likely to be minimal.

Mycorrhizae play key roles in the functioning of natural ecosystems. For example, in natural ecosystems van der Heijden *et al.* (2015) reported that up to 80% of plant N and P can be provided by mycorrhizal fungi and that the survival of many plant species depends on mycorrhizae. The ramifications of the fungal hyphae in the soil can effectively connect adjacent trees via their roots and so permit the transfer of sugars, phytoalexins and nutrients, as described by Simard *et al.* (2012). Mother trees, which will be the largest trees, may act as central hubs for vast, below ground mycorrhizal networks. Somewhat surprisingly, mother trees may be able to support adjacent young trees or seedlings, particularly their own, by infecting them with fungi and ferrying to them the nutrients they need to grow.

VAM can improve soil structure by contributing to the formation of soil aggregates (Rillig, 2014). It has been suggested that glomalin, a protein found on the surface of VAM hyphae, is responsible for improving the stability of soil aggregates (Wright, 2009).

Inoculation with VAM, ECM and commercialization

The inoculation of soils with beneficial mycorrhizal fungi to enhance crop growth has attracted attention. However, as discussed by Verbruggen *et al.* (2013), there are many basic problems to be overcome if inoculations of VAM are to be successful and persist over several seasons. Firstly, it is difficult to culture VAM to produce an inoculum, whereas most ECM can be readily cultured. Other problems include the compatibility of the introduced species with the host plant, the ability of the fungus to fit in and compete with other soil organisms and the method of introducing the fungus.

Much of the work on inoculation of mycorrhizae has been carried out in controlled conditions using sterilized soils and growing media. Typical of such studies were those of Ozdemir *et al.*, 2010) in Turkey working with rooted grapevine cuttings in a growing medium. They found that inoculation with VAM (*Glomus* spp.) promoted shoot and root growth and increased leaf P and Zn contents.

Relatively little of the work on mycorrhiza has been carried out in the field but there are exceptions. Ortega *et al.* (2004) reported that mycorrhizal associations are more likely to occur on vegetation in undisturbed sites than in disturbed sites. This led them to compare the growth of mycorrhiza-inoculated and non-inoculated *Pinus radiate* over two years; shoot growth was improved by mycorrhizae, especially under dry conditions.

It has been suggested that mycorrhizae could increase the efficiency of use of P fertilizers and the uptake of otherwise unavailable soil P, and so avert what is called in some quarters the P crisis when reserves of P run out (Roy-Bolduc and Hijri, 2011). In a Ted Talk in 2013, Hijri showed pictorially the benefits of mycor–rhizal inoculation on sorghum and soybeans in Canada.

Cassava (*Manihot esculenta*) is known to be a crop that responds to VAM in sterile soil but highly variable results have been obtained in the field, pointing to the need to identify the best fungus. This led to Ceballos *et al.* (2013) to examine the merits of *Rhizophagus irregularis* produced by an *in vitro* culture system with carrot roots. Inoculating cassava with this fungus increased yields at two sites in Colombia; however, the high cost of the inoculum precluded any economic benefit.

Penicillium bilaiae has been used as the active ingredient in JumpStart, a seed–applied product, made by Novozymes Bioag Group (www.novozymes.com). It has been sold in Canada since 1990. The same product is marketed in Argentina, Australia, Brazil and USA. In Europe, Syngenta is developing the product aimed at arable crops. JumpStart is said to increase the efficiency of phosphate fertilizer and to improve access to soil P. No claims are made about any other nutrients.

Mycorrhiza and organic farming

Many potential benefits of VAM in relation to organic farming practices were listed by Gosling *et al.* (2006) in their detailed review; benefits include improved nutrition, enhanced resistance to soil–borne pests and disease, improved resistance to drought, tolerance of heavy metals and better soil structure.

Colonization by VAM is often higher on organic farms than conventional farms and there is some evidence of an increase in species diversity of VAM on organic farms (Ryan and Tibbett, 2008).

It is thought that organic farming is likely to be less damaging to soil fungi because of the exclusion of water-soluble fertilizers, of biocides and the use of diverse rotations. It is suggested that this favours the VAM inoculum in soils and enhances colonization. However, the evidence is far from clear. Gosling *et al.* (2006) pointed out that the evidence suggests that despite the use of generally favourable management practices, organic systems do not always have large, diverse or efficient VAM communities. Although VAM might be able to substitute for some fertilizer and possibly biocide inputs in low–input organic systems there is little evidence for increased yields associated with high rates of VAM colonization. VAM only seem to thrive in infertile conditions.

Fungi and uptake from minerals

The ability of fungi to break down minerals in the soil is intriguing not only from the point of view of soil genesis but also the possibility of using this property to enhance nutrient uptake, especially from minerals of low solubility in arable soils. The latter aspect could be of great significance to organic farmers.

Bonneville *et al.* (2009) showed that ECM were able, over nineteen weeks, to extract K from biotite flakes when growing in association with *Pinus sylvestris* and when the fresh biotite was the main source of K. Chemical dissolution followed physical distortion of the biotite lattice. Later work by Bonneville *et al.* (2011) using a scanning X-ray spectroscopy technique quantified the changes in elemental composition (Si, O, K, Mg, Fe and Al) at the fungus/biotite interface. Large amounts of K (50–65%), Mg (55–75%), Fe (80–85%) and Al (75–85%) were removed in the upper 40 nm of biotite immediately underneath the fungal hyphae while Si and O were unchanged.

The biological weathering of minerals from basalt, granite and quartz was investigated by Quirk *et al.* (2012) as part of their studies on the evolution of trees and the release of calcium to sequester CO_2 in marine carbonates. Their study involved assessing weathering by hyphal colonization after five months and by chemical analysis after fourteen months of crushed rock particles (0.25–1.00 mm) in mesh bags placed in soil under mature gymnosperm and angiosperm trees supporting VAM and ECM. After five months, fungal hyphae associated with VAM and ECM tree roots preferentially colonized basalt over granite and quartz. Hyphae from ECM on angiosperms were more effective colonizers of both basalt (calcium-rich) and granite (calcium-poor) than VAM. Calcium mobilization, from minerals such as plagioclase in basalt, followed the same pattern with ECM being more effective than VAM.

Their findings are compatible with the notion that mycorrhizal-driven weathering may have originated hundreds of millions of years ago on gymnosperms (VAM) in the Carboniferous period and that it increased from arbuscular mycorrhizal to the later, evolved ECM fungi on angiosperm hosts in the Cretaceous period.

Wallander (2000) showed in a pot experiment with *Pinus sylvestris* seedlings that ECM were able to obtain P from apatite, a calcium fluoro-phosphate, to the benefit of the growth and P content of the seedlings. Blum *et al.* (2002) postulated from their studies on calcium budgets in forest ecosystems that dissolution mainly by ECM from apatite represents an important source of calcium and that calcium so released could be absorbed directly via the hyphae.

Work at the Max Planck Institute in Germany (Schulze-Lefert and O'Connell, 2016) indicated that much more remains to be learnt about the association between plant roots and fungi after studying the effects of the normally pathogenic fungus *Colletotrichum tofieldiae* when growing on the roots of *Arabidopsis thaliana*. When P supply is limiting, the plant tolerates the presence of the fungus, which behaves like an ECM and is able to improve P absorption. However, when there is sufficient available P the plant triggers its immune system in defence against the pathogenic fungus.

Bacteria in the rhizosphere

The rhizosphere being rich in root exudates such as amino acids and sugars is an ideal home for bacteria. This narrow zone of soil is influenced by the growth and activity of the root system (Dobbelaere *et al.*, 2003) and is likely to contain ten to 100 times more bacteria than the bulk of the soil. Some of these bacteria may themselves exude compounds that act as plant growth promoters and biocides (Beneduzi *et al.*, 2012).

The recent finding, in laboratory studies in Canada, of beneficial bacteria that can promote crop growth and reduce dependence on commercial fertilizers indicates that this relatively new area of research is flourishing (Eastman *et al.*, 2014; Ze-Chun Yuan, 2016). The bacteria work by enhancing root size, thereby increasing nutrient uptake. Three bacteria with beneficial potential have been identified. One bacterium, *Paenibacillus polymyxa*, has also been found to fix nitrogen and produce growth-promoting hormones. Additionally, this bacterium can produce antimicrobial chemicals that can kill pathogens. The fact that the genes responsible for phytohormone synthesis and P solubilization have been identified indicates that improvements via genetic modification are being considered.

There is growing interest in unravelling the influence of microorganisms in the human gut on several aspects of health. It seems likely that parallel work on the behaviour of organisms in the rhizosphere and of mycorrhiza will lead to new ways of improving crop nutrition, especially under low-input systems.

The enhancement of the activity of microorganisms in the rhizosphere could be valuable. However, inoculation programmes with P solubilizing bacteria and mycorrhizal fungi have had only limited success (Khan *et al.*, 2007). New or foreign microorganisms introduced into an established soil ecosystem are likely to find it difficult to succeed.

Ahemad and Kibret (2014), in their detailed review on the mechanisms and applications of plant growth-promoting rhizobacteria, listed many examples of plant growth-promoting rhizobacteria tested for efficacy on various crop types in both laboratory and field trials; success in the field is awaited.

Summary

1. Apart from gradual advances in all areas of agricultural science there are two potential developments that could have major impacts on food production. The first is improving the efficiency of photosynthesis and the second is extending the biological nitrogen fixation capability to non-leguminous staple food crops.

2. Both advances are likely to require the transfer of genes responsible for the generally more photosynthetically efficient C4 plants to C3 plants and for controlling the efficiency of photosynthesis via rubisco and the photoprotection mechanism.

3. It is unlikely that traditional breeding techniques, including mutagenesis, would find genes responsible for symbiotic nitrogen fixation in crops other than legumes.

4. Historically crop improvements have come from selecting plants that best suited the local environment. Changes in

the traits with traditional breeding could only come about by the remixing of the existing gene pools in parent plants. If the genes were not there for a particular trait then success was impossible.

5. The finding that exposure to radiation and some chemicals could alter the DNA and so create mutants opened the door in the 1940s by effectively creating new genes. This process of mutagenesis has played a major part in the production of most of the crop varieties grown across the world.

6. Mutagenesis is essentially a random, imprecise process. Together with beneficial mutations, other changes will have been made to the DNA of which very little was known unless they had obvious detrimental effects.

7. Crops produced via mutagenesis are not regulated, which contrasts markedly with the regulations that govern crops produced by the direct and more precise genetic engineering technology.

8. Genetically modified crops have been generated by the insertion of desirable genes into the DNA in the cells of an elite host plant. The genes selected for transfer may come from a related or unrelated species.

9. The conferring of tolerance of GM crops to particular herbicides arose because herbicides were an integral and primary part of the procedure. The addition of resistance to insects followed with the breeding of Bt crops.

10. The advantages, particularly of GM maize, soya bean, cotton and canola, have been quickly appreciated by farmers in many countries. In 2015 GM crops were grown on 180 million ha by 18 million farmers in twenty-eight countries. On average growing GM crops has reduced chemical pesticide use by 37%, increased crop yields by 22%, and increased farmer profits by 68%.

11. The organic movements together with Greenpeace quickly objected to GM crops in the 1990s before much was known about them. Their opposition was so vehement that they destroyed GM experiments designed to answer questions about their safety and environmental impact.

12. Opposition was initially based on the precautionary principle that not enough was known and the claim that the procedure was not natural. The fact that mutagenesis was unnatural was not appreciated. Opposition by the organic movements is beginning to focus on the way GM crops are employed. However, some objectors still believe that GM food poses health risks.

13. The dominance of a few agrochemical companies who supply both the seed and the associated herbicide is feared by the objectors.

14. Marker-assisted breeding uses genome analysis to identify genes linked to the desired traits and by doing so reduce the time taken in traditional breeding for the stable appearance of the traits. It is not an alternative to GM but it is a powerful tool.

15. The latest genetic procedure being used by the plant breeder is gene editing. Gene sequences can be altered at precise locations with the effect of modifying their expression and also of turning existing genes on or off. The capability of gene editing to perform targeted, highly efficient alterations of the genome sequence and gene expression will undoubtedly impact crop breeding.

16. There are signs, not universal, that gene editing may be accepted by parts of the organic movement.
17. The fact that traditional breeding even when assisted by mutagenesis is a long process taking several years makes genetic engineering, whether by gene transfer or by editing, very attractive. When a gene sequence for a particular trait does not exist then gene transfer will be necessary but when it exists gene editing may turn out to suffice.
18. It is now known that genetic engineering by gene transfer can be said to be a natural phenomenon, having occurred during the evolution of sweet potato.
19. Mycorrhizae have been of interest to the organic movement from the earliest days when it was postulated that the presence of mycorrhizae on roots provided a route whereby health-promoting factors associated with compost/manure could reach man via crops and livestock. There is still no evidence of such an effect.
20. Most agricultural crops are mycorrhizal. One notable exception is the brassica family.
21. Mycorrhizae are very important in natural ecosystems and probably evolved alongside their plant hosts.
22. Mycorrhiza can play an important role, especially in soils of low fertility and in low-input farming systems.
23. The main benefit from mycorrhizae is improved absorption of the P (which is relatively immobile in the soil), an attribute associated with the extension of the effective absorptive area of the roots by the fungal hyphae.
24. Other benefits include increased absorption of Cu and Zn. Improved water utilization is probably due to improved soil structure and not to water uptake.
25. Inoculation with cultured mycorrhizae is an attractive proposition but is difficult to achieve.
26. Mycorrhizae should be of great interest to organic farmers as they provide routes for several benefits, which may also include increased resistance to soil-borne pests and diseases and enhanced tolerance to heavy metals.
27. Despite the favourable organic practices there is little evidence for increased yields resulting from VAM colonization of organic crops.
28. The ability of mycorrhizae to mobilize nutrients such as P and Ca from minerals points the way for developments that could lead to greater use of water-insoluble minerals and fewer man-made fertilizers.
29. The possible benefits of the high concentrations of bacteria in the rhizosphere look promising and await investigation.

References

Ahemad, M. and Kibret, M. (2014) Mechanisms and applications of plant growth promoting rhizobacteria: Current perspective. *Journal of King Saud University – Science* 1, 26, 1–20. Available at www.sciencedirect.com/science/article/pii/S1018364713000293 Available at http://dx.doi.org/10.1016/j.jksus.2013.05.001

Andersen, M.M., Landes, X., Xiang, W., Anyshchenko, A., Falhof, J., Østerberg, J.T., Olsen, L.I., Edenbrand, A.K., Vedel, S.E., Thorsen, B.J., Sandøe, P., Gamborg, C., Kappel, K. and Palmgren, M.G. (2015) Feasibility of new breeding techniques for organic farming. *Trends in Plant Science*, 7, 426–434.

Balfour, E.B. (1940) The Haughley Research Farms. Proposal for an Experiment in Soil Fertility in Its Relationship to Health.53pp. Available at https://www.cabdirect.org/cabdirect/abstract/19412200222

Batista, R., Saibo, N., Lourenço, T. and Oliveira. M. M. (2008) Microarray analyses reveal that plant mutagenesis may induce more transcriptomic changes than transgene insertion. *PNAS* 105, 9, 3640–3645.

Beneduzi, A., Ambrosini, A. and Passaglia, L.M.P. (2012) Plant growth-promoting rhizobacteria (PGPR): Their potential as antagonists and biocontrol agents. *Genetics and Molecular Biology* 35, 4: 1044–1051.

Blum, J.D., Klaue, A., Nezat, C.A, Driscoll, C.T., Johnson, C.E., Siccama, T.G., Eagar, C., Faheyk, T.J. and Likens, G.E. (2002) Mycorrhizal weathering of apatite as an important calcium source in base-poor forest ecosystems. *Nature*. 417, 6890, 729–31. Available at www.esf.edu/efb/mitchell/Class%20Readings/Nat.417.729.731.pdf

Bonneville, S., Smits, M.M., Brown, A., Harrington, J., Leake, J.R., Brydson, R. and Benning L.G. (2009) Plant-driven fungal weathering: Early stages of mineral alteration at the nanometer scale. *Geology*. 37, 615–618, doi: 10.1130/G25699AS.

Bonneville, S., Morgan, D.J., Schmalenberger, A., Bray, A., Brown, A., Banwart, S.A., Benning, L.G. (2011) Tree-mycorrhiza symbioses accelerate mineral weathering: Evidences from nanometer-scale elemental fluxes at the hypha-mineral interface. *Geochimica et Cosmochimica Acta*, 75, 22, 6988–7005. doi: http://doi.org/10.1016/j.gca.2011.08.041.

Boris, L. (2012) The French ban on GMOS declared illegal by the French Council of State. Yet, the Interdiction will be perpetuated, Ministers said. *Journal of Regulation* II-12. Available at http://thejournalofregulation.com/en/article/ii-121-the-french-ban-on-gmos-declared-illegal-by-

Brutnell, T.P., Wang, L., Swartwood, K., Goldschmidt, A., Jackson, D., Zhu, X.G., Kellogg, E, Van Eck, J. (2010) *Setaria viridis*: a model for C4 photosynthesis. The *Plant Cell*. 22, 8, 2537–44. doi: 10.1105/tpc.110.075309.

CABI (2016) Africa Soil Health Consortium. Let's grow better beans. Available at http://africasoilhealth.cabi.org

Catchpole, G.S., Beckmann, M., Enot, D.P, Mondhe , M., Zywicki, B., Taylor, J., Hardy, N., Smith, A., King, R.D., Kell, D.B., Fiehn, O., and Draper, J. (2005) Hierarchical metabolomics demonstrates substantial compositional similarity between genetically modified and conventional potato crops. *Proceedings of the National Academy of Sciences US A*. 102, 14458–14462.

Ceballos, I., Ruiz, M., Fernández, C., Peña, R., Rodríguez. A and Sanders, I.R. (2013). The *in vitro* mass-produced model mycorrhizal fungus, *Rhizophagus irregularis*, significantly increases yields of the globally important food security crop cassava. *PLoS ONE* 8, 8. e70633. doi: 10.1371/journal.pone.0070633.

China (2012) Chinese Use Space Radiation to Mutate Food Crops. Available at www.spacesafetymagazine.com/space-hazards/radiation/chinese-space-radiation-mutate-food-crops/o

Conner, A.J., Glare, T.R. and Nap, J-P. (2003) The release of genetically modified crops into the environment Part II.

Overview of ecological risk assessment *The Plant Journal* 33, 19–46. Available at https://www.ncbi.nlm.nih.gov/pubmed/12943539

Davey, T. (2016) Artificial Photosynthesis: Can We Harness the Energy of the Sun as Well as Plants? Future of life institute (2016) Available at http://futureoflife.org/2016/09/30/artificial-photosynthesis

Dobbelaere, S., Vanderleyden, J. and Okon, Y. (2003) Plant growth-promoting effects of diazotrophs in the rhizosphere. *CRC Critical Reviews in Plant Sciences*, 22, 107–149.

Eastman, A.W., Heinrichs, D.E. and Yuan, Z.C. (2014) Comparative and genetic analysis of the four sequenced *Paenibacillus polymyxa* genomes reveals a diverse metabolism and conservation of genes relevant to plant-growth promotion and competitiveness. *BMC Genomics*. 15. 851. doi: 10.1186/1471-2164-15-851.

FAO/IAEA (2011). Plant Mutation Breeding and Biotechnology. Available at www.fao.org/3/a-i2388e

FAO/IAEA (2016) Mutant Variety Database. Available at https://mvd.iaea.org (Lagoda, IEAE, Austria, 2016, personal communication)

FAO/AEA (2016a) Mutation Induction for Breeding. Atoms for peace, health and prosperity.

FoE (2016) Friends of the Earth. Farming for the Future: Organic and Agroecological Solutions to Feed the World. Available at www.db.zs-intern.de/uploads/1466576808-FOE_Farming_for_the_Future_Final.pdf

Farmonline (2016) Greens mull rethink on GM crop policy. Available at www.farmonline.com.au/news/agriculture/general/politics/greens-mull-rethink-on-gm-crop-policy/2750469.aspx?storypage=1

Faunce, T. (2011) Global artificial photosynthesis project: a scientific and legal introduction. *Journal of Law and Medicine* 19, 2, 275–81.

FiBL (2016) An absolute no-go – a European organic opinion leader goes CRISPR. Available at http://ludgerwess.com/organic-crispr

France (2016) French GMO cultivation ban declared illegal yet again. Available at www.europabio.org/news/french-gmo-cultivation-ban-declared-illegal-yet-again

Gosling, P., Hodge, A., Goodlass, G. and Bending, G.D. (2006) Arbuscular mycorrhizal fungi and organic farming. *Agriculture Ecosystems and Environment* 113, 1–4, 17–35. doi: 10.1016/j.agee.2005.09.009.

Harrier, L.A. (2002) The arbuscular mycorrhizal symbiosis: a molecular review of the fungal dimension. *Journal of Experimental Botany* 52, 1, 469–478.

Hyperphysics (2016) Georgia State University. Available at http://hyperphysics.phy-astr.gsu.edu/hbase/biology/phoc.html

IFOAM (2002) Position on Genetic Engineering and Genetically Modified Organisms. Available at www.ifoam.bio/sites/default/files/page/files/ifoam-ge-position.pdf

IFOAM (2016) Public Consultation on the position of IFOAM – Organics International on Genetic Engineering and Genetically Modified Organisms. Available at www.ifoam.bio/en/news/2016/02/26/public-consultation-position-ifoam-organics-international-genetic-engineering-and

IRC (2014) Second Green Revolution seeks to leave no farmer behind. IRRI Annual Report. Available at http://irri.org/news/media-releases/second-green-revolution-seeks-to-leave-no-farmer-behind

ISAAA (2014) ISAAA Brief 49-2014. Global Status of Commercialized Biotech/GM Crops: 2014. Brief 49-2014: Executive Summary. Available at http://isaaa.org/resources/publications/briefs/49/executivesummary/default.asp

ISAAA (2017) USDA proposes revisions on GE crop regulations. Available at www.isaaa.org/kc/cropbiotechupdate/article/default.asp?ID=15110

James, C. (2002) World-wide deployment of GM crops: – aims and results – state of the art. In: Discourse on Genetically Modified Plants. Bad Neuenahr, Germany. Available at www.gruenegentechnik.de/Doku_Fachtagung/james_engl.pdf.

James, C. (2014) Global Status of Commercialized Biotech/GM Crops: 2014. ISAAA Brief No. 49. International Service for the Acquisition of Agri-Biotech Applications: Ithaca, NY. Available at www.isaaa.org/resources/publications/pocketk/16

James, C. (2015) 20th Anniversary (1996 to 2015) of the Global Commercialization of Biotech Crops and Biotech Crop Highlights in 2015. ISAAA Brief No. 51. ISAAA: Ithaca, New York. Available at http://isaaa.org/resources/publications/briefs/51/executivesummary/default.asp

JCAP (2016) Joint Center for Artificial Photosynthesis. Available at http://solarfuelshub.org

Keen, A. (2008) The Cult of the Amateur. How blogs, MySpace, YouTube and the rest of today's user-generated media are killing our culture and economy. Nicholas Brealey Publishing.

Khan, M. S. Zaidi, A. and Wani, P.A. (2007) Role of phosphate-solubilizing microorganisms in sustainable agriculture – A review. Agronomy for Sustainable Development, Springer Verlag/EDP Sciences/INRA, 2007, 27 (1), 29–43. Available at https://hal.archives-ouvertes.fr/hal-00886352/document

Kinver. M. (2013) UK bans sale of five invasive non-native aquatic plants. BBC News 29 January. Available at www.bbc.co.uk/news/science-environment-21232108

Klümper, W. and Qaim, M (2014) A Meta-Analysis of the Impacts of Genetically Modified Crops. *PLoS ONE* 9, 11: e111629. doi: 10.1371/journal.pone.0111629

Kromdijk, J., Głowacka, K., Leonelli, L., Gabilly, S.T., Iwai, M., Niyogi, K.K. and Long, S.P. (2016) Improving photosynthesis and crop productivity by accelerating recovery from photoprotection. *Science* 354, 6314, 857–861. doi: 10.1126/science.aai8878.

Kyndt, T., Quispe, D., Zhai, H., Jarret, R., Ghislain, M., Liu, Q., Gheysen, G. and Kuze, J.F. (2015) The genome of cultivated sweet potato contains *Agrobacterium* T-DNAs with expressed genes: An example of a naturally transgenic food crop. *Proceedings of the National Academy of Sciences* 112, 18, 5844–5849. doi: 10.1073/pnas.1419685112.

Le Page, M. (2016) Trials planned for GM super wheat that boosts harvest by 20%. *New Scientist* November. Available at https://www.newscientist.com/article/2111377-trials-planned-for-gm-superwheat-that-boosts-harvest-by-20/#.WBy4lumZfYY.twitter

Ledford, H. (2016) Titanic clash over CRISPR patents turns ugly. Accusations of impropriety feature in escalating dispute *Nature* 537, 460–461. doi: 10.1038/537460a.

Lin, M.T., Occhialini, A., Andralojc, P.J.,

Parry, M.A.J. and Hanson, M.R. (2014) A faster Rubisco with potential to increase photosynthesis in crops. *Nature* 513, 7519, 547–550. doi: 10.1038/nature13776.

Liran, O., Semyatich, R., Milrad, Y., Eilenberg, H., Weiner, I. and Yacoby, I. (2016) Microoxic niches within the thylakoid stroma of air-grown *Chlamydomonas reinhardtii* protect [FeFe]-hydrogenase and support hydrogen production under fully aerobic environment. *Plant Physiology* 172, 1, 264–271. doi: http://dx.doi.org/10.1104/pp.16.01063.

Loria, K. (2016) What do GMO-free zones mean for farmers and food? Available at https://www.geneticliteracyproject.org/2016/11/28/are-gmo-free-zones-an-organic-industry-marketing-ploy

Mader, P., Edenhofer, S., Boller, T., Wiemken, A. and Niggli, U. (1999) Arbuscular mycorrhizae in a long-term field trial comparing low-input (organic, biological) and high-input (conventional) farming systems in a crop rotation. *Biology and Fertility of Soils* 31, 2, 150–156.

Marschner, H. (1986) *Mineral Nutrition of Higher Plants*. Academic Press, London.

Montañez, A (2000) Overview and Case studies on Biological Nitrogen Fixation: Perspectives and Limitations. FAO. Available at www.fao.org/fileadmin/templates/agphome/scpi/SCPI_Compendium/Overview_and_Case_studies_on_Biological_Nitrogen_Fixation.pdf

Newsham, K.K., Fitter, A.H. and Watterson, A.R. (1995) Arbuscular mycorrhiza protect an annual grass from root pathogenic fungi in the field. *Journal of Ecology* 83, 6, 991–1000.

NIAB (2013) Breakthrough in wheat breeding science offers greater yields. Available at www.niab.com/news_and_events/article/282vgs.iaea.org

Ortega, U., Duñabeitia, M., Menendez, S., Gonzalez-Murua, C. and Majada, J. (2004) Effectiveness of mycorrhizal inoculation in the nursery on growth and water relations of *Pinus radiata* in different water regimes. *Tree Physiology* 24, 1, 65–73.

Ozdemir, G., Akpinar, C., Sabir, A., Bilir, H., Tangolar, S. and. Ortas, I. (2010) Effect of Inoculation with Mycorrhizal Fungi on Growth and Nutrient Uptake of Grapevine Genotypes (*Vitis* spp.) *European Journal of Horticultural Science* 75, 3, 103–110.

Pankievicz, V.C.S., do Amaral, F.P., Santos, K.F.D.N., Agtuca, B., Xu, Y., Schueller, M.J., Arisi, A.C.M., Steffens, Maria. B.R., de Souza, E.M., Pedrosa, F.O., Stacey, G. and Ferrieri, R.A. (2015) Robust biological nitrogen fixation in a model grass–bacterial association. *The Plant Journal* 81, 6, 907–919. doi: 10.1111/tpj.12777.

Quirk, J., Beerling, D.J Banwart, S.A., Kakonyi, G., Romero-Gonzalez, M.E. and Leake, J.R. (2012) Evolution of trees and mycorrhizal fungi intensifies silicate mineral weathering. *Biology Letters* 8, 1006–1011. doi: 10.1098/rsbl.2012.0503.

Rayner, M.C. (1927) *Mycorrhiza*. Cambridge University Press, Cambridge, England.

Ricken, B. and Hoefner, W. (1996) Effect of arbuscular mycorrhizal fungi (AMF) on heavy-metal tolerance of alfafa *Medicago sativa* L. and oat *Avena sativa* L. on a sewage sludge treated soil. *Zeitschrift für Planzenernahrung und Bokenkunde* 159, 189–194.

Rillig, M.C. (2014) Mycorrhizal fungi and soil structure: tales of crumbs, traits and walnuts In: *33rd New Phytologist Symposium. Networks of Power and Influence: ecology and evolution of symbioses between plants and mycorrhizal fungi.*

Available at https://www.newphytologist.
org/app/webroot/img/upload/
files/33rd%20NPS%20Abstract%20book_
final_onlineversion.pdf

Roy-Bolduc, A. and Hijri, M. (2011) The
Use of Mycorrhizae to Enhance Phosphorus
Uptake: A Way Out the Phosphorus Crisis.
Journal Biofertilizers and Biopesticides 2,104.
doi: 10.4172/2155-6202.1000104.

Russell, E.W. (1973) *Soil conditions and plant
growth*. 10th edition. Longman, London.

Ryan, M.H. and Tibbett, M. (2008) The
role of arbuscular mycorrhizas in organic
farming 188–229. In: Kirchmann, H. and
L. Bergstrom, L. (eds.). *Organic Crop
Production – Ambitions and Limitations*.
H. Dordrecht: Springer. Available at http://
link.springer.com/chapter/10.1007%2F978-
1-4020-9316-6_10#page-1

Schulze-Lefert, P and O'Connell. R.
(2016) Plants take on fungal tenants on
demand. Innate immune system of the
thale cress plant ensures a good phosphate
supply Max-Planck for Plant Breeding
Research. Available at https://www.mpg.
de/10390194/plants-symbiosis-phosphate

SFT (2016) Eat the week. Available
at http://sustainablefoodtrust.org/
articles/eat-the-week-gm-wheat/?utm_
source=SFT+Newsletter&utm_
campaign=93ddc69e1f-
Newsletter_07_10_2014&utm_
medium=email&utm_term=0_bf20bccf24-
93ddc69e1f-90440905

Simard, S.W., Beiler, K.J., Bingham,
M.A., Deslippe, J.R., Philip, L.J. and
Teste, F.P. (2012) Mycorrhizal networks:
Mechanisms, ecology and modelling *Fungal
Biology Reviews* 26, 1, 39–60.

Soil Association (2016) Stop genetic
modification. Available at https://www.
soilassociation.org/our-campaigns/
stop-genetic-modification

Soil Association (2016a) Organic standards,
farming and growing. Available at https://
www.soilassociation.org/media/1220/
sa-farming-and-growing-standards.pdf

Subramanian, K.S., Charest, C., Dwyer,
L.M., and Hamilton, R.I. (1995)
Arbuscular mycorrhizas and water
relations in maize under drought stress at
Tasseling. *New Phytologist* 129, 643–650.

Tien, T.M., Gaskins, M.H. and Hubbell
D.H. (1979) Plant growth substances
produced by *Azospirillum brasilense*
and their effect on the growth of pearl
millet (*Pennisetum americanum*). *Applied
and Environmental Microbiology* 37, 5,
1016–1024.

Trewavas, A. (2008) The cult of the amateur in
agriculture threatens food security. *Trends
in Biotechnology* 26, 9, 475–478. doi: http://
dx.doi.org/10.1016/j.tibtech.2008.06.002

Turan, B., Becker, J.P., Urbain, F., Finger,
F., Rau, U. and Haas, S. (2016). Upscaling
of integrated photoelectrochemical water-
splitting devices to large areas. *Nature
Communications* 7, 12681. doi: 10.1038/
NCOMMS12681.

van der Heijden, M.G.A., Martin, F.M.,
Selosse, M.A. and Sanders, I.R. (2015)
Mycorrhizal ecology and evolution: the
past, the present, and the future. *New
Phytologist* 205, 4, 1406 1423. doi: 10.1111/
nph.13288.

Verbruggen, E., van der Heijden,
M.G., Rillig, M.C. and Kiers, E.T.
(2013) Mycorrhizal fungal establishment
in agricultural soils: factors determining
inoculation success. *New Phytologist* 197, 4,
1104–9.

Wallander, H. (2000) Uptake of P from
apatite by *Pinus sylvestris* seedlings
colonized by different ectomycorrhizal
fungi. *Plant and Soil* 218, 1, 249–256.
Available at http://link.springer.com/art

icle/10.1023%2FA%3A1014936217105#page-2

Wickson, F., Binimelis, R. and Herrero, A. (2016) Should Organic Agriculture Maintain Its Opposition to GM? New Techniques Writing the Same Old Story, *Sustainability*, 8, 11, 1105. doi: 10.3390/su8111105.

Wright, S.F. (2009) Glomalin, Hiding Place for a Third of the World's Stored Soil Carbon. *Wild Ones Journal*. Available at http://wildones.org/download/mysteries/mysteryglomalin.pdf

Ze-Chun Yuan (2016) Bacteria discovery could enhance crop growth. Agriculture and Agri-Food Canada. Available at www.agr.gc.ca/eng/news/science-of-agricultural-innovation/bacteria-discovery-could-enhance-crop-growth/?id=1455047204614

Zhu, X.G., Long, S.P. and Ort, D.R. (2008) What is the maximum efficiency with which photosynthesis can convert solar energy into biomass? *Current Opinion on Biotechnology* 19, 2, 153–159. doi: 10.1016/j.cop.bio.2008.02.004.

Zhu, X.G., Ort, D.R., Whitmarsh, J. and Long, S.P. (2004) The slow reversibility of photosystem II thermal energy dissipation on transfer from high to low light may cause large losses in carbon gain by crop canopies: a theoretical analysis. *Journal of Experimental Botany* 55, 400, 1167–1175.

Appendix

The appendix is a supplement that expands the topics discussed in a few of the chapters, notably on the organic movements, health-promoting compounds, the toxicology of pesticides, nitrogen use efficiency and GHGs.

Chapter 5 Rudolf Steiner Agricultural Course (1924)

Rudolf Steiner's eight lectures "The Agriculture Course. The birth of the Biodynamic Method" provides the core text for understanding biodynamic agriculture. There are several editions available, but here Steiner (2004) has been used with kind permission of the Rudolf Steiner Press Ltd, London.

Notes on the eight-day Rudolf Steiner Agricultural Course in 1924

The following notes were made from Steiner (2004):

The 111 attendees all came from spiritual, scientific and anthroposophical backgrounds. There is no evidence of anyone from Great Britain attending. There were thirty women and eighty-one men from six countries: Germany (sixty-one); Poland (thirty); Austria (nine); Switzerland (seven); France (two); and Sweden (two). Thirty-eight were described as "agricultural" and of these twenty were farmers. There were nine priests, four medical doctors, three teachers, two artists and two engineers (Paull, 2011).

Although farmers had been advised to use the newly available mineral fertilizers to improve their poor yields, the impoverished state of farming precluded this and made the farmers receptive to Steiner's ideas about benefiting from cosmic forces that were available free of charge.

Lecture 1. No practical instructions or suggestions. Cosmic forces emphasized.

i. Importance of cosmic forces regarding all things, e.g. menstrual cycle and moon.
ii. Abundance of and therefore importance of silica assumed. Silicic acid as an essential constituent in medicines.
iii. All living things contain forces that come not from the Earth but from distant planets.
iv. The question is how can these forces be assisted?
v. Water contains more than oxygen and hydrogen. Connections exist between Moon and rain whereby growth is enhanced when the forces of a full Moon follow rain. The advice was to sow during a full Moon after rain.

vi. The forces of the planets vary. Saturn (revolution thirty years) and Jupiter (twelve years) affect the growth of perennials. Annuals are influenced by planets with short revolutions.
vii. Greater heat is given by firewood from trees that were planted according to the cosmic rhythms.

Lecture 2. No practical instructions. Value of manure appreciated. Cosmic forces again emphasized.

i. A farm should be a self-contained unit with the manure produced by the livestock being sufficient for the crops.
ii. The surface of the Earth should not simply be regarded as consisting of mineral and organic matter but as containing an effective astral principle as well. The soil is under the influence of the forces of distant planets, Mars, Jupiter and Saturn, whereas plant growth is affected by the planets near the Earth Moon, Mercury and Venus.
iii. Cosmic forces are stored in and relayed by the soil to plants.
iv. Root development is largely determined by the extent to which the cosmic forces have been retained by rocks and soils in which silica is seen to play an important part. Silica has the power to absorb light energy beneath the surface and then benefit plants.

v. The conditions under which the cosmic spaces are able to pour their forces down into the Earth need to be understood to determine when to cultivate.

vi. Flower colours are determined by the influence of distant planets – red by Mars, blue by Saturn and yellow by Jupiter.

vii. Cosmic forces are present, in very great excess, in *Equisetum* (horsetail) as it is rich (10–15%) in SiO_2 compared with <0.5% SiO_2 in most dicotyledonous plants.

viii. Cosmic forces can be concentrated in the roots, e.g. by growing in a sandy silica-rich soil or in the shoots.

ix. It is necessary to have the right number of cows, horses, pigs, on the farm to produce the correct amount of manure. It should not be necessary to import manure on to the farm. A farm is only healthy if it can produce its own manure.

x. The cosmic influences that are effective in a plant rise upward from the interior of the Earth. Animals eating a plant rich in such cosmic influences excrete excellent manure.

Lecture 3. No practical instructions or suggestions. N and legumes considered.

i. The importance of nitrogen was challenged. Carbon, oxygen, hydrogen and particularly sulphur were highlighted. Sulphur is seen as a mediator between the spiritual and the physical. Sulphur is the carrier of the spiritual.

ii. Today's chemists are not aware of the significance of C, O, H and S in the working of the cosmos as carriers of the Spirit.

iii. The spiritual moves through the world via the carbon-like framework of plants.

iv. The linking of C with O is mediated by N, which carries an astral spirit.

v. In man, N is essential for the life of the soul, which is the mediator between the Spirit and life.

vi. It was recognized that legumes collect nitrogen.

vii. The consideration of nitrogen led to a discussion of the various forms of manuring.

Lecture 4 Manure and the cow horn preparations 500 and 501.

i. Food is understood to be important in giving the body forces particularly quality forces.

ii. Use of man-made fertilizers as manure does not provide the earthy elements but will at best connect with the watery elements of the Earth.

iii. Composts preserve something of the ethereal and the astral elements, which are not as concentrated as in solid or liquid manure.

iv. Suggestions are given about making compost in layered heaps including adding lime to the heap.

v.   The question as to why cows have horns and other animals have antlers was discussed. It is believed that unlike the Earth, from which ethereal forces flow outwards in cows, the horns concentrate inward currents.

vi. Farmyard manure is perceived as containing and concentrating both astral and ethereal factors. The astral has been permeated with the nitrogen-carrying forces and with the ethereal by the oxygen-carrying forces.

vii. The way in which the forces in the manure can be properly harnessed was described. By burying a cow horn filled with manure in the soil over winter the forces in the food from the animal are preserved, i.e. the property of radiating back whatever is life-giving and astral. All the radiations that tend to etherealize and astralize are poured into the inner hollow of the horn.

viii. Precise instructions are given for use of the manure from the horn, based on experience at the Goetheanum; this is preparation 501. The manure from one horn diluted in half a bucket of water is sufficient for 1,350 sq m. Precise details as how to stir the mixture were given, namely stir at the edge of the pail, so that a crater is formed causing the entire contents to rotate; this is followed by reversing the direction of stirring.

ix. The cow horn manure, called spiritual manure, has good results when combined with ordinary manure.

x.   Precise instructions for making the cow horn silica preparation are given; this is preparation 502. The very finely ground silica (or feldspar) is buried over the summer in a cow horn. A tiny fragment (pea or pin's head size) is mixed in a bucket of water for one hour. Good effects are expected after sprinkling on crops especially vegetables.

The methods of making and applying the preparations have been refined over the years but the basics are as Steiner described. See Chapter 5.

Discussion after Lecture 4
The fact that Steiner talked of using buckets suggests that he had thought about the use of preparations 500 and 501 on gardens rather than farms. Steiner was keen that the stirring of the preparations was done by hand and not mechanically. He said that the results were not due to the substances as such, but to the dynamic radiant activity in the cow horn manure. The reuse of cow horns and the likelihood that the provenance of horns affects their potency was discussed. Cow horns were preferred to ineffectual ox or bull horns.

Steiner advised that cow-horn manuring is not intended as a substitute for ordinary manuring but that it should be regarded as an extra enhancing the effect of the manuring.

Lecture 5 Details of making preparations 502 to 507.

i.   Steiner showed his distaste for the studies on mineral nutrition and the use of fertilizers. He was prepared, however, to say that the Anthroposophists should try to supplement genuine modern achievements with their own spiritual concepts.

ii.   The materialistic world should be abhorred and spiritual science favoured.

iii. Steiner talked of taking forces and not substances from the farm in exported produce and emphasized the value of manure, albeit enhanced by appropriate treatment.

iv.   The addition of unspecified mineral substances to manure was not approved. In contrast it was seen as necessary to bring the manure into a condition that it is able to vitalize the soil.

v.   Silicic acid, lead and mercury were mentioned from time to time as though they are valuable but there was no understandable explanation other than that because they were present in rain they must be what nature intended should be supplied. Silicic acid, lead and mercury were seen as working homeopathically.

vi.   Homeopathy had a great appeal at the Goetheanum in Dornach where it is said that the Koliskos had placed homeopathy on a sound scientific footing and had proved that the required radiant forces in organic substances are set free by the minute quantities of homeopathic materials provided they are used in the proper way.

vii. The use of homeopathic doses in the proper way led to the preparations 502–507, which are added to manure and compost in order to supply the essential living forces.

viii. Steiner appreciated that potash is needed by crops.

ix.   The making of preparations 502–507 was described (the numbers were not used by Steiner).

x.   *Preparation 502 (based on yarrow).* Yarrow was held in high esteem by Steiner, who said that it should not be treated as a weed. There was no need to mix the yarrow preparation in the manure as the radiating power was so strong; it only needed to be inserted in a few places. Steiner refers to yarrow's homoeopathic sulphur content, saying it influenced potash but not calcium in the plant. The yarrow preparation involves placing it in a stag's bladder because the beneficial force is intensely preserved by the processes that take place between the kidneys and the bladder. The bladder is kept for a year, sometimes buried. Steiner may have known that historically yarrow had been used medicinally in relation to bladder and uterus bleeding complaints.

xi. *Preparation 503 (based on chamomile)*. Steiner advised using chamomile because of its potash and calcium content. The use of chamomile tea as a food and medicine for treating stomach and menstrual cramps led Steiner to specify that chamomile flowers must be stuffed into bovine intestines and then buried in the soil.

xii. *Preparation 504 (based on stinging nettle)*. For Steiner stinging nettle was a replacement for yarrow and chamomile if they were difficult to find, calling it the greatest benefactor of plant growth. Maybe Steiner was aware of the abundant growth of nettles on patches of fertile soil. He valued it because it carried the spiritual sulphur as well potash, calcium and iron radiations. Nettles were buried bare in the soil.

xiii. *Preparation 505. (based on oak bark)* Steiner wished to bring calcium, which he believed was connected to plant diseases, into the manure but not as an inorganic substance. He advocated the use of oak bark to provide calcium in a living form. Crushed oak bark (only outer layer) is buried in the skull of any domestic animal over winter. When added to manure the bark was intended to confer forces to combat or to arrest any harmful plant diseases.

xiv. Steiner returned to silica for his next preparation, saying that something was needed to attract the silicic acid from the cosmic environment. Steiner believed in transmutation, saying that potash, for example, transmutes into nitrogen, provided the potash is working properly.

xv. *Preparation 506. (based on dandelions)* Silica, which draws in the cosmic forces, is said to interact with potassium with the silica being transmuted into a substance that was not yet included among the chemical elements. A preparation made from dandelions if added to manure is claimed to have the ability to aid this relationship. Dandelion flowers are buried over winter in bovine peritoneum. After addition in homeopathic amounts to manure the preparation will give the soil the ability to attract just as much silicic acid from the atmosphere and from the cosmos as is required.

xvi. *Preparation 507. (based on Valerian)* An extract of crushed flowers of Valerian is diluted and added to manure. The extract has the power to stimulate the manure.

xvii. The various preparations are added to the manure in a few shallow holes along the top of the heap from which the forces radiate through the manure.

xviii. It is implicit in much of what Steiner said that he appreciated the value of manure at the same times as believing that manure could be enhanced by the preparations.

Lecture 6 Weeds, insects and diseases.

i.   Steiner reiterated that the nearby planets Mercury and Venus, and the Moon, do not act directly from the planets to the plants, but via the soil in order to influence plant growth and seed formation. In contrast, the distant planets directly influence fruit formation. (How this differs from seed–setting is not clear).

ii.   Understanding the differences between the effects of distant vs. nearby planets is essential when it comes to weed control.

iii.   Weeds and many plants are greatly influenced by the workings of the Moon. Apart from reflecting sunlight, the Moon mixes the sun's rays with its own rays, which then arrive as lunar forces. Moreover, with the Moon's rays the whole reflected cosmos comes to the Earth. All influences that pour on to the Moon are radiated out again.

iv.   In order to obtain the maximum benefit from the cosmic forces it is better to carry out agricultural operations, e.g. sowing and reaping, at the time of a full Moon. However, there is a let out in that if the operations are done at the wrong time the influences of the Moon will simply remain in the Earth until the next full Moon. (Sadly the Moon also influences weed growth).

v.   Steiner's recommendation for controlling specific weeds consisted of burning the seeds of the particular weed and then scattering the ash (in homeopathic amounts) over the weeds. The opposite force to that which was developed in the weed by the Moon-forces is concentrated in the ash, which becomes an effective weed killer. The annual sprinkling of ash from a variety of weeds for a few (four) years can eliminate weeds. The powder, sprinkled as from a pepper pot, has such a large radius of influence that it is only necessary to sprinkle a little across the field to kill all weeds.

vi.   Steiner, who did not have a farm, had not verified the weed-killing property of ash. He was sure that it would work, believing that spiritual-scientific truths are true in themselves; they do not need to be confirmed.

vii.   Deterring field mice is achieved by sprinkling over the soil the ash of burnt mouse-skin obtained when Venus is in Scorpio.

viii   Insects are also combatted by an ash prepared from the insect in question at the time when sun is in Taurus. The ash is sprinkled on the field in homeopathic amounts. Steiner strangely used nematodes as his example.

ix. In answer to a question in discussion, Steiner said that optimal timing for other insects would need to be determined.

x. With regard to plant diseases, Steiner perceived a problem insofar as plant diseases should not be possible without the presence of an astral body and plants do not possess astral bodies.

xi. An over-intense lunar influence is seen as responsible for diseases such as mildew, blight and rust. This influence can be countered by preparation 508 made from *Equisetum arvense*, which is diluted and sprinkled as a liquid in homoeopathic doses over the field. (Did Steiner get the idea for the use of *Equisetum* from its use as an herbal remedy for kidney and other complaints, dating back to Roman times or was he influenced by its high silica content and the associated cosmic forces?)

xii. Steiner agreed that all insects (including cabbage root-fly) could be controlled by sprinkling their burnt ash. He distinguished between animals without backbones, for which the ash of the whole body was sufficient, and those with backbones, for which only the skin was burnt.

xiii. In answer to a question about inorganic manures, Steiner showed an intense dislike of mineral manuring, which he said must cease altogether in time. He cited the loss of nutritive value when mineral manures are used. He was confident that artificial manures will go out of use.

Lecture 7 Farm animals and orchards. No practical suggestions.

i. Steiner discussed the growth of trees and plants. He talked about stems, blossom and fruit being united with their roots through the spirit, that is, through the ethereal.

ii. It is very difficult to understand most of this lecture. The plant is seen as having a physical body and an ethereal body surrounded by an astral cloud. Whenever the plant connects with the astral cloud (as happens in fruit formation), food is produced that will support the astral in the animal and human body.

iii. Steiner hypothesized that the roots of plants, e.g. fruit trees grown in close proximity, intertwine and form a single root system with exchange of sap and fluids. Steiner was way ahead of his time in this (see Chapter 20).

Lecture 8 General discussion on forces.

i. Steiner emphasized the importance of acquiring the necessary spiritual-scientific insight in order to be able to act with individual intelligence when adopting the new measures.

ii. Steiner had the opinion that animal and human nutrition was not understood insofar as there was much more to feeding than supplying materials for growth and activity.

iii. He asserted that a claw or a hoof must not be thought of as being formed from the feed eaten by the animal. What the animal eats is merely for the purpose of developing its inner forces of movement, so that the cosmic principles may be driven right down into the metabolic and limb system. In this way the animal gains cosmic substantiality.

iv. Steiner said the head can only assimilate the nourishment that it receives from the body if at the same time it can get the necessary forces from the cosmos. Animals should not be shut up in dark stables where the cosmic forces cannot flow towards them.

v. The concept of the farm as a living organism was developed with the aim of promoting farms as individual self-contained units. The cycling of manure was important. The use of manure from Chile (sodium nitrate) was said to impair the proper working of nature.

vi. Steiner concluded Lecture 8 by saying that what had been divulged during the course should remain confidential within the farmers' circle. The ideas will be enhanced and developed by experiments. The farmers' society (the Experimental Circle) that has been formed will decide when to publish results. Attendees were asked not to fall into the prevalent anthroposophical mistake and proclaim the new ideas from the rooftops. Steiner was aware that the anthroposophist movement had often been harmed in this way and that farmers following the ideas in his lecture might be thought of as going crazy by fellow farmers!

vii. In answer to the question: "What is the best way of combating couch-grass?" Steiner thought that the fact that seed is seldom formed will in the end eliminate the weed and that without seed there is not a real weed problem. Rampant growth was assumed to be associated with sufficient seed production.

References

Paull, J. (2011) Attending the First Organic Agriculture Course: Rudolf Steiner's Agriculture Course at Koberwitz, 1924. *European Journal of Social Sciences* 1, 1, 64.

Steiner, R. (2004) Agricultural Course. *The Birth of the Biodynamic Method, Classic Translation* by G. Adams. Rudolf Steiner Press Ltd, London, England.

Goetheanum/Section for Agriculture

The Section for Agriculture is one of eleven departments of the School for Spiritual Science founded by Rudolf Steiner, which is based at the Goetheanum in Dornach, Switzerland. The central concern of the Section for Agriculture is to contribute to finding solutions for the global challenges facing agriculture. At the same time, the Section for Agriculture is the focal point for biodynamic agriculture worldwide and has the mandate of maintaining, promoting and developing the biodynamic practices. Research, training courses, conferences, publications and the coordination of various activities connected to biodynamic farming make the Section for Agriculture a social laboratory dedicated to the active quest for sustainable, future-oriented farming (www.sektion-landwirtschaft. org). The Biodynamic Ambassadors is a program at the Goetheanum that aims to support biodynamic projects abroad by sending young adults who have completed a biodynamic training course to be ambassadors. There have been recent projects in South Africa, Morocco, Southern India and Peru.

The International Biodynamic Association (IBDA) is the worldwide union of national biodynamic associations. IBDA works in close cooperation with the Section for Agriculture at the Goetheanum on the one hand, and with Demeter International on the other hand. IBDA holds the ownership rights of the biodynamic and Demeter trademarks (www.ibda.ch).

Chapter 6 Founder and early members of the Soil Association

The backgrounds of founder and early members of the Soil Association and of several of the organizations and magazines to which they were associated are presented here in order to indicate the diversity of their activities and thinking; they were all listed as leading figures in the organic movement by Conford (2001) in his detailed biographies, to which readers are directed for more information. Floris Books are thanked for giving permission for the information from Conford (2001) being used here.

Acronyms indicate membership of a group or of writing for a magazine

FM	Founder/early member of Soil Association
CM	Council member of Soil Association
CC&C	Council for the Church and Countryside.
EM	English Mistery
EA	English Array
GS	Guild Socialism
KH	Kinship in Husbandry
NEW	New English Weekly

Individuals

Baker R. St B., A forester (CM, FM).

Barlow, K.E., A doctor (CM, FM, NEW).

Bruce, M.E., A biodynamic farmer (CM, FM).

Chapman, G., A New Zealand dentist (FM).

Coward, R., An organic farmer (CC&C).

Easterbrook, L., A biodynamic farmer and journalist (KH).

de la Mare, R., Editor, Faber and Faber (CM, FM).

Gardner, R., Influential supporter of organic farming (CM, FM, CC&C, GS, KH, NEW).

Greenwell, P. Mc., Farmer following Howard's compost instructions (CM).

Jenks, J.E.F., A farmer and journalist favouring of small-scale farming (FM, CC&C, NEW).

Viscount Lyminton, 9th Earl of Portsmouth, politically right wing. He wished for a return to self-contained mixed farming with few inputs and envisaged that the medieval Yeoman system would become the backbone of farming in the UK (CM, FM, EA, EM, KII).

Massingham, H.J., A prolific writer (FM, CM, CC&C, GS, NEW).

Pearse, I.H., With her husband Scott Williamson, the doctors who set up the Peckham centre (CM, FM).

Picton, L.J., A doctor in Cheshire who played a major part in the production of the"MedicalTestament"(FM,CC&C).

Rayner, M.C., A scientist. Influential due to her work mycorrhizae. On the advisory panel for the Soil Association.

Sanderson-Wells, T.H. A physician/surgeon (FM).

Saxon, E.J., A journalist (FM).

Shoobridge, H., An Australian hop grower who founded the Living Soil Association of Tasmania.

Lord Semphill, Connected to the organic movement via membership of the Economic Reform Club and Institute (CM, FM).

Sykes, Friend., A farmer renowned for successfully employing Howard's composting techniques when transforming a rundown farm (FM).

Lord Teviot, Soldier and politician (CM, FM).

Turner F.N., A farmer and follower of Howard (CM).

Westlake, A.T., A doctor who later farmed organically (CM, FM)

Williamson, G.S., A doctor who with his wife (Pearse, I.H.) established the Pioneer Health Centre in Peckham in 1926 (CM, FM, KH).

Wilson, R.G.M., An organic vegetable farmer who had been advised on composting by Howard (FM).

Wood, M., A biodynamic farmer (CM, FM).

Analysis

Conford (2001) identified twenty-four founder and early members of the Soil Association; they came from several parts of society, with farmers who were already farming organically and doctors concerned about the prevention of diseases predominating:

6 were doctors

6 were using compost made according to Howard

3 were farming according to the bio-dynamic methods of Steiner

4 were writers/journalists

2 wished for a return to old rural life styles

2 had extreme right wing tendencies
1 was a scientist

Groups and magazines
Council for the Church and Countryside

Many of the Soil Association's early members were members of the Council who believed that God had created man to be dependent on nature's laws that it was vital to establish a rural/natural life style if society was to be healthy and stable. They emphasized the return of all wastes to the soil.

English Mistery

English Mistery (an old word for a guild) was a secret, far right, political group that was in favour of bringing back the feudal system. It was ultra-royalist and anti-democratic, attracting landowners, political figures and intellectuals; it spawned the *English Array*.

English Array

English Array was active from 1936 to the early months of the Second World War advocating "back to the land". Its objective was to turn the clock back to feudal times.

Guild Socialism

Guild Socialism was an early 20th century English socialist group advocating state ownership of industry with control and management by guilds of workers.

Kinship in Husbandry

Kinship in Husbandry was a group of lovers of rural life who were opposed to the modernizing of agriculture. Their aim was to create a forum in which members could share their experiences in organic farming. They drew ideas mainly from Howard, McCarrison, Stapledon and Wrench. In several ways, it was the precursor of the Soil Association.

New English Weekly

The New English Weekly was initially a political and literary review that became a major vehicle for transmitting organic ideas. During the 1940s most people involved in the organic movement wrote for the weekly.

Individual supporters of the mainstream organic movements

Albrecht, W. A. (1888–1974) was Head of the Soils Department at the University of Missouri who had concluded from his studies that there was a direct link between soil quality, food quality and human health. He warned of the dangers of industrial agriculture.

Lord Bledisloe (1867–1958) was a distinguished agriculturist serving as Parliamentary Secretary to the Ministry of Food and as Governor of New Zealand (1930–1935). He was very sympathetic to the views of the organic school and assisted in raising funds for the Haughley experiment. He was also Chairman of the Committee of the Lawes Agricultural Trust.

Bromfield, L. (1896–1956) was an American journalist and ultimately an organic farmer in Ohio who supported Rodale. Bromfield was an early proponent of organic and self-sustaining methods, and was one of the first to stop using pesticides.

Elliot, R.H. (1837–1914) anticipated many of the organic movement's concerns, including those about new artificial fertilizers. He farmed in India, Ireland and Scotland and advocated building up the organic content of soil by planting a mixture of grasses and other, particularly deep-rooted plants.

Lord Northbourne (Walter James 1896–1942) experimented with biodynamic methods having studied agriculture at Oxford. He was the first to use the term "organic farming" in his book, *Look to the Land* (Northbourne, 1940). In it he foresaw a contest of organic versus chemical farming and a clash of world views that could last for generations. Rather than the practicalities of organic farming, Lord Northbourne's book presents the philosophy, the rationale, and the importance of organic farming. His ideas on organic farming spread and were quickly taken up in the USA and Australia in particular.

Reference

Conford, P. (2001). *The Origins of the Organic Movement*. Floris Books, Edinburgh.

Chapter 12 Manures and Slurries, Composition

Reference

Chambers, B., Nicholson, N., Smith, K., Pain, B., Cumby, T. and Scotford, I. (2001). ADAS Booklet 1. Managing Livestock Manures – Making better use of livestock manures on arable land. ADAS

Table 12a   Amounts of collectable excreta produced by livestock during the housed period Chambers *et al.* (2001)

Animal/Age	Body weight kg	% of year housed	Daily kg	Annual t
Dairy cow	650	50	64	11.6
Dairy cow	450	50	42	7.6
Beef cattle/1–2 years	400	66	26	6.2
Beef cattle/0.5–1.0 years	180	50	13	2.4
Pigs/11–23 weeks	35–105	90	4.5	1.5
1000 Layer hens	2,200	97	11.5	4.1
1000 Broilers	2,200	76	60	16.5

Gleadthorpe Research Centre. Available at www.rothamsted.ac.uk/sites/default/files/Booklet1ManuresArableLand.pdf

Chapter 14 Benefits of fruit and vegetables and supplementation with dietary antioxidants

Cancer

An early review of about 200 epidemiological studies showing the benefits of fruit and vegetables in reducing cancer was carried out by Block *et al.* (1992).

Cardiovascular disease and cancer

The notion that fruit and vegetables are likely to have beneficial health effects came from a meeting in California in 1991 when around twenty fruit and vegetable companies met with the US National Cancer Institute and the non-profit group Produce for Better Health (US, CDC, 2005). The outcome was that Americans were encouraged to eat at least five portions (80 g) of fruit and vegetables every day. This was and still is good advice because of the enhanced intake of minerals, vitamins and fibre without too many calories. It has been wrongly assumed that the five-a-day recommendation was based on firm evidence. The advice has been remarkably successful and spread to many countries, some of which have programmes of eight-a-day. The World Health Organization recommends daily eating a minimum of 400 g of fruit and vegetables

Table 14a   Review of epidemiological studies on cancer showing protection by consumption of fruits and vegetables (Block *et al.*, 1992)

Cancer site	Fraction of studies showing significant cancer protection
Epithelial, Lung	24/25
Oral	9/9
Larynx	4/4
Oesophagus	15/16
Stomach	17/19
Pancreas	9/11
Cervix	7/8
Bladder	3/5
Colorectal	20/35
Miscellaneous	6/8
Hormone-dependent	
Breast	8/14
Ovary/endometrium	3/4
Prostate	4/14
Total	129/172

to lower the risk of serious health problems, such as heart disease, stroke, type 2 diabetes and obesity.

There have been several huge long-term epidemiological studies in the USA and in Europe in which very clear evidence has been gained showing positive effects of consuming fruit and vegetables on health.

Bazzano *et al.* (2002), in a nineteen-year study in the USA on 9,608 healthy adults, found that consuming three helpings per day of fruit and vegetables compared with one was associated with reductions in stroke incidence of 27%, in stroke-related deaths of 42% and of 27% in cardiovascular disease mortality. Increasing the intake of fruit and vegetables (from 250 g/day to 400/500 g/day) in Denmark was estimated to increase life expectancy by 0.8 and 1.3 years, respectively (Gundgaard *et al.*, 2003). In addition, it was estimated that 19–32% of cancers could be prevented.

Bradbury *et al.* (2014) reported on the current findings of the European Prospective Investigation into Cancer and Nutrition (EPIC) study on the associations between fruit, vegetable and fibre consumption and the risk of fourteen different cancers that began in 1992. More than 500,000 participants from ten European countries were involved. The risk of cancers of the upper gastrointestinal tract was found to be inversely linked with fruit intake but was not associated with vegetable intake. The risk of colorectal cancer was inversely associated with intakes of total fruit and vegetables and total fibre, and the risk of liver

cancer was also inversely linked with the intake of total fibre. The risk in smokers of cancer of the lung was inversely associated with fruit intake but was not associated with vegetable intake. There was a borderline inverse association between fibre intake and breast cancer risk. For the other nine cancer sites studied (stomach, biliary tract, pancreas, cervix, endometrium, prostate, kidney, bladder, and lymphoma) there were no reported significant links with intakes of total fruit, vegetables, or fibre.

A meta-analysis of sixteen cohort prospective published studies (lasting five to twenty-five years) by Wang *et al.* (2014) provided further evidence of the ability of fruit and vegetables to reduce cardiovascular disease, and cancer mortality. More than 800,000 adults were involved in the USA, Europe and Asia. High consumption of fruit and vegetables was significantly associated with a lower risk of all causes of death. There was a threshold around five servings of fruit and vegetables a day, after which the risk of death from all causes was not further reduced. They also found that eating more fruit and vegetables was mainly effective in reducing the incidence of deaths due to cardiovascular disease and strokes rather than with affording protection against cancer.

A meta-analysis by workers at Imperial College London indicates the health benefits (relating to heart attack, stroke, cancer and early death) of consuming more than 400 g per day fruit and vegetables, namely 800 g, i.e. equivalent to eight portions; this is due to be published in the *International Journal of Epidemiology* in 2017 and contrasts with the meta-analysis by Wang *et al.* (2014).

There is no doubt about the health-promoting capacity of fruit and vegetables, particularly on cardiovascular disease and some cancers, but the precise reasons are not clear. Most attention has been focused on the antioxidants vitamin E, ß-carotene/vitamin A, and vitamin C (Wojcik *et al.*, 2010). The possible involvement of fibre and of the micronutrient selenium, which is an essential component of the enzyme superoxide dismutase, should not be ignored.

Benefits of compounds found in fruit and vegetables

Antioxidants

Antioxidants are compounds that have the capacity to scavenge, by providing electrons to, the highly reactive free radicals, e.g. hydrogen peroxide, superoxide and super-hydroxide, which are produced by the body during normal metabolism and which are reactive, being short of electrons. The free radicals can cause oxidative damage (by stealing electrons) to cellular DNA, protein and lipids, resulting in the initiation or development of numerous diseases such as cancer, cardiovascular diseases, type 2 diabetes mellitus, rheumatoid arthritis, Alzheimer's disease, Parkinson's disease, and eye diseases such as cataracts and age-related macular degeneration. All these diseases have been linked to attacks by free radicals (US Dept. of Health,

2013). Antioxidants have attracted a lot of media attention as being the key to many health issues so it is not surprising that studies such as Baranski *et al.* (2014) focused on them.

The human body makes its own antioxidants (e.g. uric acid, superoxide dismutase, glutathione peroxidase) and would be expected to benefit from dietary antioxidants, notably vitamins A, C and E and polyphenols. Vitamin A is synthesized in the body from β-carotene in fruit and vegetables, otherwise it is obtained as preformed vitamin A in animal products. Vitamin C is a cofactor in several enzymatic reactions. Vitamin C is made internally by almost all organisms with a few exceptions, notably humans and other apes. The antioxidant effects of vitamin C have been demonstrated in many *in vitro* experiments but it may be that its antioxidant activity is not as important as its enzyme activity. Citrus fruits are a particularly rich source of vitamin C but it is also found in many other fruits and vegetables in amounts that vary with varieties. Apple skin and the adjacent tissues are rich in vitamin C, containing two to four times the amount in the flesh.

Vitamin E comprises a group of compounds that include both tocopherols and tocotrienols. Of the different forms of vitamin E, γ-tocopherol is the most common. Vitamin E has many biological functions, the antioxidant function being the best known. Vitamin E is found in a wide variety of foods but particularly in plant oils, such as soya, corn and olive oil. Other good sources include nuts, seeds and wheat germ.

Polyphenols (compounds with several hydroxyl groups on an aromatic ring) commonly occur (maybe several thousand) in higher plants, and several hundred in edible plants. They are generally involved in defence against ultraviolet radiation and pathogens (Manach *et al.*, 2004). Bioavailability appears to vary widely between the different polyphenols, which makes it important to identify the polyphenols when assessing health benefits based on plant composition. The grouping of all phenolic compound together as having health benefits may well be misleading and lead to wrong conclusions. For example, just because polyphenols have antioxidant activity does not automatically make them all useful; e.g. flavonoids are phytooestrogens and will induce mutations in rodents at high concentrations.

Whether the differences in the flavonoid content of crops, such as apples, are significant with regard to human metabolism is likely to depend on several complicating factors, including absorption in the gut, which as pointed out by Lotito and Frei (2006) may be increased by uric acid – the levels of which may be influenced by the fructose content of apples; it was the fact that flavonoids are poorly bioavailable and reach only very low concentrations in human plasma after the consumption of flavonoid-rich foods that prompted this work. Fruit and beverages such as tea, red wine, and coffee constitute the principal dietary sources of polyphenols, but vegetables, leguminous plants, and cereals are also good sources, containing up to several g/kg. Polyphenol concentrations in foods vary according to genetic and environmental factors (Manach *et al.*, 2004).

Investigation of the reasons for the beneficial effects of fruit and vegetables have followed two routes: firstly the administration of synthetic antioxidants and secondly prospective studies in which intake of dietary antioxidants has been related to cardiovascular health and cancer.

Effects of supplementation with synthetic antioxidants

Long-term clinical trials involving a combined total of more than 100,000 people with vitamins C and E supplements have so far failed to prove their efficacy in reducing the risk of developing cardiovascular diseases and cancer and in prolonging life. Of particular note are the Women's Antioxidant Cardiovascular Study, the Women's Health Study, the Physicians' Health Study and the Selenium and Vitamin E Cancer Prevention Trial. There have been trials that have linked supplements with increased incidence of cancer, e.g. vitamin E and prostate cancer.

The Women's Antioxidant Cardiovascular Study found no beneficial effects of vitamin C (500 mg/day, vitamin E (600 IU every other day), or of β-carotene (50mg every other day) supplements on cardiovascular events (heart attack, stroke or death from cardiovascular diseases) or the likelihood of developing diabetes or cancer in more than 8,000 female health professionals, aged 40 years or older, who were at high risk for cardiovascular disease (Cook *et al.*, 2007).

The Women's Health Study, which included almost 40,000 healthy women at least 45 years of age, found that vitamin E supplements did not reduce the risk of heart attack, stroke, cancer, age-related macular degeneration, or cataracts. (Lehesranta *et al.*, 2007).

The Physicians' Health Study II, which included more than 14,000 male physicians aged 50 or older, found that neither vitamin E nor vitamin C supplements reduced the risk of major cardiovascular events (heart attack, stroke or death from cardiovascular diseases), cancer, or cataracts (Sesso *et al.*, 2008). Somewhat worryingly, vitamin E supplements were associated with an increased risk of haemorrhagic stroke in this study.

The Selenium and Vitamin E Cancer Prevention Trial (SELECT), a study of more than 35,000 men aged 50 or older, found that selenium supplements, taken alone or together with vitamin E, did not prevent prostate cancer. A 2011 updated analysis from this trial, based on a longer follow-up period, concluded that vitamin E supplements increased the occurrence of prostate cancer by 17% in men who only received the vitamin E supplement (Dunn *et al.*, 2010).

In contrast, the Age-Related Eye Disease Study (AREDS) found that a combination of antioxidants (vitamin C, vitamin E, and β-carotene) reduced the risk of developing the advanced stage of age-related macular degeneration by 17% but did not help to prevent cataracts or slow their progression. The fact that the addition of zinc and of lutein and zeaxanthin (two carotenoids found in the eye) had benefits when used in combination with one of the vitamins indicates the complexity of the situation (US Dept. of Health, 2013).

Wojcik, *et al.* (2010) reviewed the use of vitamin C alone, or in combination with other antioxidants (vitamin E, β-carotene, selenium and zinc) in twelve epidemiological studies. These studies followed promising results from cell culture and animal studies on the cardioprotective effect of vitamin C. However, there was no clear evidence of any protective effect of vitamin C (maximum 500 mg/day) on cardiovascular diseases.

Researchers at the Cleveland Clinic USA carried out a meta-analysis of seven large randomized supplementation trials with vitamin E (more than 80,000 patients) given alone or in combination with other antioxidants, and eight trials with β-carotene (138,000 patients) that were followed for up to twelve years. The doses of vitamin E ranged from 50–800 international units and for β-carotene, 15–50 mg (Vivekananthan, *et al.*, 2003). Vitamin E did not lower mortality compared to control treatments, and it did not significantly decrease the risk of cardiovascular death or stroke. As a result it was concluded that the routine use of vitamin E could not be supported. Supplementation with β-carotene led to a small but statistically significant increase in all-causes of death and a slight increase in cardiovascular-related deaths, a clear indication that supplementation should also be discouraged.

The absence of positive responses to vitamin C, E and β-carotene supplements together with the risk that high doses of vitamin E may increase the risk of prostate cancer and that β-carotene may increase the risk of lung cancer in smokers and in cardiovascular-related death mean that high doses of these vitamins cannot be recommended.

The finding of adverse effects of vitamin supplements does not mean that eating food rich in antioxidants should be discouraged. The results of the prospective supplementation trials with synthetic antioxidants may need to be treated with caution as they employed doses far higher than nutritional recommendations and for long periods. Moreover, vitamins C and E and β-carotene may only make up a relatively small contribution to the antioxidant capacity of fruit and vegetables with their hundreds of different antioxidants. It is possible that antioxidants act together and it is their collective effect that is important.

Dietary antioxidants and health

Arts and Hollman (2005) and Hollman *et al.* (2011) reviewed the evidence supporting the proposition that the beneficial effects of polyphenol-rich foods on cardiovascular health are due to their antioxidant activity. However, they were unable to establish a causal relationship between the total antioxidant content of the diet and cardiovascular health. Although some polyphenol-rich foods exert beneficial effects on some biomarkers of cardiovascular health, there was no evidence that this was caused by an enhanced antioxidant capacity. Moreover, they questioned whether a direct antioxidant effect of polyphenols in the diet could occur because the concentrations found in blood are low compared with other antioxidants. Polyphenols are metabolized following ingestion, which can quickly lower their antioxidant activity.

Following the previous inconclusive studies on dietary antioxidant intake, including β-carotene, vitamin C and vitamin E, and breast cancer risk Nagel *et al.* (2010) examined the data (over nine years and on more than 7,500 cases of breast cancer) provided by the European Prospective Investigation into Cancer and Nutrition (EPIC). Overall, dietary intake of β-carotene, vitamin C and E was not related to breast cancer risk in either pre- or post-menopausal women. However, in some instances with post-menopausal women, a weak protective effect between dietary β-carotene and vitamin E and breast cancer risk was indicated.

Zamora-Ros *et al.* (2013) found a huge variation in the intakes of phenolic acids by adults across Europe, from about 1,000 mg/day in Denmark to about 200 mg/day in Greece. Coffee was the main source of phenolic acids and accounted for 55–81% of the total intake, followed by fruits 2–18%, vegetables (1–7%) and nuts (<1–5%).

Serafini *et al.* (2012) investigated the link between the total dietary antioxidant content and gastric cancer in the European Prospective Investigation into Cancer (EPIC) study (twenty-three centres, ten European countries involving more than 500,000 subjects mostly 35–70 years old started in the 1990s). They found that dietary antioxidant intake is associated with a reduction in the risk of gastric cancer. There appeared to be a threshold effect at higher levels of total antioxidant intake that suggests that care might need to be taken to avoid overloading with antioxidants. The effect of dietary antioxidants was more evident in subjects where specific risk factors linked to oxidative stress (e.g. smoking) were present. The Serafini results contrasted with the review by Bjelakovic *et al.* (2004 and 2007) of fourteen randomized trials (170,525 participants) that found no evidence that antioxidant supplementation with β-carotene, vitamin A, vitamin C, vitamin E and selenium prevented gastric cancer. On the contrary, antioxidant supplementation increased overall mortality, raising strong concerns about the use of antioxidants supplementation for gastric cancer prevention. The supply of dietary antioxidants from different sources of fruit and vegetables may be better than using synthetic supplements for reducing the risk of gastric cancer.

The very large number of secondary metabolites that possess health-promoting potential will make it difficult to identify precisely the active compounds in fruit and vegetables. It may be that some aspect of organic farming can have positive effects on the relevant compound(s). Meanwhile, it is to be noted that there are only two indications from epidemiological studies that consumption of organic food can benefit health, namely links with a lower incidence of non–Hodgkin lymphoma and of pre-eclampsia (related to organic vegetables but not fruit) (see Chapter 10).

References

Arts, I.C.W. and Hollman, P.C.H. (2005) Polyphenols and disease risk in epidemiologic studies. *American Journal of Clinical Nutrition*. 81, supplement, 317S–25S. Available at http://ajcn.nutrition.org/content/81/1/317S.full.pdf

Barański, M., Średnicka-Tober, D., Volakakis, N., Seal, C., Sanderson, R., Stewart, G.B., Benbrook, C., Biavati, B., Rembiałkowska, E., Markellou, E., Giotis, C., Gromadzka-Ostrowska, J., Skwarło-Sońta, K.,Tahvonen, R., Janovská, D., Niggli, U., Nicot, P. and Leifert, C. (2014) Higher antioxidant and lower cadmium concentrations and lower incidence of pesticide residues in organically grown crops: a systematic literature review and meta-analyses. *British Journal of Nutrition* 112, 5, 794–811. doi: 10.1017/S0007114514001366.

Bazzano, L.A., He, J., Ogden, L.G., Loria, C.M., Vupputuri, S., Myers, L. and Whelton, P.K. (2002) Fruit and vegetable intake and risk of cardiovascular disease in US adults: the first National Health and Nutrition Examination Survey Epidemiologic Follow-up Study. *American Journal of Clinical Nutrition* 76, 1, 93–99.

Bjelakovic, G., Nikolova, D., Simonetti, R.G. and Gluud, C. (2004) Antioxidant supplements for prevention of gastrointestinal cancers: a systematic review and meta-analysis. *Lancet* 364, 9441, 1219–1228.

Bjelakovic, G., Nikolova, D., Gluud, L.L., Simonetti, R.G. and Gluud, C. (2007) Mortality in randomized trials of antioxidant supplements for primary and secondary prevention: systematic review and meta-analysis. *Journal of the American Medical Association* 297, 8, 842–857.

Block, G., Patterson, B. and Subar, A. (1992) Fruit, vegetables and cancer prevention: a review of the epidemiologic evidence, *Nutrition and Cancer* 18, 1, 1–29. Available at http://dx.doi.org/10.1080/01635589209514201

Bradbury, K.E., Appleby, P.N. and Key, T.J. (2014) Fruit, vegetable, and fiber intake in relation to cancer risk: findings from the European Prospective Investigation into Cancer and Nutrition (EPIC). *American Journal of Clinical Nutrition* 100, Supplement 1394S–398S. doi: 10.3945/ajcn.113.071357.

Cook, N.R., Albert, C.M., Gaziano, J.M., Zaharris, E., MacFadyen, J. Danielson, E., Buring, J.E. and Manson, J. (2007) A randomized factorial trial of vitamins C and E and beta carotene in the secondary prevention of cardiovascular events in women: results from the Women's Antioxidant Cardiovascular Study. *Archives of Internal Medicine.* 167, 15, 1610–1618. doi: 10.1001/archinte.167.15.1610.

Dunn, B.K., Richmond, E.S., Minasian, L.M., Ryan, A.M., and Ford, L.G.. (2010) A nutrient approach to prostate cancer prevention: The Selenium and Vitamin E Cancer Prevention Trial (SELECT). *Nutrition and Cancer* 62, 7, 896–918. doi: 10.1080/01635581.2010.509833.

Gundgaard, J., Nielsen, J.N., Olsen, J. and Sørensen, J. (2003) Increased intake of fruit and vegetables: estimation of impact in terms of life expectancy and healthcare costs. *Public Health Nutrition* 6, 1, 25–30.

Hollman, P.C., Cassidy, A., Comte, B., Heinonen, M., Richelle, M., Richling, E, Serafini, M., Scalbert, A., Sies, H. and Vidry, S. (2011) The biological relevance of direct antioxidant effects of polyphenols for cardiovascular health in humans is not established. *Journal of Nutrition.* 141, 5, 989S–1009S. doi: 10.3945/jn.110.131490.

Lehesranta, S.J., Koistinen, K.M., Massat, N., Davies, H.V., Shepherd, L.V., McNicol. J.W., Cakmak, I., Cooper, J., Lück, L., Kärenlampi, S.O. and Leifert, C. (2007) Effects of agricultural production systems and their components on protein profiles of potato tubers. *Proteomics* 7, 4, 597–604. doi: 10.1002/pmic.200600889.

Lotito, S.B. and Frei, B. (2006) Consumption of flavonoid-rich foods and increased plasma antioxidant capacity in humans: cause, consequence, or epiphenomenon? *Free Radical Biology and Medicine.* 41, 12, 1727–1746.

Manach, C., Scalbert, A., Morand, C., Rémésy, C. and Jiménez, L. (2004) Polyphenols: food sources and bioavailability *American Journal of Clinical Nutrition* 79, 5, 727–747. Available at http://ajcn.nutrition.org/content/79/5/727.full

Nagel, G., Linseisen, J., van Gils, C.H., Peeters, P.H. and 44 colleagues (2010). Dietary beta-carotene, vitamin C and E intake and breast cancer risk in the European Prospective Investigation into Cancer and Nutrition (EPIC). *Breast Cancer Research and Treatment* 119, 3, 753–765. doi: 10.1007/s10549-009-0444-8.

Sesso, H.D., Buring, J.E., Christen, W.G., Kurth, T., Belanger, C.J., Bubes,

V., Manson, J.E., Glynn R.J. and Gaziano, J.M. (2008) Vitamins E and C in the Prevention of Cardiovascular Disease in Men: The Physicians' Health Study II Randomized Trial. *Journal of the American Medical Association*. 300. 18, 2123–2133. doi:10.1001/jama.2008.600.

Serafini, M. Jakszyn, P., Luján-Barroso, L. and 48 EPIC co-authors (2012) Dietary total antioxidant capacity and gastric cancer risk in the European prospective investigation into cancer and nutrition study. *International Journal of Cancer*. 131, 4, E544–54. doi: 10.1002/ijc.27347.

US Dept. of Health (2013) Antioxidants and Health: An Introduction National Center for complementary and integrative health US Dept. of Health and Human Services. Available at https://nccih.nih.gov/health/antioxidants/introduction.htm#safety

US, CDC (2005) US National Cancer Institute National Institutes of Health. UD Dept. of Health and Human services, Centers for disease control and prevention. Available at https://www.cdc.gov/nccdphp/dnpa/nutrition/health_professionals/programs/5aday_works.pdf

Vivekananthan, D.P., Penn, M.S., Sapp, S.K., Hsu, A and Topol, E.J. (2003) Use of antioxidant vitamins for the prevention of cardiovascular disease: meta-analysis of randomised trials. *Lancet* 361, 9374, 2017–2023. doi: http://dx.doi.org/10.1016/S0140-6736(03)13637-3639.

Wang, X., Ouyang, Y., Liu, J., Zhu, M., Zhao, G., Bao, W. and Hu, F.B. (2014) Fruit and vegetable consumption and mortality from all causes, cardiovascular disease, and cancer: systematic review and dose-response meta-analysis of prospective cohort studies. *British Medical Journal* 349, g4490. doi: http://dx.doi.org/10.1136/bmj.g4490

Wojcik, M., Burzynska-Pedziwiatr, I. and Wozniak, L.A. (2010) A Review of Natural and Synthetic Antioxidants Important for Health and Longevity. *Current Medicinal Chemistry*, 2010, 17, 28, 3262–3288.

Zamora-Ros, R., Rothwell, J.A., Scalbert, A. and 41 EPIC co-authors (2013) Dietary intakes and food sources of phenolic acids in the European Prospective Investigation into Cancer and Nutrition study. *British Journal of Nutrition* 110, 8, 1500–1511. doi: 10.1017/S0007114513000688.

Chapter 15 LD_{50} and HERP (Human Exposure dose/Rodent Potency dose)

LD_{50}

All chemicals have the potential to cause harm depending on the amounts involved. Toxicological tests permit assessments of carcinogenicity, teratogenicity, mutagenicity, etc., and the comparison of the relative potential for harm.

The LD_{50} as determined in rodent studies provide a useful way of comparing different chemicals (Table 15a). The LD_{50} is the dose that is lethal for 50% of animals consuming or exposed to the chemical; it is normally determined on rats or mice and is expressed as the quantity of the chemical per kg live weight of the animal.

For comparison (Table 15a), the LD_{50} for aspirin is 200 mg/kg, for vitamin A 2,000 mg/kg, for common salt (NaCl) 3,000 mg/kg and for sugar (glucose) 30,000 mg/kg.

Table 15a The seventeen most deadly compounds (RSC)

	LD$_{50}$ mg/kg
Botulinum toxin A	3×10^{-8}
Tetanus toxin A	5×10^{-6}
Diphtheria toxin	3×10^{-4}
Dioxin*	3×10^{-2}
Muscarine**	2×10^{-1}
Bufotoxin***	4×10^{-1}
Sarin	4×10^{-1}
Strychnine ****	5×10^{-1}
Soman*	6×10^{-1}
Tabun*	6×10^{-1}
Tubocurarine chloride****	7×10^{-1}
Rotenone****	3
Isoflurophate*	4
Parathion*	4
Aflatoxin B1	10
Sodium cyanide*	15
Solanine	42

* Synthetic.
** Lethal ingredient in deadly mushrooms.
*** In venom secreted by European toad.
**** Extracted from plants.

HERP (Human Exposure dose/Rodent Potency dose)

The relative cancer risks to humans of different chemicals can be assessed by expressing the human intake (as mg/kg body weight per day) of a given chemical as a percentage of the amount (also mg/kg) that caused cancer in 50% of the test animals by the end of a lifetime of consumption (the TD$_{50}$ value for cancer). These values (Table 15b) were first calculated by Ames and Gold (2000) and Gold *et al.* (2002) and are known as the HERP value, the Human Exposure dose/Rodent Potency doses. HERP values allow a comparison of different chemicals. The larger the HERP value the potentially more harmful the chemical is to humans.

The fact that a substance is near the top of the list does not mean that it will cause cancer, it simply indicates its relative toxicity. DDT is the first pesticide in the list. The caffeic acid content of coffee makes it potentially fifty times more hazardous than DDT.

It is noticeable that the carcinogenic compounds found in many foodstuffs are more hazardous than synthetic pesticides, so much so that if foodstuffs were regulated with the same criteria used for synthetic chemicals many compounds and foods would not be approved. Moreover, it is clear that the risks posed by synthetic pesticides are extremely small in comparison with the small risk posed by many of the foods we eat. In America, Scheuplein (1991) estimated that of the 200,000 food-related cancers, 99% were from traditional foods; 2,000 were associated with spices, 400 with additives and twenty with pesticides (a similar number applied to food preparation and drug residues).

If fruit and vegetables contain only a few molecules of a pesticide (e.g. amounts below the detectable limit) then the HERP values will be infinitesimally lower than lindane in Table 15b.

Chemicals in roasted coffee

Thirty chemicals of the more than 1,000 found in roasted coffee have

Table 15b Ranking of possible carcinogenic hazards base on calculated values for HERP using American dietary data (Ames and Gold, 2000; Gold *et al.*, 2002)

Human Dose	Dose of Carcinogen[+]	HERP%
Phenobarbital, 1 sleeping pill	Phenobarbital 60 mg	12.0
Alcoholic drinks	Ethyl alcohol 22.8 ml	3.6
Beer 229 g	Ethyl alcohol 11.7 ml	1.8
Wine 20.8 g	Ethyl alcohol 3.67 ml	0.6
Conventional Home Air	Formaldehyde 598 µg	0.4
Coffee 11.6 g	Caffeic acid 20.8 mg	0.1[++]
Lettuce (14.9g)	Caffeic acid 7.9 mg	0.04
Orange Juice 138 g	*d-* Limonene 4.28 mg	0.03
Tomato 88.7 g	Caffeic acid 5.46 mg	0.03
Coffee 11.6 g	Catechol 1.16 mg	0.02
Mushrooms 2.55 g	Hydrazines	0.02
Apple 32.0 g	Caffeic acid 3.4 mg	0.02
Celery 14 g	Caffeic acid 1.5 mg	0.007
Coffee 11.6 g	Furfural 783 µg	0.006
Coffee 11.6 g	Hydroquinone 290 µg	0.005
Saccharin daily	Saccharin 7 mg	0.005
Carrot 12.1 g	Aniline 624 µg	0.005
Potato 54.9 g	Caffeic acid 867 µg	0.004
Coffee 11.6 g	4–Methylcatechol 378 µg	0.002
Carrot 12.1 g	Caffeic acid 374 µg	0.002
Ethylene thiourea*	Ethylene thiourea 9.5 µg	0.002
DDT USA before 1972 ban**	DDT 13.8 µg	0.002
Pear 3.7 g	Caffeic acid 270 µg	0.001
Plum 1.7 g	Caffeic acid 235 µg	0.001
Bacon 19 g	Diethylnitrosamine 19 ng	0.001
Celery 14 g	8-methoxypsoralen 8.5 µg	0.0004
Carbaryl USA average**	Carbaryl 2.6 µg	0.0003
Toast 79 g	Urethane 948 ng	0.00008
Parsnip 48.8 g	8-methoxypsoralen 1.4 µg	0.00006
Lindane USA average**	Lindane 32 ng	0.000001

+ This value is divided by 70 kg (average adult weight) and then expressed as a percentage of the intake (as mg/kg body weight per day) by the test animal that caused cancer in 50% of the tested animals (TD$_{50}$ value).
* A breakdown product of some fungicides.
** Insecticide.
++ This means that the caffeic acid content in 11.6 g of coffee provides one thousandth of the amount of caffeic acid that would cause cancer in 50% of rodents.

been tested toxicologically. Gold *et al.* (2002) rated twenty-one as carcinogens, namely: Acetaldehyde, benzaldehyde, benzene, benzofuran, benzo (a) pyrene, caffeic acid, catechol, 1, 2, 5, 6-dibenzanthracene, ethanol, ethylbenzene, formaldehyde, furan, furfural, hydrogen peroxide, hydroquinone, isoprene, limonene, 4-methylcatechol, styrene, toluene and xylene.

Eight chemicals were not carcinogens, namely acrolein, biphenyl, choline, eugenol, nicotinamide, nicotinic acid, phenol and piperidine. Caffeine could not be classified.

Toxicology of Phytoalexins – Natural pesticides

Gold *et al.* (2002) reviewed the data on the toxicology of phytoalexins of which 53% (n=38) were found to be rodent carcinogens namely: Acetaldehyde methylformylhydrazone, allyl isothiocyanate, arecoline. HCl, benzaldehyde, benzyl acetate, caffeic acid, capsaicin, catechol, clivorine, coumarin, crotonaldehyde, 3,4–dihydro coumarin, estragole, ethyl acrylate, N2-γ-glutamyl-p-hydrazinobenzoic acid, hexanal methyl formylhydrazine, p-hydrazinobenzoic acid.HCl, hydroquinone, 1-hydroxy anthraquinone, lasiocarpine, d-limonene, 3-methoxycatechol, 8-methoxypsoralen, N-methyl-N-formylhydrazine, α-methylbenzyl alcohol, 3-methylbutanal methylformylhydrazone, 4-methylcatechol, methyl eugenol, methylhydrazine, monocrotaline, pentanal methylformylhydrazone, petasitenine, quercetin, reserpine, safrole, senkirkine, sesamol and symphytine.

The rodent carcinogens are found in: absinthe, allspice, anise, apple, apricot, banana, basil, beet, broccoli, Brussels sprouts, cabbage, cantaloupe, caraway, cardamom, carrot, cauliflower, celery, cherries, chili pepper, chocolate, cinnamon, citronella, cloves, coffee, collard greens, comfrey herb tea, corn, coriander, currants, dill, eggplant, endive, fennel, garlic, grapefruit, grapes, guava, honey, honeydew melon, horseradish, kale, lemon, lentils, lettuce, liquorice, lime, mace, mango, marjoram, mint, mushrooms, mustard, nutmeg, onion, orange, oregano, paprika, parsley, parsnip, peach, pear, peas, black pepper, pineapple, plum, potato, radish, raspberries, rhubarb, rosemary, rutabaga, sage, savory (herb), sesame seeds, soybean, star anise, tarragon, tea, thyme, tomato, turmeric, and turnip.

A total of 47% of the phytoalexins tested were rated (Gold *et al.* 2002) as non-carcinogens: (n= 34) namely: Atropine, benzyl alcohol, benzyl isothiocyanate, benzyl thiocyanate, biphenyl, d–carvone, codeine, deserpidine, disodium glycyrrhizinate, ephedrine sulphate, epigallocatechin, eucalyptol, eugenol, gallic acid, geranyl acetate, β–N–[γ–l(+)–glutamyl]–4-hydroxymethylphenylhydrazine, glycyrrhetinic acid, p-hydrazinobenzoic acid, isosafrole, kaempferol, dl–menthol, nicotine, norharmane, phenethyl isothiocyanate, pilocarpine, piperidine, protocatechuic acid, rotenone, rutin sulfate, sodium benzoate, tannic acid, 1-trans-δ9- tetra hydro cannabinol, turmeric oleoresin and vinblastine.

References

Ames, B.N. and Gold, L.S. (2000) Paracelsus to parascience: the environmental cancer distraction *Mutation Research* 447, 3–13. http://toxnet.nlm.nih.gov/cpdb/pdfs/ Paracelsus.pdf

Gold, L.S., Slone, T.H., Manley, N.B. and Ames, B.N. (2002) Misconceptions About the Causes of Cancer 1415–1460. In: Paustenbach, D. (ed.) *Human and Environmental Risk Assessment: Theory and Practice* New York: John Wiley and Sons.

RSC Natural or man-made chemicals. Available at www.rsc.org/learn-chemistry/content/filerepository/CMP/00/000/185/naturalnotes_tcm18-115179.pdf

Chapter 16 Nitrogen use efficiency

Lassaletta *et al.* (2014) calculated nitrogen use efficiency NUE (as N in harvested product as a percentage of N applied from combined fertilizer, manure and symbiotic N fixation) for 124 countries over fifty years from 1961 based on FAO data (FAOSTAT).

Table 16a Nitrogen use efficiency and total N input kg/ha/year (Lassaletta *et al.*, 2014) Average values for the first and final four years

Country	1961–1964		2005–2009	
	NUE %	N input kg/ha	NUE %	N input kg/ha
Africa				
Angola	101	9	69	17
Benin	83	10	132	15
Botswana	17	74	10	110
Burkina Faso	76	13	80	23
Burundi	112	16	86	19
Egypt	63	100	44	472
Ghana	79	8	92	16
Kenya	87	18	51	41
Malawi	112	15	64	39
Nigeria	92	13	97	23
Senegal	58	32	68	28
South Africa	73	28	46	122
Tanzania	91	11	85	20
Uganda	92	11	84	20
Zambia	86	15	47	58
Asia				
Bangladesh	55	56	43	223
Cambodia	43	41	71	62
China	84	39	29	343
India	53	31	33	130
Indonesia	65	30	42	98
Iran	68	27	41	115
Japan	49	156	41	214
Malaysia	44	26	11	135

Table 16a   Continued

Country	1961–1964		2005–2009	
	NUE %	N input kg/ha	NUE %	N input kg/ha
Pakistan	53	30	23	194
Philippines	50	34	56	79
Europe				
Austria	51	94	71	131
Belgium	39	296	53	251
Denmark	48	141	69	170
Finland	33	100	58	120
France	45	97	68	157
Germany	37	149	67	193
Hungary	49	63	67	101
Ireland	46	168	24	561
Israel	46	82	24	233
Italy	54	58	55	139
Netherlands	16	464	32	371
Norway	19	255	22	326
Poland	52	63	47	131
Portugal	38	43	47	82
Spain	50	53	46	101
Sweden	46	115	48	188
United Kingdom	29	225	51	240
South America				
Bolivia	57	26	60	72
Brazil	67	26	54	151
Chile	87	33	29	267
Colombia	42	41	18	216
Costa Rica	24	64	14	177
Cuba	22	66	26	60
Peru	42	67	39	127
North America				
Canada	191	13	80	79
Mexico	66	33	50	101
USA	67	65	66	172
Oceania				
Australia	72	32	53	55
New Zealand	14	276	13	778

Reference

Lassaletta, L, Billen, G., Grizzetti, B., Anglade, J. and Garnier, J. (2014) 50 year trends in nitrogen use efficiency of world cropping systems: the relationship between yield and nitrogen input to cropland. *Environmental Research Letters* 9, 10. doi: 10.1088/1748-9326/9/10/105011. Supplement http://iopscience.iop.org/1748-9326/9/10/105011/media/erl502906suppdata2annex.pdf

Chapter 17 Greenhouse gas emissions by country

References

WRI (2010) List of countries by greenhouse gas emissions. Available at https://en.wikipedia.org/wiki/List_of_countries_by_greenhouse_gas_emissions

WRI (2014) World Resources Institute. Climate Analysis Indicators Tool (CAIT) 2.0: WRI's climate data explorer. http://cait.wri.org

Table 17a Total greenhouse gases in countries emitting more than 1% global emissions in 2012 and agricultural emissions $CH_4 + N_2O$ Tg CO_2 equivalent (WRI, 2010 and 2014)

Country/Region	Total 2012[*]	Agricultural 2014[+]
World total	43,286	5,247
European Union	4,399	588 [**]
Argentina	338	112
Australia	648	142
Brazil	1,013	442
Canada	714	62
China	10,975	712
France	457	72
Germany	887	61
India	3,014	627
Indonesia	761	166
Iran	715	35
Italy	465	30
Japan	1,344	21
South Korea	693	13
Mexico	724	85
Russia	2,322	92
Saudi Arabia	527	7
South Africa	463	30
UK	553	45
USA	6,235	351

[*] WRI (2014) excluding those related to land use change and forestry.
[+] FAOSTAT.
[**] All Europe.

Organic Products

Apart from organic food and feed produced on farms, a wide range of organic products are marketed, including the following:

Baby clothes
Baby powder
Baby laundry detergents
Beer
Chocolate
Coconut oil
Cotton shirts and sheets
Cotton lingerie
Cosmetics, anti-aging skin care creams, bath salts, body oils, lipstick eye shadow, mascara and skin cleansers
Furnishing fabrics
Hamburgers
Herbs and spices
Olive oil
Omega-3 oil
Shampoo
Sleep balm
Tampons
Tea
Vodka
Water (USA)
Wine

Extension of Organic activities
Ethical investments

Financial planners are known to suggest that organic farming should be considered alongside animal welfare, AIDS research and disaster relief when advising ethical investments. In contrast, investors are advised to avoid investments related to pesticides, which are listed with the arms trade, gambling and pornography.

Forest management

The Soil Association (UK) became involved with forest management in 1992 with the setting up of a Forestry Programme with a view to extending its philosophy of sustainable resource use and certification expertise to the area of forestry and timber. The programme is now called Soil Association Woodmark. The main aim is to minimize illegal logging.

Glossary

Acceptable daily intake

Acceptable Daily Intake (or ADI) is the amount of a specified substance (e.g. pesticide residue) in food or drinking water that can be ingested daily over a lifetime without an appreciable health risk. It is normally based on one hundredth of the NOAEL (the minimal intake level at which no effect is found on the most sensitive test animal).

Acid deposition also known as acid rain

Acid deposition involves the precipitation on to soils and vegetation of various acidic compounds, particularly man-generated nitrogen oxides and sulphur dioxide from the atmosphere. Nitrogen oxides can also be produced naturally by lightning strikes, and sulphur dioxide is produced by volcanic eruptions.

Acidification

The process of making soils and water bodies more acid through the inputs of acidifying chemicals. Increased acidity can alter lake habitats and the growth of terrestrial vegetation. Acidification is used to describe the loss of nutrient bases (calcium, magnesium, potassium) from the soil, through leaching, and their replacement by acidic elements (hydrogen and aluminium). Pollutant deposition can enhance the rate of acidification.

AHDB

Agriculture and Horticulture Development Board.

Agroecology

Agroecology is the science of applying ecological concepts and principles to the design and management of sustainable agroecosystems. It includes the study of the ecological processes in farming systems such as: nutrient cycling, carbon cycling/sequestration, water cycling, food chains, lifecycles, herbivore/predator/prey/host interactions and pollination.

Alkaloid

Alkaloids are a diverse group of organic chemicals that in addition to carbon, hydrogen and nitrogen may also contain other elements such as oxygen and sulphur. More than 3,000 different types of alkaloids have been identified in more than 4,000 plant species. Alkaloids are also produced by a large variety of organisms, including bacteria, fungi and animals. Well-known alkaloids include caffeine, morphine, codeine, strychnine, quinine, ephedrine, cocaine and nicotine.

Ammonia

A reactive nitrogen form (NH_3), which is colourless gas with a pungent odour. A product of biological nitrogen fixation, it can be synthesized using the Haber–Bosch process. Most industrially synthesized ammonia is used in the manufacture of synthetic fertilizers.

Ammonium

Ammonium is a reactive nitrogen form (NH_4^+), which is a constituent of many synthetic fertilizers, such as ammonium nitrate and ammonium sulphate. In the soil it is nitrified to NO_3^- by bacteria.

Anthroposophy

Anthroposophy was the brainchild of Rudolf Steiner (1861–1925), who combined his understanding of science and philosophy following studies in Vienna. He was influenced by the philosophy of Goethe. The result was Anthroposophy and the founding in 1913 of an Anthroposophical Society at the Goetheanum in Switzerland. Anthroposophy is very much a personal philosophy that permits the individual to find his way under the influence of spiritual nature of the universe in many walks of life, not just agriculture and education.

Antioxidant

An organic compound that can limit the formation, and thereby counteract the damaging effects of, oxygen free radicals that lack electrons. Common antioxidants include vitamins C and E, ß-carotene, N-acetylcysteine, selenium and zinc.

Biodiversity

Biodiversity is the variability among living organisms, particularly of plants and animals in the biosphere. Diverse populations of plants and animals are valuable in preserving the integrity of natural ecosystems.

Biological nitrogen fixation (bnf)

BNF is the transformation of gaseous nitrogen into usable, or reactive, forms of nitrogen, by microbes, often existing in a symbiotic relationship with leguminous plants. Microorganisms that are able to fix nitrogen are called diazotrophs.

Broadbalk

See Rothamsted.

CABI

CABI (Centre for Agriculture and Biosciences International) is an international not-for-profit organization that provides information and applies scientific expertize to solve problems in agriculture and the environment. CABI evolved in 1986 from CAB, the Commonwealth Agricultural Bureaux.

Carbon dioxide (CO₂)

Carbon dioxide is a naturally occurring gas produced during respiration (by both animals and plants); it is also generated by the burning of fossil fuels and of biomass, and also from industrial processes (e.g. cement production). It is the principal anthropogenic greenhouse gas (GHG) that affects the warming of the Earth. It is the reference gas against which other GHGs are measured.

Carbon dioxide capture and storage

This is a process whereby a relatively pure stream of carbon dioxide from industrial and energy-related sources is separated, conditioned, compressed, and transported to a storage location for long-term isolation from the atmosphere.

Carbon sequestration

Carbon sequestration is the deliberate removal of carbon dioxide from the atmosphere and of storing it in a place (a sink) where it will remain; main examples are soil organic matter and natural vegetation, particularly trees.

Cation exchange capacity (CEC)

The CEC is a measure of the amount of cations that can be held on negative sites on a material and is expressed as milli equivalents (meq) per 100g. The CEC of humus, which may range from 250 to 400 meq/100 g, is much higher than those of clay minerals, which vary from about 10 meq/100 g (kaolinite) to 25–40 meq/100 g (illite) and 60–100 meq/100 g (montmorillonite). Soils with high organic matter contents will have high CEC.

Chemical

A chemical is a substance containing one or more elements (from the periodic table of more than 100 elements). All organic chemicals contain carbon most often in combination with hydrogen with or without oxygen, nitrogen, sulphur, phosphorus, and other elements. Organic compounds exist as either carbon chains or in ring form.

Chlorofluorocarbons (CFCs)

A chlorofluorocarbon is an organic compound that contains chlorine, carbon, hydrogen, and fluorine. CFCs have many uses including refrigeration, air conditioning and aerosol propellants. As they are not destroyed in the lower atmosphere, CFCs drift into the upper atmosphere where, given suitable conditions, they break down the protective ozone (O₃) layer. It is one of the greenhouse gases covered under the 1987 Montreal Protocol, following which manufacture was phased out.

Chloroplasts

Chloroplasts are the small green organelles in plants and algae where photosynthesis takes place. Chloroplasts are rich in chlorophyll and other pigments. Chloroplasts, like mitochondria in animals, contain their own DNA.

CO_2-equivalent concentration

The amounts of different GHG expressed as the equivalent amount of CO_2 that would have the same effect on heat trapping. The CO_2-equivalent concentration permits a comparison of the contributions made by different GHG in a mix.

Cisgenic modification

Cisgenic modification involves the insertion into the DNA of genes from the same species as the plant being modified.

Compost

Compost is produced by the aerobic decomposition of biodegradable organic materials.

Conventional and industrial agriculture

Conventional agriculture is here understood to be the system of agriculture that has evolved and has adopted many of the advances of agricultural science. When conventional agriculture involves large, heavily mechanized farms dependent on many inputs the term "industrial agriculture" may be more appropriate.

Cover crops and catch crops

Crops grown after the main crop has been harvested to provide ground cover rather than leaving a bare soil. As well as protecting the soil from erosion, leaching losses, especially of nitrogen, will be minimized and the recycling of nutrients promoted. Leguminous cover crops are often grown between cash crops in the rotation with the principal objective of building soil fertility. When cover crops are grown specifically as fertility builders they will commonly be called green manure crops; they will be incorporated into the soil before sowing the next crop. Organic matter and N levels will be increased when a leguminous crop is grown. Catch crops are normally fast-growing crops that can be grown between successive crops or at the same as a cash crop, e.g. between the rows.

Denitrification

Denitrification is the microbial regeneration in soils and water bodies of gaseous nitrogen and of nitrous oxide (N_2O) from nitrate, nitrite and nitric oxide (NO).

Dobson unit

A Dobson unit is the basic measure used in ozone thickness research. One Dobson Unit (DU) is 0.01 mm thickness at STP (standard temperature and pressure). Ozone thickness is expressed in terms of its physical thickness in the Earth's atmosphere. It is very thin with a normal range being 300–500 Dobson units.

Digestate

Digestate is produced by anaerobic digestion of biodegradable organic materials and includes liquids and fibrous materials separated after digestion.

Ecosystem

An ecosystem is a community of living organisms interacting with each other and with their environment.

Eutrophication

Eutrophication is the process of increasing nutrient inputs, especially compounds of nitrogen and phosphorus, into a body of water from anthropogenic sources; the result is an overgrowth of organisms, particularly algae, and the subsequent depletion of oxygen, leading to hypoxic or anoxic conditions.

Gene editing (CRISPR-Cas9)

Gene editing is dependent on firstly recognizing and understanding the activity of gene sequences (the function of the CRISPR bit) and secondly on the ability to cut and/or affect the genome at precise locations (using the Cas9 bit, which is basically an enzyme). The gene sequences at targeted locations may be replaced by sequences brought with the CRISPR-Cas9 package or they may be turned on or off.

Genetic modification

The modification of an organism's DNA by the direct introduction of novel genes or by gene editing.

Genome

The genome is an organism's complete set of DNA, including all of its genes. Each genome contains all of the information needed to build and maintain the organism. In humans, a copy of the entire genome – more than 3 billion DNA base pairs of nucleotides – is contained in all cells that have a nucleus. It is the order of the four nucleotide bases that make up the sequences in the DNA, namely adenine, cytosine, guanine and thymine, that determine the formation of messenger RNA and the subsequent generation of enzymes outside the nucleus.

Global warming potential (gwp)

Greenhouse gases differ markedly in their capacity to trap heat in the atmosphere. For example, the GWP for methane over 100 years is twenty-five times that of CO2 and for nitrous oxide it is 298 times. This means that emissions of 1 million t of methane and nitrous oxide respectively is equivalent to emissions of 25 and 298 million t of carbon dioxide.

The amounts of the various GHG are expressed as CO_2 equivalents.

Greenhouse gases (GHG)

GHG are gases in the atmosphere that trap heat re-radiating from the surface of the Earth which thus contribute to the warming of the Earth. The main greenhouse gases are water vapour, carbon dioxide, methane, nitrous oxide, ozone and several chlorofluorocarbons. The chlorofluorocarbons are entirely anthropogenic, while the other gases originate from both natural and anthropogenic sources.

Green Revolution

The Green Revolution is the name given to the reduction of famines in many countries, notably India, following the breeding of new varieties of wheat, rice and maize, which started in the 1950s. One plant breeder who is famous for his involvement in the Green Revolution is Norman Borlaug, who in the 1940s developed a strain of wheat that could resist diseases, was short-strawed (and so less prone to wind damage), and produced large seed heads and high yields.

Haber–Bosch process

The Haber–Bosch process is the industrial procedure for the production of ammonia from nitrogen and hydrogen when reacted at high pressures and temperature (400–650°C) using iron as a catalyst. It is named after its inventors, the German chemist Fritz Haber and Carl Bosch, an industrial chemist, who developed the process at the beginning of the First World War.

Hazard

A hazard is a potentially damaging situation or set of circumstances. In agriculture it is often thought of as being associated with a chemical or compound that may cause injury, or harm people or the environment

HERP … human exposure dose/rodent potency dose

HERP is a measure/Index of possible hazard where human exposure (daily lifetime dose in mg/kg) is expressed as a percentage of the rodent TD_{50} (toxic dose affecting 50% of the population) dose (in mg/kg) for a given carcinogen.

IAASTD (2005–2007)

International Assessment of Agricultural Knowledge, Science and Technology for Development was an international collaborative exercise involving FAO, WHO and others aimed at evaluating the relevance, quality and effectiveness of agricultural knowledge, science, and technology.

IFOAM

The International Federation of Organic Agriculture Movements Organics has, since 1972, occupied the central position and as the international umbrella organization of the organic world.

Homeopathic treatment

Treatment of diseases based on administration of remedies prepared through successive dilutions of a substance that in larger amounts produces symptoms in healthy subjects similar to those of the diseases themselves. The dilutions are such that the homeopathic dose is unlikely to contain any molecules of the original substance; the impression in the water of the substance is said to be effective.

Industrial agriculture

A form of agriculture that is capital-intensive, substituting human activities and animal labour by machinery, fertilizers and pesticides. The term is often used pejoratively.

Integrated pest management (IPM)

A procedure for integrating and applying practical management methods to manage insect populations so as to keep pest species from reaching damaging levels while avoiding or minimizing the potentially harmful effects of pesticides on humans, non-target species and the environment. Beneficial predators or parasites that attack pests are encouraged.

John Innes Centre

The John Innes Centre is an independent facility for research and training in plant science and microbiology located in Norwich, Norfolk, England. The JIC works with the Sainsbury Laboratory, also in Norwich, which specializes in research on plant microbe interactions.

Kyoto Protocol

The Kyoto Protocol, an international treaty committed 192 countries to reduce greenhouse gas emissions, based on the premise that (a) global warming exists and (b) human-made CO_2 emissions have caused it. The Kyoto Protocol was adopted in Japan in 1997 and came into force in 2005. The GHG included carbon dioxide, methane, nitrous oxide and sulphur hexafluoride (SF_6) a very potent GHG with a warming potential 23,900 times that of CO_2. The Protocol was updated by the Doha amendment in 2012; as of 2016 the amendment has been accepted by sixty-six countries.

LD_{50}

The lethal dose 50% is the amount of an ingested substance that kills 50% of a test animal; it is usually expressed in mg of substance per kg of body weight.

Leaching

Leaching is the washing out of soluble ions and compounds by water draining through the soil profile.

Legumes

Legumes are plants which form a symbiotic relationship with rhizobia bacteria that can fix nitrogen from the atmosphere in nodules growing on the roots. Legumes are valuable in being able to replenish the levels of reactive nitrogen in the soil when grown in a crop rotation sequence or in pastures e.g. clovers.

Lodge

Lodging of cereals is the breaking of the stem under the influence of storms. It is exacerbated by excessive N application and dense stands. Short-strawed wheat and rice varieties, which are less liable to lodge than tall varieties, played a major part in the Green Revolution.

Maximum residue levels (MRL)

The Maximum Residue Levels for a given pesticide is the maximum residue levels likely to be found in a crop or food item after a given pesticide has been properly applied. The MRLs are not related to health risks. When a MRL is exceeded an assessment of risk is made by considering the lifetime intake in relation to the ADI. The MRLs are set to ensure consumption does not exceed the ADI even if everything consumed contains the permitted MRL. In Europe the MRL is now used as a legally binding level that should not be exceeded; this had led to a confusing and unsatisfactory situation because the ADI is not considered when assessing risk.

Methane

A greenhouse gas (CH_4) that is twenty-five times more effective at trapping heat in the atmosphere than CO_2. Methane is the major constituent of natural gas; it is formed biogenically in anaerobic environments. It is also a major product of enteric fermentation in the stomach of ruminants.

Methaemoglobinemia

A disorder commonly referred to as "blue baby syndrome" is characterized by the presence of a higher than normal level of methaemoglobin in the blood. Methaemoglobin is an oxidized form of haemoglobin that has almost no affinity for oxygen, resulting in almost no oxygen delivery to the tissues; formation is promoted by high levels of nitrate and by bacterial contamination in drinking water.

Mineralization

Mineralization is the process whereby soil microorganisms break down organic matter and convert it into inorganic forms that can be absorbed by plants.

Mutagenesis

The creation of mutations in plant materials – a method used in plant breeding whereby random mutations are induced in plant DNA by exposure to ionising radiation, notably gamma rays, or to chemicals. Mutagenesis is perceived as a driving force of evolution.

Mycorrhizae

Mycorrhizae are fungi that form associations with plant roots and rely on the host plant to supply their energy needs. There are four kinds, namely the arbuscular mycorrhiza (VAM), ectomycorrhiza

(ECM), orchid mycorrhiza and ericoid mycorrhiza, of which VAM and ECM are important in agriculture. VAM live inside the roots and form arbuscules in the cortical cells. ECM grow on the surface of the root. In both types hyphae grow out into the body of the soil, effectively increasing the surface area for nutrient absorption; a feature that is likely to be more important for slowly diffusing nutrients such as P.

Nanotechnology

Nanotechnology involves the manipulation of materials to create structures and systems at the scale of atoms and molecules. Nanoparticles have at least one dimension of less than 100 nanometers (nm) ($1,000 \times 10^6$ nm = 1 meter, 1,000 nm = 1 micron). Red blood cells measure 7,000 to 10,000 nm across. Nanomaterials include nanoemulsions and nanostructures such as nanocapsules, nanotubes, fullerenes (buckyballs) and quantum dots (semi-conductors) and nanowires.

NIAB

The National Institute of Agricultural Botany, Cambridge, UK, has since 1919 been a centre of excellence in the breeding and development of new varieties of crops as well as in promoting improved crop husbandry.

Nitrate

Nitrates are a reactive form of nitrogen (NO_3^-). All nitrate salts are soluble in water and are very mobile in soils due to the paucity of positively charged binding sites. Nitrates are the most important crop nutrient, but high concentrations in waters can cause eutrophication.

Nitric oxide

A reactive form of nitrogen (NO) formed mainly in combustion processes and also emitted during nitrification and denitrification processes in soils.

Nitrification

Nitrification is a multi-stage microbial conversion of ammonium cations to nitrate anions. NH_4 is oxidized firstly to nitrite NO_2 (by *Nitrosomonas*) and then to NO_3 (by *Nitrobacter*).

Nitrite

Nitrites are a reactive form of nitrogen (NO_2^-). They have a high oxidative potential and are used in cured meats. Nitrite is typically much less abundant than nitrate and in high concentration is toxic to humans.

Nitrogen dioxide (NO_2)

NO_2 is an oxide of nitrogen formed mainly by combustion processes where N in fuel is oxidized or gaseous nitrogen is oxidized at high temperatures.

Nitrogen oxides (NO_x)

There are two reactive forms of nitrogen known collectively as NO_x, namely

NO (nitric oxide) and nitrogen dioxide (NO_2). These oxides, which are mainly formed during combustion, promote the formation of smog.

Nitrogen use efficiency (NUE)

The ratio of nitrogen input to output of a system (e.g. soil, plant, farm animal or farm). Various techniques can be employed to increase nitrogen use efficiency in crop and livestock systems.

Nitrous oxide (N₂O)

N_2O is a reactive nitrogen form, also known as laughing gas. It is mainly formed during microbial denitrification processes in soils and waters acting on NO_3 derived from fertilizers or organic matter. It is also emitted by combustion and other industrial processes. N_2O is a greenhouse gas that is 298 times more effective at trapping heat in the atmosphere than CO_2.

Nutrient cycling

Nutrient cycling is the movement of nutrients through an ecosystem.

Ozone (O₃)

Ozone the triatomic form of oxygen (O_3) is a gaseous atmospheric constituent. In the troposphere, it is created both naturally and by photochemical reactions involving gases resulting from human activities. Tropospheric O_3 acts as a GHG. It is created in the troposphere by the action of ultraviolet radiation and molecular oxygen (O_2).

Particulate matter (PM)

PM is microscopic solid or liquid matter suspended in air. Sources of particulate matter can be man-made or natural. They can impact climate and can adversely affect human health. Increasing attention is being given to particulate matter with a diameter smaller than 2.5 micron ($PM_{2.5}$), a serious pollutant. Other pollutants of concern are $PM_{1.0}$ and PM_{10}. Combustion of diesel fuel is a major source of particulates.

Perfluorocarbons (PFCs)

PFCs are included in the Kyoto Protocol but they are not relevant in agriculture. PFCs are by-products of aluminium smelting and uranium enrichment.

Pesticides

Initially the term pesticide was used to denote a compound that killed insect pests but it is now used to encompass a wide range of biologically active compounds, notably fungicides and herbicides but also miticides, molluscicides and rodenticides.

Phenolic compounds

More than 4,000 phenolic compounds may exist in plants. They have an aromatic benzene ring structure with one or more hydroxyl groups. The simplest is phenol (C_6H_5OH). Phenolic compounds are classified as simple phenols or polyphenols based on the number of ringed phenol units combined in the molecule. Phenolic phytoalexins play a major part in plants, mainly in providing protection against disease and insect stress. Phenolics play an important role in lignin formation.

Phytoalexins

Phytoalexins are secondary metabolites generated by plants in response to stress, insect attack and disease organisms that can protect the plant against particular insects or pathogens. These metabolites are collectively known as phytoalexins, or as natural pesticides. The name is derived from the Greek "warding-off (alexein) agents" in plants. Phytoalexins have been found in toxicology tests not only to be carcinogens (probably 50–60%), but also teratogens, oestrogen mimics, sterility inducers, nerve toxins, goitrogens and chromosome breakers.

Phytoanticipins

Phytoanticipins are similar to phytoalexins insofar as they are defence compounds but they differ in being present before any challenge by a pathogen or pest. Phytoanticipins occur in a wide variety of crops, including brassicas, currants, lupins and tomato.

Planetary boundaries

The concept of Planetary Boundaries aims at defining the environmental limits within which humanity can safely operate. Limits have been set, somewhat arbitrarily, for nine systems, including N fixation, atmospheric CO_2, ozone, extinctions and changes in land use.

Precautionary principle

The precautionary approach to risk management is invoked to justify an action relating to a suspected risk of a substance or action causing harm to the public, or to the environment, in the absence of scientific consensus that the action is not harmful. For example, the United Nations Framework Convention on Climate Change stipulated that precautionary measures should be taken to anticipate, prevent, or minimize the causes of climate change and mitigate its adverse effects and that lack of full scientific certainty should not be used as a reason to postpone such measures.

Reactive nitrogen (Nr)

Nr includes all biologically and chemically active nitrogen compounds in the atmosphere and biosphere. It includes inorganic forms of nitrogen, such as ammonia (NH_3) and ammonium (NH_4^+), nitric oxide (NO), nitrogen dioxide (NO_2), nitric acid (HNO_3), nitrous oxide (N_2O), and nitrate (NO_3^-), and organic compounds such as urea, amines, proteins and nucleic acids. Di-nitrogen gaseous N is not reactive.

Respiration

Aerobic respiration is the process in both animals and plants whereby glucose is broken down in the presence of oxygen and catalysed by enzymes to generate energy held in the bonds of adenosine tri-phosphate (ATP) and CO_2 as a waste product. The energy released by respiration via ATP is used to make large molecules from smaller ones. In plants, for example, sugars, nitrates and other nutrients are converted into amino acids to form proteins: In anaerobic respiration oxygen is absent and lactic acid is produced as well as ATP. Anaerobic respiration occurs in many environments, including freshwater and marine sediments and soils.

Rhizosphere

The rhizosphere is the soil region immediately surrounding plant roots that is normally rich in microorganisms that benefit from root exudates of sugars and amino acids.

Risk

Risk is an assessment of the probability of an identified hazard having a deleterious effect. For example, when assessing the risk of the harm that could be caused by ingesting pesticide residues found to be toxic in animal studies it is necessary to estimate the actual amounts of the pesticide that would need to be consumed before any harm was caused and compare them with actual intakes.

Rothamsted

In 1843 Lawes and Gilbert set up what are commonly referred to as the classical experiments on Lawes' Rothamsted Estate. The continuous wheat trial on the field named "Broadbalk" is the most famous but other similar fertilizer/manure experiments were carried out on barley, root crops and permanent pasture, all on heavy soils. Parallel long-term experiments were later set up in 1876 on light-textured soils at Woburn.

Rubisco

Rubisco (ribulose-1,5-bisphosphate carboxylase/oxygenase) is an enzyme that is vital for the capture of CO_2 in photosynthesis. All carbon present in organic material exists because of the activity of rubisco.

Ruminants

Ruminants are mammals with a four-chambered complex stomach that digest plant-based food by initially softening it within its first stomach, then regurgitating the semi-digested mass, now known as cud, and chewing it again. Breakdown of the cellulose is aided by bacteria, which generate CH_4.

Sulphur hexafluoride (SF_6)

SF_6 is a very potent GHG covered under the Kyoto Protocol. SF_6 is largely used in heavy industry to insulate high-voltage equipment and to assist in the manufacturing of cable-cooling systems and semi-conductors. It has no relevance with regard to agriculture.

Tg teragram, Gg gigagram and Mg megagram

Teragram = 1×10^{12} g.
1 Teragram = 1 million t
Gigagram = 1×10^{9} g.
1 Gigagram = 1 thousand t
Megagram = 1×10^{6} g.
1 Megagram = 1 t

Terpenoids

Terpenoids, which occur in all plants (more than 22,000 found), are a diverse class, derived from five-carbon iso-prene units assembled and modified in many ways. About 60% of known natural products are terpenoids, making them the largest class of natural compounds. Plant terpenoids are used extensively for their aromatic qualities and play a role in traditional herbal remedies.

Tonne, ton

1 tonne (t) = 1,000 kg

1 ton = 1,016 kg. Historical uses of ton have been retained.

Transgenic modification

Transgenic modification is genetic modification that involves the insertion of genes from a different species to the plant being modified. The novel genes exists in all cells and are passed through to progeny.

Woburn

See Rothamsted.

Urea

Urea is an organic compound with the chemical formula $(NH_2)_2CO$ containing 46% N. Urea is widely used in fertilizers as a concentrated source of nitrogen. Despite losses of NH_3 on application, urea is the preferred N source in many countries because, being rich in N, transport costs per unit N are the smallest.

Epilogue

If this epilogue is one of your first ports of call it is hoped that it will lead you to locating the evidence and story behind the following comments.

Consideration of organic and conventional farming revealed several key areas concerning health, food composition, natural plant pesticides, the inadequacies of assessments of the risks of pesticide residues, the three sources of crop N, the importance of distinguishing between recycled N in manure and that from biological and factory fixation of N, the inability of organic farming to feed the world, the degradation of soils in regions where little fertilizer N is applied, the dominant and important position of small-scale farmers in global food production and the prospects of advances in agricultural science.

Health

Apart from pre-eclampsia during pregnancy there is little evidence linking consumption of organic food with better health. Proponents of organic crops rely on the composition of organic food to indicate their health-promoting superiority. The basic difficulty is that of identifying the beneficial compounds from a cast of hundreds. There can be no guarantee that all organic crops will exhibit the claimed superior composition.

Pesticides

Pesticide residues are commonly believed to be harmful without any consideration of the amounts consumed, which are dwarfed (15,000 times) by the amounts of the equally toxic phytoalexins, the natural plant pesticides, consumed in all fruit and vegetables. Greater amounts of harmful natural chemicals are consumed in two cups of coffee than of pesticide residues eaten in one year in fruit and vegetables.

Nitrogen

Nitrogen is the key plant nutrient, the availability of which is the major factor limiting crop growth. There are three basic sources of available soil nitrogen, namely biological N fixation, man-made

N fertilizers and the mineralization of soil organic matter. Before the advent of N fertilizers virtually all the nitrogen was derived from biological N fixation. Manure and composts do not generate nitrogen, they recycle it.

Livestock farming leads to the generation of more of the greenhouse gases methane and N_2O than arable farming reliant on fertilizers. Organic farming is a greater contributor of GHG than conventional insofar as it largely depends on the recycling capabilities of livestock. Minimal nitrate leaching, which varies with the amount and timing of application of N sources, can be more easily achieved when fertilizers rather than manure are used.

There is a need to increase the efficiency of use of both inorganic and organic sources of N.

The beneficial effect of nitrates on cardiovascular health indicates that attitudes to and legislation regarding nitrate in drinking water need to be re-examined.

Inability of organic farming to feed the world

Despite the fact that it is accepted that biological N fixation on its own would only be able to support a global population of around 3 billion, it is thought by some that organic farming has the capacity to feed the world. If the whole world stopped eating livestock products there is a possibility that organic farming could suffice. The global consumption of livestock products is, however, expected to increase.

Misunderstandings about organic productivity are often due to the fact that allowance is not made for the land used by legumes to fix N either on the farm or elsewhere. Provided sufficient nutrients, especially N, are supplied at the right time by manure, yields will be comparable to those obtained by applying fertilizers. The problem is the land area and the number of animals required to generate and recycle the manure or the land area required for fertility-building crops are unattainably large.

In practice organic yields fall well short of conventional yields.

Soil organic matter

Soil organic matter plays a very important part in maintaining good soil structure and particularly soil fertility with its ability to provide N when broken down and of humus to retain cations.

Soil degradation

Fertilizer N inputs are much lower in Africa than in other regions. It is worrying that it would appear that the N in soil organic matter is effectively being mined in many African countries. Fertility-building legume crops and fertilizers will be required to reverse the degradation.

Farm size

Small-scale farmers in developing countries are responsible for feeding most people in the world. In contrast, in developed countries large-scale farming dominates. Achieving long-term sustainable global food production will largely depend on how small-scale farmers can cope; they will need many inputs including fertilizers and there will be limited room for the less productive, land-demanding organic practices. Man-made fertilizers will necessarily be preferred to fertility-building legume crops to provide N in countries (e.g. China) where there is little opportunity to expand the arable area. In developed countries with few food security problems there will be opportunities for organic farming.

Agricultural science

Globally over the years conventional farming has benefited by adopting the findings of agricultural science, particularly in breeding, crop nutrition and crop protection. Massive population growth has been facilitated. All indications are that further advances in the pipeline will help to sustain adequate food production.

Index